Axel Rossmann

Aircraft Turbine Engine Safety

Problem Oriented Engine Technology for Professionals

Translated from the German by Oleg Benesch

- Examiners
- Consultants
- Public Authorities and
- Agencies
- Operators
- Quality Assurance
- Designers
- Service and Maintenance

Volume 1B
Foreign Object Damage(FOD), Bird Strikes,
Erosion: Particles, Drop Impact/Rain, Cavitation.
Corrosion: Stress Corrosion Cracking (SCC),
Hot Gas Corrosion/Sulfidation, Hydrogen Embrittlement.

This volume contains over 200 illustrated pages with about 300 diagrams and sketches, as well as detailed descriptions.

turbo consult
Axel Rossmann

1st Edition
(Series 1.0 Letter)

Contact: www.turboconsult.de or fax (Germany) (0044) (0)8131 50 50 67

ISBN 3-00-005842-7

Foreword

The safety of aircraft engines has reached a high degree. However, the ever-increasing number of aircraft necessitates continual effort in order to reduce the absolute number of engine problems. This book is intended to give professionals who deal with the safety of turbine aircraft engines an overview of engine-related problems. The reader should become able to recognize the complex relationships that are typical for the subject matter, and be able to ask the right questions, provide correct answers, and be able to analyze these properly.
In my opinion, there is no better approach than the analysis and evaluation of incidents relating to safety, since the maxim „learn from damage“ still holds true. I have asked myself how it might be possible to impart this experience, which can only be gained from actual practical work in the field. Therefore, I have attempted to take examples from technical literature and analyze and explain them, relying on many years of practical experience and an understanding of the technical backgrounds.
This particular type of book was chosen in order to increase the effectiveness of the learning progress, present the complex relationships in an understandable manner, and motivate the reader. It relies to a large degree on visual depictions and is similar to overhaul manuals used in turbine engine-related fields. Concise explanations and labels based on my experience and evaluations of the subject matter of diagrams are intended to pique the readers interest. This may be the first time that the familiarity of current generations with comics as an explanatory media has been applied to a comprehensive technical text.
The possibility, that I may have misinterpreted documents in individual cases and/or that flawed or vague data may have influenced the analysis of factors, cannot be absolutely ruled out. In case the reader has any evidence pertaining to actual cases dealt with in this book, I will do my best to incorporate this into later editions.

I thank my wife Dr. Daniela Rossmann for her years of patient understanding and support, and also for her many helpful recommendations concerning design. This book would never have been possible without her support.

Axel Rossmann

I am especially indebted to the following individuals for technical consultations and corrections

Dipl.-Ing. (TU) Hagen P. Hanser (emphasis: understanding of engine technology, technical terminolgy)
Dipl.-Ing.(FH), Michael Müller (emphasis: damage and damage analysis)
Dipl. Ing. Dipl.-Wirtsch.-Ing. Helmut Schubert (emphasis: literature selection and procurement)
Dipl.-Ing. Karl Wolf, (emphasis: clear and understandable presentation, technical terminology)
Reinhold Gräter (emphasis: materials science)

Contents

Foreword

1. Introduction

2. Trends and Statistical Observations

3. Superordinate Issues

4. Aircraft Accident Investigation

4.1 Reports
4.1.1 Recommendations for Evaluating Reports

4.2 Operating Data
4.2.1 Recommendations for Incorporating Operating Data

4.3 Reconstruction
4.3.1 Recommendations for Reconstructing Damage Processes

4.4 Reproduction
4.4.1 Recommendations for Reproducing Damage

4.5 Damage Sequences

5. External Factors

5.1 Weather
5.1.1 Rain
5.1.1.1 Measures Against Problems due to Water Ingestion

5.1.2 Hail
5.1.2.1 Measures Against Problems due to Hail

5.1.3 Lightning
5.1.3.1 Measures Against Lightning Damage

5.1.4 Ice
5.1.4.1 Measures Against Ice Damage

5.1.5 Surrounding Temperatures
5.1.5.1 Measures Against Damage due to Surrounding Temperatures

5.1.6 Turbulence
5.1.6.1 Measures Against Damage due to Turbulence

5.2 Foreign Object Damage (FOD)

5.2.1 Damage due to (non-biological) Foreign Objects
5.2.1.1 Fundamentals and Damage Mechanisms
5.2.1.2 Ingestion of Foreign Objects
5.2.1.3 Foreign Object Types
5.2.1.4 Measures Against Foreign Object Damage

5.2.2 Bird Strikes
5.2.2.1 Bird Strike Damage
5.2.2.2 Bird Strike Damage Processes
5.2.2.3 Constructive and Technological Influences on Bird Strike Behavior
5.2.2.4 Measures Against Bird Strike Damage

5.3 Erosion

5.3.1 Erosion Mechanisms

5.3.2 Erosion Damage

5.3.3 Measures Against Erosion Damage

5.4 Corrosion

5.4.1 Corrosion without Mechanical Loads
5.4.1.1 Corrosion Fundamentals
5.4.1.2 Corrosion Damage
5.4.1.3 Measures Against Corrosion Damage

5.4.2 Stress Corrosion Cracking (SCC)
5.4.2.1 Fundamentals of Stress Corrosion Cracking
5.4.2.2 Damage due to Stress Corrosion Cracking

5.4.3 Corrosion Fatigue
5.4.3.1 Fundamentals of Corrosion Fatigue
5.4.3.2 Damage due to Corrosion Fatigue
5.4.5.3 Measures for Preventing Corrosion Fatigue Damage

5.4.4 Hydrogen Embrittlement
5.4.4.1 Fundamentals of Hydrogen Embrittlement
5.4.4.2 Damage due to Hydrogen Embrittlement
5.4.4.3 Measures Against Damage due to Hydrogen Embrittlement

5.4.5 Hot Gas Corrosion
5.4.5.1 Fundamentals of Hot Gas Corrosion
5.4.5.2. Damage due to Hot Gas Corrosion
5.4.5.3 Measures Against Damage due to Hot Gas Corrosion

5.5 Special Media (Unusual Influences)
5.5.1 Measures for Preventing Unusual Influences

Index 1A

5.2 Foreign Object Damage (FOD/OOD)

5.2.1 Damage from Non-Biological Foreign Objects

5.2.2 Bird Strikes

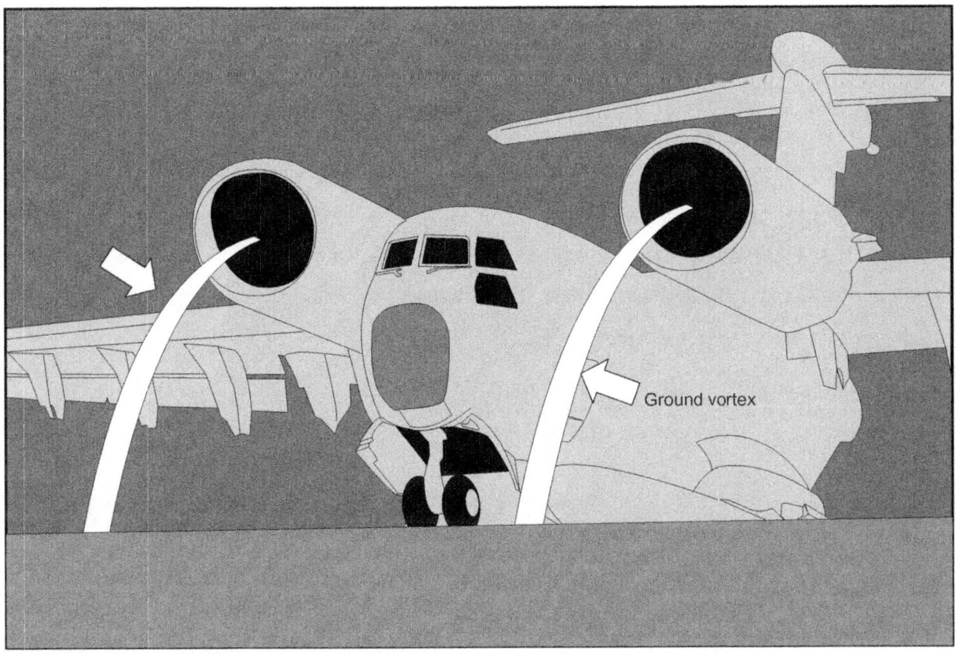

There are several possible answers to the question, "What is a foreign object that can cause engine damage?" For this reason, this chapter uses the term "**foreign object**" to include everything that enters the engine from outside (**F**oreign **O**bject **D**amage = FOD) or originates in the engine itself (**O**wn **O**bject **D**amage = OOD or **D**omestic **O**bject **D**amage =DOD) and is not a part of the normal inlet air or the specified auxiliary materials. Particles that contaminate the oil flow, as well as erosive (Chapter 5.3) and corrosive (Chapter 5.4) media in the air flow are also considered foreign objects. Bolts that come loose inside the engine are also foreign objects.

In the broadest sense, weather influences such as ice, hail, and rain which were dealt with separately in Chapter 5.1 can also be considered to be foreign objects.

The classification of foreign objects into "non-biological foreign objects" and "**bird strikes**" is arbitrary and was chosen because massive, hard foreign objects and foreign objects of technical origin can be differentiated from bird strikes and weather factors.

If corrosion and erosion are also included under FOD, then FOD is a deciding factor for engine cost, exceeding over 50% of the total purchasing cost in military machines, for example (Ill. 5.2.1.1-1). In most cases parts with FOD that exceeds specified limits (given in maintenance manuals and overhaul handbooks, etc.) must be reworked, and if this is no longer possible, replaced.

The increased use of electric and electronic engine parts such as regulators increases the danger of **short circuits caused by extremely fine fragments of carbon fiber**. These are created during burning or reworking (cutting) of fiber-reinforced parts such as fuselage sheeting (Ref. 5.2.1-1).

Ingested **dust** can cause **erosive wear** in the compressor, as well as block up cooling air ducts in the hot parts, leading to dangerous temperature increases.

However, some FOD is reversible and can be removed fairly simply. This includes **fouling** of the compressor blading through oil and dust, which sticks to the blades and can greatly decrease compressor performance. Washing and cleaning procedures that remove these coatings are a necessary part of routine maintenance.

Bird strike is one of the most feared types of FOD and is treated in a separate chapter (Chapter 5.2.2) due to its unique effects. However, even insects can have a similarly unallowable, if less spectacular, effect on engines. If insects carried by the airflow become stuck on the blading upon impact, it results in so-called "**insect roughness**" which can have a considerable effect on compressor performance. For example, in one case a small 100 kW gas turbine ingested a mosquito swarm which worsened the surge limit of the radial compressor so quickly that compressor surges made further operation impossible and required immediate cleaning.

Illustration 5.2.1.1-1: *These diagrams depict an older (Ref. 5.2.1-2) statistical evaluation of* ***military engine overhauls*** *(percentages are rounded). The top diagram refers to a power-shaft engine type for helicopters. The bottom diagram covers a wide range of helicopters and winged aircraft used by the US Navy. If one goes by the broad definition of FOD used here, including* ***erosion*** *and* ***corrosion*** *damage, then this type of damage is the reason for more than 50% of overhauls. The notion that helicopters are highly susceptible to FOD is plausible, since they are operated in an environment with a large amount of foreign objects.*

If one considers the total number of engines in the lower diagram then, in addition to the portion explicitly labeled as FOD, one could certainly include some of the cases from the "decreased performance" section (fouling, erosion, corrosion). This would mean that in all the studied military aircraft, roughly 30% of cases necessitated an overhaul due to FOD.

External Factors
Foreign Object Damage: Fundamentals

Reasons for engine overhauls according to old, but interesting, statistics.

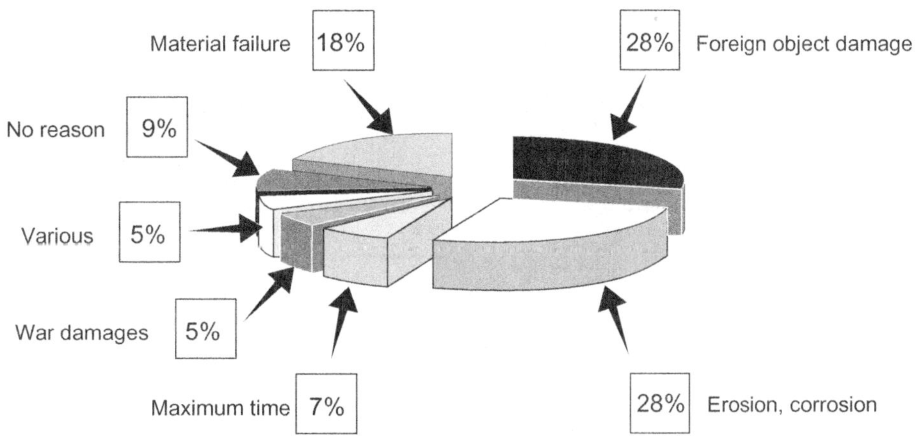

Typical military helicopter engine

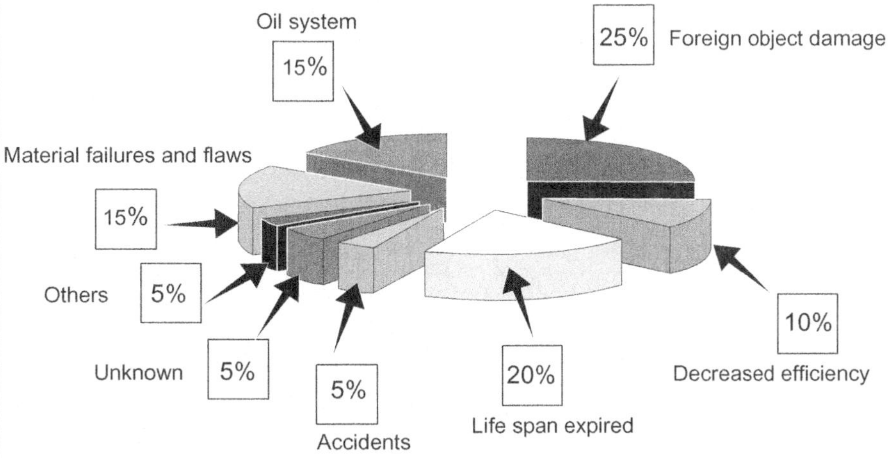

Various engine types from military airplanes and helicopers

Ill. 5.2.1.1-1

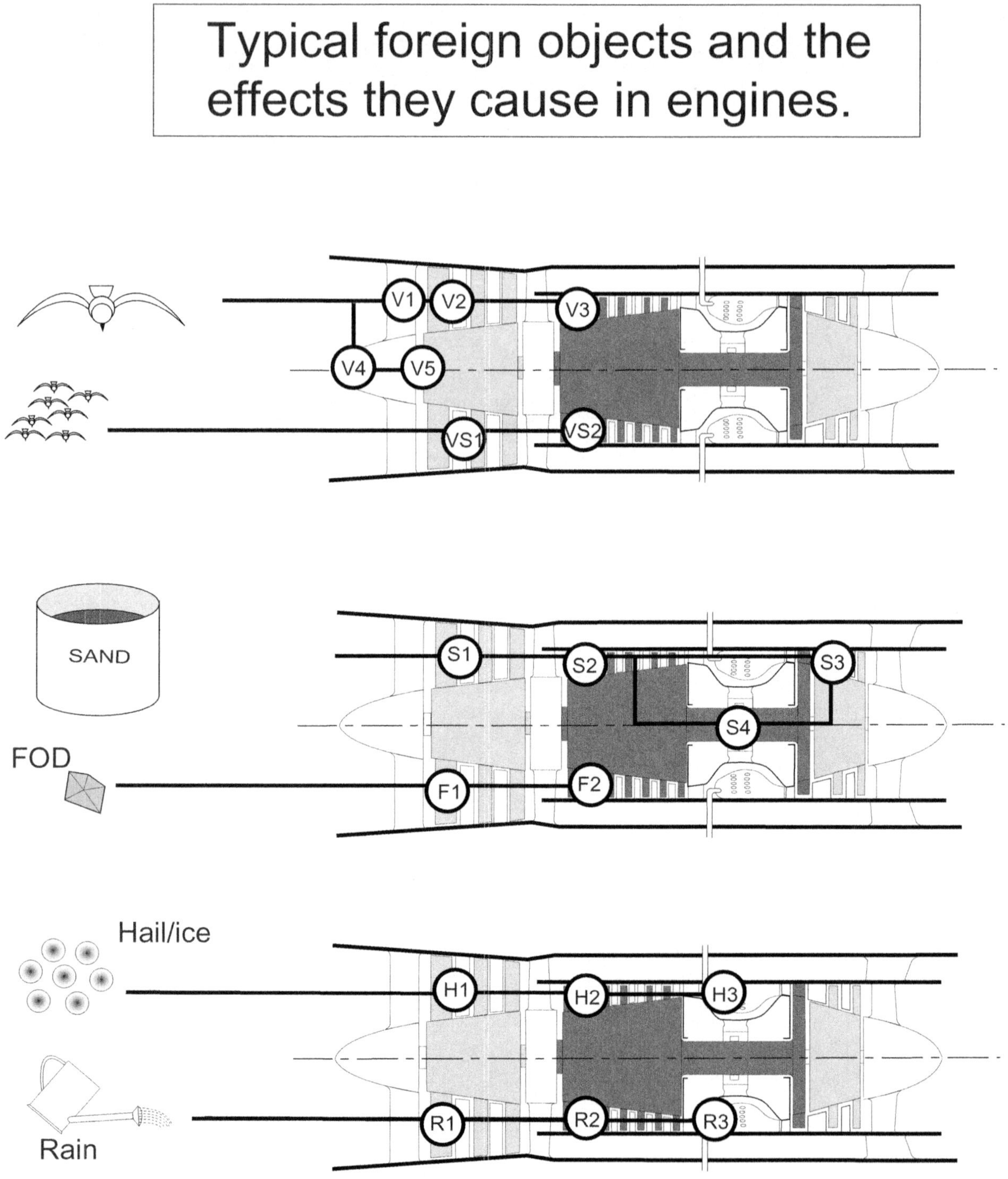
Typical foreign objects and the effects they cause in engines.
V1
V2
V3
V4
V5
VS1
VS2
SAND
S1
S2
S3
S4
FOD
F1
F2
Hail/ice
H1
H2
H3
Rain
R1
R2
R3

Ill. 5.2.1.1-2

Illustration 5.2.1.1-2: *This diagram describes the damages characteristic of typical foreign objects (see Chapter 5.1.1).*

Larger birds:

*- Intake guide assembly/intake **(front) bearing (V1)**: This type of construction is used in some military and older civilian aircraft types (Ill. 5.2.2.3-7). Parts of this assembly are stressed either by direct hits or indirectly through the inlet cone when it is struck. Experience has shown that this configuration has certain weaknesses when compared with configurations with rotating nose cones and no **intake guide vanes**.*

*- **Blading of the fan** and low-pressure compressor **(V2)**: Powerful forces on the blading cause damage including blade failure. The guide vanes are particularly susceptible to this (Ill. 5.2.2.2-7) since they typically have thin profiles and are subject to high speed impact from large bird pieces that pass through the fan whole. However, the risk of damage to rotor blading is greater at slower flight speeds (**low speed impact**, see Chapter 5.2.3), because it must put more energy into accelerating the ingested bird mass, creating large bending force.*
*Adjustable/**variable guide vanes** are at risk for damage to the adjustment mechanism and/or closing of the vanes.*

*- High-pressure compressor blading **(V3)**: An example of a dangerous situation is **bird parts collecting** on a housing strut ahead of the compressor and then striking the relatively filigreed compressor blading as a large mass (Ill. 5.2.2.3-11).*

*- **Spinner** (rotating nose cone) **(V4)** : Imbalances caused by deformation of metallic spinners, fragments of fiber-reinforced synthetic spinners.*

*- **Main bearings (V5)**: Short-term overstressing of the main bearing through **large axial forces** (depending on the stiffness of the spinner, etc.) and/or dynamic overstress caused by **imbalances** (results: fractures, fatigue).*

Flock of birds (small birds):

*- **Guide vanes** of the low-pressure compressor **(VS1)**: Even small birds can cause dangerous damage to smaller engines, especially if these are of a very filigreed type and/or made from particularly sensitive materials such as fiber-reinforced synthetics or aluminum alloys.*

*- **High-pressure compressor** blading **(VS2):** Damage can occur if large amounts of bird matter collect on struts or guide vanes and enter the compressor as a large mass (Ill. 5.2.2.3-11). For this reason, the first high-pressure compressor rotor stage must be given special consideration.*

Sand and dust:

*Fan and low-pressure compressor **(S1)**: **Erosions** wear, roughing, and small notches on the blading.*

*-High-pressure compressor **(S2)**: The **centrifuging of the sand** causes most of the sand to stay out of the high-pressure compressor in bypass engines. However, erosive wear, roughing, and small notches on the blading (especially near the tips) will occur. Erosion of the inner housing wall, especially of soft abradable coatings, will increase the clearance gap at the blade tips.*

*- **Labyrinths (S4)**: Erosion of labyrinths in the cooling air flow.*

*- Cooled hot parts **(S3)**: **Blockages in the cooling air** ducts causes overtemperatures with considerable reductions in the life span; damaging reactions (hot gas corrosion,*

sulfidation) with the inner and outer surfaces of the hot parts.

Small rocks and split:

*Fan, low-pressure compressor (**F1**) and high-pressure compressor (**F2**); small **notches** that can usually be reworked.*

***Hail and ice** (see Chapters 5.1.2 and 5.1.4):*

*Damage to the fan (for example, exit guide vanes made from Al alloys or **fiber-reinforced plastics**) and low-pressure compressor (**H1**); because the ice particles melt before impact, no damage is to be expected in the high-pressure compressor. However, a compressor stall (**H2**) due to the created water and/or unstable combustion or flame-out are possible results (**H3**). These effects are especially dangerous in hail, because hailstones are more massive than **rain drops** and less likely to be diverted into the bypass duct, causing a larger amount of water to enter the high-pressure area.*

***Rain and spray:** Rain can cause **rain erosion**, i.e. rain drop impact in the fan (**R1**). While microscopic traces of rain erosion can be detected on titanium blades, there does not seem to be any acute risk of damage. Noticeable damage is possible on the surfaces of coated parts or those made from **fiber-reinforced plasics**. In the high-pressure compressor and combustion chamber, the same damage can occur as with hail ingestion (**R2** and **R3**), usually in combination with hail, since rain is more likely to be directed through the bypass duct away from the high-pressure section.*

***Illustration 5.2.1.1-3:** This diagram shows the influence of FOD on operating behavior, i.e. the efficiency of the engine components and the engine as a whole. These are usually damage mechanisms with a long incubation period, such as erosion, buildup of deposits, or wear.*
The creeping worsening of engine performance at the very least increases the probability of hot part damage (higher temperature levels required to achieve the necessary power) and, under poor operating conditions, the risk of compressor stalls.

External Factors
Foreign Object Damage: Fundamentals

Typical important influences that affect the efficiency of the main components of a run-in engine.

Influences on the fan
- Erosion of rotor blades (inlet edge, chord length)
- Foreign object strikes, heavily faded notches
- Rounded edges at blade tips
- Eroded or dulled stator vane edges
- Angle deviations in the inlet guide vanes and stator vanes
- Roughness on the blades

Low-pressure compressor
- Rubbing at the blade tips
- Erosion at the leading edge of the rotor blades
- Foreign object damage on rotor and stator blades (leading and trailing edges)
- Soiling of the rotor and stator
- Dulling of the stator blade leading edges
- Angle deviations in the stator vanes
- Leakage flow from labyrinth seals
- Roughness on the blades
- Leakage from the bleeder system

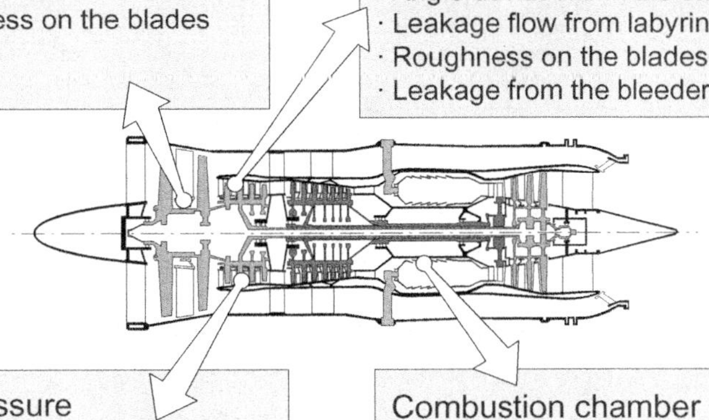

High-pressure compressor
- Erosion on the stator vanes and shortening of the rotor blade chords
- Changed leading and trailing edge angles and edge radii of stator and rotor blades
- Blade roughness/soiling
- Leakages at the:
  bleeder system
  stator and rotor blade tips
  labyrinth seals
  housing flange
- Rotor blade length:
  overly large tip clearances

Combustion chamber and turbine
- Coking of the injection nozzles
- Crack development in combustion chamber walls
- Bent and burned stator vanes
- Repair- and rework-induced deviations from original dimensions of inlet guide vanes
- Erosion of the rotor blades, roughness and burn marks
- Leakages from flange around the high pressure turbine
- Leakages at blade shrouds
- Wear and damage to the gas seals

Ill. 5.2.1.1-3

Illustration 5.2.1-3: *Foreign objects and* ***ingested contaminants*** *can affect almost all components in the gas flow in many different ways, changing their* ***operating characteristics*** *and worsening the overall . The direct relationship is often* ***not immediately discernible*** *since there is a chain of effects that mutually influence each other. For example, the failure of a* ***hot part*** *may be causally related to erosion wear on seals in the cooling air ducts.*
Mostly failure mechanisms which need longer pereiods of time, like ***erosion****, the formation of* ***deposits*** *or* ***wear processes*** *are concerned,*

5.2.1 Damage from Non-Biological Foreign Objects

5.2.1.1 Fundamentals and Damage Mechanisms

5.2.1.2 Intake of Foreign Objects

5.2.1.3 Types of Foreign Objects

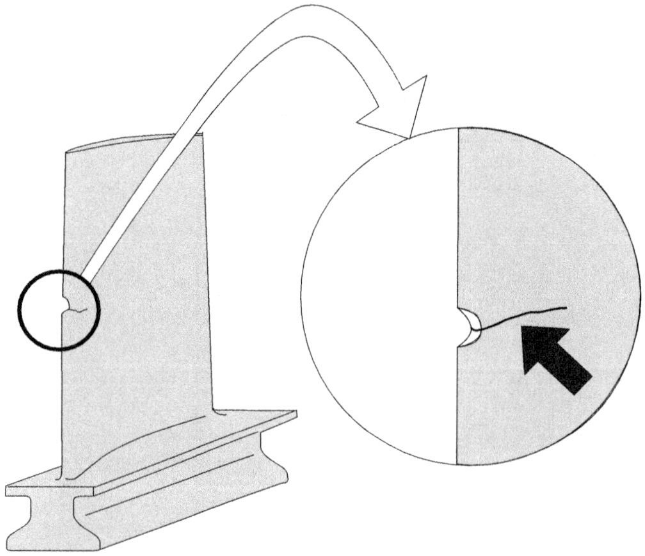

This chapter deals with **damage caused by the impact** of large, hard bodies or foreign **objects of technical origin** (plastic sheeting, cleaning rags, etc.). It covers both the **ingestion** of external foreign objects and internal foreign objects (engine parts or fragments).

The **FOD** described in this chapter is primarily caused by mechanical overstressing of engine parts, which can have many different **damage mechanisms**.

There is a special focus on the possibility of foreign objects being sucked into the engine, since an understanding of this process is a prerequisite for effective prevention of foreign objects entering the engine.

5.2.1.1 Fundamentals and Damage Mechanisms of Foreign Object Damage

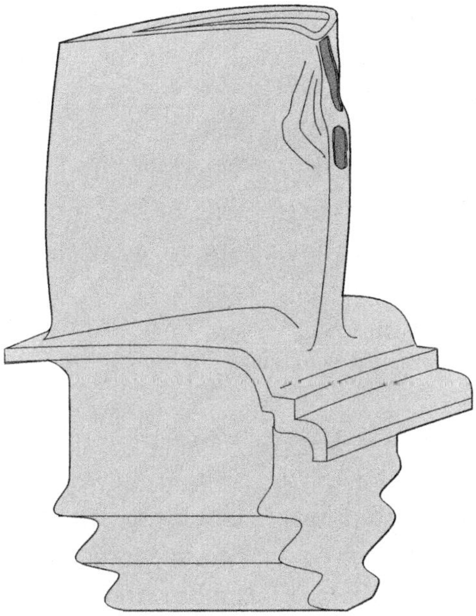

Foreign object damage through **mechanical overstress** can occur in many different ways. The following possibilities are given for FOD to **blading**, which is an engine part typically affected by FOD (see Ill. 5.2.1.1-4):

- Elastic deformation (mostly bending) on the **entire part** (of a blade, etc):
This can result in **contact with other parts** with relative movement and consequential damages such as wear, fractures, and the ignition of fires and dust explosions.

- **Local plastic deformation** (**notches**, distortion):

The creation of notches and marks (for example, on bolt threads, Ill.5.2.1.3-3) creates immediate **danger of dynamic fatigue** overstress. This can occur if the normal dynamic operating loads, which are below the fatigue strength of the undamaged part and would not affect it, act on an area where a notch has induced a tension increase.

A further damaging influence could be **local tension residual stress** induced by the plastic deformation. These act to increase the mean stress, further decreasing the part's dynamic strength. During impact the material around the **plastic deformation** becomes permanently stretched, while the surrounding material expands elastically. When the load is relaxed (after the impact), the tension is reversed, causing compressive stress in the stretched impact area that balance with the tension stress in the surrounding material.

- Plastic deformations across large part sections:

A typical example is the **bending of a blade**. Consequences include imbalances, aerodynamic interference, failure, rubbing-induced vibration, or flow interference/blockage. These increase the likelihood of LCF and HCF dynamic fatigue fractures.

- Excavation of material (notches, grooves, Ill. 5.2.1.1-10) :

If **hard partic**les strike the blades near the edges with sufficient impact speed, they can “dig out“ material and create notches. Strafing impacts result in cut-like grooves that usually correspond to the movement of the foreign object in the direction of the flow, i.e. across the blading‘s flexural modes, promoting dynamic fatigue cracks.

- **Crack initiation** and overload fractures:

At **critical impact speeds** (Ill.5.2.1.1-10) crack initiation will most likely occur from the impact notch, seriously lowering the dynamic strength of this area. Larger foreign objects can overstress the entire part and cause overload cracks and unstable crack growth. This type of crack growth, which is similar to a tear in a plastic sheet, has been observed especially in notched (cracked) **titanium alloy blades** under high tensile stress. If the crack exceeds the **critical fracture toughness** (k_{IC}), it will cause a brittle overload fracture.

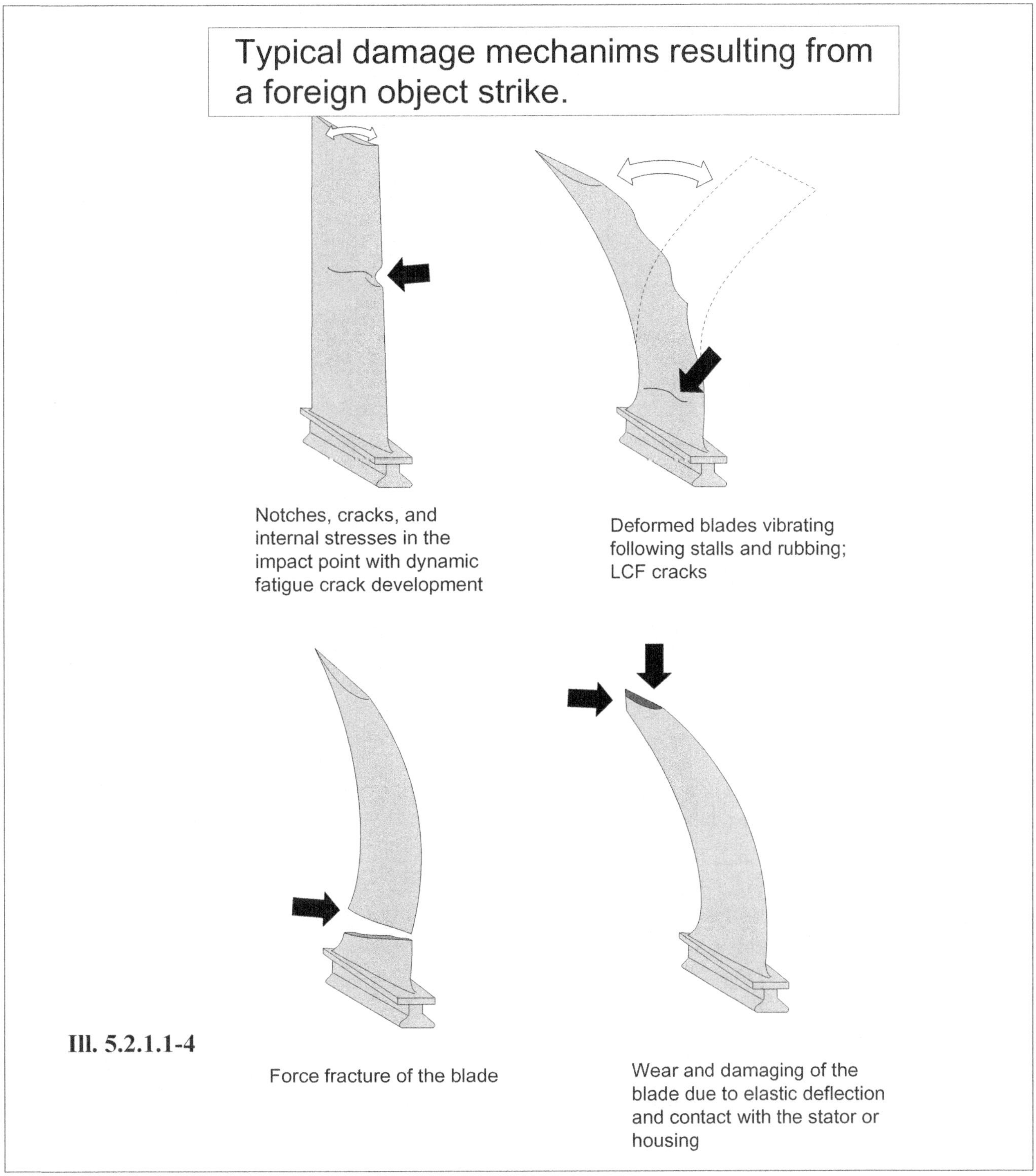

Ill. 5.2.1.1-4

Illustration 5.2.1.1-4: *Foreign object strikes can* ***damage rotor blades*** *in different ways. If the blade does not fracture immediately due to overstress, then the biggest danger is from a dynamic* ***fatigue crack*** *from increased dynamic loads (rubbing, stall) and/or from increased tensile stress at the impact notch (see the introduction to this chapter).*

Blade fractures in the fan and low-pressure compressor due to small notches caused by foreign object impacts.

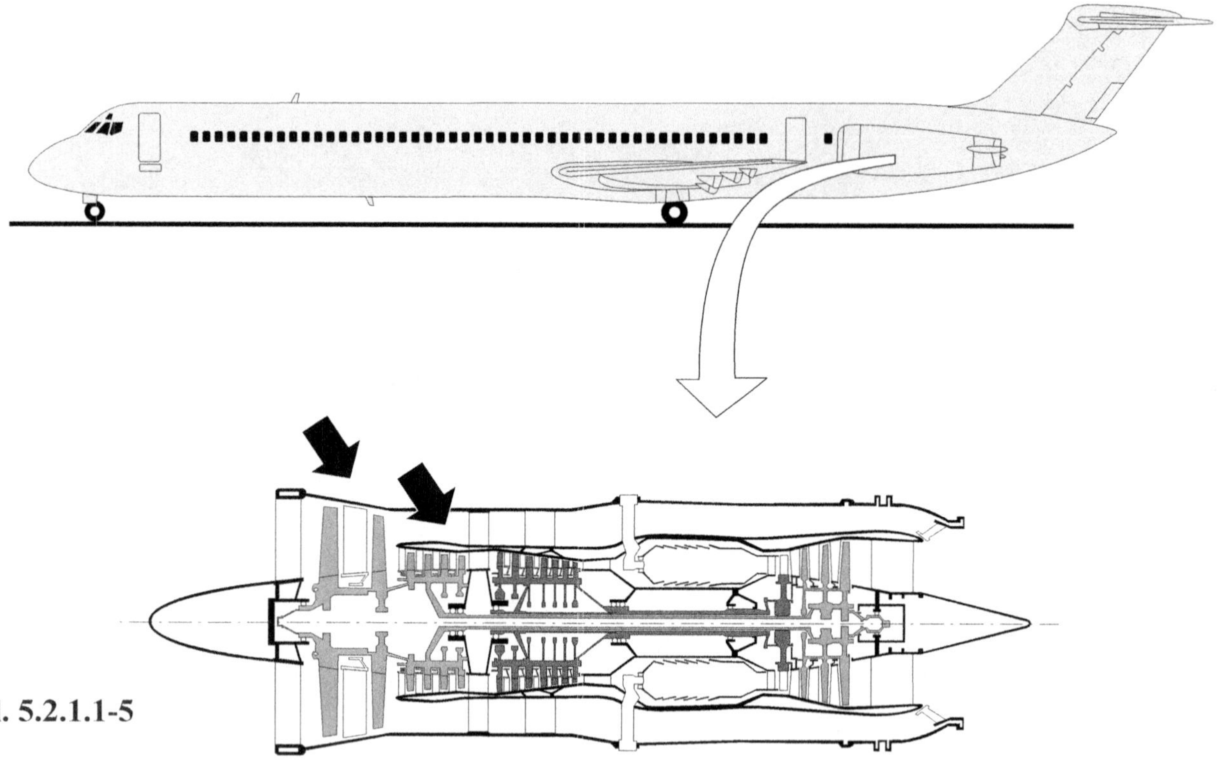

Ill. 5.2.1.1-5

Illustration 5.2.1.1-5 *(Ref. 5.2.1-3): This is a diagram depicting the events in Example 5.2.1.1-1 (possible variations depending on the engine and aircraft types). It shows the primary zones for blade failures due to dynamic* ***fatigue fractures*** *originating in notches (FOD, also see Ill. 5.2.1.1-6). It is interesting that this foreign object problem especially affects* ***engines mounted on the rear*** *fuselage. Evidently this configuration promotes foreign object ingestion. Possible reasons for the high probability of FOD are objects thrown up by the* ***landing gear*** *(Ill. 5.2.1.2-1) and the possibility that ice from the fuselage or wing could have been carried into the engine by the air flow (also see Ill. 5.1.4-2).*

Example 5.2.1.1-1 (Ref. 5.2.1-3):

Excerpt: "..Optimized airfoils, seals, burners etc. combined with increasing focusing on weight reduction, lead to a optimum design that by nature will be much more vulnerable to foreign object/debris...What first lead ....to a chronic lack of spare engines caused by many unscheduled removals due to FOD...

A even more serious experience, severely affecting flight safety, was the alarming increase in compressor blade fractures. In particular, blades in the fifth and seventh stage of the Low Pressure Compressor (LPC) fractured due to fatigue, in most cases originating in impact damages (small nicks) from foreign objects.

In many cases these blade fractures caused severe damage downstream the engine, thus requiring a complete engine overhaul...

A rather new flight safety issue brought up as a result of FOD, is the stress concentration nicks are causing on fan-blades.

Just in the ...18 months the industry has experienced 4-5 fan-blade fractures, all originating in nicks/cracks caused by foreign objects...

Stress analysis conducted by the engine manufacturer has revealed that the stresses in the fan blades, in particular close to the blade root (see Ill. 5.2.1.1-7), has found to be in the upper region of what was predicted..."

Comments: Possible conditions for the damage are shown in Ill. 5.2.1.1-5. It seems especially important that blade improvements undertaken to increase performance evidently caused this increase in blade failures due to the improved bladings greater sensitivity to FOD.

Fatigue fracture of a fan blade shortly after a FOD during start.

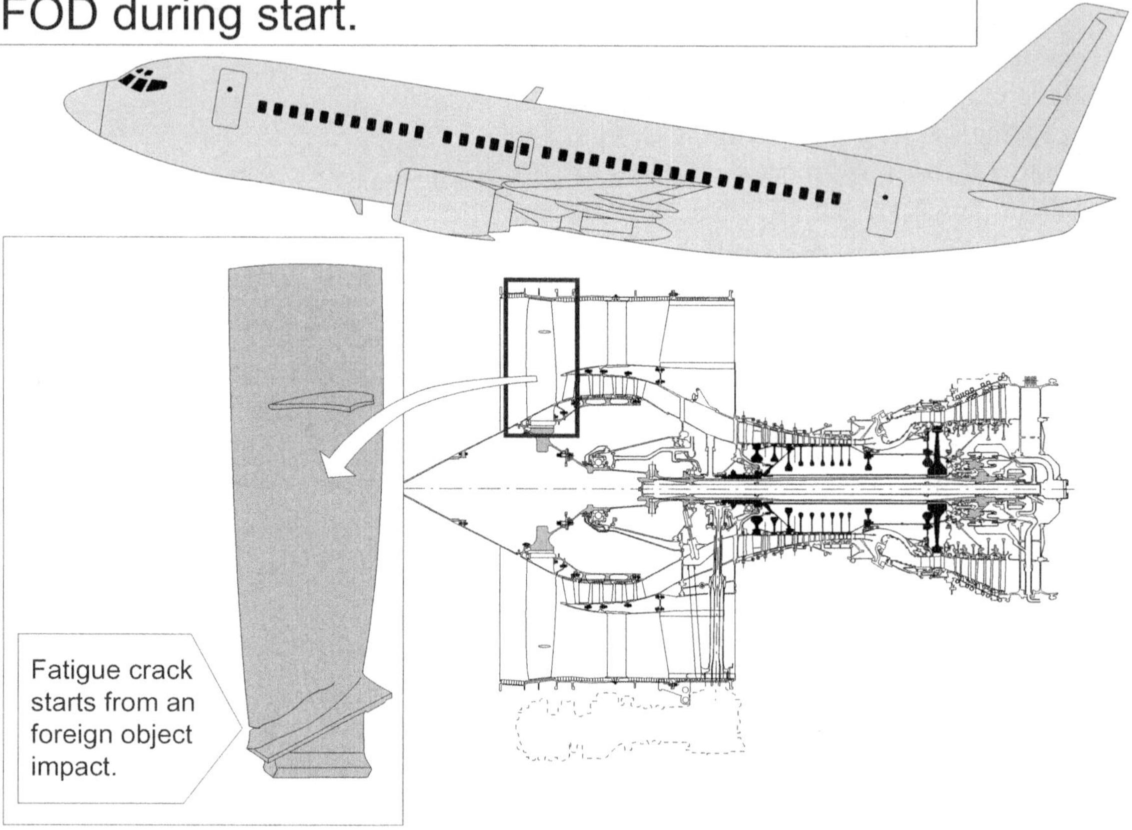

Example 5.2.1.1-2 (Ref. 5.2.1-4):

Excerpt: "...that the fan blade failure ...was caused by foreign object damage. The fan blade, thought to have been damaged during the take-off run...failed at around 7000ft (2100m) and broke off close to the root, causing 'considerable secondary damage to the adjoining blades'."

Comments: This case shows that even the fan blades of modern engines with optimized damage resistance can suffer dynamic fatigue fractures due to foreign object strikes. The relatively short operating time between the foreign object strike and the blade failure is astonishing. This indicates relatively high dynamic loads acting on the blading.

At this also point further incidents at the same type of aeroengine and aircraft (volume 3 Ill. 12.6.3.1-9 and 12.6.3.3-5).

llustration 5.2.1.1-6: FOD in the compressor *is always a potential threat for the aircraft. Extensive damage that makes itself known through the operating behavior of the compressor is easily observable is unpleasant and possibly very expensive. However, damage that occurred at a far earlier time than the ultimate extensive consequential damage can be extremely dangerous. Typical examples of times prone to this type initial damage include a* ***delivery flight*** *or a flight after an overhaul, which experience has shown to be accompanied by an increased*

External Factors
Foreign Object Damage: Fundamentals

The danger posed by a notch also depends on its location on the blade and the blade geometry.

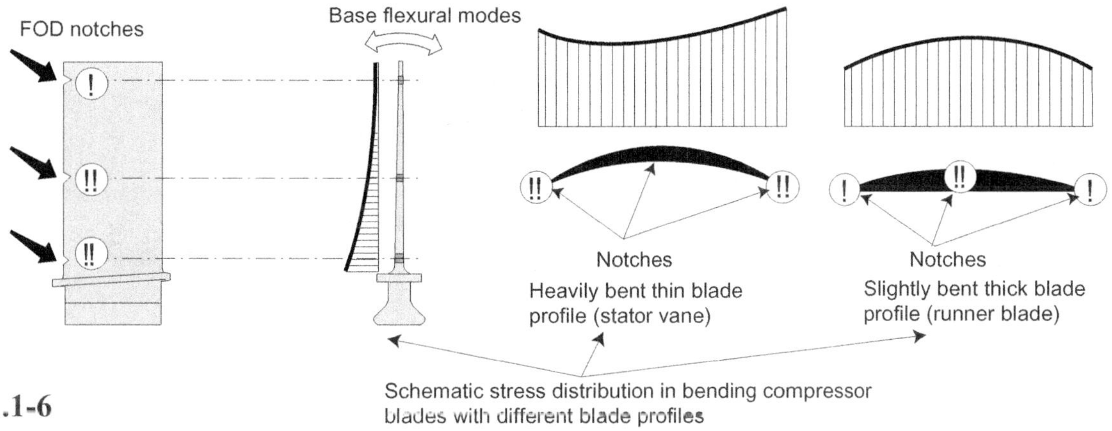

Ill. 5.2.1.1-6

risk of FOD. Notches and nicks mean considerably higher local tension (notch effect). The unavoidable dynamic fatigue stress on the blading is less than the fatigue limit of the undamaged parts and can be tolerated indefinitely. Therefore, it does not present any danger of crack initiation to parts without notches, but the ***notch effect*** *would increase this stress to dangerous levels. Because the vibrations usually only occur for a short time during flight through resonances, these load changes accumulate until a crack is formed and the blade fractures. In* ***titanium compressor blades*** *at the rear area of the compressor, there is the additional danger of a* ***titanium fire*** *occurring, even if this is fairly unlikely. This type of damage should be avoidable with the aid of regular* ***boroscope inspections*** *and a standard for evaluating the findings. It has only been recently that potentially fast and inexpensive techniques have been introduced that* ***smooth out notches*** *without having to open the housing. This type of procedure requires a suitable cutting technique (also see Ill. 5.2.1.1-9). A suitable cutting tool is fed through the boroscope opening. The reworked area might then be hardened through* ***laser peening*** *in a similar manner.*

Dangers at blisks and blings by mistuning *develop during* ***seemingly small changes of the natural frequency of blades****. Such are just to be expected from a* ***rework****. So the concerned or* ***neigbored*** *blades can be excited during operation to* ***multiple increased vibrations*** *(violume 4 chapter 16.2.2.9). For this reason the* ***rework standards****/specifications of inserted blades must not simple adopted. Instead for every tyspe of blisk rework limits, based on investigations and experience, must be developed.*

The danger posed by a notch also depends on its location on the blade and the blade geometry.

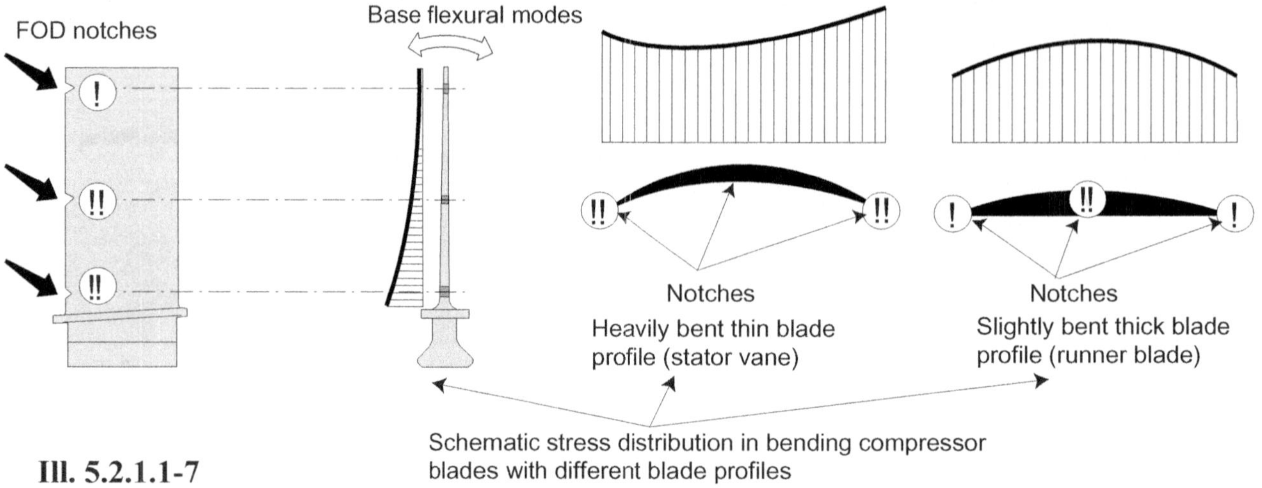

Ill. 5.2.1.1-7

Illustration 5.2.1.1-7: *The promoting effect* ***foreign object impact notches*** *have on dynamic fatigue crack initiation in blades also depends on the* ***location of the notch on the blade****. Since the most common type of vibration is the flexural mode, notches in the root platform are especially dangerous, as the highest dynamic stresses can be expected in this area (see Example 5.2.1.1-1). An* ***'Eiffel tower-shaped' blade*** *cross-section can reduce this tendency. The profile of the blade also has an effect. Very curved, thin blades of the type used in stators are especially sensitive to damage on the* ***blade edges****. Thicker, less curved blade profiles such as those of rotor blades are also sensitive to notches towards the middle of the profile.*

Typical characteristics and appearances of small foreign object impacts.

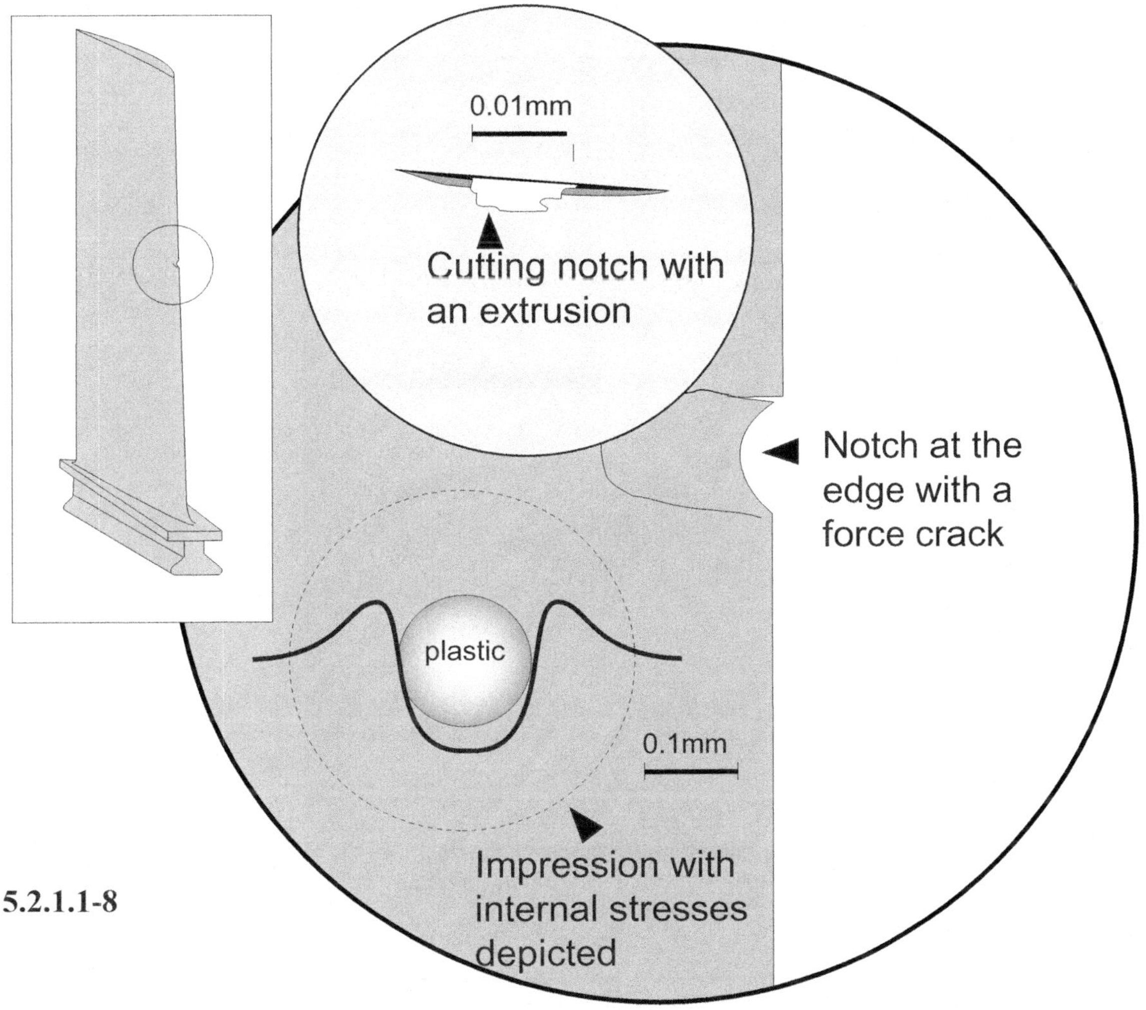

Ill. 5.2.1.1-8

Illustration 5.2.1.1-8: *A notch is not simply a geometric change that affects the dynamic* ***fatigue strength*** *through its shape. Other important factors include its location in relation to the greatest dynamic loads (with cut-like notches, for example), the size and distribution of residual stresses induced by the plastic deformation, any overload fractures or cracks (Ill. 5.2.1.1-10), as well as material changes such as* ***embrittlement*** *or ruptures of hard surface coatings such as oxide layers. In the cut-like notch in the above diagram, there is an extrusion similar to those that are occasionally observed on a microscopic scale in surfaces subject to extreme dynamic loads. The extrusion consists of material that has been pressed out of the notch through the "****fulling****" process.*

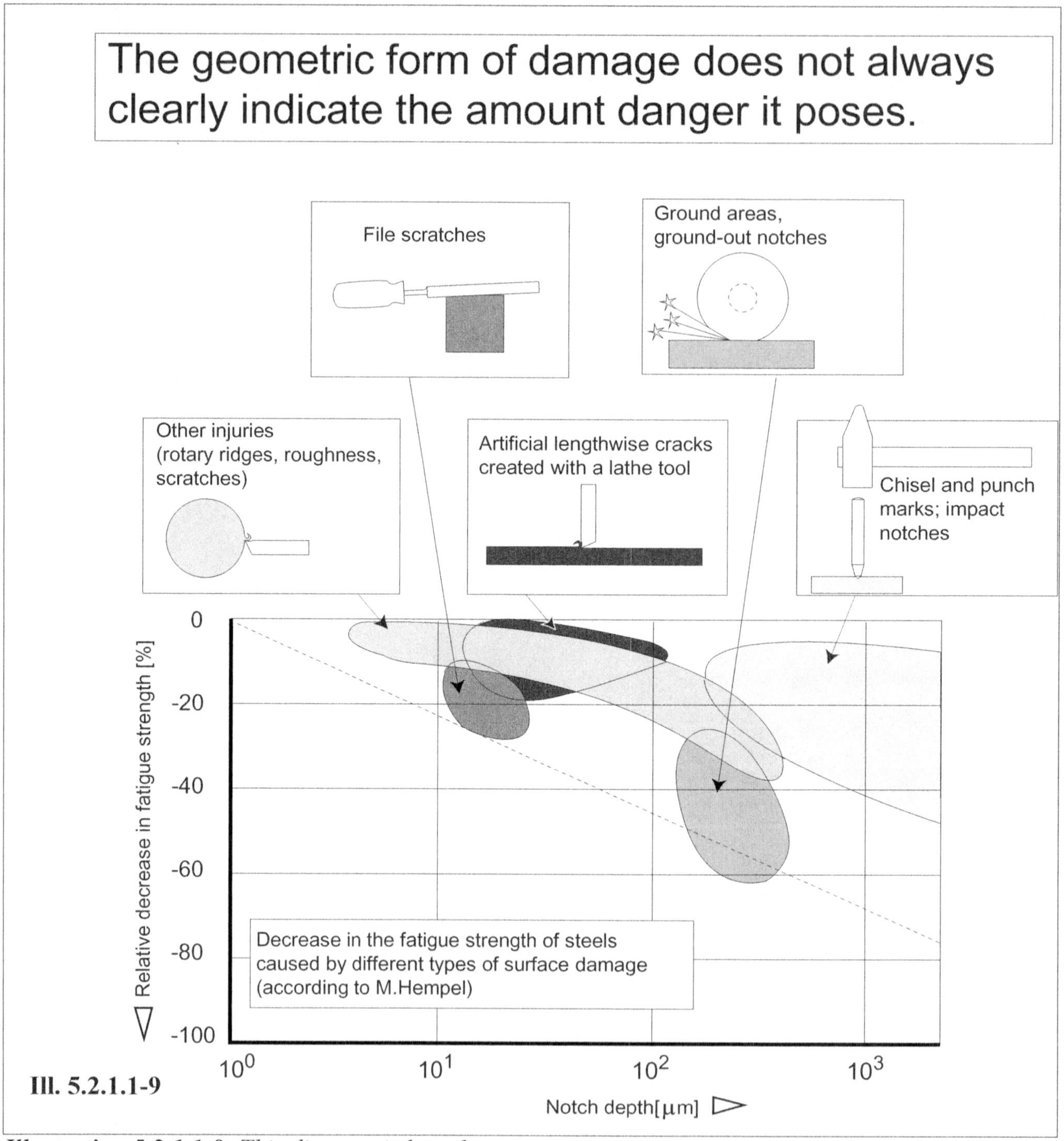

Ill. 5.2.1.1-9

***Illustration 5.2.1.1-9:** This diagram is based on Ref. 5.2.1-5 and shows the effect various **kinds of notches** and surface damage have on the dynamic strength of steel samples. It can be assumed that similar effects would be observed in titanium alloys and superalloys. These "injuries" are especially dangerous to **titanium alloys** due to their well-known high sensitivity to sharp notching. On the other hand, the hardness and relaxation at relatively low operating temperatures should create other focal points of the damaging influences. Unfortunately the influence of the impact speed when the creating the notches cannot be deduced from the diagram, since this*

continued on following page

External Factors
Foreign Object Damage: Fundamentals

Also in the turbine a small FOD respectively OOD can cause big secondary failures.

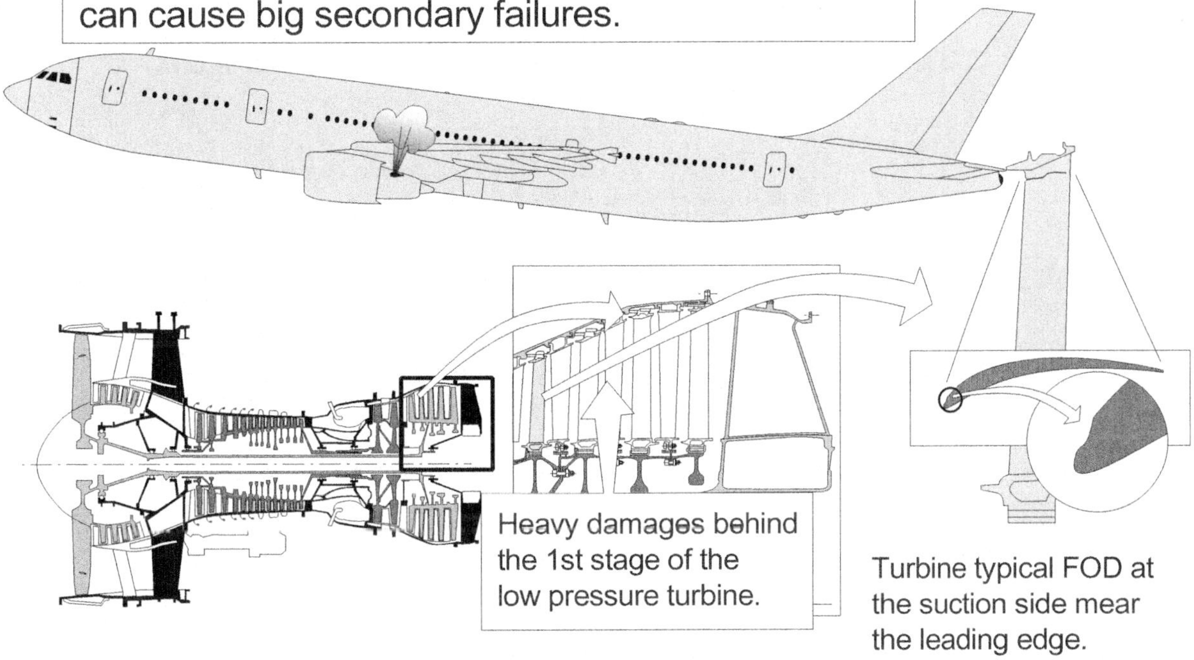

Example 5.2.1.1-3 (Ref. 5.2.1-8):

Excerpt: "...(the large twin-jet commercial aircraft) experienced an uncontained low pressure turbine (LPT) blade fracture in the....engine installed on the No1, left, position, shortly after takeoff...The flight crew reported that as the aircraft was climbing through 1000 feet, they heard a "soft thunk" and the No 1 engine spooked down to about idle. The flight crew also reported that the No.1 engine remained running until they shut down after landing.

...The disassembly of the engine.. .(showed). The LPT case had six holes, including one that was 17.375 inches (about 45 cm) long circumferentially and 2 inches wide axially, in the plane of the LPT Stage 4 rotor. A T4.9 thermocouple probe was found with the tip broken off just inboard of the LPT case flowpath wall. There was one LPT stage 1 blade that had blade tip broken off and the remaining downstream LPT stages had damaged the blades that was progressively worse to the rear of the engine...Examination of the T4.9 probes, including the probe with the missing tip, was inconclusive because of the oxidation and corrosion on the fracture surfaces. The examination of the LPT Stage 1 blade **revealed a fatigue fracture that originated from an impact mark on the convex side of the airfoil just in back of the leading edge**."

Comments: This is the typical impact point of foreign objects on turbine rotor blades as described in Ills. 5.2.1.1-11 and 5.2.1.1-12. It is surprising that a seemingly harmless blade fracture led to an 'uncontained failure'.

OOD = Own Object Damage = damage, caused from **aeroengine own foreign objects**

Small cracks originating in FOD can damage the dynamic strength of blades more than expected.

Impact crater of spherical FOD

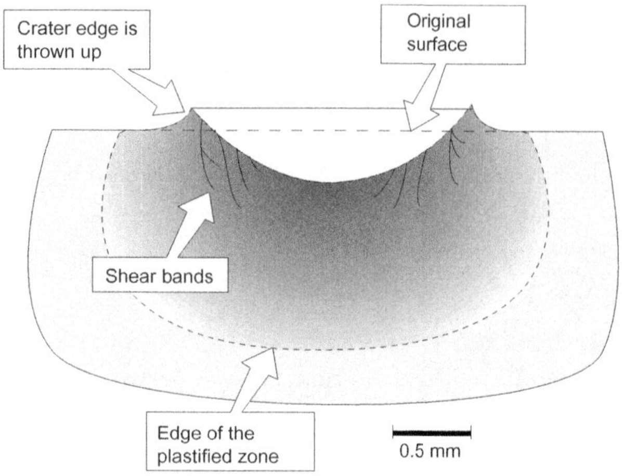

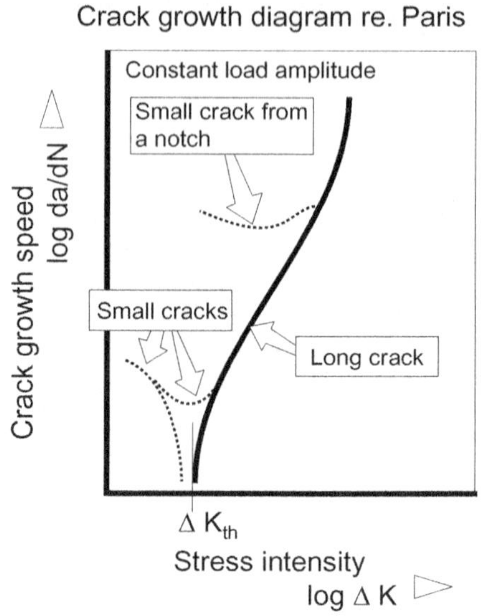

Ill. 5.2.1.1-10.1

continued from page 5.2.1.1-17

question was not addressed as with was in "Foreign Object Strikes on Bladings".
However, several important insights can be gained:
While the permanent strength loss increased along with the depth and sharpness of the notch as expected, these geometric factors were frequently less important than the influence of ***the way in which the damage*** *occurred. For example, ground areas or notches that have been* ***ground*** *can be especially prone to dynamic crack initiation. However,* ***hammered*** *or* ***chiseled*** *notches affect the fatigue strength of materials that are ductile or not sensitive to notching much less than they affect the fatigue strength of extremely hard, brittle materials.*

Illustration 5.2.1.1-10.2 *(Ref. 5.2.1-9):* ***HCF-dynamic fractures*** *in compressor blades are often caused by* ***foreign object strikes.*** *Crack progression data is usually taken from samples with cracks that extend through the entire cross-section. These data do not necessarily apply for damages caused by small foreign object strikes (diagram), which may be only tenths of a millimeter across (Ills. 5.2.1.1-8 and 5.3.2-2).*
Shear bands on the surface of impact craters have only been observed at middle impact speeds. High impact speeds caused raised crater edges. Heavy impact damage such as this results in a considerable loss of dynamic strength.
Cracks of the order of magnitude of the microstructure (grain size) or of the plastic zone at the crack tip can grow at much greater speeds than larger cracks at the same tension intensity. *Small* ***cracks can grow even below the threshold*** *value ΔK_{th} (diagram). This must be taken into account when determining the dynamic strength of blades during design.*

The impact speed of particles has a characteristic effect on the resulting damage.

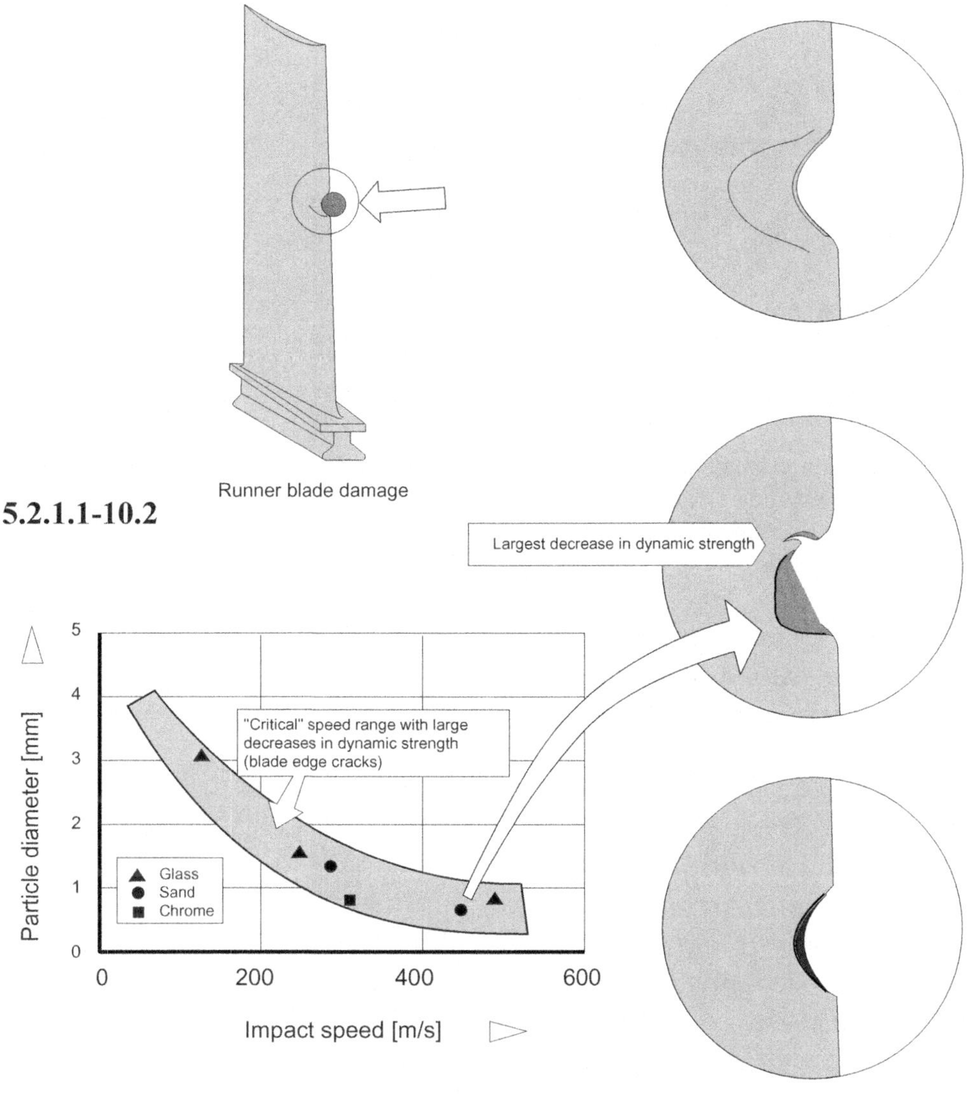

Ill. 5.2.1.1-10.2

Illustration 5.2.1.1-10.2 *(Ref. 5.2.1-6): Interesting material behavior was reported in tests in which spherical particles of various sizes and materials were fired at different* ***speeds*** *against blade and sample edges made from a high-strength* ***titanium alloy****.*

*At relatively low impact speeds, largely independent of the foreign object material (!), bulge-like plastic deformations (top right diagram) with little influence on the dynamic strength of the part At high impact speeds the edge was punched out cleanly (bottom right diagram), also with only a moderate decrease in dynamic strength. The greatest losses of dynamic strength occurred at a transitional impact speed (****critical impact speed****), which was only slightly dependent on the particle diameter, inside of which the edge cracked.*

External Factors
Foreign Object Damage: Fundamentals

Rotor blades in compressor and turbine are hit from foreign objects at different sides of the blade.

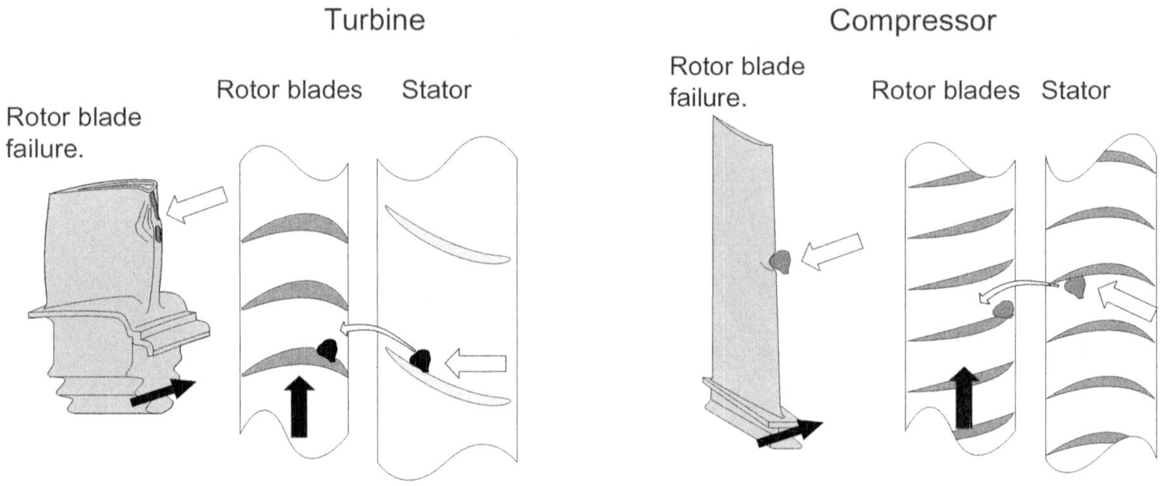

In the turbine the impact of foreign objects (e.g., coke particles or spalled particles of thermal barrier doatings) are located typically at the suction sice of the rotor blade.

In the compressor, foreign objects usually strike the pressure sid of the rotor blades.

Ill. 5.2.1.1-11

Illustration 5.2.1.1-11: *The* ***impact direction*** *of foreign objects is* ***different in compressors and turbines****. This has a considerable effect on the location of blade leaf erosion and impacts on the blade edges.*

If one assumes that a larger foreign object (particles larger than about 0.01 mm usually follow the gas flow without being deflected) does not follow an ideal path and travels at a lower speed than the flow speed due to its aerodynamic profile, then this particle will strike the rotating blade against the direction of rotation.

For this reason, foreign object impacts in compressors usually occur from the pressure side. Exceptions may occur if the foreign object is deflected.

In the turbine, the foreign object will strike the suction side of the blade at its leading edge.

Failures/damages at turbine blades caused from impact of spalled thermal barrier particles.

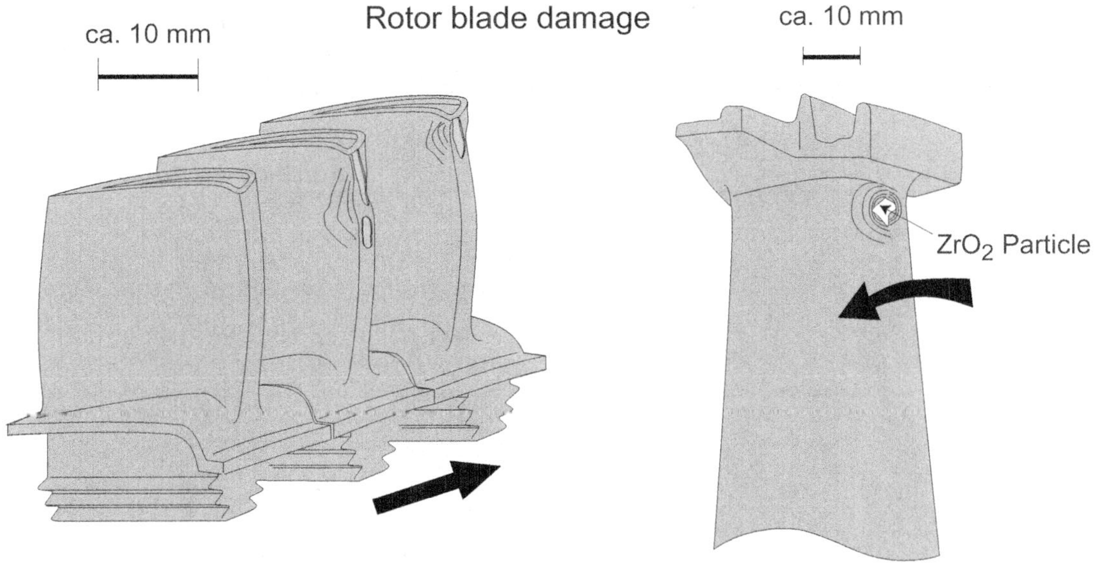

Ill. 5.2.1.1-12

Illustration 5.2.1.1-12: *This diagram shows typical* ***FOD to turbine rotor blades****. The top diagram depicts damage to heavily cooled blades from a small turbine after the impact of carbon from the combustion chamber. Due to the internal cooling ducts, the blade walls are only a few tenths of a millimeter thick and are subject to operating temperatures of up to 1100 °C at the leading edge (see Ref. 5.2.1-7), evidently making them extremely sensitive to seemingly harmless foreign objects such as carbon particles. The impacts were typically on the* ***suction side*** *of the blades.*

The bottom diagram depicts the back (suction) side of a turbine rotor blade which has been struck by a ***ceramic particle from the thermal barrier*** *of either the combustion chamber or a front blade stage. In this case the local plastic deformation led to crack initiation in the relatively brittle protective oxide layer (****Al-diffusion layer****).*

Illustration 5.2.1.1-13 *(Ref. 5.2.1-10): Displayed are features around a foreign object impact at the* ***physical vapour deposited (PVD) thermal barrier coating*** *(TBC) of a turbine blade. The material structure of such TBCs is characterized ba a* ***columnar structure*** *(volume 3, Ill. 11.2.3.1-4). Typical features at the impact region are:*
- ***Permanent indentation****. From it can be concluded at the delamination above the undercoating /adherence coating and so enables the* ***evaluation of the deterioration dimension respectively of a possible spalling****.*
- ***Compacted zone around the indentation****. It is also involved at the introduction of internal stesses into the coating. These increase with the thickness of the coating. In the zone lack the distances between the columnar crystals. It mshows a field of* ***concentric cracks****, which can penetrate to the surface (Ill. 5.2.1.1-12). The compression effect may be especially distinct at PVD coatings. This could be a reason for the comparatively good erosion resistance of this coating type. Erosion does than happen, if the amount of particles is not sufficient to produce a crack formation which leads to the spalling of the coating.*
- ***Crack formation in shear bands*** *runs inclined from the compressed zone to the adhesion coating. Above the adhesion coating these cracks deflect as* ***delamination cracks****. A further crack growth during the operation is supported from induced internal stresses of the impact. Delamination cracks determine the* ***size of a possible spalling****.*
The influence of the base material at the crack formation seems at sufficiens stiffness of the bearing cross section rather minor.
The delamination can be suppressed with an optimizing of the ***coating properties at operation temperature. The hardness must be lowered and the „toughness“*** *(yielding)* ***increased****. This can be achieved e.g., with an increase of the porosity.*

Typical influences at the deterioration/damage of a thermal barrier coating by FOD.

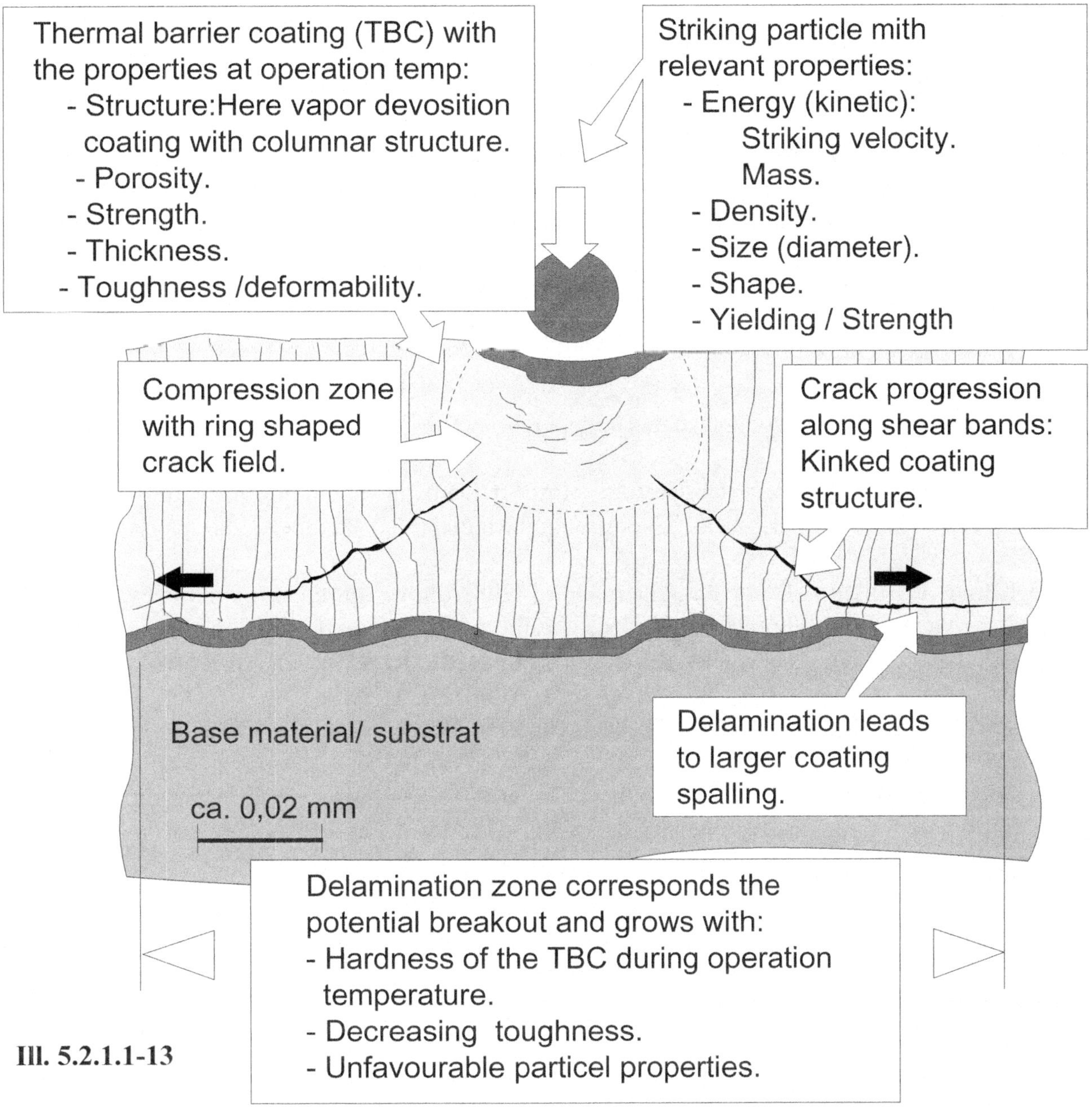

Ill. 5.2.1.1-13

*However with this the erosion strength an drop unacceptable. For a coating specific **threshold of the foreign object energy** for delamination conditions can be specified.*

Chapter 5.2.1.1 References

5.2.1-1 "Carbon Fiber Hazard Concerns NASA", periodical "Aviation Week & Space Technology" March 1979, pages 47-50.

5.2.1-2 J.E. Flynn, "Requirements For Advancement Of Technology, Gas Turbine Engine Monitoring Instrumentation", Report-NAPTC-ATD-163, 1968, pages 43-46.

5.2.1-3 P. Stokke, "Erosion, Corrosion and Foreign Object Damage Effects on Gas Turbines", AGARD-CP-558, Proceedings of the Conference "Erosion, Corrosion and Foreign Object Damage Effects in Gas Turbines", Rotterdam, The Netherlands, 25-28 April 1994, Chapter 7, pages 2 and 3.

5.2.1-4 "Next Generation 737 CFM56 failure was a 'one-off', says CFMI", periodical "Flight International, 1-7 July, 1998, page 12.

5.2.1-5 M. Hempel, "Beeinflussung der Dauerschwingfestigkeit metallischer Werkstoffe durch den Oberflächenzustand", Klepzig-Fachberichte, October 1963, pages 371-382.

5.2.1-6 T. Nocholas, J.P. Barber, R.S. Bertke, "Impact Damage on Titanium Leading Edges from Smann Hard Objects", periodical "Experimental Mechanics", October 1980, pages 357-364.

5.2.1-7 P. König, T. Miller, A.Roßmann, "Damage of High Temperature Components by Dust-Laden Air", AGARD-CP-558, Proceedings of the Conference "Erosion, Corrosion and Foreign Object Damage Effects in Gas Turbines", Rotterdam, The Netherlands, 25-28 April 1994, Chapter 25.

5.2.1-8 NTSB Identification DCA96SA026, Index for Jan 1996.

5.2.1-9 R.O. Ritchie, B.L. Boyce, J.P. Campbell, O. Roder, A.W. Thompson, W.W. Milligan, "Thresholds for high-cycle fatigue in a turbine engine Ti-6Al-4V alloy", periodical "International Journal of Fatigue" 21, 1999, pages 653-662.

5.2.1-10 X.Chen, R.Wang, N.Yao, A.G.Evans, J.W.Hutchinson, R.W.Bruce, „Foreign object damage in a thermal barrier system: mechanisms and simulations", Elsevier „Materials Science & Engineering" A352 (2003) 221-231.

5.2.1.2 Intake of Foreign Objects

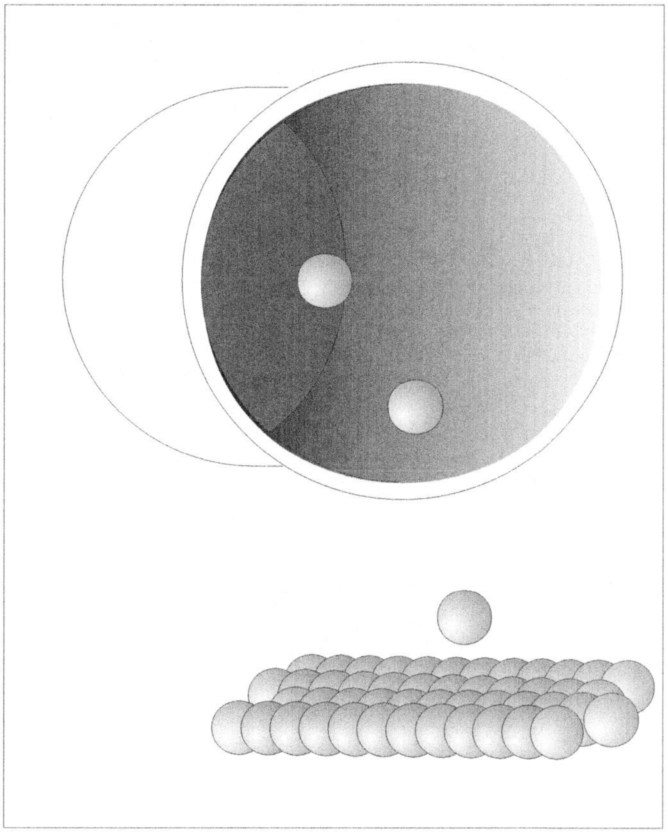

Engines can **ingest foreign objects** in the air. Fairly common foreign objects include objects on the **runway** thrown up by the **landing gear** or parts of the landing gear itself, such as tire pieces (Ills. 5.2.1.2-1 and 2). There are attempts to minimize this risk through suitable constructive measures. There is also a risk when using **thrust reversers**, since these may throw up particles and objects that get dangerously close to the engine intakes (Ill. 5.2.1.2-3).

It seems less natural that **even large compact objects** such as **rocks** or **metal** objects (bolts, etc.) can be **sucked up from the ground**. The powerful suction needed for this is created by (ground-) **vortexes** (Ills. 5.2.1.2-5 to 8). In order to minimize this risk, modern commercial aircraft with large fan engines on the wings with their intakes relatively near the ground (Ill. 5.2.1.2-7) are outfitted with constructive measures to prevent vortexes from forming (Ill. 5.2.1.2-9).

Some Russian aircraft have a very effective method of **preventing the ingestion of foreign objects** during takeoff and landing (Ill. 5.2.1.2-10). They are outfitted with a shutter mechanism in the intake duct that can be closed when the aircraft is near the ground.

The trend towards higher efficiency and more concentrated performance necessitates not only RPM increases, but also optimization of the **blade shapes**. This puts greater mechanical (static and dynamic) and aerodynamic loads on the blades. The trend towards supersonic shapes leads towards sharper and thinner edges. This means that, compared with older engines, smaller damages (roughing, profile changes, deformation) already have an unallowably strong effect and can no longer be tolerated (Ref. 5.2.1.2-2).

An unnoticed damage by an ingested tire fragment of the main landing gear caused an emergency landing.

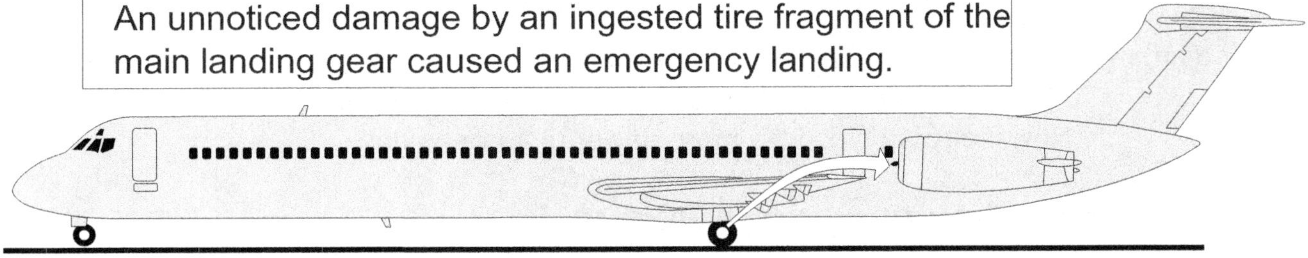

Example 5.2.1.2-1 (Ref. 5.2.1.2-1):

Excerpt: "...The take off roll progressed normally until a point 4,200 feet from initiation. The crew reported a vibration and the left engine power diminished. The take off was aborted, but the aircraft speed at the time of the abort and runway distance remaining was insufficient to avoid departing the end of the runway. The aircraft became mired in the rain soaked terrain up to the axle. Post-incident examination of the **left main gear tires indicated a failure of the carcass**....The cause of the initial power loss on the left engine was from the ingestion of the tire debris."

Comments: This case involved a small cargo aircraft with two engines with small bypass ratios **on the rear fuselage** (similar to Ill. 5.2.1.2-1 top). Experience shows that this engine configuration seems to promote the ingestion of foreign objects that have been thrown up or broken off (example 5.2.1.2-2).

The below shown case is connected with a tire failure which was triggered by a sharp foreign object on the runway,.**and caused in the critical start phase** (Ill. 2-13) **the loss of thrust**. Additionally it came to a fuel fire, obviously because a large piece of the tire sprang a leak at the wing tank. Nobody in the aircraft survived the following catastrophic crash.

In this case a tire fragment from the main landing gear acted catastrophic. Also an aeroengine was concerned.

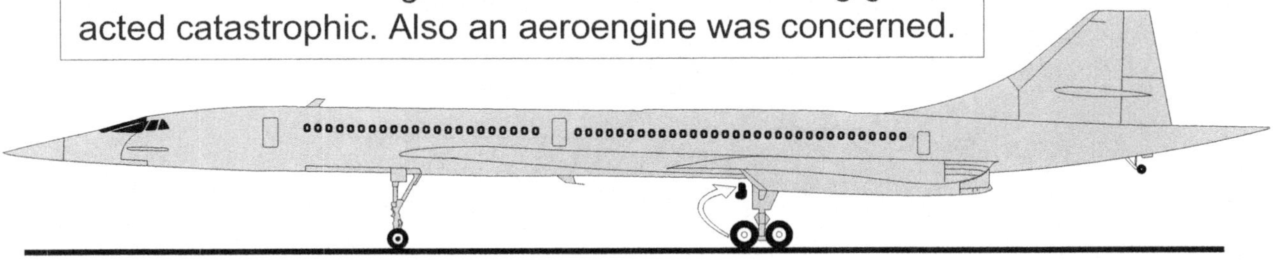

External Factors
Foreign Object Damage: Intake

Both aeroengine types look similar, anyway the frequency of the foreign object damages in the aeroengines is markedly different. Reason is the higher performance, respectively the bigger intake diameter/cross section for the higher flow-rate.

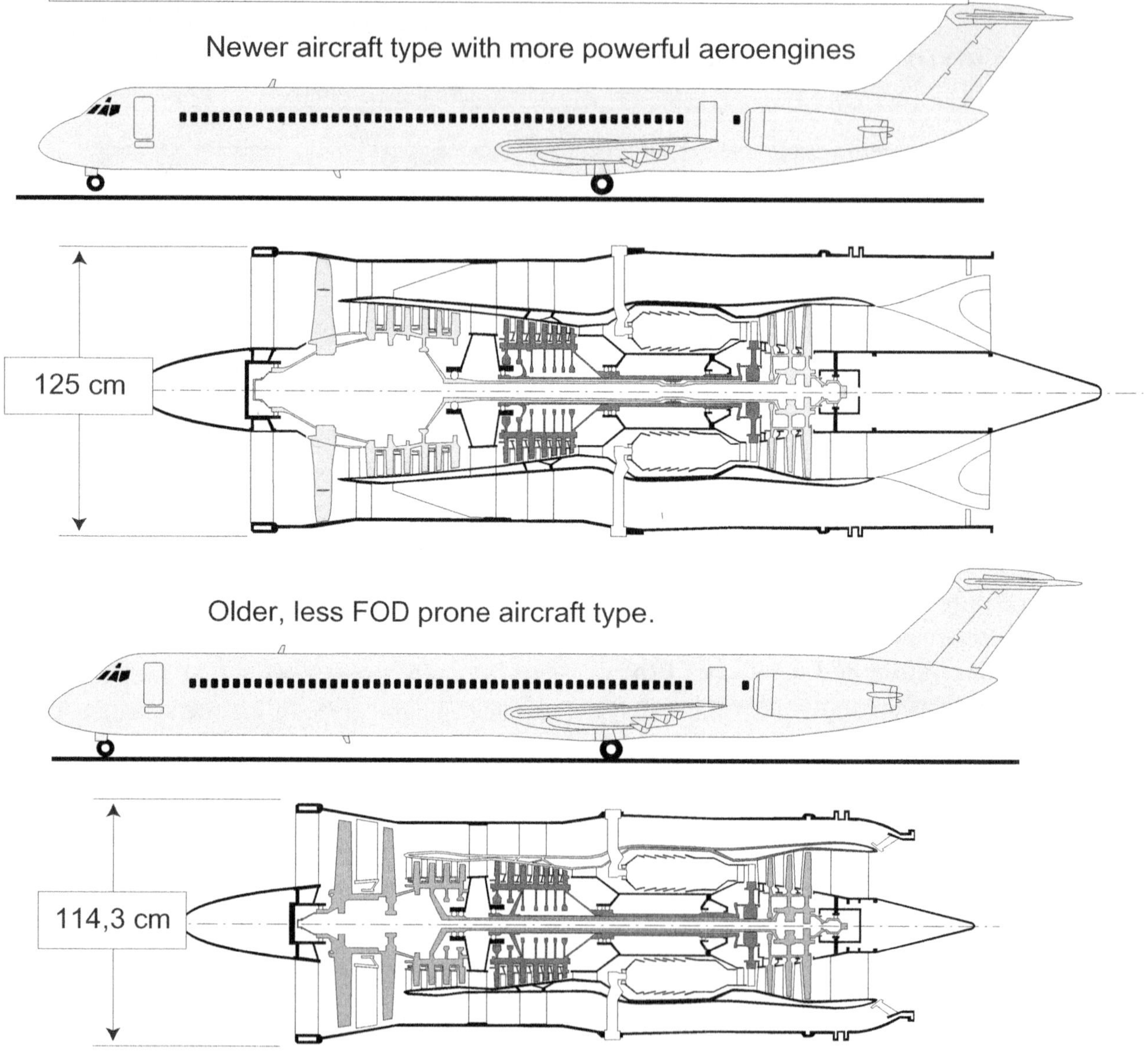

Example 5.2.1.2-2 (Ref. 5.2.1.2-2):

Excerpt: "...Two obvious contributors to the increased FOD-rate in the (transport aircraft with two engines at the end) were increased airflow and approximately 30% larger air inlet compared with (a former type)."

Comments: The probability of an increased FOD risk with larger inlet diameters is also taken into account in the rating regulations (see table in chapter 5.1.1).

External Factors
Foreign Object Damage: Intake

Many different engine configurations can all promote FOD.

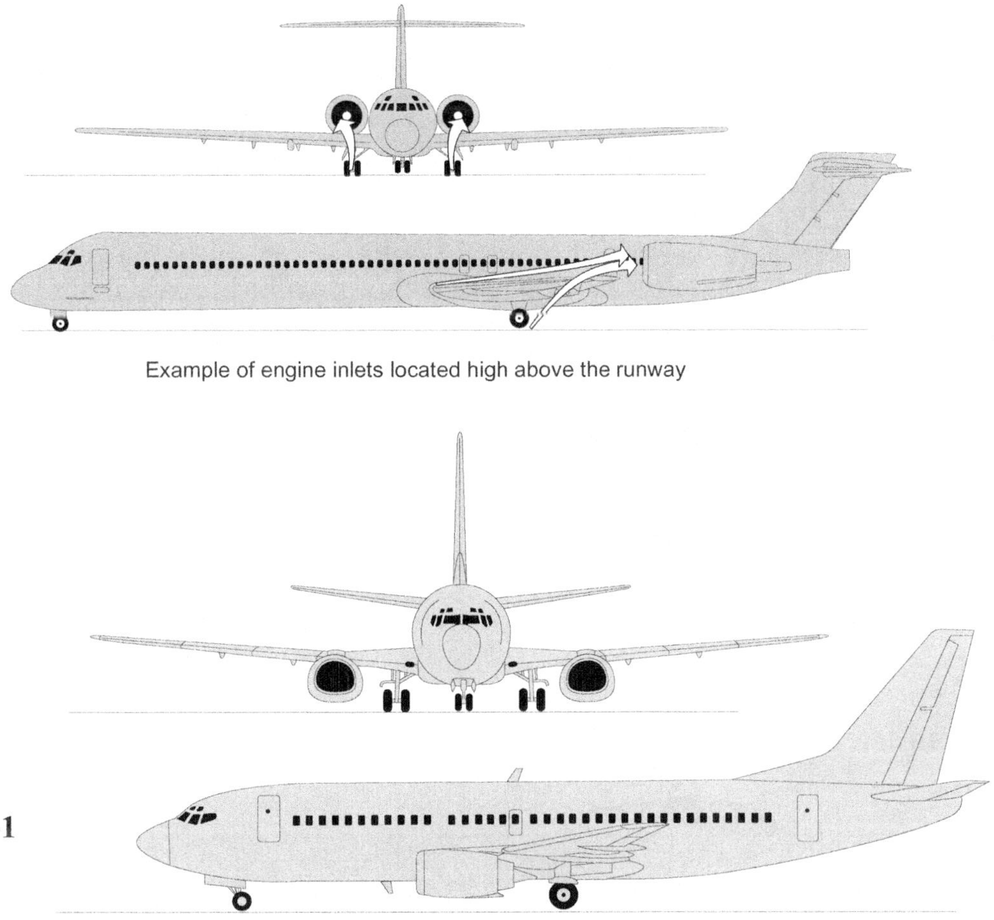

Example of engine inlets located high above the runway

Ill. 5.2.1.2-1

Example of engine inlets located low down close the runway

***Illustration 5.2.1.2-1:** The **location of the engines** affects the probability of foreign objects entering the engine (also see Ill. 5.2.1.1-5). In general, the probability of foreign objects being sucked in by the engine from the ground increases, the lower the engine inlet is to the ground. The probability also increases with the size of the inlet (Example 5.2.1.2-2).*
*However, experience has shown that this is only true for engines that are located on the wing in a similar fashion. **Engines located on the rear of the fuselage** at a relatively high position have a high rate of FOD (Ref. 5.2.1.2-4). Ice breaking off from the fuselage or **broken parts** of the fuselage (fragments of the radome, windshield wipers, etc. Example 5.1.2-1) can easily be sucked up by the engine. Even foreign objects **thrown up by the main landing gear** can threaten engines located at the rear of the aircraft (Example 5.2.1.2-1).*

The position of the nose wheel relative to the engine inlet is decisive for the sensitivity to FOD.

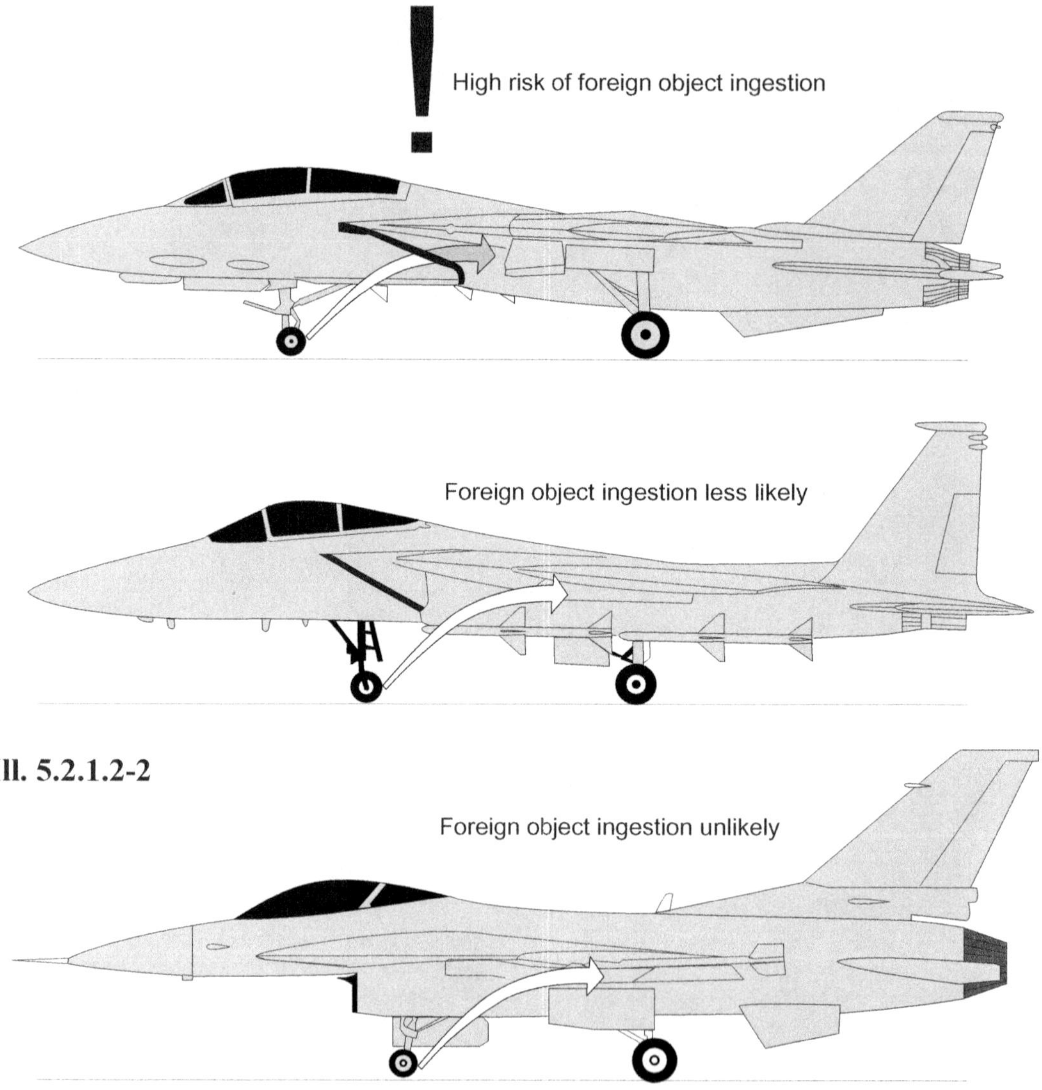

Ill. 5.2.1.2-2

Illustration 5.2.1.2-2: *In **fighter aircraft**, the front wheel is often the cause of foreign objects being thrown up into the engine. The tendency to ingest foreign objects depends on the relationship between the inlet size (diameter, diagonal, cross-section) and the height of the inlet above the ground (Ill. 5.2.1.2-6). The greater this relationship is, the more probable foreign object ingestion becomes. For this reason, the **engine inlets underneath the fuselage** of modern fighter aircraft must be critically considered (see Ills. 5.2.1.2-5 to -8).*

Protective covers on the nose wheel of a fighter aircraft for takeoffs from unpaved runways.

Ill. 5.2.1.2-2.1

Illustration 5.2.1.2-2.1 *(Ref. 5.2.1.2-2): Fighter planes that are intended to take off from unpaved runways are commonly outfitted with deflectors or protective covers on the front landing gear (the diagram shows a Russian design).*

Coating break-outs of the runway caused a surge in one aeroengine and a dangerous aborted take-off.

Example 5.2.1.2-3 (Ref. 5.2.1.2-3):

Excerpt: „.......(The aircrafttype with two rear engines) aborted a take off...after having a **compressor stall on the right engine**. The aircraft returned to the terminal and deplaned normally...The aircraft received only minor damage...The examination of the engine revealed it had ingested bituminous-polymer sealant that had been recently applied on the runway."

Comments: Evidently the **poorly adhering sealant** was thrown up by the landing gear and ingested by the engine. It is surprising that the persons responsible for the condition of the runway overlooked this flaw. Probably it was not a stall at some blades, but a surge.

External Factors
Foreign Object Damage: Intake

Measures to minimize the FOD frequency

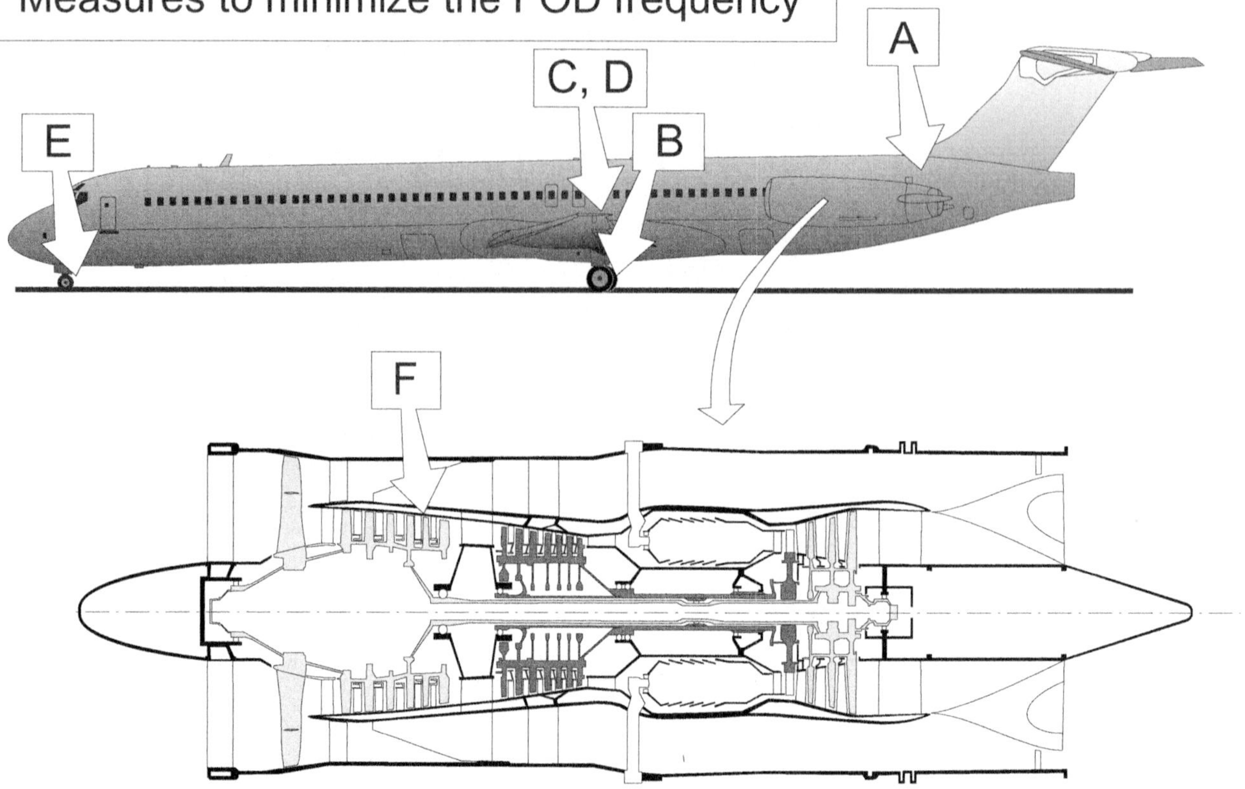

Example 5.2.1.2-3 (Ref. 5.2.1.2-2):

Excerpt: "...the following modifications released in the 1988 to 1990 timeframes:

- repositioning (rotation) the thrust reverser doors.(„**A**")
- improved main wheel water/debris deflector („**B**").
- blocking plate between flap and wing trailing edge.(„**C**")
- debris curtain on wing aft spar.(„**D**").
- improved nose gear water deflector („F").

In addition the engine manufacturer... introduced new improved fifth stage (compressor) blades („**F**") and a new sixth stage anti flutter bleed system to reduce seventh stage blade flutter and - induced stresses.

...all modifications mentioned (were implemented) without seeing the anticipated improvements on the engine deterioration rate.

The only aircraft **modification that has proven to reduce the amount of debris entering the engines are the improved debris deflector on the main wheels**, even though the improvement is marginal.

Comments: This example shows the difficulty of preventing the ingestion of foreign bodies when the basic principle of the **engine location** is unsuitable for this requirement. Even effective **deflectors on the wheels** require a great deal of knowhow. Russian fighter aircraft landing gear have been the subject of much successful pioneering work in the area (Ill. 5.2.1.2-2.1)..

Under poor conditions, even engines located high up on aircraft can suck up large amounts of dust.

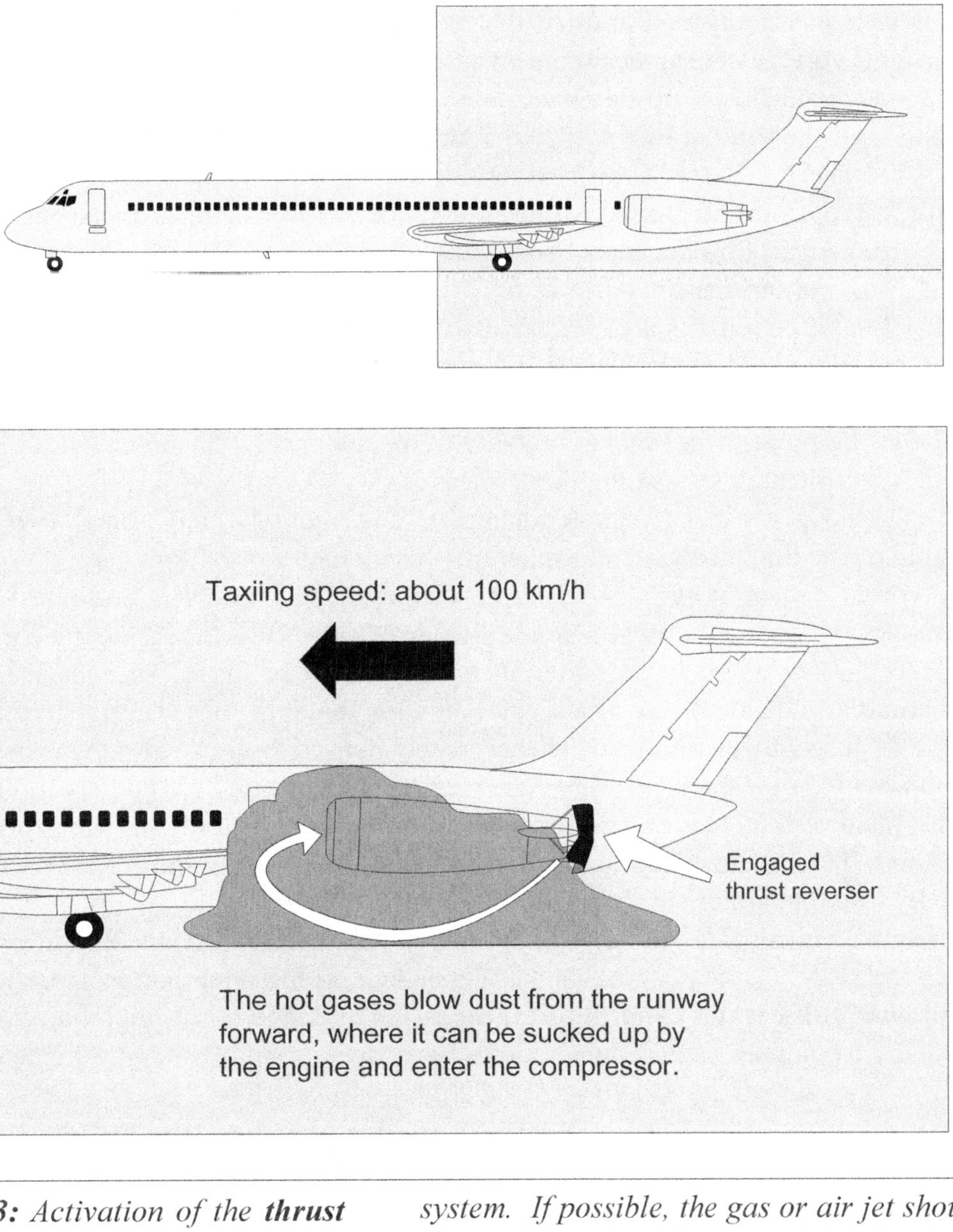

Ill. 5.2.1.2-3

Illustration 5.2.1.2-3: *Activation of the* ***thrust reverser*** *can throw up foreign* ***objects from the runway*** *that are then ingested by the engine (Refs. 5.2.1.2-2 and 5.2.1.2-3). This is also dependent on the direction of the gas jet, i.e. the deflection system. If possible, the gas or air jet should be directed away from the engine inlet. The location of the engines on the side of the fuselage at the rear limits possible options more than the location of engines on the wing does.*

Formation of an inlet vortex in front of engines

A vortex consists of rapidly whirling air that creates a tube-like **underpressure zone**. In nature, vortexes occur as tornados. A comparison can also be made with the whirlpool formed by fluid draining through a hole (Ill. 5.2.1.2-4). In aircraft, vortexes often form around wing edges and can be recognized by condensation in these areas.

The larger the inlet of an engine, and the closer its center is **located to the ground**, the higher the probability of a vortex forming which could suck up dangerous foreign objects. This risk increases with higher bypass ratios and/or inlets that are located to the ground (Refs. 5.2.1.2-5 and 5.2.1.2-6).

Ground vortexes are undesirable in aircraft. The underpressure in the airflow due to the high rotation speed can **throw up dense objects such as bolts and stones from the ground and carry them through the tube of underpressure that is the vortex** (Ref. 5.2.1.2-6). The throwing up of the foreign objects is impulsive. They may include objects from inside cracks on the runway, from the edge of the runway, or from soft ground. Vortexes do not always form. Instead, they require **specific conditions** such as sufficient air flow speed and prevailing wind (strength and direction with regard to engine axis; Ill. 5.2.1.2-8). For this reason, taxi and takeoff procedures are often important factors for the frequency of FOD. If these conditions are met, for example, through an increase in engine performance, then a vortex can suddenly form (Ref. 5.2.1.2-6). It spreads from the engine inlet down to the ground. The larger the diameter of the vortex, the heavier the foreign objects it can suck up (Ill. 5.2.1.2-6, Ref. 5.2.1.2-6). The power of a vortex is related to the cube of its diameter. The following example from Ref. 5.2.1.2-5 illustrates the strength of vortexes:

An engine inlet with a two meter diameter, the center of which is about two meters off the ground, is capable of sucking up a concrete sphere with a diameter of up to 30 cm!

A vortex can be **recognized** as a hose-like shape if there is **condensation** around it due to temperature or pressure drops. Unlike the outside of the vortex, the inside is clear without condensation (Ref. 5.2.1.2-7). This gives the vortex its typical hose-like shape. This shape is created by the lower relative humidity inside the vortex, a corresponding tendency for the moisture to evaporate (steam is clear!), and the centrifugal force which acts on condensed droplets, moving these out of the inner zone. On damp runways the underpressure zone near the ground can be seen in splashing water if the vortex is over a puddle. **The vortex and its end can move very quickly on the ground and also in other directions. This means that maintenance personnel near running aircraft are at risk.** This is especially true for fighter aircraft on aircraft carriers.

Engines with sufficiently high performances can show pronounced vortex formation around the inlet (Ill. 5.2.1.2-5). If the engine intake (dependent on the inlet diameter) is sufficiently **near to the ground, then this creates conditions for ingestion** of dangerous foreign objects from the ground. The trends in both commercial and fighter aircraft are towards smaller height/diameter ratios and therefore towards greater sensitivity to FOD (Ill. 5.2.1.2-8).

! The suction zone must be clean!
(even compact objects such as pebbles and wires can be sucked up by engines).

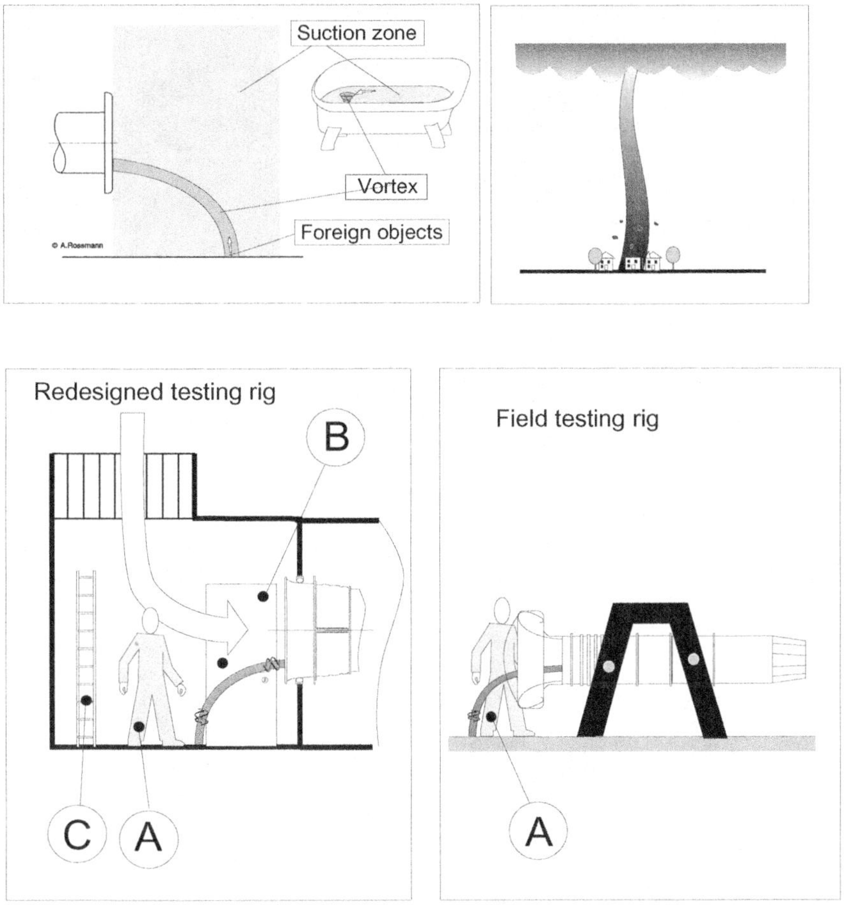

Ill. 5.2.1.2-4

Illustration 5.2.1.2-4 *(see the previous page): The top two diagrams explain the concept of a* ***vortex*** *with the aid of commonly known natural phenomena.*
The bottom diagrams depict phenomena that were observed during tests (Ref. 5.2.1.2-5). These ***phenomena*** *are important for understanding the various factors in the surrounding environment that can have an influence on* ***vortex formation****. These factors should be understood by maintenance crews and testing rig operators in order to prevent accidents or damage.*
For example, it was observed that vortexes became stronger when holding one's hand between the inlet and outlet of the fan of a testing assembly, thus affecting the intake flow. Merely opening the door to the testing chamber or movement by the testing personnel near the test assembly can create vortexes that suck up foreign objects.
Therefore it must be assumed that ***interferences*** *such as people (A), doors (B), and machines (C) in an enclosed testing rig will* ***influence the ingestion of foreign objects****. This may also be the case when* ***maintenance personnel*** *are active near an engine on an aircraft or field testing rig.*

Ingestion of foreign objects due to a vortex (rotating underpressure)

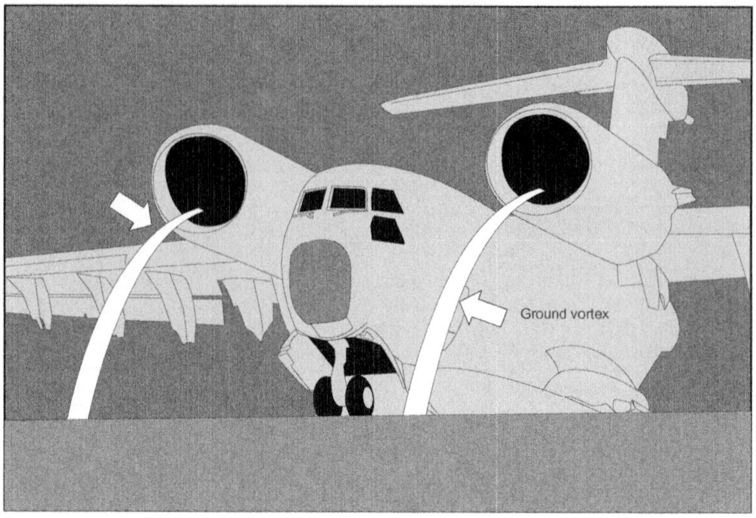

Ill. 5.2.1.2-5

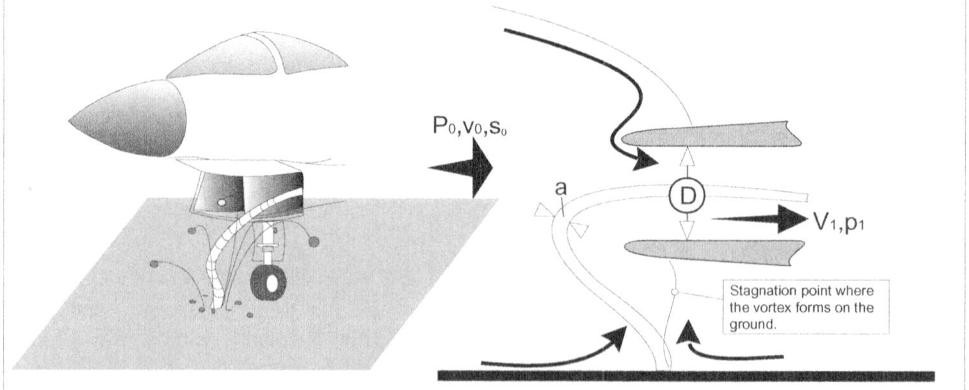

***Illustration 5.2.1.2-5:** The top diagram depicts obvious vortexes formed by a military (testing) STOL cargo aircraft (Ref. 5.2.1.2-7). It is surprising that the engines, which are **located high on the wing**, can create underpressure tubes that reach all the way to the ground and increase the risk of FOD. This is especially disconcerting in the depicted case, since this aircraft is designed for especially short runways, which are not usually as debris-free as longer standard runways.*

*The lower diagram shows the **parameters that influence vortex formation** as explained in Refs. 5.2.1.2-6 and 5.2.1.2-12. The values indicated by "0" are for the surrounding wind conditions, those marked "1" refer to the flow conditions in the inlet. The mass of the foreign object that could potentially be sucked in is proportional to the cube of the **vortex diameter** "a".*

This diagram is intended to give an impression of the capability of engines to suck up foreign objects.

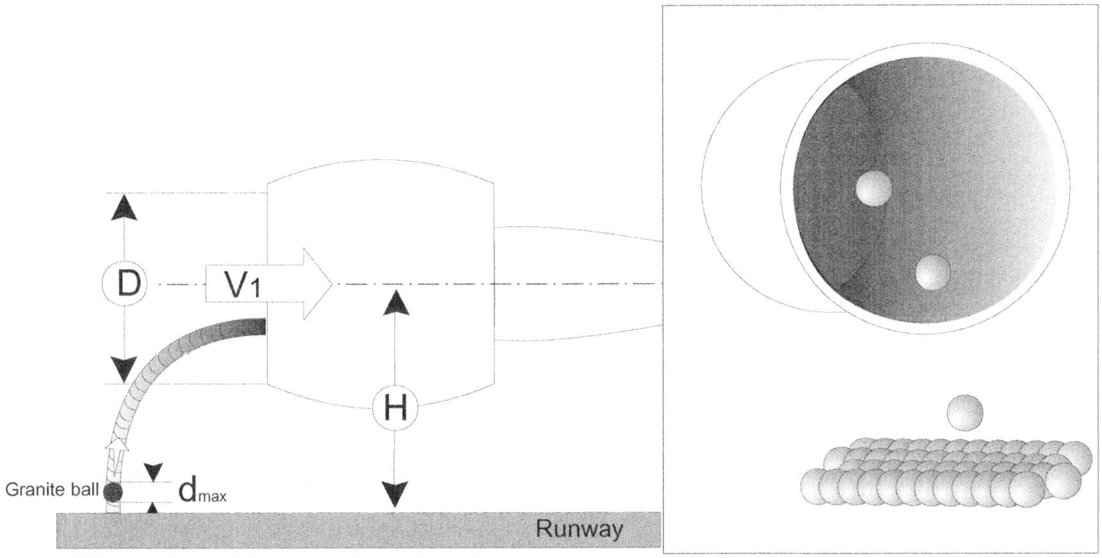

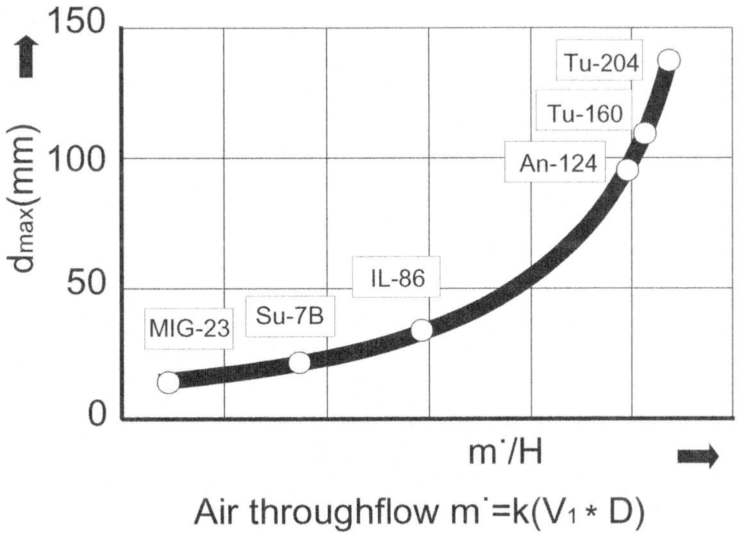

Ill. 5.2.1.2-6

Illustration 5.2.1.2-6: *The top right diagram shows a test assembly described in Ref. 5.2.1.2-5 with a H/D of 1.2, with an inlet diameter (D) of 30 cm sucking up 25 mm diameter glass spheres. The lower diagram is based on Ref. 5.2.1.2-8, and refers to the ratios shown in the top left diagram, giving the* ***maximum ingestible foreign object sizes*** *(granite spheres) and the corresponding (Russian) aircraft engine types. As the diagram shows, in extreme cases rocks with diameters of considerably more than 10 cm can be sucked up, depending on the air flow rate.*

The relationship H/D is an important value for the tendency of an engine to suck up foreign objects.

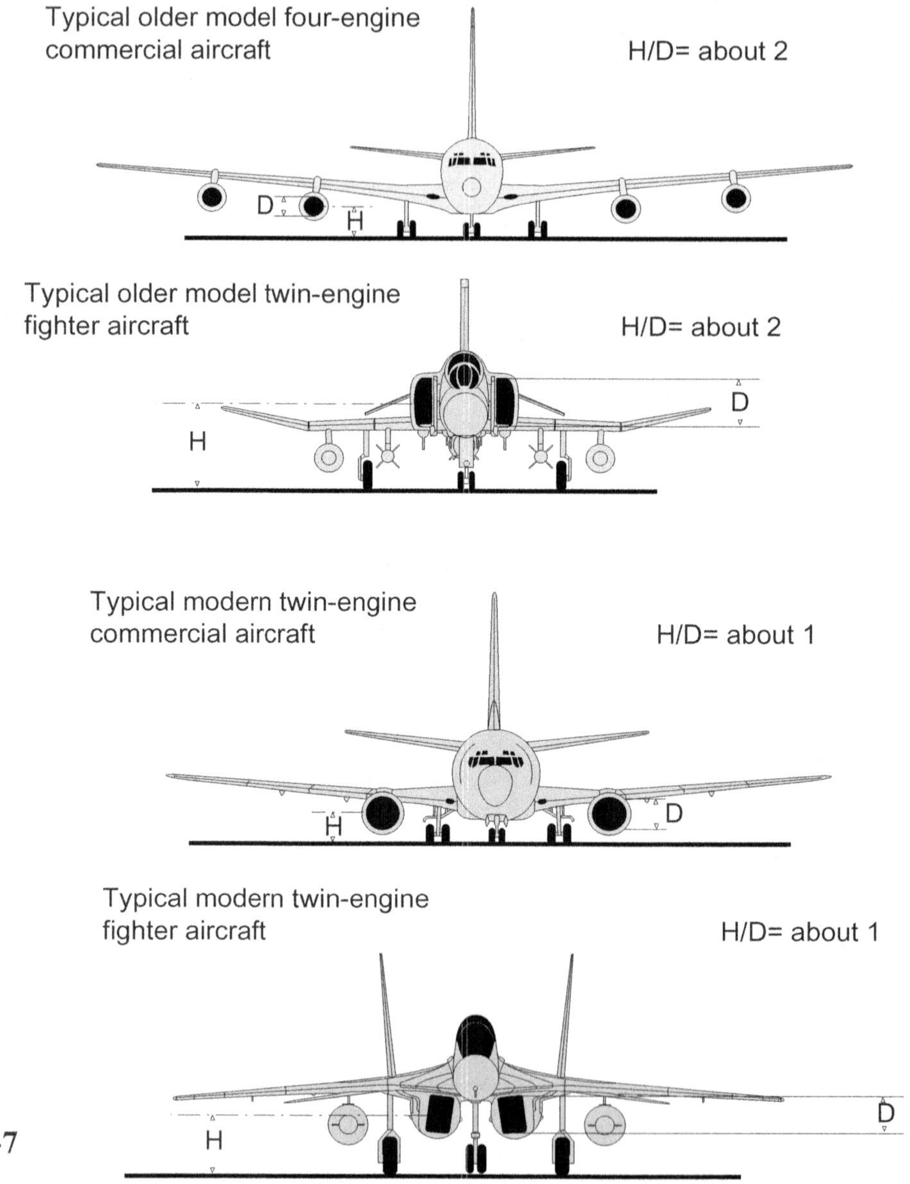

Ill. 5.2.1.2-7

Illustration 5.2.1.2-7: *The ratio of the* ***height of an inlet above the ground to the inlet size*** *(i.e. air flow rate) is important for the probability that an engine on the ground will suck up foreign objects. The smaller this ratio is, the greater the probability of FOD occurring. The recent trend for both commercial aircraft and fighter aircraft is towards smaller H/D ratios. This has led to, for example, the following special measures in inlet of the fighter aircraft depicted below: a* ***shutter mechanism*** *(Ill. 5.2.1.2-10) closes the inlet during start-up, and the engine takes in air through openings on top of the wing during this phase.*

External Factors
Foreign Object Damage: Intake

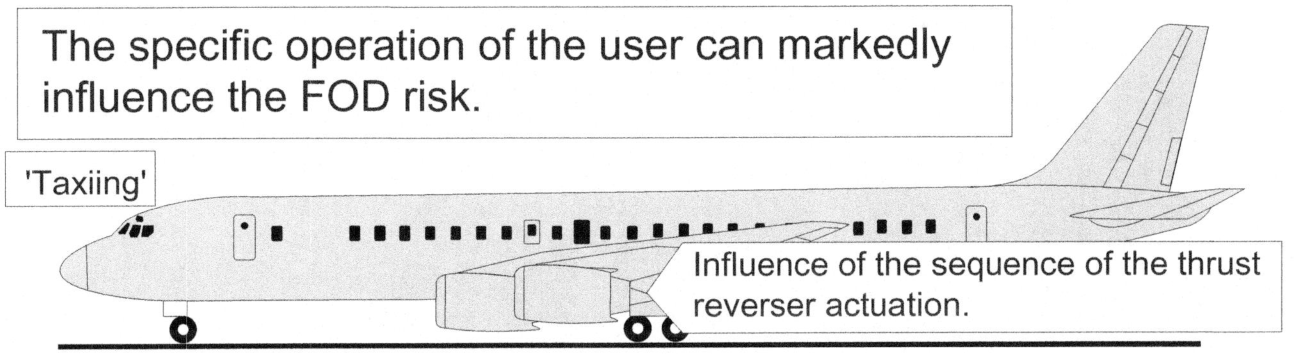

Example 5.2.1.2-5 (Ref. 5.2.1.2-9):

Excerpt: ".....engines that power the carrier's reengined (Aircraft)...have encountered sand and particle ingestion problems.

Changes are being considered seriously by the carrier and the manufacturer... in a continuing analysis of the engines...Aim of the redesign and the vortex dissipater would be to reduce particle ingestion that has been causing high levels of erosion and has necessitated a higher engine removal rate 'than we would like to see'...

The vortex dissipater under consideration would be a tube inserted at the bottom of the cowling that would blow the engine bleed air downward to disrupt vortices that develop under the engine.

... the ingested dirt was **clogging high-pressure turbine blade cooling holes**.

..**taxiing procedures**...probably helped cause more vortices and contributed to the particle ingestion problem.

The procedure borrowed from the carrier's previous practices....involved the activation of **thrust reversers on outboard engines t**o slow taxiing speeds.

(The carrier)... has revised the procedure to use the inboard engines only, one engine at a given power setting at idle and the other at varying speeds."

Comments: The affected aircraft type is an older four-jet commercial aircraft type, outfitted (reengined) with **modern fan engines with large bypass ratios**. It is interesting that improper operation of the thrust reversers changed the flow direction and promoted vortex formation (see Ill. 5.2.1.2-8), leading to an increased risk of FOD.

External Factors
Foreign Object Damage: Intake

Influence of wind on vortex formation

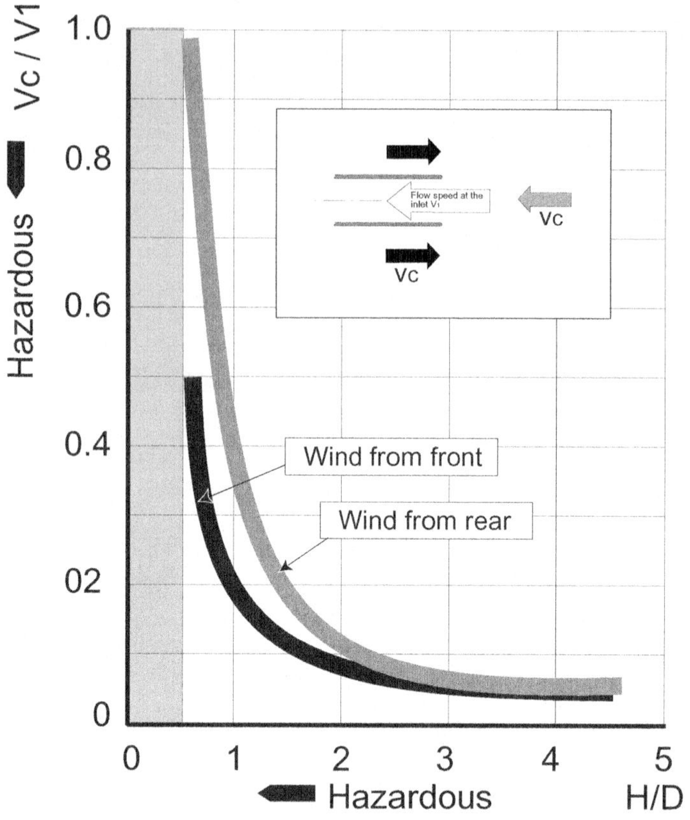

Vc = Speed of side wind
V1= Wind speed at inlet
H = Height of the center of the inlet above the ground
D = Inlet diameter

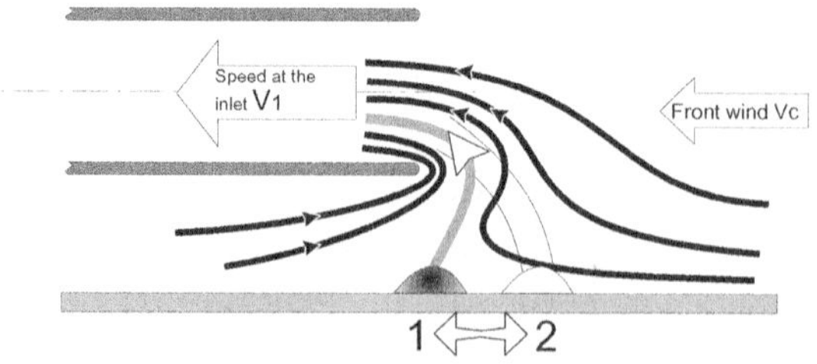

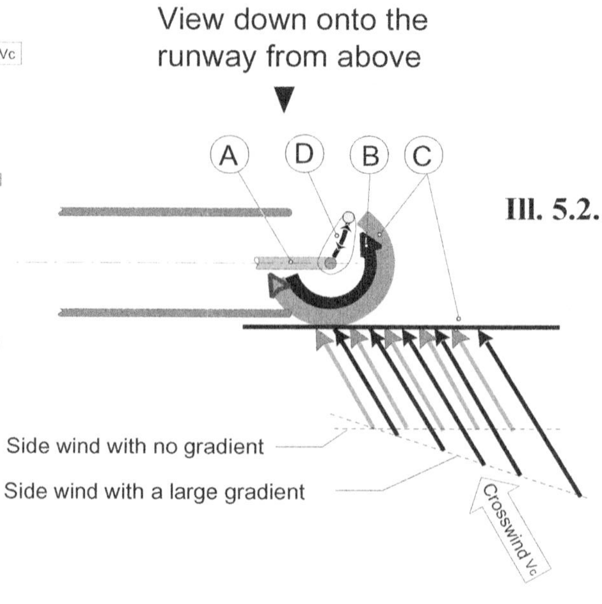

Ill. 5.2.1.2-8

The highest possibility of vortex formation and the accompanying danger of foreign object ingestion is when the v_1 is large and the V_c is small.

***Illustration** 5.2.1.2-8 (Ref. 5.2.1.2-5): The **wind conditions** around the inlet play an important role in **vortex formation**. This effect increases with smaller H/D ratios and weaker side winds (top diagram). The range for vortex formation lies below the curves for headwind (top curve) and tailwind (bottom curve). Therefore, with small H/D ratios, vortexes can form even with considerably stronger side winds. **The most dangerous situation is to operate a high-performance engine in a windless environment**. Since the most powerful suction effect of the vortex is at "A" in the range of the stagnation flow line, the strength of the headwind is especially important in determining its location. With strong headwind, such as during taxiing, the stagnation flow line is shifted backwards ("1" in the bottom left diagram) and loses its contact with the ground. When there is no wind, it is located in front of the air inlet ("2").*

*During operation of **thrust reversers** while taxiing, the forward-directed flow can reduce the vortex-preventing headwind and additionally contribute to the throwing-up of any foreign objects on the ground, thus increasing the risk of foreign object intake (see Example 5.2.1.2-4).*

The side wind gradient affects the direction of rotation of the vortex (bottom left diagram).

External Factors
Foreign Object Damage: Intake

In dusty environments, it may be necessary to implement specific measures to prevent vortexes from forming.

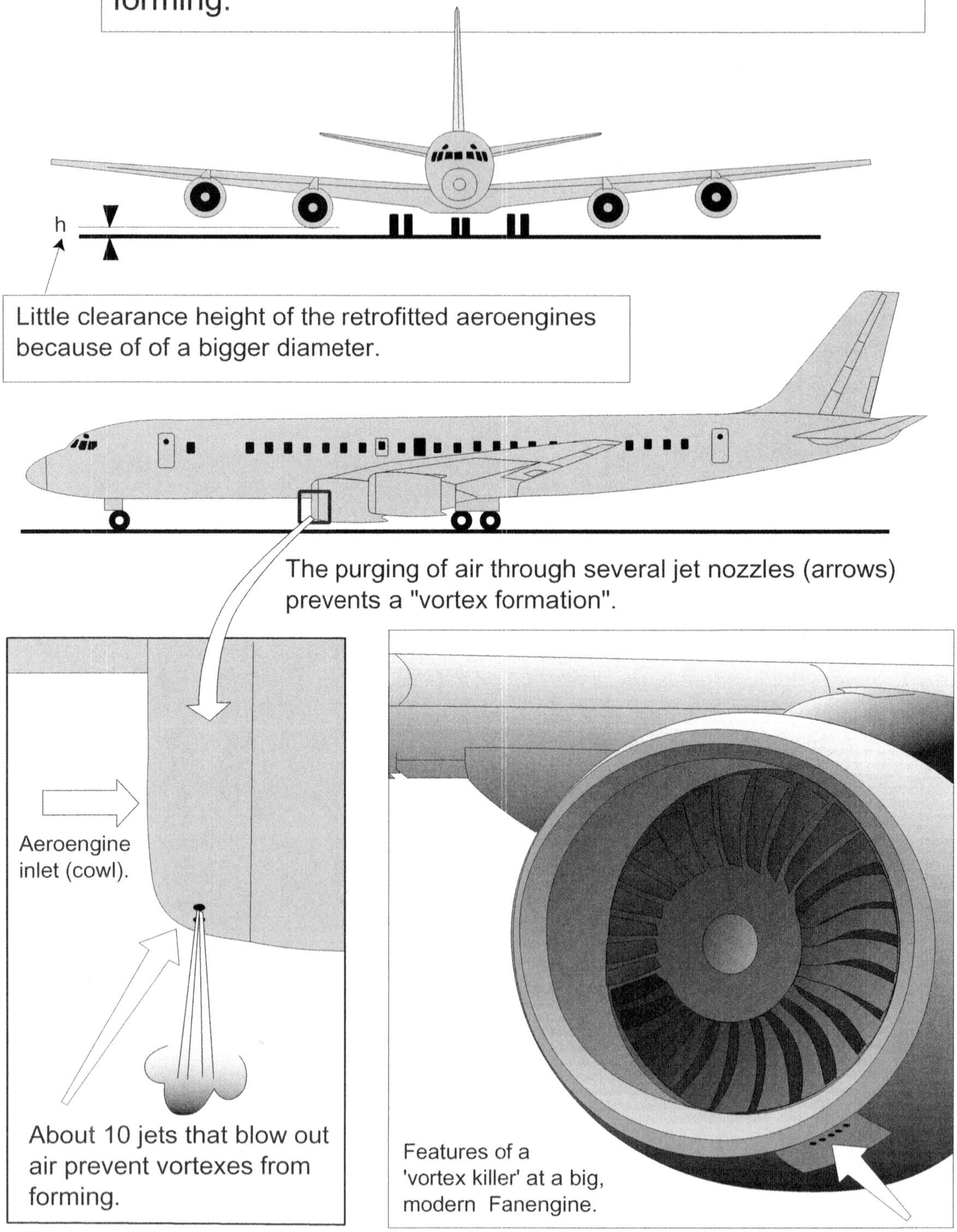

Ill. 5.2.1.2-9

The importance of foreign object problems can be seen in the elaborate constructive measures in inlet ducts that are designed to reduce the FOD danger.

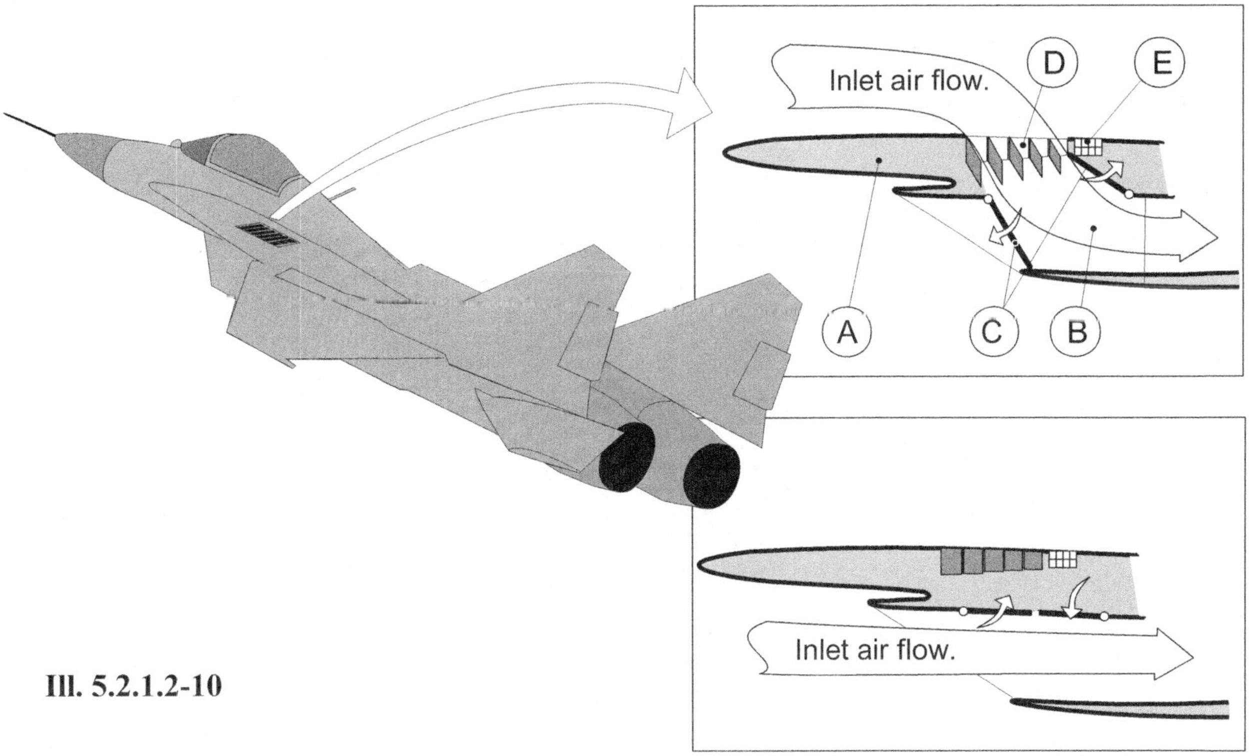

Ill. 5.2.1.2-10

Illustration 5.2.1.2-9: *There are various possibilities for* ***suppressing the formation of a dangerous vortex****. These include air* ***jets around the inlet that blow air radially outward*** *(Example 5.2.1.2-4, Refs. 5.2.1.2-9 and 5.2.1.2-10) or small wings that mechanically prevent foreign objects from entering the engine. However, evidently none of these measures is completely reliable (Ref. 5.2.1.2-5). Because sufficiently strong headwind prevents vortexes from forming and thus minimizes the risk of FOD,* ***rapid forward movement seems the safest measure at high engine performance****. This requires progressive power increases in the engines during takeoff. The degree to which this can even be realized depends on the specific takeoff conditions.*

Illustration 5.2.1.2-10: *This Russian fighter aircraft type has a special* ***shutter mechanism in the inlet duct*** *(also see Ill. 5.2.2.3-3) of the engine (Refs. 5.2.1.2-12 and 5.2.1.2-13) because the large inlet openings are located very low on the underside of the fuselage (also see bottom of Ill. 5.2.1.2-7), which would otherwise increase the probability of foreign objects being sucked up from the runway. Because this aircraft is designed for takeoff from makeshift runways, this risk is especially high. The function of the complex air intake system is depicted in the two bottom diagrams:*
From start-up on the ground (top diagram) until the landing gear leave the ground, the main air inlet "C" is closed except for three permanently open vents in the main shutter, which are

The complex air inlets with adjustable devices are a source of FOD in modern fighter aircraft.

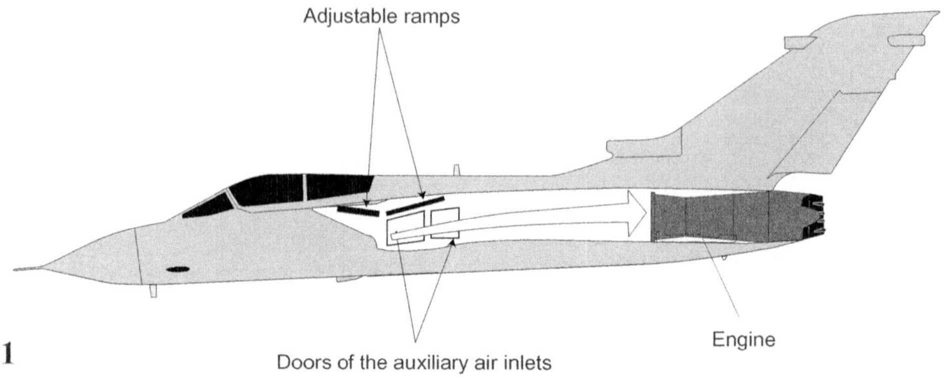

Ill. 5.2.1.2-11

protected with blind-like metal strips. The necessary air flow is drawn through a secondary inlet "D" on top of the wing "A", which is opened through a shutter system.
During flight at subsonic speeds (bottom diagram) the main air inlets are opened and the secondary inlet "D" is closed by the shutter system "C".
During supersonic flight the shutter system "C" is partially closed as an inlet ramp, and the excess air is blown off through the overflow openings "E".
The shutter system closes again during landing as soon as the oil shock struts of the main landing gear are compressed. This system has evidently proven itself during operation and makes possible the extremely low air inlets that are advantageous in flight situations with a high angle of attack.

Illustrations 5.2.1.2-11: *The air inlets of supersonic aircraft, especially modern fighter aircraft, have complex* ***adjustable shutter and ramps systems****. These systems consist of components such as hinges, actuators, and parts made from metal sheeting attached with fastening devices such as bolts and rivets. These present many sources for possible FOD to the engine. Internals such as those described in Ill. 5.2.1.2-10 should further increase this risk.*
Testing of these systems during inspection and maintenance work may require tools such as flashlights that may accidentally be left in ducts and inlets with poor visibility, later resulting in extensive damages.

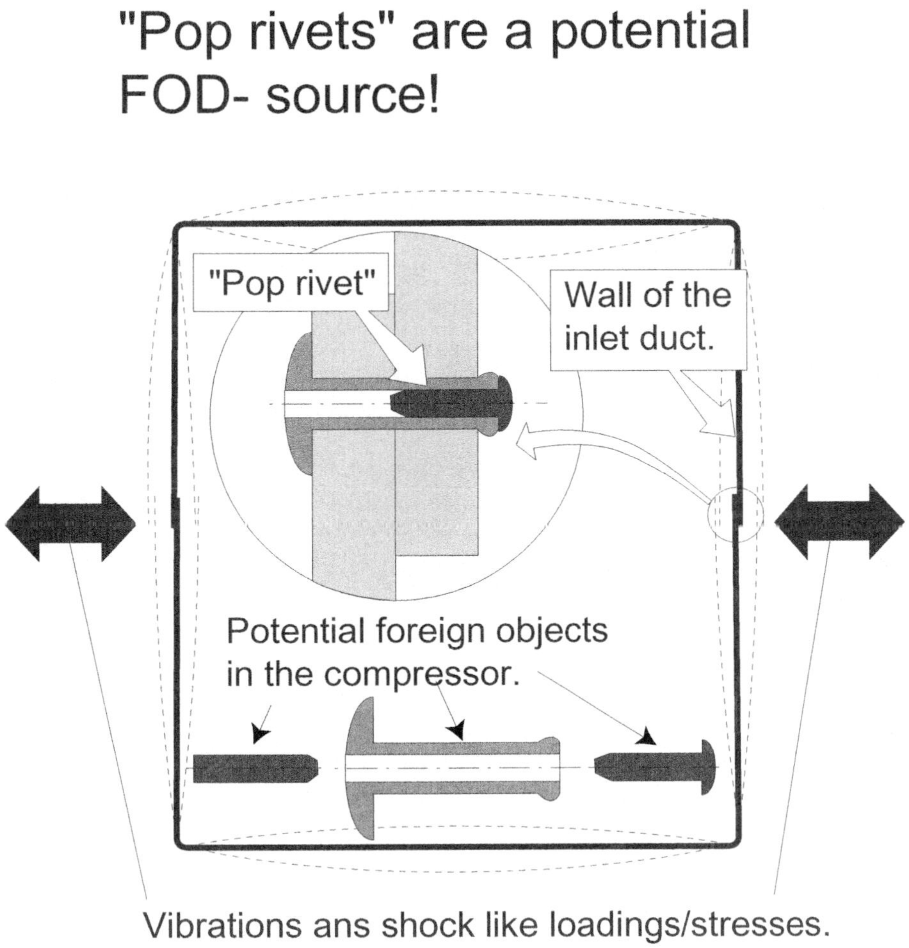

Ill. 5.2.1.2-11.1

Illustration 5.2.1.2-11.1: *During surge of the aeroengine a sudden deceleration of the air flow in the inlet duct can occur. Thereby develop intense shock waves (**„hammer shock“**, see volume 3, Ill. 11.2.1.1-17 and Ill. 11.2.1.1-18). This pressure shocks load the whole structure of the inlet duct very high. As especially dangerous has shown the **loosening and separating** of so called **„pop rivets“** (blind rivets) in **provisional inlet ducts** of test rigs in the development phase. These foreign objects produce very expensive and dangerous failures/damages at the blading of the compressor.*

Chapter 5.2.1.2 Reference

5.2.1.2-1 NTSB Identification MKC89IA024, microfiche number 37058A, Index for Nov. 1988.

5.2.1.2-2 P. Stokke, "Erosion, Corrosion and Foreign Object Damage Effects on Gas Turbines", AGARD-CP-558, Proceedings of the Conference "Erosion, Corrosion and Foreign Object Damage Effects in Gas Turbines", Rotterdam, The Netherlands, 25-28 April 1994, Chapter 7.

5.2.1.2-3 NTSB Identification SEA98WA086, microfiche number 37058A, Index for June 1998.

5.2.1.2-2 T.L. Alge, J.T. Moehring, "Modern Transport Engine Experience With Environmental Ingestion Effects", AGARD-CP-558, Proceedings of the Conference "Erosion, Corrosion and Foreign Object Damage Effects in Gas Turbines", Rotterdam, The Netherlands, 25-28 April 1994, Chapter 9.

5.2.1.2-5 D.E. Glenny, "Ingestion of debris into intakes by vortex action", U.D.C. No. 621-757: 533.697.2:532.527, N.G.T.E. Peyestock, G.P. 1114, December 1968.

5.2.1.2-6 C.L. Bore, "Scaling Laws For Vortex Induced Debris Ingestion Into Air Intakes", Research Note BAe-KRS-N-GEN-274, March 1983.

5.2.1.2-7 "AMST: Tactical Airlift Into the 21st Century", Boeing Co. and not ascribable literature source of a contribution to the proceedings.

5.2.1.2-8 A.K. Ivanyushkib, E.V. Pavlyukov, "Aerodynamic Problems of Propulsion System Operation Safety", Proceedings of the " Aircraft Flight Safety Conference", Zhukovsky, Russia, August 31-September 5, 1993, pages 148-162.

5.2.1.2-9 J. Jeffrey, M. Lenorovitz, "CFM56 Powerplant Fixes Based on DC-8 Operations", periodical "Aviation Week & Space Technology", February 14, 1983, page 32.

5.2.1.2-10 "Delta Weighs Changes to CFM56-2", periodical "Aviation Week & Space Technology", February 14, 1982, page 31.

5.2.1.2-11 V.I. Vasiliev, "Vortex Ingestion of Foreign Objects in Gas Turbine Engines of Aircraft", Proceedings of the " Aircraft Flight Safety Conference", Zhukovsky, Russia, August 31-September 5, 1993, pages 163-167.

5.2.1.2-12 MIG-29/Fulcrum, periodical "Soldat und Technik, 8/1988, page 480.

5.2.1.2-13 periodical "Military Technology, MILTECH", 4/87, pages 124-128.

5.2.1.3 Foreign Object Types

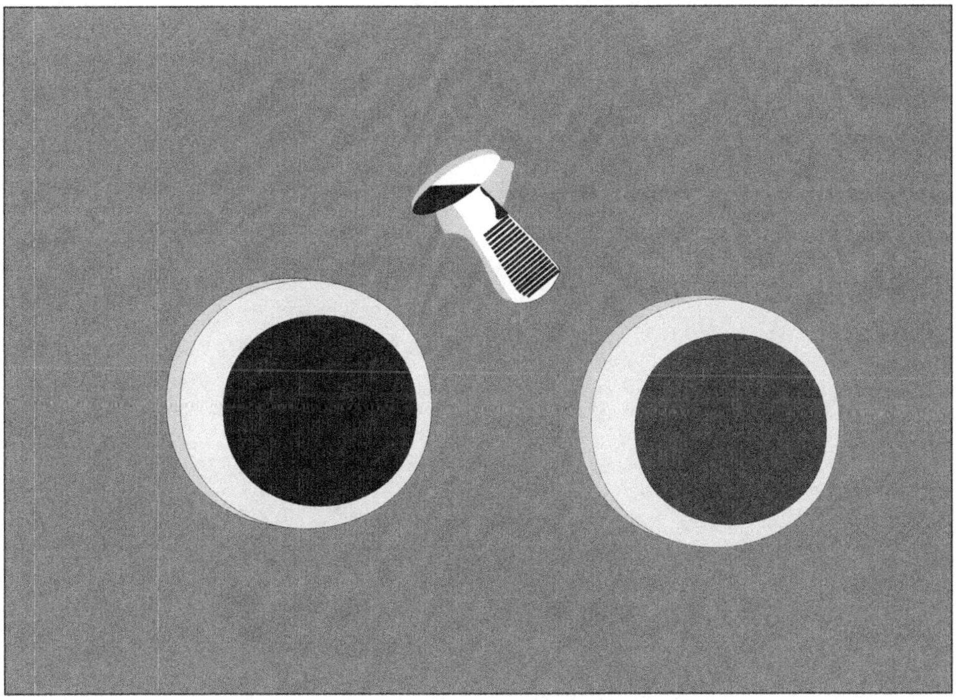

It is of fundamental importance to determine what **kind of foreign object** was the cause of engine damage. **Identification of the foreign** object enables conclusions as to its origin and is a prerequisite for designing appropriate **preventive measures** which, in turn make possible targeted measures for avoiding future damage.

Even though virtually everything found in and around engines has been found inside as a foreign object (Ill. 5.2.1.3-1), **fastening devices** such as bolts (Examples 5.2.1.3-1), nuts, shims, and rivets or parts thereof are found most frequently.

The probability of foreign objects remain in the inlet area is certainly connected with the degree to which the inside of the engine is visible from the outside. In fighter aircraft, especially, the engines are integrated into the fuselage and require long, often curved inlet ducts with complicated internal structures (such as shutters and flow guide vanes; Ill. 5.2.1.2-11), in which **tools or fastening devices** can easily "hide". This type of problem is very unlikely to occur in the easily observable inlets of nacelle engines from modern commercial aircraft with large bypass ratios.

Engines in military use seem to be especially prone to FOD in comparison with civilian aircraft. This is probably due to the types of foreign object that are characteristic for military uses. These include parts of **projectiles** (shell housings, etc.) and rockets, as well as ricochets (see Examples 5.2.1.3-2) that were ingested into the engine during shooting practice. However, even flight missions that occur primarily at **low altitudes** considerably increase the FOD risk, especially the probability of **bird strike** (Ills. 5.2.2.1-2 and 5.2.2.1-3).

External Factors
Foreign Object Damage: Foreign Objects

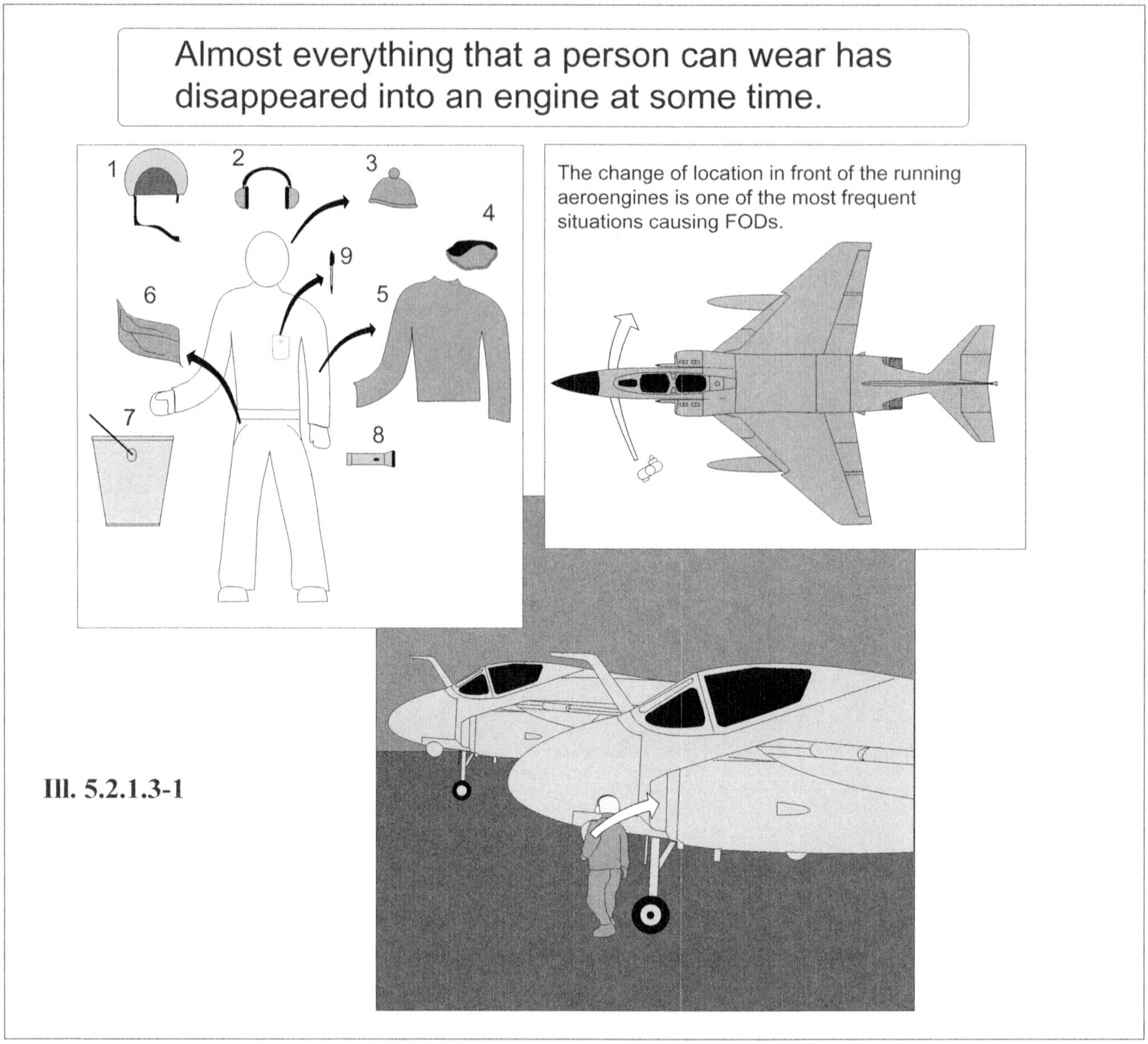

Ill. 5.2.1.3-1

Illustration 5.2.1.3-1: *Almost everything that is found around engines has been found inside them at some point. This diagram shows* ***typical human-related foreign objects****. There have been documented cases in which engine damage was traced back to these objects. Understandably, there is an especially high risk of the maintenance personnel's equipment being sucked in when they pass by the engine inlet area. The vortex formation described in Chapter 5.2.1.2 is also important in relation to this type of foreign object ingestion. Vortexes can also suck in apparently securely fastened objects such as articles of clothing. There have been reported cases of FOD due to safety helmets "1", ear protection "2", hats "3", a hood attached with a zipper "4" , and jackets "5".*

Personal equipment *that has been sucked out of pockets includes pens "9" and cleaning rags "6". Even a bucket "7" was torn from the hand of a maintenance worker and sucked into the engine.*

With flashlights "8" and tools the greater risk is that they are accidentally forgotten in the engine during inspection of the inlet duct.

External Factors
Foreign Object Damage: Foreign Objects

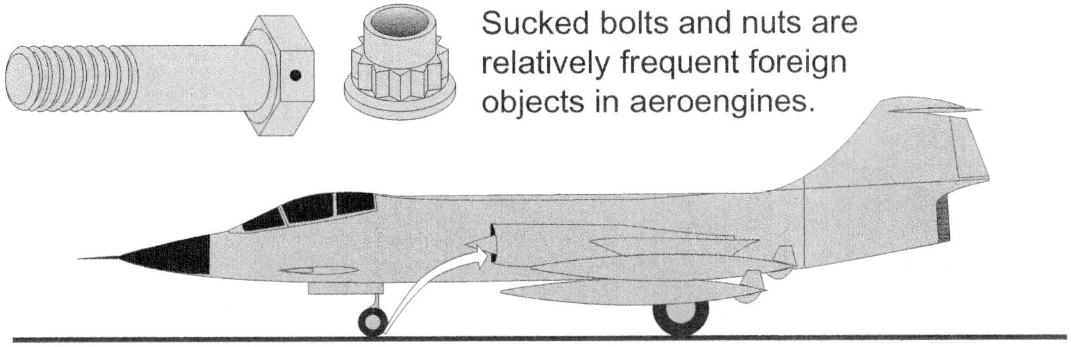

Example 5.2.1.3-1.1 (Ref. 5.2.1.3-1):

Excerpt: "...shortly after takeoff engine failure due to foreign object intake (unidentifiable bolt) followed by a crash..."

Example 5.2.1.3-1.2 (Ref. 5.2.1.3-2):

Excerpt: "......during approach (to the airport) engine failure due to damage from an ingested bolt...crash landing."

Comments: This case concerns a single-jet fighter aircraft. Bolts are *the* typical foreign object. There are several similar cases listed in the reference literature. It would be important to determine whether the bolts had been sucked up off of the runway or if they had originated in the aircraft itself.

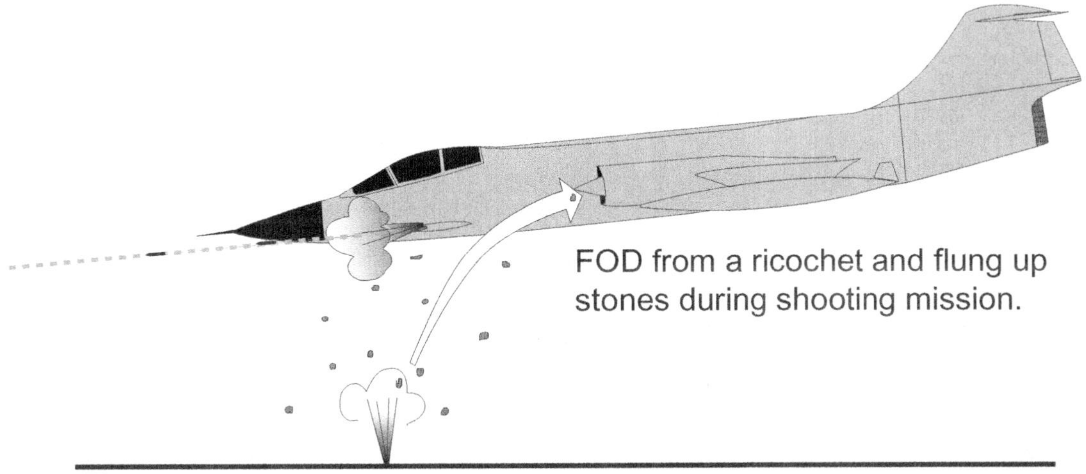

Example 5.2.1.3-2.1 (Ref. 5.1.2.1.3-3):

Excerpt: "...during shooting, engine failure due to ricochets into the fuselage and engine compressor; aircraft crashed."

Example 5.2.1.3-2.2 (Ref. 5.2.1.3-4):

Excerpt: "...during air-to-surface shooting, aircraft crashed following engine failure due to a compressor stall caused by foreign object ingestion."

Comments: Typical "military FOD"; it is worth noting that **missed shots from guns or rockets** can evidently ricochet so far from the target that they endanger the aircraft, i.e. engine. There are frequent reports of this type of case.

External Factors
Foreign Object Damage: Foreign Objects

Maintenance caused foreign objects.

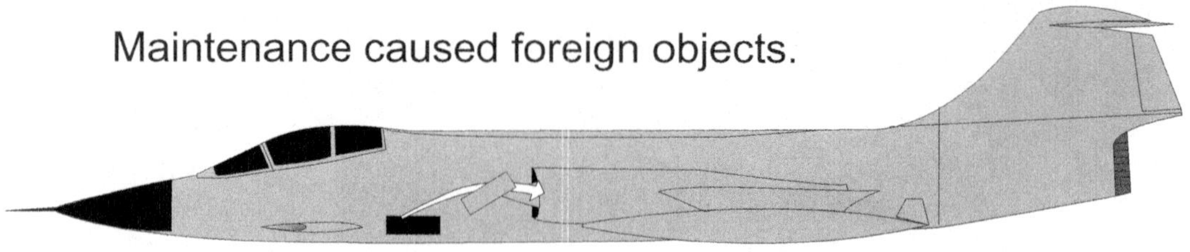

Example 5.2.1.3-3 (Ref. 5.2.1.3-5):

Excerpt: "...crash due to engine damage following the loss of a maintenance flap on the left front fuselage."

Comments: A typical case of FOD caused by an aircraft's own parts.

Forgotten maintenance tools.

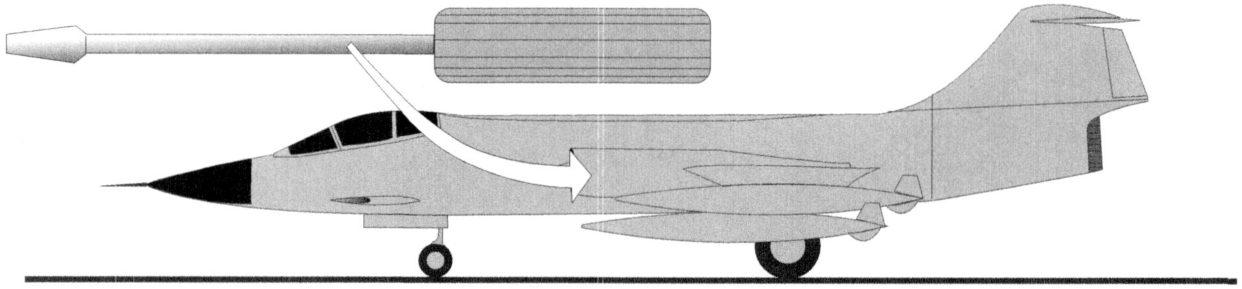

Example 5.2.1.3-4 (Ref. 5.2.1.3-6):

Excerpt: "...immediately after takeoff engine failure due to a compressor stall; cause: a **screw driver** in the intake duct!"

Comments: Forgotten tools in the engine are not so rare as causes of catastrophic engine damage that leads to an aircraft crash, as it did in this single-jet type. For this reason, extreme care must be taken during maintenance to ensure that no objects are left **in the inlet duct in front of the engine**. Because of the darkness here are **pocket lamps** necessary, which in ofher cases already have been forgotten. With this in mind, after the end of maintenance work the tool set used should be carefully checked to ensure that all of the tools used are present and accounted for (Volume 5 Ill. 19.2.2-5).

External Factors
Foreign Object Damage: Foreign Objects

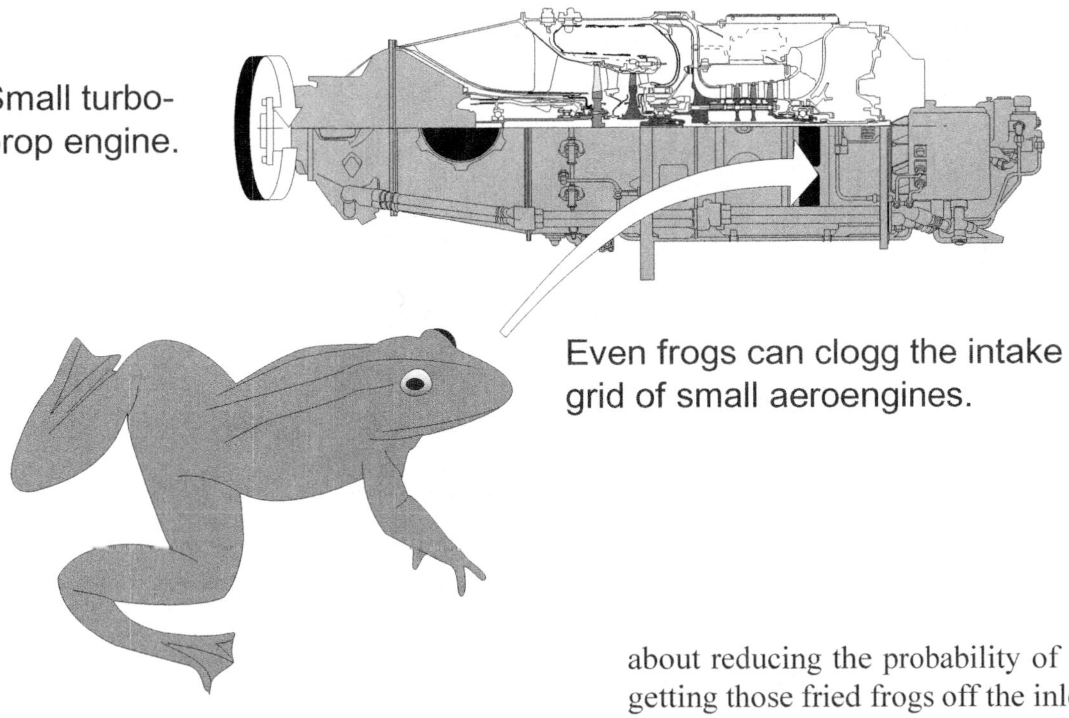

Birds are not the only animals that cause FOD. There is also the following bizarre example:

Example 5.2.1.3-5 (Ref. 5.2.1.3-8):

Excerpt: "...Then there was the case of a couple of frogs that jumped just at the wrong moment...the (small engine type) is really good about reducing the probability of ingestion but getting those fried frogs off the inlet screen was a real challenge.

Comments: Even relatively little foreign objects are dangerous for smaller engines. A frog in the engine can completely destroy the compressor. Here this was prevented by an intake grill. Hovever, thereby the cross section is clogged and so a compressor surge can be triggered.

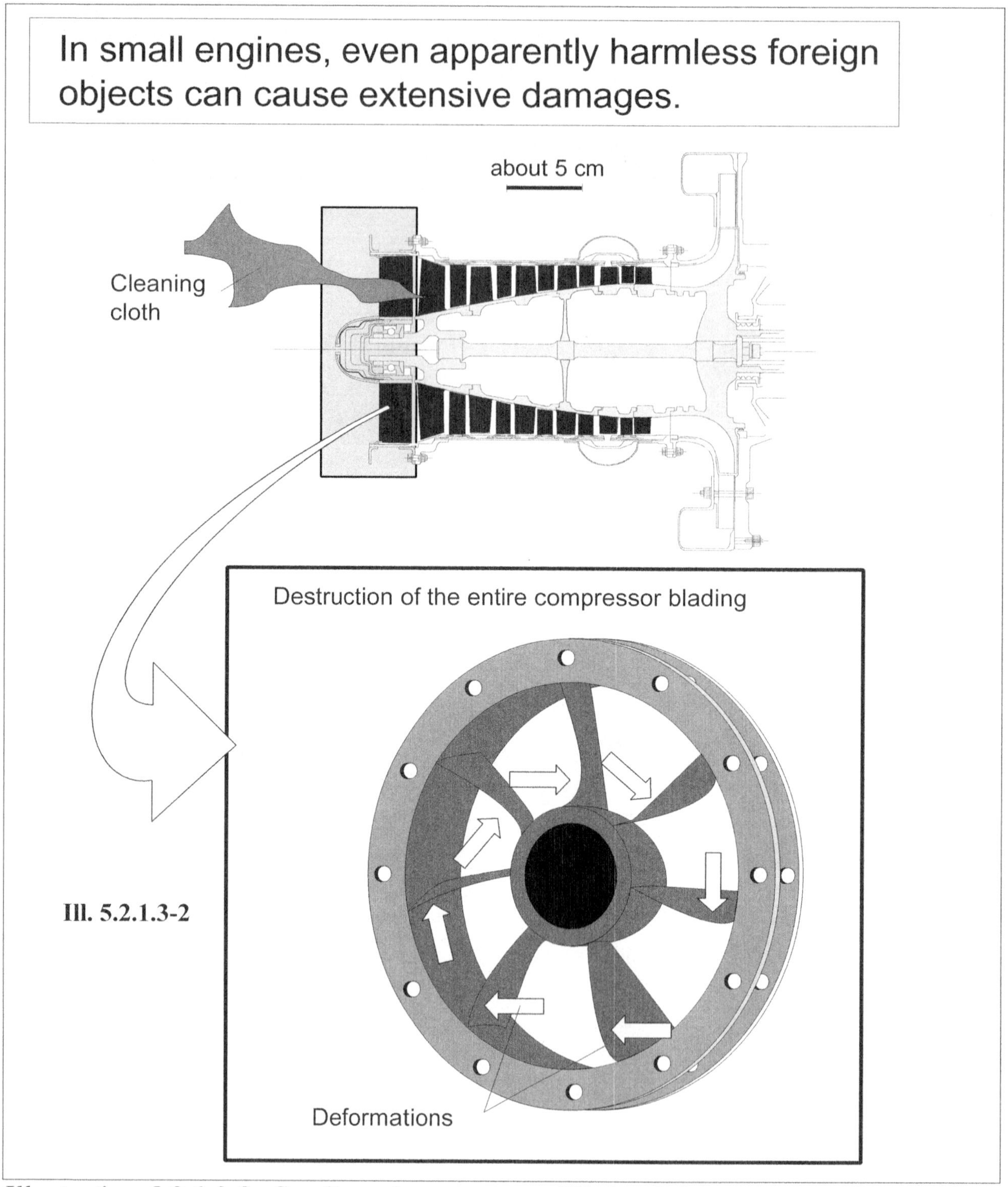

Ill. 5.2.1.3-2

Illustration 5.2.1.3-2: Small engines *are especially sensitive to apparently harmless foreign objects due to their filigreed, thin blading. For example, in the case depicted here a* ***cleaning rag*** *totally destroyed the compressor of a 400 kW class helicopter engine. It cannot be said with absolute certainty that a paper tissue would not cause unallowable deformation to the blading of a small engine.*

Foreign objects often reveal their origin through characteristic impressions and fragments.

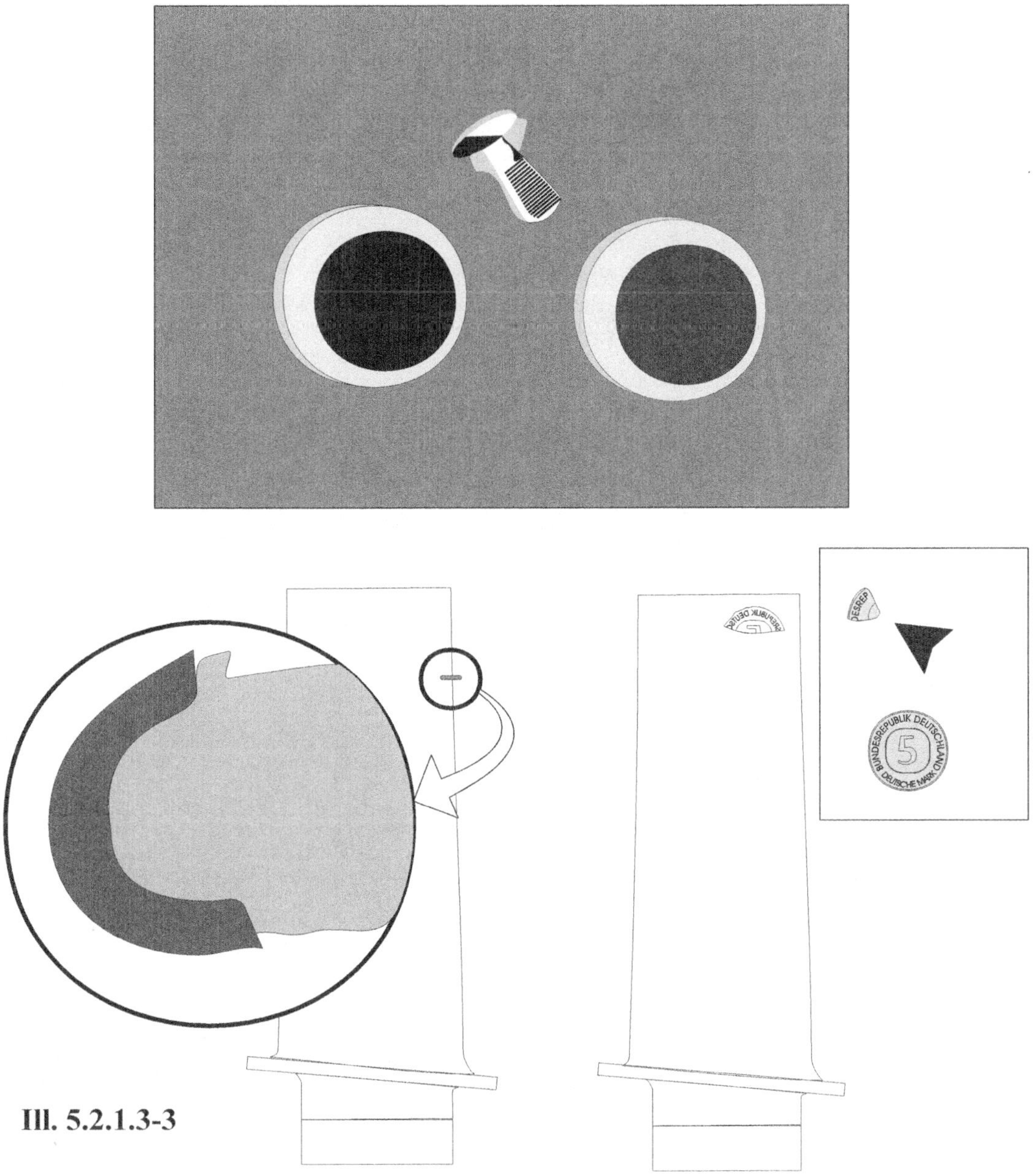

Ill. 5.2.1.3-3

Illustration 5.2.1.3-3: *Often* ***parts of the foreign object (pieces of shims, nuts, wires, etc.) remain in the blading.*** *Analysis of the material and geometry of the foreign objects can indicate their origin.*

If this is inconclusive, there are other important characteristics such as the ***geometric contours of the impact marks****, the evaluation of which is made easier if there is FOD to multiple blades. A nick with a crescent shape and smooth contour*

is typical for a smooth, cylindrical foreign object. This assumption is supported if the blade edges of subsequent stages show smaller or larger nicks of increasingly irregular shape. These nicks were created by the already ***deformed foreign objects and/or its fragments****.*
It is also possible that the trailing edges of the blading will have impact marks. This is the case if, for example, the blades of the ***following stage threw the foreign object forward again*** *(Ill. 5.2.1.3-7). This is a relatively frequent occurrence. The top diagram shows a bolt imprint on the inside of a light metal compressor housing. Evidently the bolt was thrown free from the compressor or was flung radially outward by the rotor blades.*
Many other foreign objects also leave ***characteristic imprints****, even if they are made from a softer material than the blades. For example, the coin left a clearly identifiable imprint on a hard Cr-steel blade.*
Unique patterns such as ***bolt thread imprints*** *indicate not only the type of foreign object, but measurement of the imprint can show whether it was a* ***metric or non-metric thread****, which might indicate whether the foreign object came from within the engine itself or from a different source.*
Analyzable foreign object fragments can even be found outside of the compressor in the gas duct. For example, the cooling air ducts in older combustion chambers are a good place to look for foreign object residue.

! A prerequisite for targeted prevention of foreign objects is identifying them.

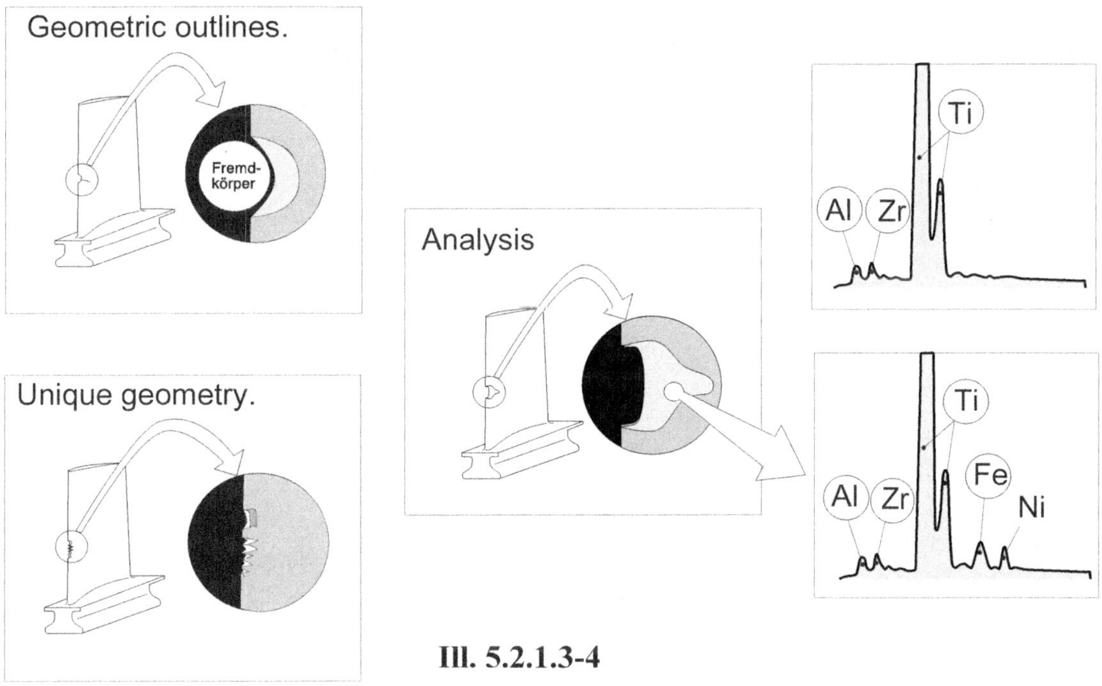

Ill. 5.2.1.3-4

Illustration 5.2.1.3-4 *(Ref. 5.2.1.3-9): In order to* ***specifically prevent*** *further foreign object damage, the source of foreign objects must be determined with as much certainty as possible. A great deal of experience and specific knowledge of the engine and periphery (inlet duct with internals, etc.) is necessary for* ***identifying foreign objects****. It is extremely important to determine whether the foreign object came from the engine itself (through a loose part, etc.) or was itself consequential damage (delaminated coatings, etc.). The foreign object may also have entered during* ***maintenance*** *or* ***mounting work*** *(forgotten or overlooked nuts or the ends of fuse wires, etc.).*

There are often ***remnants of the foreign object*** *left in the blading which can be analyzed to determine their source.*

Metallic foreign objects leave traces in the impact area through friction which can be ***EDX analyzed*** *with an REM in order to determine their composition.* ***Comparing*** *these results with analyses of undamaged sections (base material composition) will indicate the material of the foreign object and therefore its origin.*

If it small FOD is suspected of having caused serious compressor damage, then this can be confirmed by analysis of the dynamic fatigue fracture on the blade which initiated the damage *(if it is still analyzable)****.***

***Illustration 5.2.1.3-5** (see following page): Damage statistics show that the likelihood of damage increases considerably immediately after an overhaul. This behavior can be seen in the typical damage history throughout the operating time (**bathtub curve**, bottom diagram). The increased damage rates in new parts and after overhauls are referred to as **"infant mortality", "wear-in", or "burn-in".** This damage also includes foreign object damage caused by forgotten tools, fastening devices, auxiliary materials, etc.*

Experience has shown that during normal work on a standing engine (top left diagram) there is a special danger of foreign objects such as fuse wire trimmings falling into the uncovered engine. This kind of foreign object can slide along the conical compressor rotor hub and fall between the disks. The disk band stops the foreign objects from falling out when the engine is turned over. Then during the test run the foreign object is flung into the blading where it can create dangerous nicks.

In engines with curved inlets, such as are not unusual in helicopters (Ill. 5.2.1.3-6), objects that are set down or fall into the inlet can only be found by feeling around for them.

*The increased damage rate of the wear-in phase levels off into a constant rate (**random failures**) throughout the estimated safe life span.*

*The increase in the damage rate near the end of the life span is referred to as "**wear-out**". It is caused by normal aging processes such as fatigue, corrosion, and wear.*

*The bathtub curve is closely related to the **"overhaul philosophy"**, i.e. the determined overhaul intervals. If this is **"conflict oriented"**, then the first overhauls will be set early in the engine's safe life span in order to ensure that the engines are in optimal condition in case of a conflict. On the other hand, a **"peace oriented"** overhaul philosophy recommends that overhauls be done based on the condition of the engine parts, i.e. at the beginning of the wear-out phase.*

External Factors
Foreign Object Damage: Foreign Objects

The assembly process as a source of foreign objects, also forming a bathtub curve.

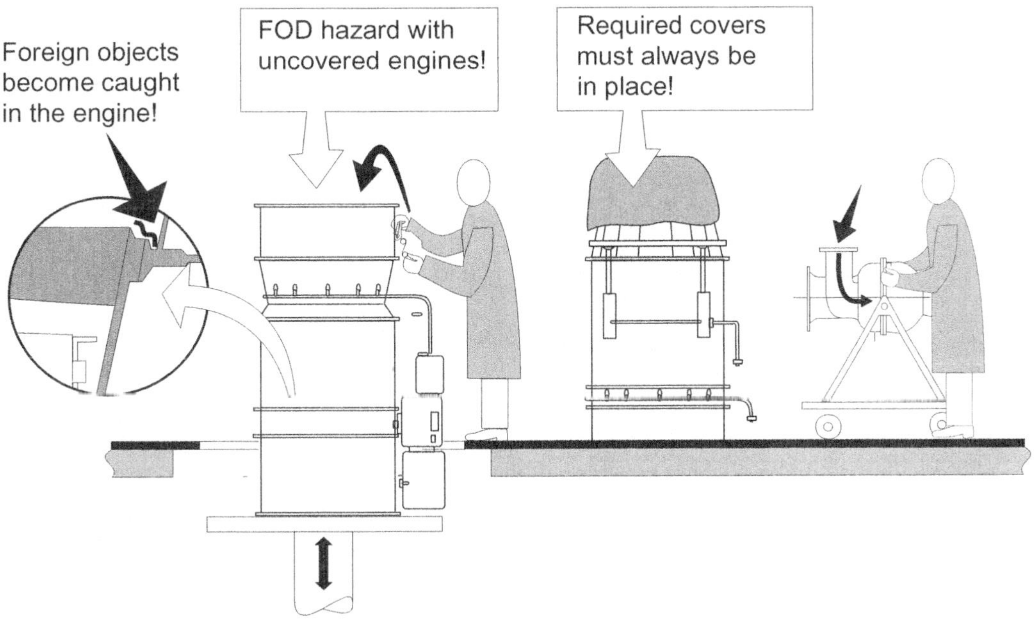

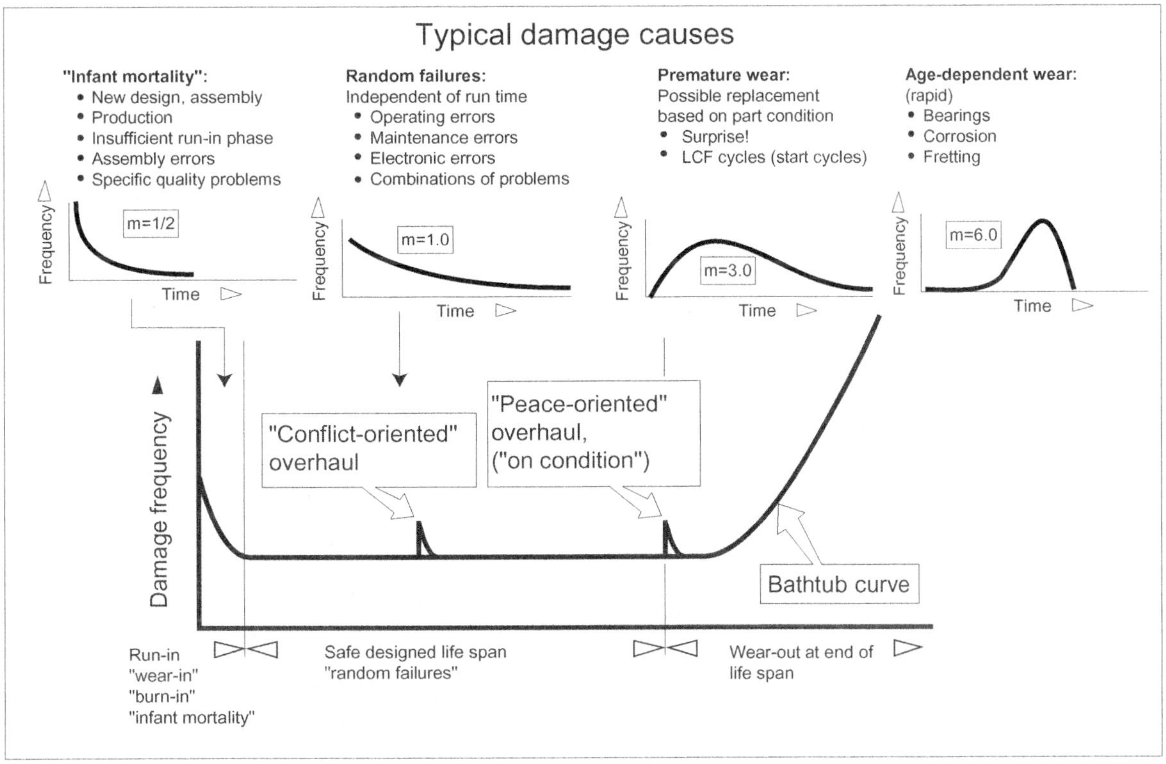

Ill. 5.2.1.3-5

explanation of diagram on previous page

Poorly visible air intake ducts must be especially closely inspected for foreign objects after any assembly work.

It is difficult to see inside the curved air intake duct, increasing the probability that foreign objects may remain in it after assembly work.

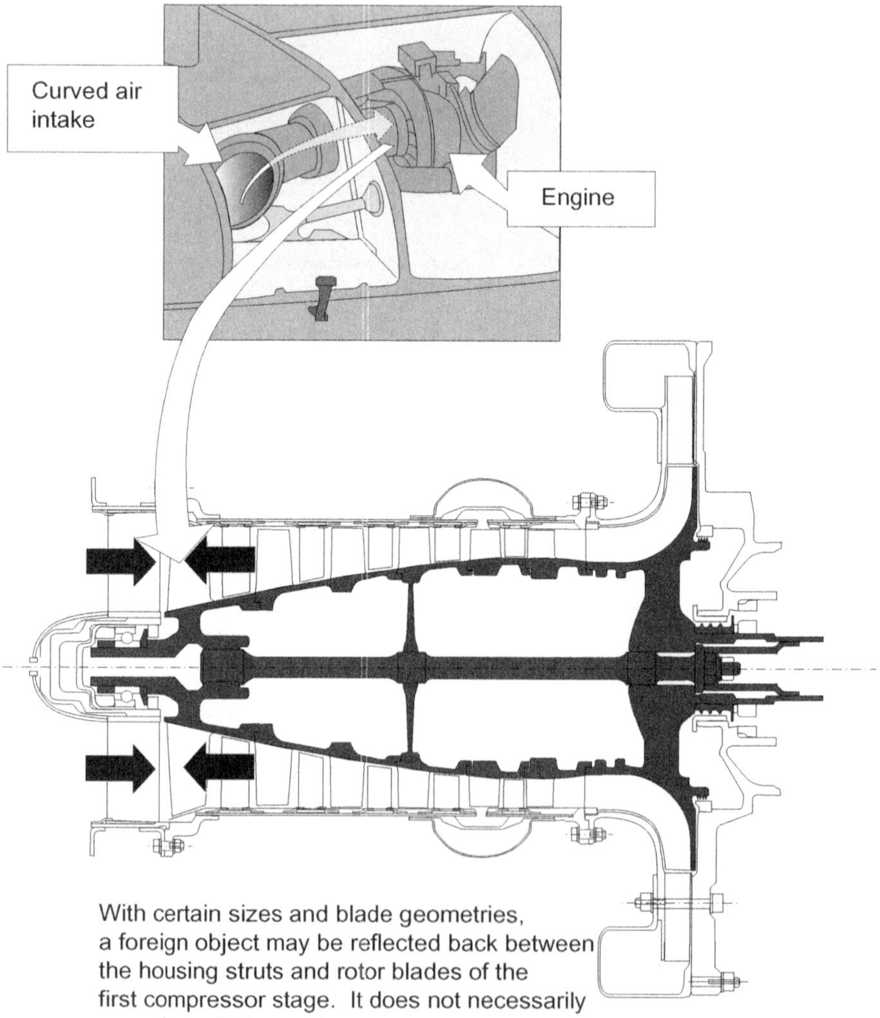

With certain sizes and blade geometries, a foreign object may be reflected back between the housing struts and rotor blades of the first compressor stage. It does not necessarily pass through the engine.

Ill. 5.2.1.3-6

***Illustration 5.2.1.3-6: Helicopters** often have curved **air inlet ducts** with relatively small diameters (top diagram). It is difficult to see into these ducts. In the case described, a foreign object was obviously **left in the duct during mounting**. It finally entered the engine after about an hour of flight. The flow combined with the normal vibrations during operations and maybe even special movement by the helicopter carried the foreign object into the engine inlet. The foreign object then **oscillated between the first compressor rotor stage and the inlet housing** (bottom diagram), causing extensive damage in this area (Ill. 5.2.1.3-7).*

External Factors
Foreign Object Damage: Foreign Objects

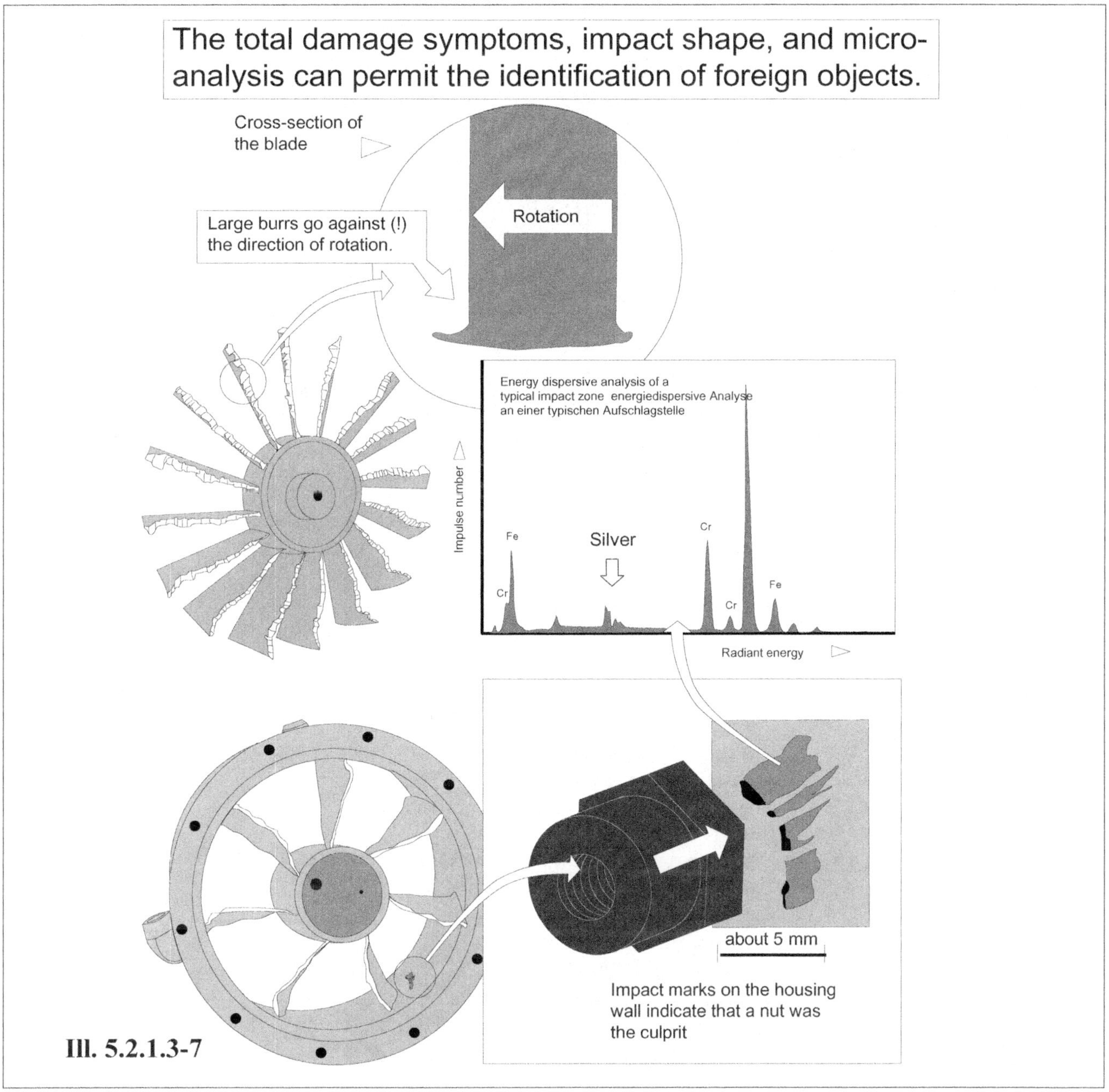

Ill. 5.2.1.3-7

Illustration 5.2.1.3-7: *The foreign object that caused this FOD (see Ill. 5.2.1.3-6) was identified to a high degree of certainty by* ***typical characteristics****.*

The damage to the rotor disk in the first stage and the inlet struts (right diagrams) indicated a foreign object that was too large to fit through the rotor blades. The burrs on the leading edges of the rotor blades (top detail) go against the direction of rotation, which was observed in parallel cases to be typical for a foreign object oscillating against the direction of flow (Ref. 5.2.1.3-11). The ***geometry of the impact marks*** *(bottom right detail) indicated a six-edged foreign object, and the high silver content at the impact areas indicated that the foreign object was silver-plated.*

The combination of these characteristics was typical for a silver-plated nut with a diameter of about 8 mm. This type of nut is used in several places on the helicopter fuselage.

Typical causes of foreign objects in the air and oil systems.

Maintenance:
- Unsuitable remnants from cleaning, such as liquids and abrasive particles
- Wear products from seals
- Excess sealing paste

Operation:
- Labyrinth wear products and spalled armor coatings
- Wear products and chunks broken out of rub coatings near the blade tips
- Chips from rubbing
- Fatigue spalling from pitch surfaces e.g. from bearings and gears

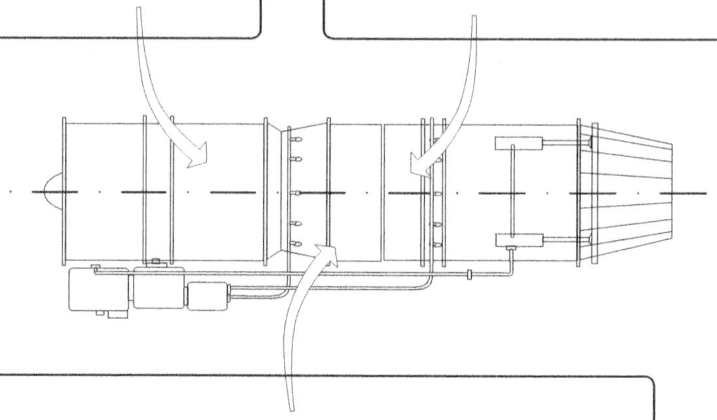

Production and repairs:
- Abrasive blasting media (Al_2O_3 crystals)
- Materials from shot peening (steel balls)
- Chips from structural movements (burrs)
- Seal material removal (e.g. from O-rings)
- Excess sealing paste
- Remnants from cleaning baths
- Chips from machining

Ill. 5.2.1.3-8

Illustration 5.2.1.3-8 *(see previous page): Many different particles from a wide variety of sources can* ***contaminate the flow in the air and oil systems****. They can occur at all stages of an engine's life. In many cases they can potentially cause extensive consequential damages such as* ***bearing track fatigue*** *(Example 5.2.1.3-7) or catastrophic* ***labyrinth failure*** *due to self-increasing rubbing. These foreign particles can primarily be prevented by a knowledgeable, careful, and conscientious crew and stable, safe processes during maintenance, manufacture, repair, and installation.*

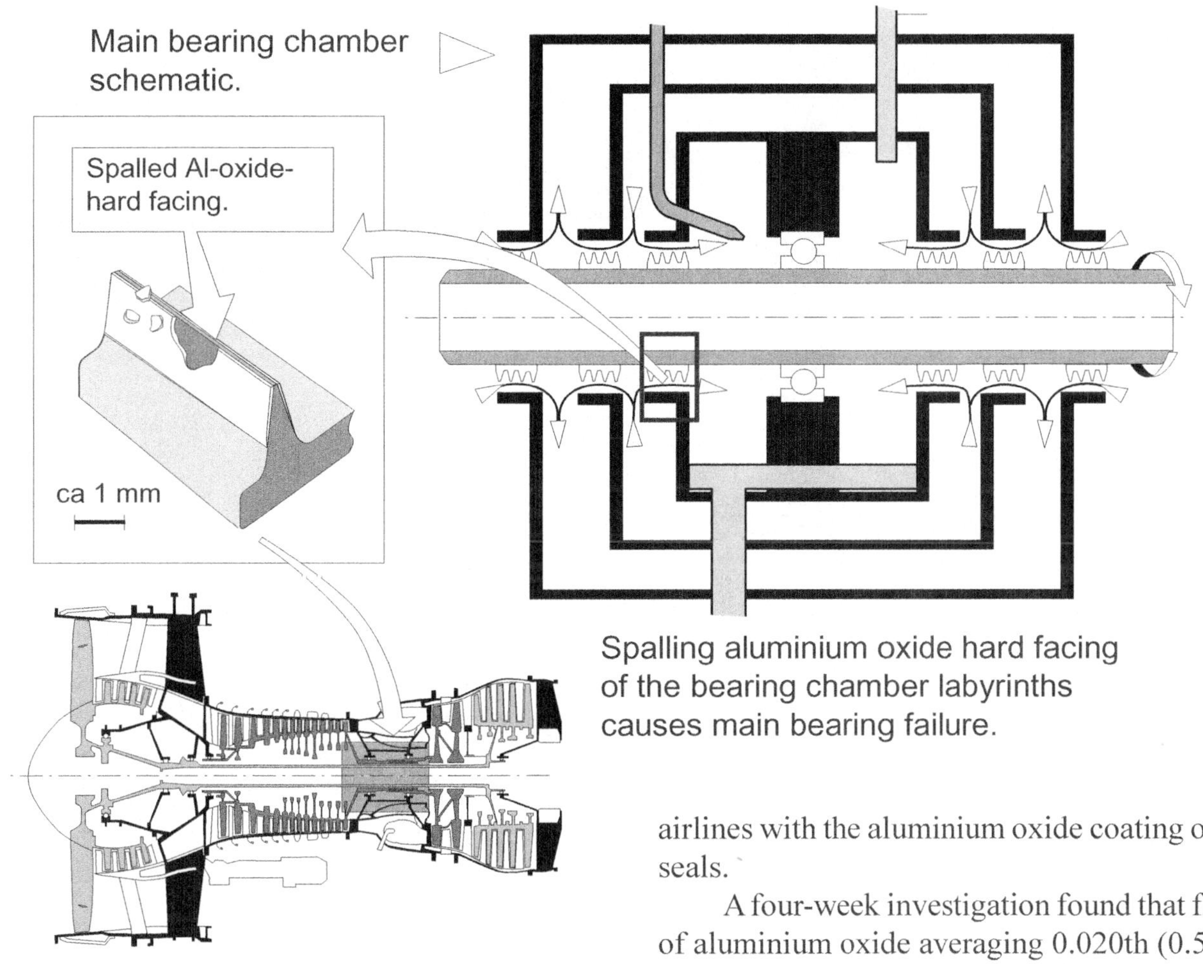

Example 5.2.1.3-6 (Ref. 5.2.1.3-10):

Excerpt: "Combination of aluminium oxide contamination and bearing loads has been identified as the cause of bearing wear found on some....engines in service...the electron-microscope examination of No.4 and No.5 roller bearings found evidence of **aluminium oxide entering the bearing area and causing premature spalling of bearing surfaces**.

Aluminium oxide, an abrasive material, had been used...to coat air seal teeth adjacent to the bearing area in the pressurized cavity in the oil system. **The material facilitates mating of air seal surfaces during initial run-in of the engine**.

...The engines ranged 40-4000 hr.use. A total of 18 engines have been delivered to the airlines with the aluminium oxide coating on the seals.

A four-week investigation found that flakes of aluminium oxide averaging 0.020th (0.5mm) of an inch size were coming off the knife edges of the seals, which keep the oil system pressurized and entering the bearing cavity area...

(the engine manufacturer) also sent samples of the flakes to both airlines for comparisons.

To rectify the problem, (the engine manufacturer) is stripping the aluminium oxide coating from the affected area in all ...engines, and bearing and seal clearances are being increased to provide a higher tolerance to contaminants."

Comments: The pressure distribution in the bearing chamber should normally be designed in a way that ensures that wear products and spalled bearing seal material are blown out of the bearing chamber and cannot enter into the bearing (volume 5 Ill. 23.1.2-4).

External Factors
Foreign Object Damage: Foreign Objects

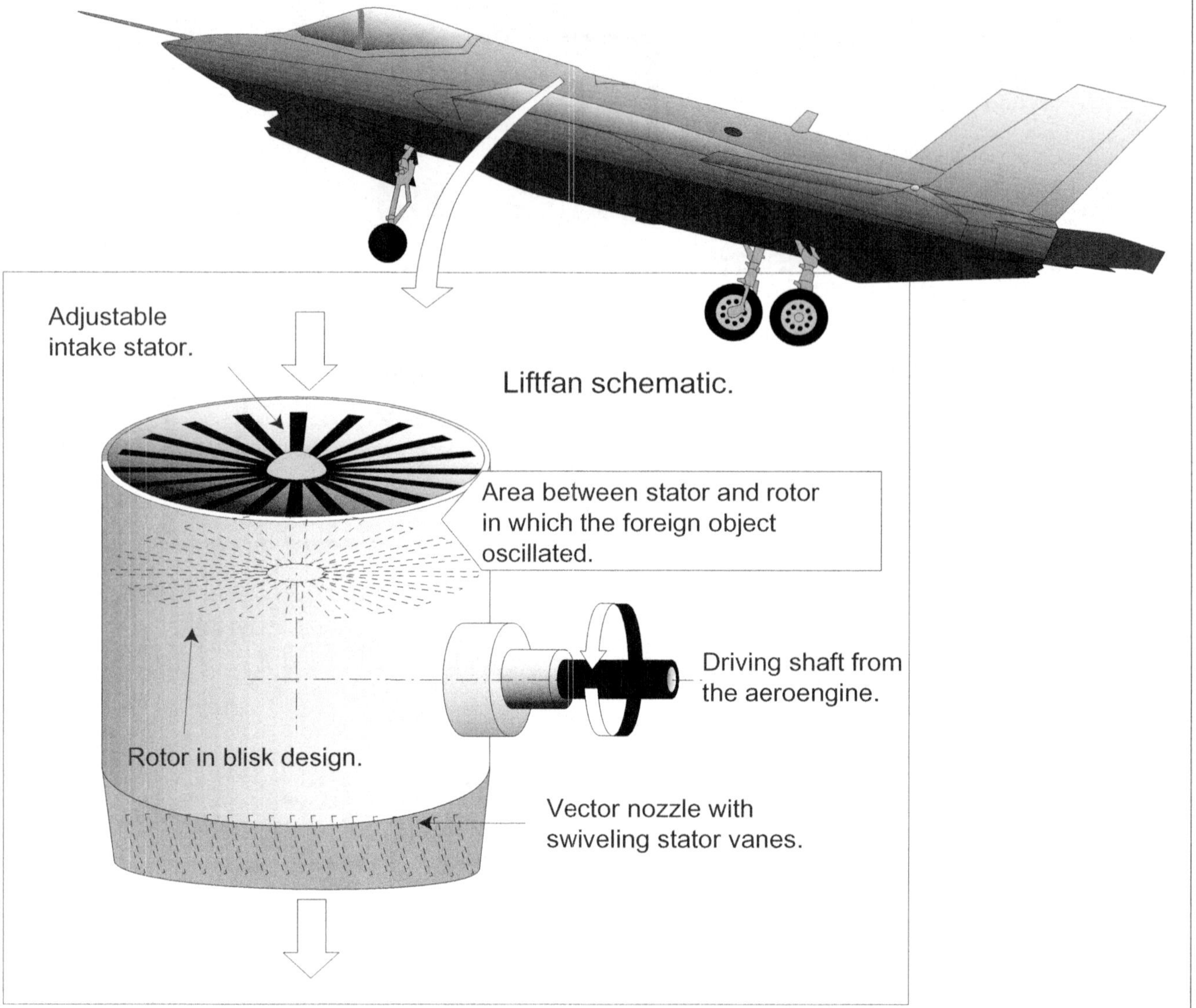

Example 5.2.1.3-8 (Ref. 5.2.1.3-11):

Excerpt: "...(the aircraft developer) has suffered another in a series of minor bug nagging setbacks with the lift fan specially built for the short take off, vertical landing version of its Joint Strike fighter demonstrator aircraft....a part apparently vibrated loose from somewhere within the lift fan assembly during testing. **The part then ricocheted around the spinning unit** with enough force to knock metal filings off it and possibly from moving elements of the lift fan."

Comments: This case seems to be similar to that shown in Ill. 5.2.1.3-7, in which a **foreign oblect was repeatedly ricochetted/thrown around** by the spinning rotor and caused extensive damage.

5.2.1.4 Measures Against Foreign Object Damage

(Measures against bird strike damage are covered in Chapter 5.2.2.4)

• **Identifying foreign objects** is a prerequisite for specific preventive measures. Foreign objects can be identified by the geometry of the impact marks and a comparative microanalysis of the impact area (residue) and an undamaged surface.

• **The periphery of running engines** must be kept clean. Loose dense objects such as pebbles can even be sucked out of cracks and other apparently unreachable areas (or even out of running joints in the runway in the case of vortex formation!).

• **Cleanliness** and **no loose part** in the suction area of testing rigs. This includes cleaning rags and tools.

• Use of the **prescribed tools** and following the usual installation rules such as holding the ends of fuse wires when trimming them; confirming that all of the tools are present and accounted for immediately after finishing work.

• **Not temporarily setting down** auxiliary materials, tools, or replaced parts in the suction area, especially the inlet duct.

• **Threaded fasteners** and **rivets** must be inspected for completeness and they must sit tightly! If fastening devices such as pull rivets are used, the number of pull rods must be complete. When loosening fused bolts the removed fuse wires must be disposed of immediately.

• **During assembly/installation** work, especially on vertically standing engines, the compressor inlet must always be properly covered. Particles such as fuse wire trimmings cannot necessarily be removed by turning the engine and shaking it.

• The location of the engines on the aircraft should minimize the probability of foreign object ingestion. This refers to both **sufficient height of the engine above the runway** as well as the possibility of the **landing gear tires** throwing up foreign objects.

• **Inlet/intake designs** that allow air to be sucked in from less FOD-prone areas (such as on top of the wings of fighter aircraft) when the aircraft is near the ground.

• With long inlet ducts with complex configurations such as **shutter systems**, which are common in supersonic fighter aircraft, the designer must ensure that no parts of the **adjustment mechanism can come loose**.

• **Mud guards for the hull landing gear** of fighter aircraft; cargo aircraft with engines located on the rear fuselage and high sensitivity to FOD should be examined to see if a similar type of protection can be affixed to the landing gear.

• Proper **use of thrust reversers** in aircraft with engines in locations that are prone to FOD

• Following regulations that give the **minimum distance which persons should be away** from the inlet of the engine when it is running; **securing objects which are usually carried loosely** in clothing pockets such as cleaning rags, pens, and screwdrivers.

• There must be openings in the engine housing suitable for **boroscopic inspections** of the entire blading for foreign object strikes, as well as for doing limited reworking (fading out, hardening, etc.) without opening the housing.

• Using the most **impact and notch resistant materials** and coatings available; sufficient FOD behavior (such as a tolerable drop in dynamic strength) must be verified especially in brittle and hard blade coatings (armor, erosion protection).

• Engines located close to the runway may require **measures against vortex formation** (**'vortex killer'** Ill. 5.2.1.2-9).

• **Solid design of the blading**, which is especially sensitive to FOD. This includes relatively thick leading edges, a relatively straight profile (not curved) and a low dynamic basic load.

Optimal design of the inlet duct to the core engine can considerably improve an engine's resistance to FOD.

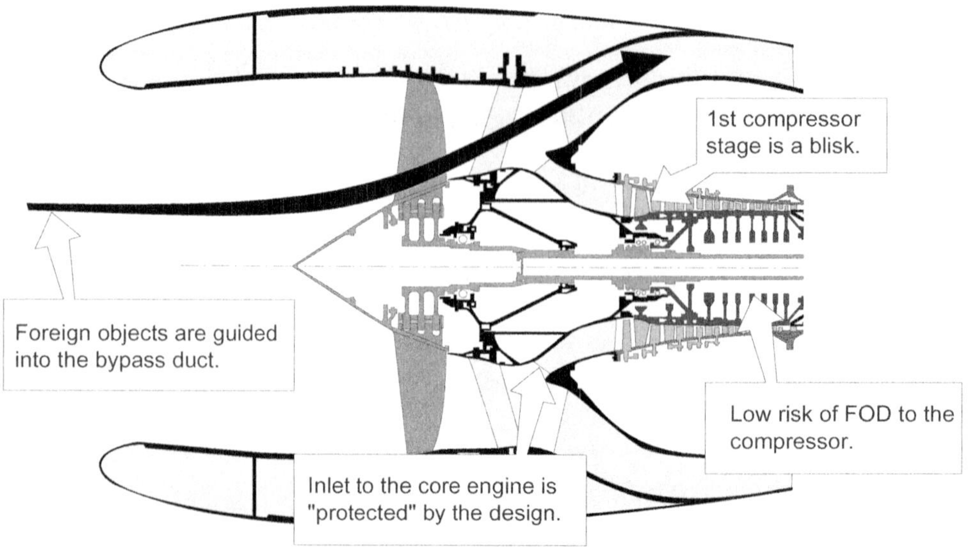

Ill. 5.2.1.3-9

External Factors
Foreign Object Damage: Foreign Objects

• Proper **design of the engine inlet** (also see Ill. 5.2.2.3-4) decreases the FOD danger to the core engine decisively. This also made possible the use of a blisk in the core engine compressor of a large serial engine, which has been successful for a long time (the engine in Ill. 5.2.1.3-9 is from General Electric Corporation).

• The fins of **labyrinth seals and the seal coatings** of bearing chamber seals should not contain any hard and abrasive elements (such as ceramic armor) which could cause damage if they entered the bearing (Example 5.2.1.3-7).
The **pressure ratios in and around the bearing chamber** must be designed so that pressure is lower outside of the bearing chamber, ensuring that wear products are blown out.

• The number of compressor rotor blades should be even so that no imbalances occur when **smoothing out a FOD nick** on a blade; if the number is even, simply smoothing out the opposite blade will restore the balance. This procedure has proven itself in practice. However, designers often dislike even numbers of blades because these increase the probability of interference with the guide vanes and also the danger of the blading reaching resonant frequency.

• Sufficient **inspection openings** and good optical access to the blading.

• Easy **replaceability of the blades without complex installation** work or extensive disassembly of the part groups. This can be achieved through constructive measures such as axially split housings and suitable blade root fastening mechanisms.

• During testing rig operation, especially with **field testing rigs**, it must be ensured that no particles such as **rust** or flaking coatings come off of inlet ducts and tunnels. Any screens in front of the inlet ducts should be inspected for worn wires, since pieces of wire could break off due to a dynamic fatigue fracture.

• If possible, engines should not be run on the ground (for example, in field testing rigs or test runs) if there are any loose **dusts or chemicals in the near vicinity** (several hundred meters). Paint fumes, man-made fertilizer dust, and other materials have been shown to be damaging to compressors and hot parts.

• During manufacture and repair it must be ensured that any **manufacturing residue** such as metal shavings or auxiliary materials such as blast medium or cleaning fluids is completely removed from the engine parts. Oil and filters should be inspected for contaminants, especially in the first few hours of operation.

• Only the required amount of **auxiliary materials** such as **sealants** or **lubricants** should be used during maintenance and overhauls. Care must be taken not to press larger, excessive amounts into the inside of engine parts where they might block up the oil system or cause other problems.

Chapter 5.2.1.3 References

5.2.1.3-1 G. Fischbach, "916 Deutsche F-104 Starfighter, ihre Bau- und Lebensgeschichten", page 662.

5.2.1.3-2 G. Fischbach, "916 Deutsche F-104 Starfighter, ihre Bau- und Lebensgeschichten", page 595.

5.2.1.3-3 G. Fischbach, "916 Deutsche F-104 Starfighter, ihre Bau- und Lebensgeschichten", page 92.

5.2.1.3-4 G. Fischbach, "916 Deutsche F-104 Starfighter, ihre Bau- und Lebensgeschichten", page 307.

5.2.1.3-5 G. Fischbach, "916 Deutsche F-104 Starfighter, ihre Bau- und Lebensgeschichten", page 303.

5.2.1.3-6 G. Fischbach, "916 Deutsche F-104 Starfighter, ihre Bau- und Lebensgeschichten", page 593.

5.2.1.3-7 NTSB Identification MKC89IA024, microfiche number 37058A.

5.2.1.3-8 D.A. Lombardo, "Controlling Foreign Object Damage", periodical "AM" , April 1997, pages 27-31.

5.2.1.3-9 P. König, A. Roßmann, "Ratgeber für Gasturbinen-Betreiber" Vulkan Publishing Essen, page 57.

5.2.1.3-10 "Investigation Team Identifies Causes of CF6-80 Problem", periodical "Aviation Week & Space Technology", February 7, 1983, page 32.

5.2.1.3-11 D.A. Fulghum, "JSF Lift Fan Has Another Glitch", periodical "Aviation Week & Space Technology", September 11, 2000 page 37.

5.2.2 Bird Strikes

5.2.2.1 Bird strike damage

5.2.2.2 Bird strike damage mechanisms

5.2.2.3 Constructive and technological influences on bird strike behavior

5.2.2.4 Measures against bird strike damage

The **term "bird strike"** refers to the process of a bird entering an aircraft engine, including the impact of the bird against inlet and engine components.

Damages to engines caused by bird strikes can vary greatly (Ill. 5.2.2.1-7). Usually, the **compressor blading** is affected. The allowable amount of damage caused by bird strikes should be specified in the **certification and rating regulations** for each engine type. **Flocks of smaller birds** up to about 80 grams must generally be tolerated by the engine without any unallowable damage or changes to operating behavior. The blading must not be permanently deformed or suffer impact nicks that would reduce the dynamic strength considerably. Even if **larger birds** are ingested, at the very least it should be guaranteed that a controlled **shutdown of the engine** is possible. The engine will usually have to create a minimum amount of **power** over several minutes after the bird strike. During this period, it must be ensured that no consequential damages such as blade failures occur due to dynamic fatigue, etc. If blades are seriously deformed, vibrations will most likely occur due to stalls and flutter (Ill. 5.2.2.1-15, Ref. 5.2.2.1-16). In case of a blade failure, under no circumstances may fragments escape from the engine or put unallowably high stress on the engine suspension or nacelle. Blade failures with extreme imbalances can cause **titanium fires** in the high-pressure compressor, causing serious and extensive damage in the core engine resulting in engine failure within a few seconds.

External Factors
Bird Strikes: Damage

Heavy bird strikes will often result in damage to the **nose cone** and inlet housing struts, depending on the design of the engine. Powerful axial forces caused by the stiffness of the struck parts and/or large imbalances and overload the main bearings and make them fail.

Bird strikes are one of the most common causes of engine damage. Especially **military aircraft** that operate at relatively low altitudes are prone to bird strike damage. This can be exacerbated if flight paths are in the area of migratory bird routes. Bird strikes are one of the leading causes of single-jet **fighter aircraft** crashes.

An important parameter for the stressing of the engine components by the ingested bird(s) is the **flight speed**, which is usually equal to the speed at which the bird strikes the engine, since the bird‘s airspeed is negligible compared with the speed of the aircraft.

The sensitivity of engines to bird strikes also depends on a large number of **constructive factors** inside the engine. These include not only the location of the engine on the aircraft, but also the engine materials and the design of engine components.

Certification of engines for civilian and military applications includes verifications (through testing and design analysis) of acceptable behavior in case of bird strikes (see Chapter 5.1.1). This verification is generally conducted by using an air-pressure cannon to fire actual birds into the engine at the required speed, while the engine is running at a specified performance level. The engine must then show a specified operating behavior depending on the size and number of the birds. For example, the performance may only decrease a certain percentage, or in extreme cases with large birds engine failure must be controllable and any engine part fragments must be contained.

The procedures used in these **bird strike tests** was developed over a long period of time. It was discovered, for example, that birds cannot be simulated by plastic bags filled with water of equal weight, since the air pockets inside birds and their unique shape are less damaging. For this reason, during the engine development stage a special **substitute mass** is used that has sufficient air pockets to realistically simulate the shockwaves created in the foreign object and engine component by the impact.

These tests are very elaborate and a determining step of the certification process that must often be repeated after **constructive changes.** In this manner, the use of fiber-reinforced synthetic materials in large fan blades failed due to the poor bird strike behavior of this material combined with the design standards at the time.

5.2.2.1 Bird Strike Damage

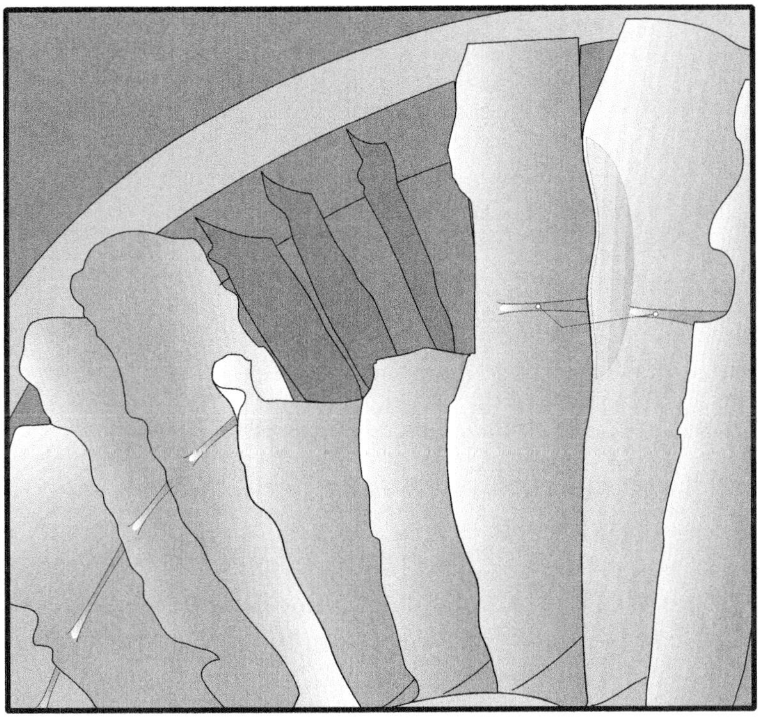

The common perception that birds are a relatively soft and yielding foreign object does not apply to the **impact conditions** in an engine due to the high typical impact speeds. A bird with a mass of a few hundred grams can not only seriously damage the compressor blading, but it may even overload solid components such as housing struts.

Statistical data for bird strikes:

The statistics that describe the frequency of bird strikes in engines are heavily scattered, and in some cases even contradict each other.

This is due partly the large differences in the **definition of bird strikes**; for example, whether this term includes mere bird impact, noticeable damage, damage in need of repair, or even aircraft accidents. This is further complicated by the fact that various bases are used for comparisons:

Aircraft movement: Takeoff + landing = 2 aircraft movements

Aircraft flight: 1 takeoff + 1 landing, but no indication of the number of engines involved

Engine flight: Each flight of each engine is counted separately (1 takeoff + 1 landing);

Furthermore, in many countries there are either no available data or only rough estimates, especially for military aircraft.

Civilian flight:

There is comprehensive data available from various sources. The following conclusions can be made on basis of this:

- Birds strike several hundred aircraft per 10^6 flights.

- In roughly 3/4 of all cases there is no damage to the aircraft.

- In about 20% of all bird strikes on aircraft than one or more engines are affected. However, it is very rare that multiple engines are struck simultaneously.

- Only less than 10 % of all bird strikes on aircraft, i.e. 35% of bird strikes on engines result in engine damage that requires immediate repair.

Simultaneous damage to multiple engines making continuation of flight impossible is an extremely rare occurrence in civilian cargo aircraft. In one case in the mid-1970s, a large, three jet, passenger aircraft was destroyed due to consequential damages resulting from several birds weighing roughly 2.5 kg each being ingested into one of the large-bypass fan engines. In general, after this type of incident, the certification regulations are changed accordingly (made stricter) in order to decrease the probability of similar damage occurring in future aircraft types.

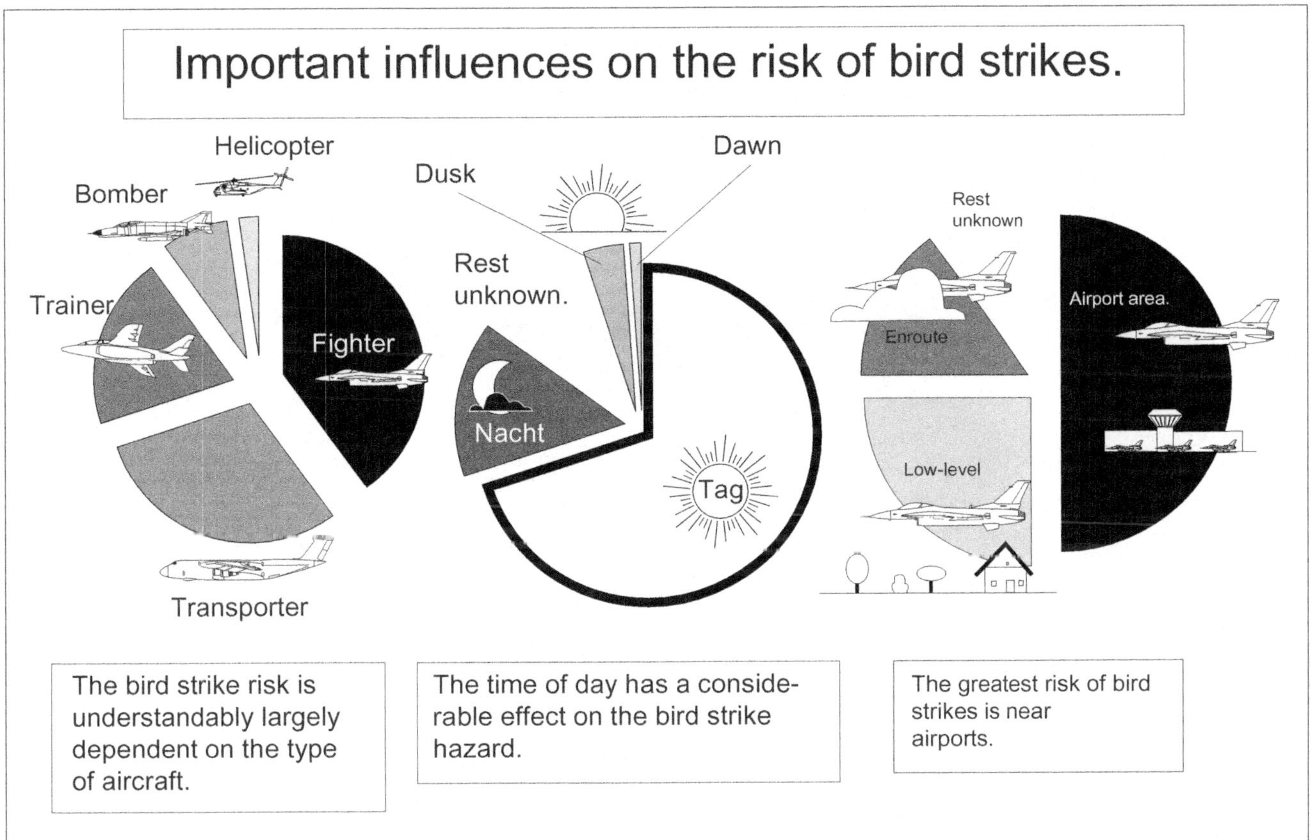

Ill. 5.2.2.1-1

Illustration 5.2.2.1-1 *(Ref. 5.2.2.1-1): This* ***statistical analysis*** *is based on military aircraft in the early 1980s. The greatest bird strike risk was in* ***fighter aircraft****, followed by* ***large cargo aircraft*** *(top diagram). The relatively large percentage of missions that fighter aircraft fly at* ***low altitudes*** *should increase the probability of bird strikes considerably (see Ill. 5.2.2.1-2).*
Large cargo aircraft have a large engine ***intake surface area****, which should also be a reason for increased bird strikes.*
Since most bird species are active during the ***daytime****, it is natural that the greatest risk of bird strikes was during this time (middle diagram). However, it is interesting that there were also a considerable number of bird strike incidents registered at* ***night****.*

By far the greatest bird strike risk is ***around airports*** *(bottom diagram) which is partly due to the low flight altitudes in this area. However, this also means that a large number of birds are evidently present around airports. This is the reason for the extensive effort to remove birds from areas near airports, which continues today with limited success.*
Bird strike is relatively rare in helicopters, *even though these operate primarily at low altitudes (top diagram). One explanation for this is the relatively low flight speeds, combined with possible deflection of the birds through the angled or vertical rotor wash.*

The risk of bird strikes increases extremely during flight missions at low altitudes.

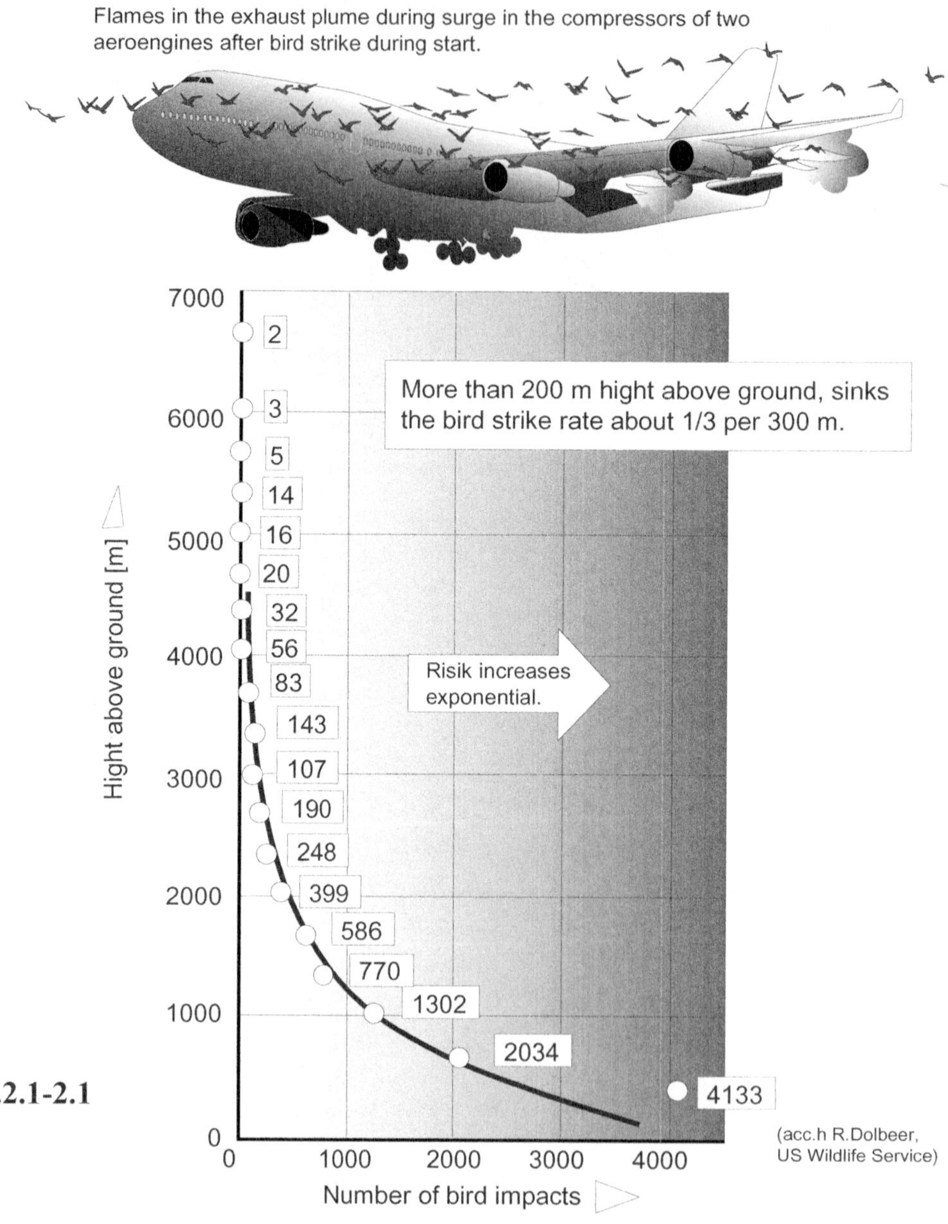

Ill. 5.2.2.1-2.1

Illustration 5.2.2.1-2.1 *(Refs. 5.2.2.1-2 and 5.2.2.1-3): The greatest number of bird strike damages occurs at* ***altitudes*** *below 500 meters. This is correlated with the corresponding* ***flight situations****, i.e. takeoff, landing, and approach flight (Ill. 5.2.2.1-3). The documented cases involving fighter aircraft show a similar distribution. This is due to both the preferred low flight altitudes of birds, as well as the fact that flight missions of modern fighter aircraft are often at low altitudes below radar height, even during cruising flight.*

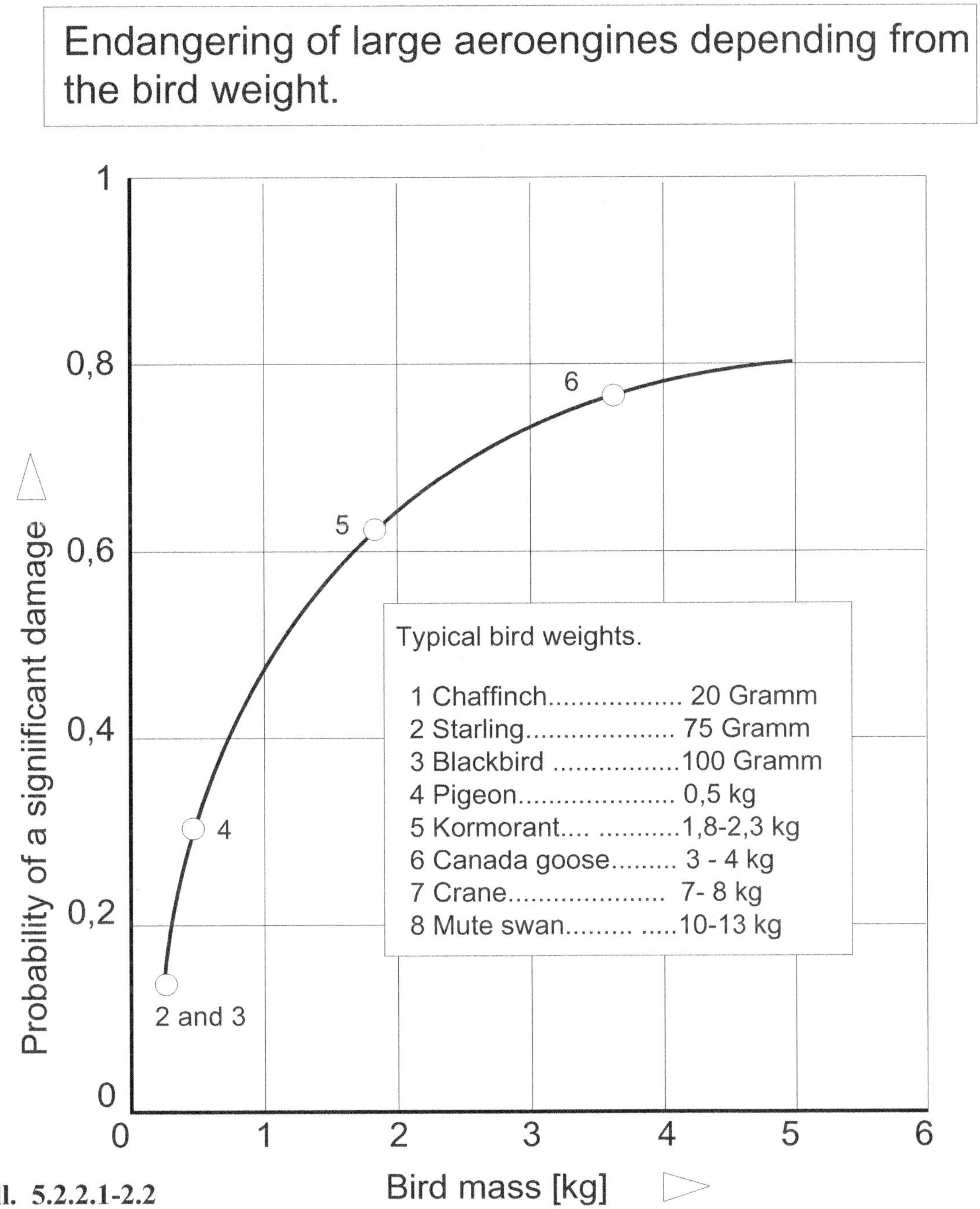

Ill. 5.2.2.1-2.2

Illustration5.2.2.1-2.2 *(Ref. 5.2.2.1-20): This diagram from a statistik (DOT/FAA/CT-91/17) of the U.S. authority FAA betrays a clear* ***connection between bird weight and 'significantem' damage*** *at the aeroengine. The definition of the damages is not distinct. Concerned may be damages which make a repair necessary (Ill. 5.2.2.1-6.6). Interresting is the markedly increase of the damage risk up to about 2 kg bird mass. Even birds with aroud 100 g are obviously in the position to cause bigger damages. Thereby it can be supposed, that it's about several birds (flocks, Ill. 5.2.2.1-6.1) and not about single exemplars.*

External Factors
Bird Strikes: Damage

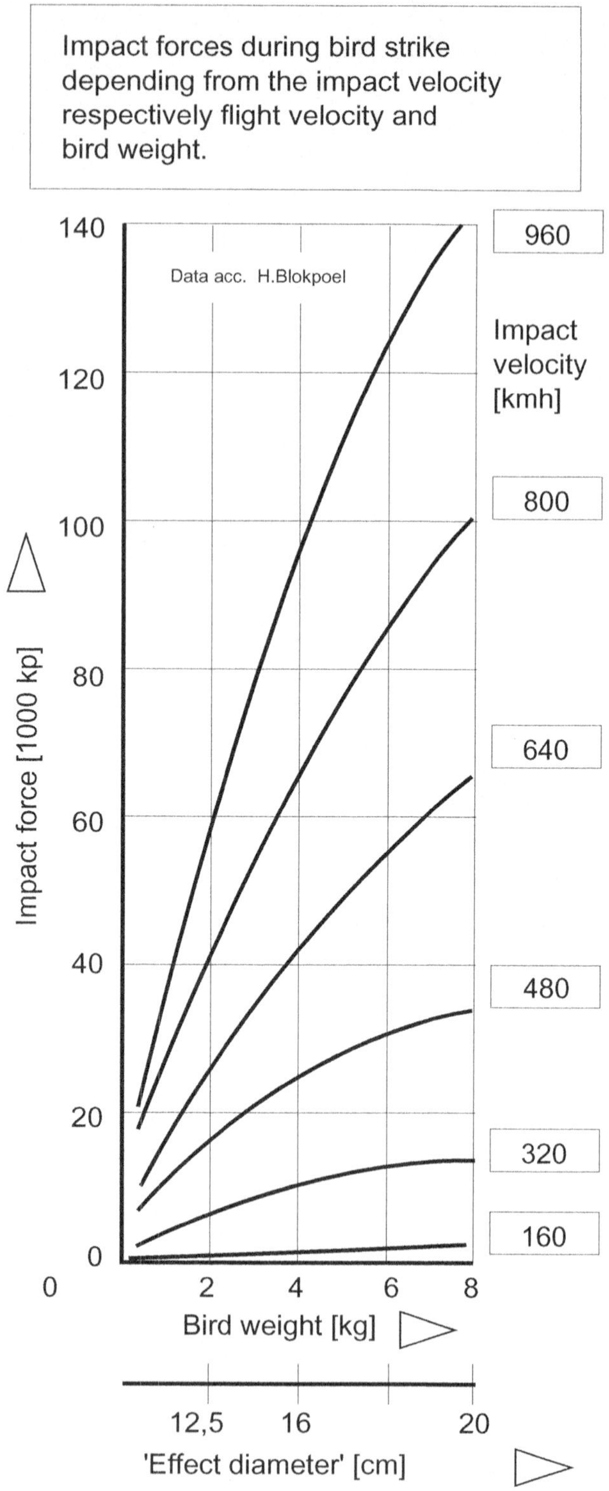

Ill. 5.2.2.1-2.3

Illustration 5.2.2.1-2.3 *(Ref. 5.2.2.1-12): This concerns theoretical estimated* ***impact forces*** *(Ill. 5.2.2.1-2.4). Considered is a bird impact at the outer skin of an aircraft fuselage. Anyway, at least* ***magnitude and tendencies*** *may also apply for aeroengine components like fan blades, casing struts and nosecone as well as the intake duct. The forces in the chart will heavy scatter in the real case. This has several reasons. The calculations where based on an impact angle of 90°. With this the real* ***inpact angle*** *in the individual case was not considered. An angle of 30° may be more realistic, at least for the fuselage surface. This would besect the stated impact forces. With the introcuction of an „* ***bird diameter*** *“ (effective diameter) this inaccuracy shall be a little reduced. The forces can be also a multitude higher, depending from the supposed ‘****bird consistence****’. Contrary a* ***flexibility of the impacted surface*** *will act diminishing.*

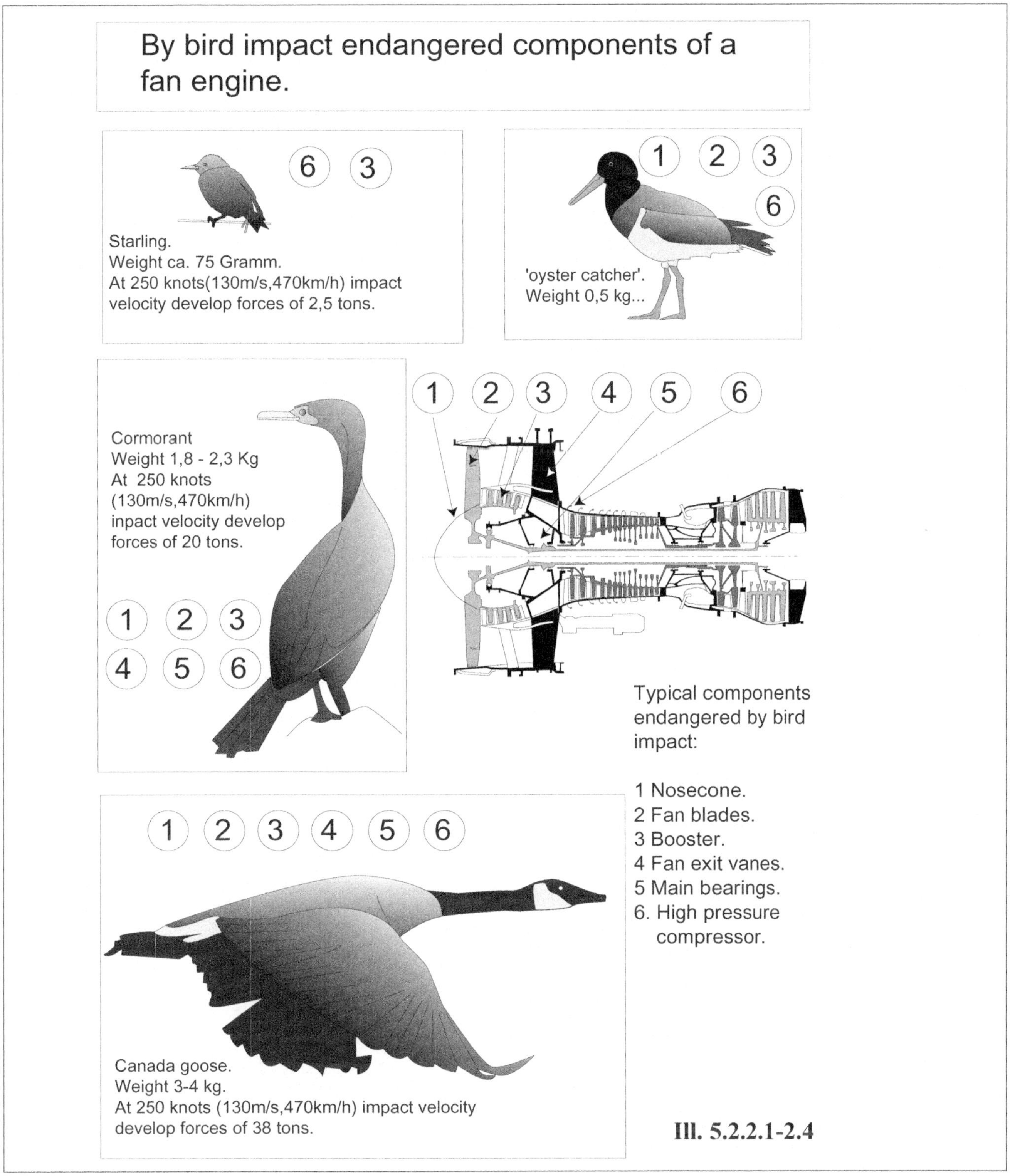

Ill. 5.2.2.1-2.4

Illustration 5.2.2.1-2.4 *(Ref. 5.2.2.1-18): In this picture will be tried to assign typical, for the bird impact outage endangered aeroengine components,* ***sizes of birds.***

Birds around 100g *can already represent a danger. This is especially to expect for* ***flocks.*** *They can lead to changes at/in the 'booster' (3)* ***(deformations of blades, cross section reduction)*** *which trigger a* ***surge*** *(Ill. 5.2.2.1-3).*

External Factors
Bird Strikes: Damage

This is especially risky because bird impacts occur during the start phase at full aeroengine power.
*The rotorblades of the 1st stage high pressure compressor (6) after a '**swan neck channel**' ('S-channel') are from experience also endangered by overload from small birds. These blades are located behind the booster (Ill. 5.2.2.3-12) or at three shaft engines between intermediate pressure compressor and high pressure compressor (Ill. 5.2.2.3-11). Here the danger exists, that **bird rests accumulate at the struts in the channel** and then, as a bigger mass, act at the filigree rotorblades as **low speed impact** (Ill. 5.2.2.2-8).*
*Even small single birds can destroy **sondes**/probes (intake temperature, pressure/flow velocity) which extend into the flow channel and 'consternate' control units and the cockpit instruments.*

***Birds up to 1 kg** are above the problems with smaller birds already in the position to produce alarming failures with a **direct impact**. This is also true for the **nosecone** (rotating or fixed, „1"). Such made of **fibre reinforced plastic** had to be suitable strengthened (example 5.2.2.1-2). Also big **fan blades** (2) from a high strength titanium alloy can be already so **deformed**, that a replacement is necessary (Ill. 5.2.2.1-6.6). At smaller aeroengines a **blade fracture** is absolutely possible.*
*Also indirect heavy aeroengine failures/damages can be triggered already from birds of this weight class. This is the case if **cowls/shrouds** of the fuselage or the nacelle respectively the **intake channel** to the aeroengine are ripped off (Ill. 5.2.2.1-9.1, Ill. 5.2.2.1-9.2 and Ill. 5.2.2.1-14). The secondary failures are as well heavy mechanical damages of the blading as also vibration overload from compressor surge and/or flow disturbances.*

***Birds up to 2,5 kg** can produce impact forces in the range of 10-tons. With this the danger exists, that already **struts in the intake to the front bearing chamber** break (Ill. 5.2.2.1-6.4.1). A direct hit at the spinner respectively the **low pressure rotor shaft** can destroy mainbearings. In such a case high rotor unbalances and catastrophic secondary failures must be expected. Occures an abrupt braking of the rotor by intense rubbing, also in the rear aeroengine area catastrophic failures by overloads like the fracture of casings are possible..*
*Heavy rubbing processes and/or the fracture of **oil flown through casing struts** can cause **fire** in the **intake region** of the aeroengine (Ill. 5.2.2.1-6.4.1). Thereby obviously also **plastic components can be ignited** and **Al honeycomb structures** of the lining from the nacelle can be molten.*

***Birds more than 2,5 kg** let expect the already discussed damages/failures in a far disastrous dimension (Ill. 5.2.2.1-6.4.1).*

External Factors
Bird Strikes: Damage

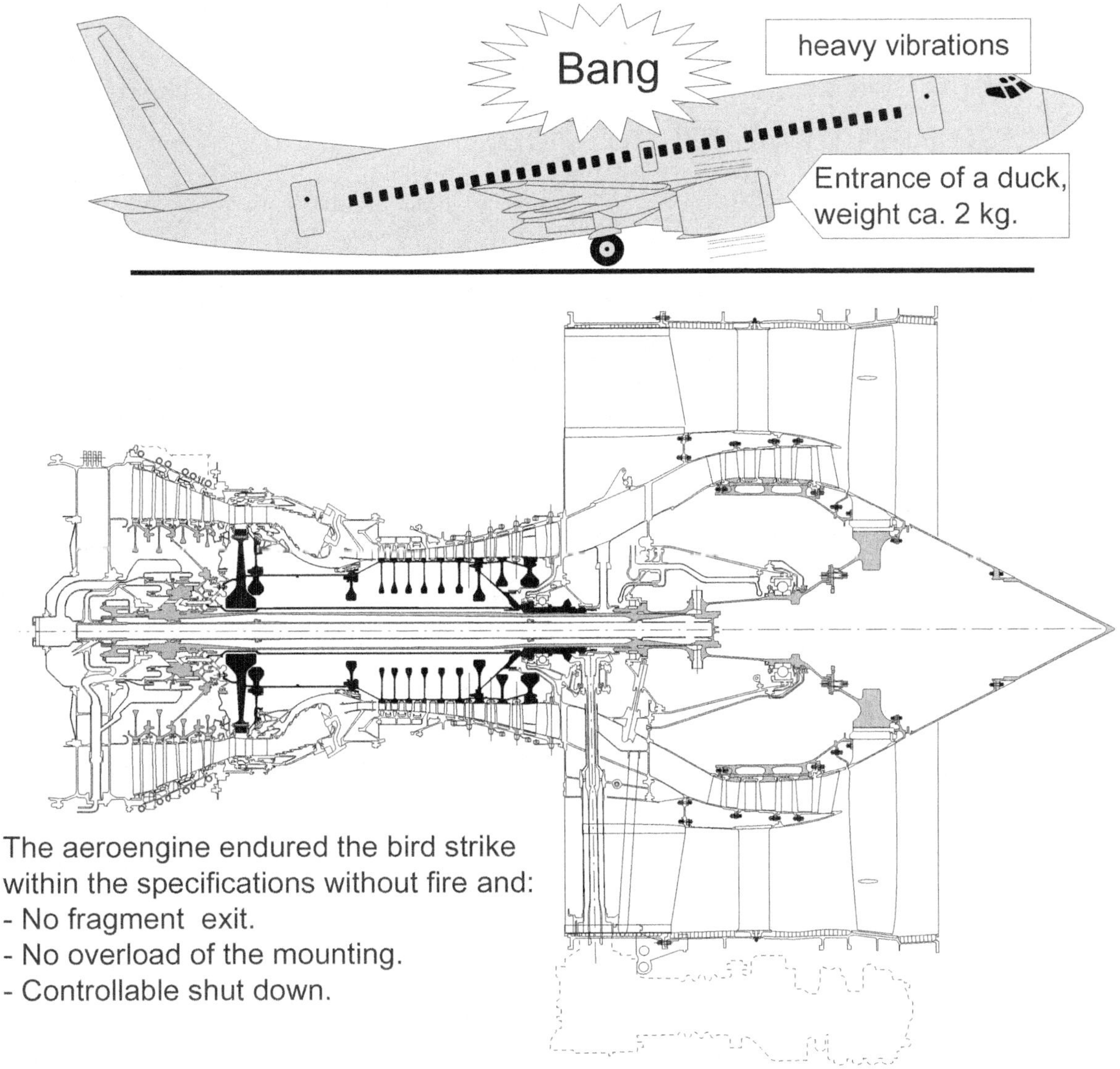

Example 5.2.2.1-1 (Ref. 5.2.2.1-4):

Excerpt: "...During rotation on the takeoff roll the right engine of (a two engine transport aircraft), ingested a bird. Several fan blades broke, however there was no fire and the engine continued to run until shut down by the flight crew... Feathers removed from the engine belonged to a bird with ...an average weight of **3.8 pounds**. The engine performed within the limits of its certification criteria."

Comments: The bird strike during takeoff is a "classic" bird strike situation (Ill. 5.2.2.1-3). The pilot heard a loud bang and felt heavy vibrations. Inspection of the feathers and remains of the bird that were found in the fan, compressor, and combustion chamber showed that it was a duck. The given value is the average weight of that species of duck. With this large bird, the contained failure of fan blades is allowable. The special stressing of the fan rotor blades is to be expected during low speed impacts such as this one (also see Ill. 52.2.1-6.5.1 and chapter 5.2.2.2).

'Plastic spinner' must be safe agaunst bird impacts.
Fragments must not develop.

Fragments of the fan blades and the rotating nose cone of glass fibre reinforced plastic, get into the core engine.

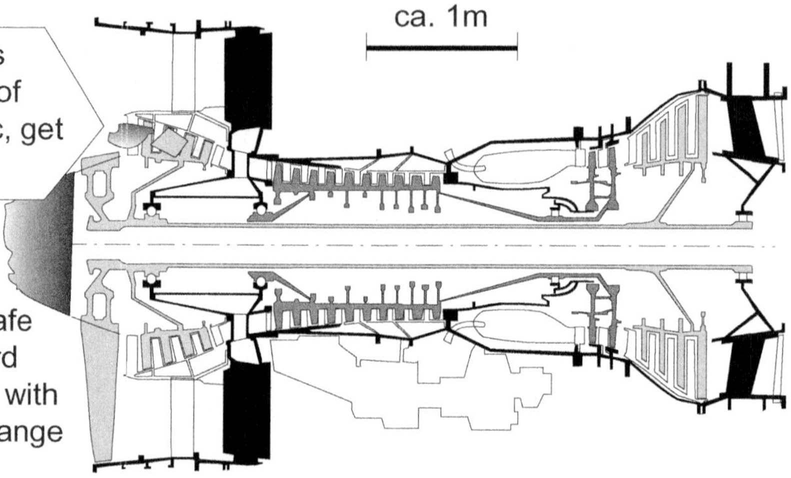

'Plastic spinner' must be sufficient safe against high impact forces during bird impact. In the shown case of a duck with 0,8 kg weight, the forces are in the range of several tons.

Example 5.2.2.1-2 (Ref. 5.2.2.1-5):

Excerpt: "A 28 ounce (about 900g) duck was ingested into a ...turbofan engine of a (4 engine aircraft) on takeoff, impacting and **breaking the 10 ply fiberglass fan spinner. Downstream engine destruction followed.** Engine manufacturer and FAA was aware of a number of previous bird-strike failures with this spinner. New and stronger 20 ply spinners are being phased in; however thin spinners are still in use..."

Comments: This case involves a passenger aircraft with four large fan engines with high bypass ratios. The pilot was able to shut down the engine in a controlled manner and land the aircraft safely. The damage investigation showed that the engine damage had been caused by **fragments of the spinner** (rotating nose cone) **and fan blades.**

At the concerned aircraft type obviously spinners have shown in an early version during bird strike as not sufficient against the high impact forces (Ill. 5.2.2.1-2.3). After this the **number of glass fiber layers was doubled.**

Dangerous takeoff abortion after the entrance of a relative large bird into the aeroengine.

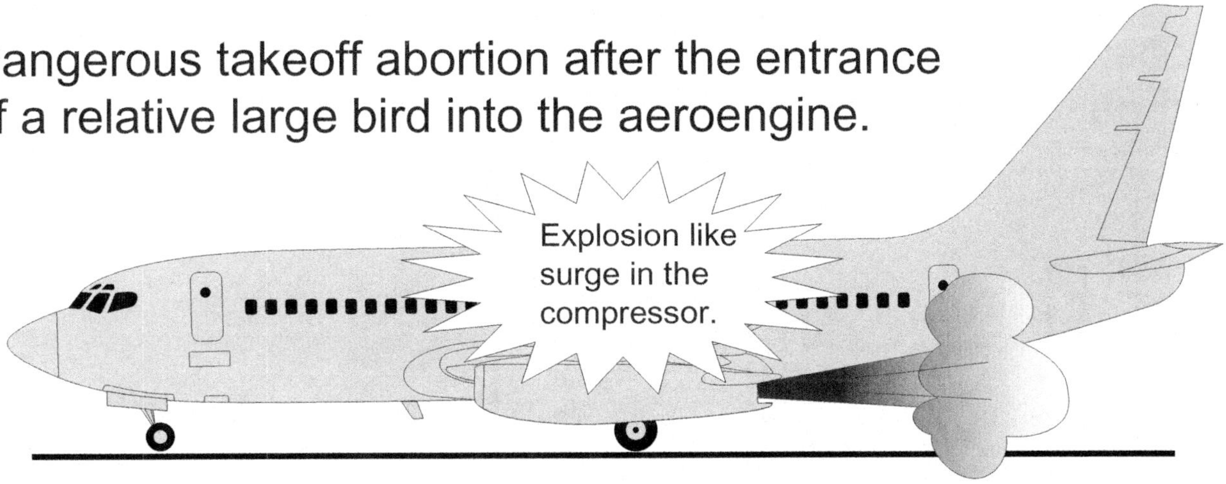

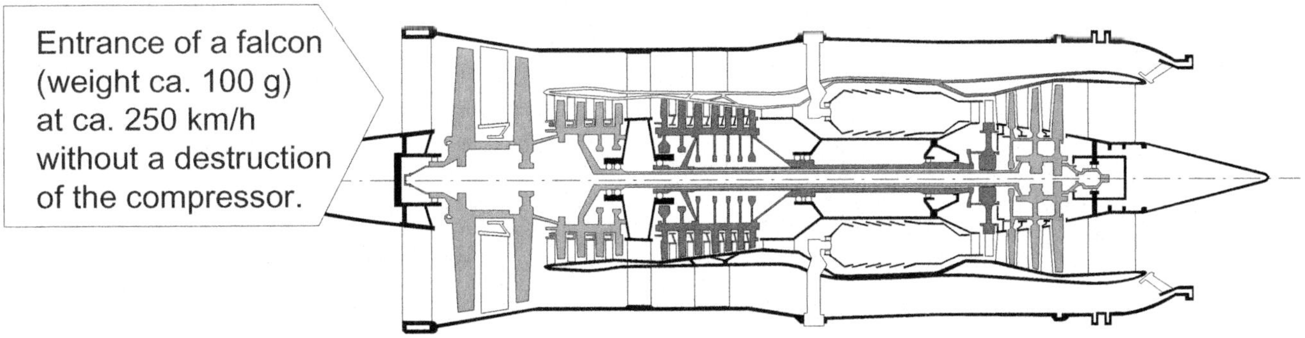

Beispiel 5.2.2.1-3 (Lit. 5.2.2.1-6):

Example 5.2.2.1-3 (Ref. 5.2.2.1-6):

Excerpt: "During the takeoff roll after V1 was called by the officer, a bird was ingested in the left engine.

...according to the captain, he observed an object flash past the airplane. The first officer reported he saw a bird on the right of the nose of the airplane. Shortly afterwards he heard a loud 'explosion'...The captain reported that the **explosion, louder than any compressor stall he had ever experienced**, created a shudder in the airplane....the airplane was subsequently stopped approximately 750 feet off the departure end of runway..."

Comments: This was a smaller twin jet passenger aircraft with two older fan engines with relatively low bypass ratio. The bird was a hawk that weighed roughly 120g. Evidently the damage to the engine was **not as serious as the noise and vibrations** of the triggered **surge** (volume 3 Ill. 11.2.1.1-7) caused by the bird strike seemed to suggest.

External Factors
Bird Strikes: Damage

The influence of the proximity to the ground is for the impact probability especially pronounced.

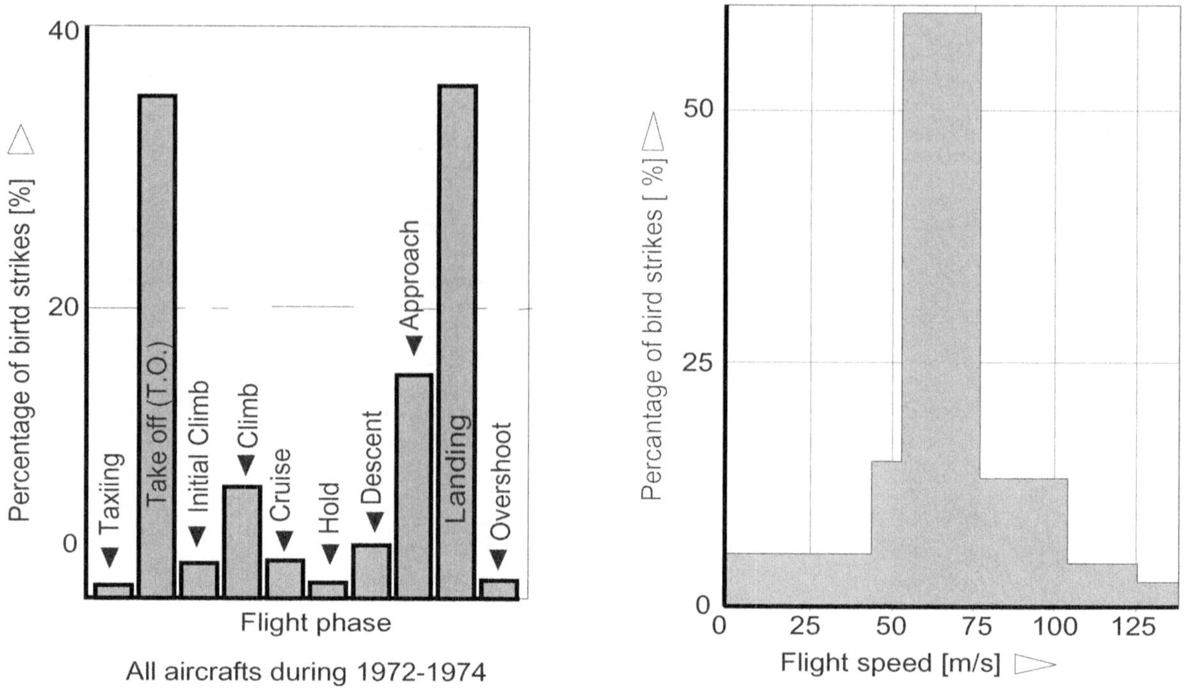

Ill. 5.2.2.1-3

Illustration 5.2.2.1-3 *(Ref. 5.2.2.1-2): In the low flight altitudes at which bird strikes are most frequent, civilian aircraft (tleft diagram) will generally flying at relatively low speeds (takeoff, landing, approach flight; right diagram). Therefore most strikes will below speed impacts that especially stress the fan rotor blades.*

Fighter aircraft are a different situation, since these usually travel at relatively high speeds at low altitudes. In this case, the resultant high speed impacts will especially stress static engine parts in the air flow such as housing struts or intake guide vanes (also see chapter 5.2.2.2).

Bird strikes in military aircraft

There are no accurate published statistics for military aircraft. In general, the situation should be similar to that of civilian aircraft, with the major difference being that there is an additional and probably dominating danger factor due to the **low altitude flights** over land and water at **high cruising speeds** (Examples 5.2.2.1-4). Also, the frequency of bird strikes is probably considerably increased by the fact that the engines of pursuit and fighter aircraft are integrated into the fuselage. If one refers to the civilian aircraft damage statistics (Ill. 5.2.2.1-5), the **fuselage** was the most frequently affected part of the aircraft. Assuming that due to the considerably smaller dimensions of military jet aircraft fuselages, roughly 50% of birds will strike the relatively large engine inlets (Ill. 5.2.2.1-13), then the bird strike rate for fighter aircraft should be roughly twice the 20% engine strike rate in civilian aircraft. The risk of total thrust loss is further increased in all single-engine aircraft. It should only be decreased relative to large civilian aircraft due to the fact that the (absolute) engine inlet surface area is considerably smaller. Ill. 5.2.2.1-5 shows that the strike rate only decreases underproportionally with smaller engine face cross-sections.

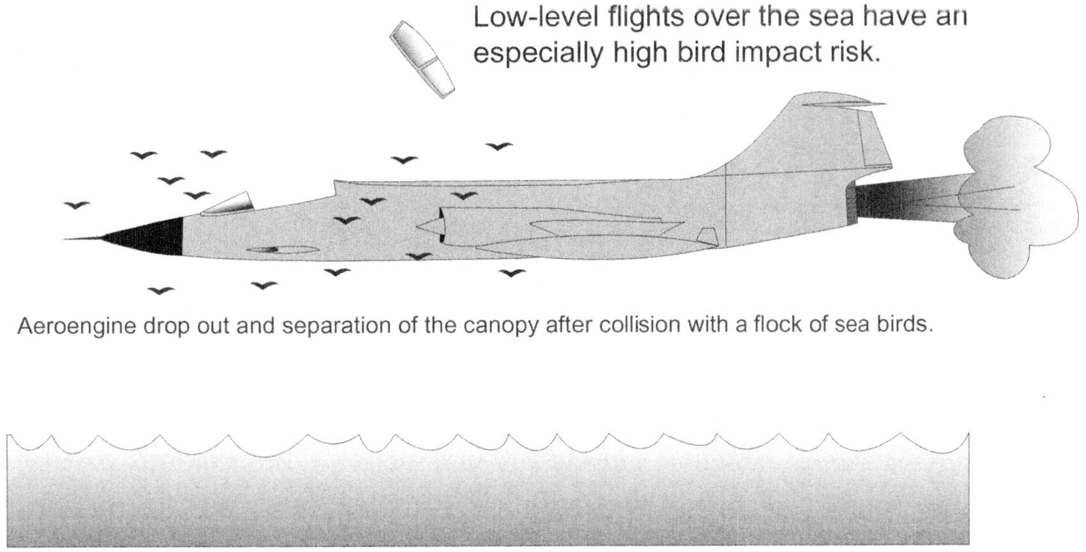

Aeroengine drop out and separation of the canopy after collision with a flock of sea birds.

Example 5.2.2.1-4.1 (Ref. 5.2.2.1-8)**:**

Excerpt: "...during low altitude flight at 50 meters over the ocean, flew into flock of birds, lost cockpit roof, engine failed, and crashed."

Example 5.2.2.1-4.2 (Ref. 5.2.2.1-7)

Excerpt: "...during low altitude flight at about 60 meters in three-aircraft formation, aircraft struck a bird and crashed into the Baltic Sea..."

Example 5.2.2.1-4.3 (Ref. 5.2.2.1-9)

Excerpt: "...during approach flight...crashed due to engine damage caused by flying into a flock of ducks..."

Comments: These are only three examples of many similar incidents, involving this single-engine fighter aircraft type. It is also typical that many of the incidents occurred at low altitude flight missions over the ocean. The (single cycle) engine was an older model with adjustable intake guide vanes. The blading of the rotor and stator was made from Cr steels.

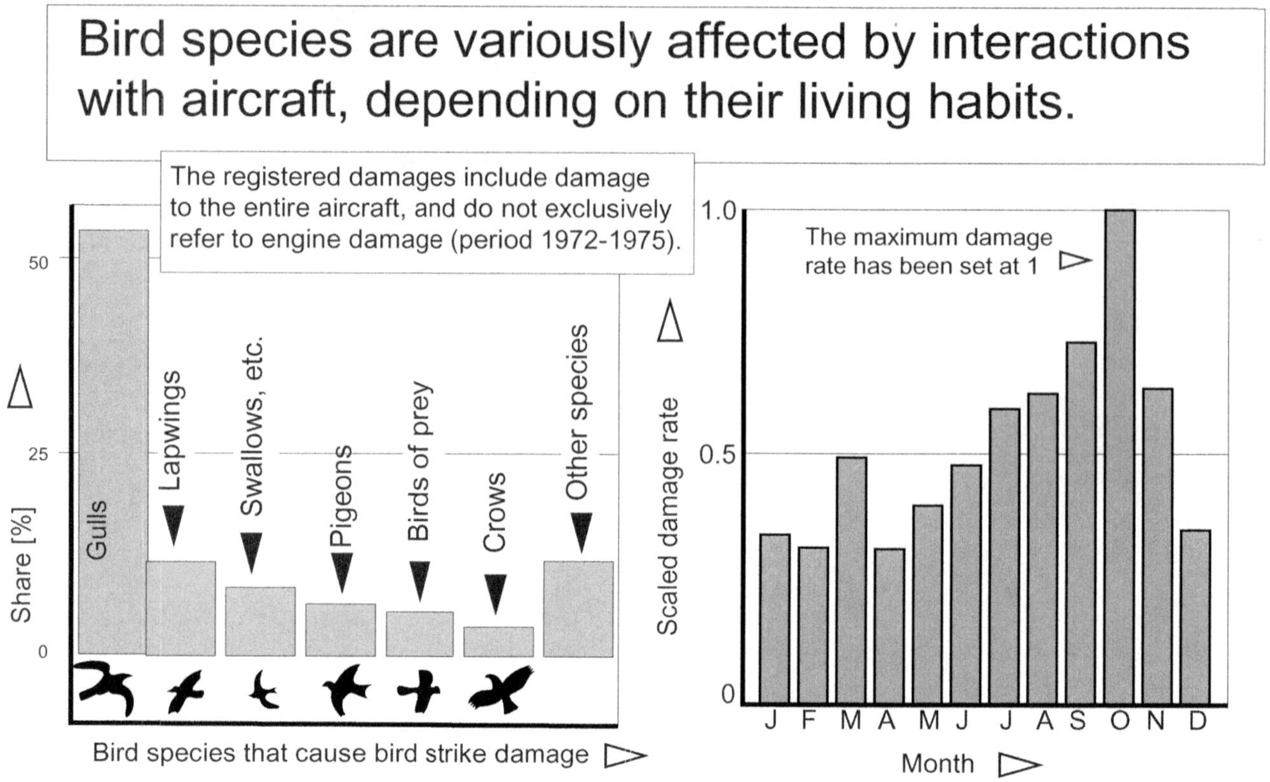

Ill. 5.2.2.1-4

Illustration 5.2.2.1-4 *(Ref. 5.2.2.1-2): The fact that in this year 50% of all bird strikes were caused by* ***gulls*** *and a further 12% were caused by* ***lapwings*** *indicates that airports and low-altitude flight paths in coastal areas and expansive wetlands are especially dangerous. Neither of these bird species is migratory. This statistic is taken from the European region.*

The number of bird strike damages in the ***nordic countries*** *seems considerably smaller than that in central Europe. This is probably due not only to the shorter stay of migratory birds in the north, but also the relatively small number of flights.*

The danger is especially great in the ***migratory bird routes*** *(Israel, etc.) and their land bridges (Denmark, Gibraltar, etc.) in the spring (May) and autumn (October), since expansive, dense flocks of birds during these times increase the probability of several birds striking one or more engines simultaneously (Ill. 5.2.2.1-6).*

Additionally, ***flocks of birds*** *often fly at greater altitudes during migration than they would otherwise, increasing the danger zone for aircraft.*

However, by far the largest influencing factor is the area ***surrounding an airport*** *(feeding areas, garbage dumps, fields, nesting areas, wetlands). In altitudes above 4500 meters there is virtually no risk of bird strikes since even large birds of prey such as eagles, condors, and vultures are rarely found at these heights.*

The frequency of bird strikes varies greatly between different areas of an aircraft.

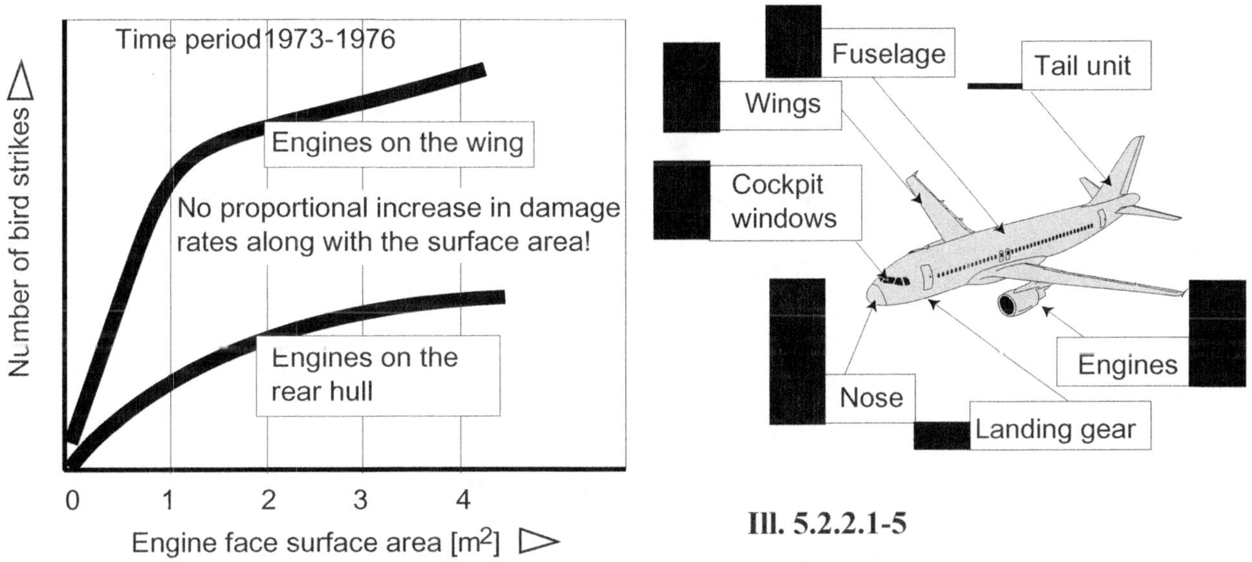

Ill. 5.2.2.1-5

***Illustration 5.2.2.1-5** (Ref. 5.2.2.1-2): The left diagram shows that with engines on the wings of large civilian cargo aircraft, the frequency of bird strikes increases considerably more slowly with **inlet cross-sections** larger than a certain size. There is no similarly clear limit in engines on the fuselage. It is not clear whether the under-proportional bird strike probability at larger inlet cross-sections is due to the introduction of modern fan engines with large bypass ratios. It is possible that the pinching of the inlet flow may have an effect on bird strikes (Ill. 5.1.1-5.1).*
The right diagram shows the distribution of bird strike damages across the entire aircraft during the early 1970s. If one adds up the fuselage components (nose, cockpit windows, hull), over half of bird strikes are on the fuselage.

Not so seldom as could be supposed: Both aeroengines suffer at the same time a bird strike.

"Multibirdimpact": In the USA 293 cases occurred in the period of time 1990 - 2003.
In 63 cases both aeroengines of twin-engined airplanes habe been hit.

Ill. 5.2.2.1-6.0

Illustration 5.2.2.1-6.0: *It would be estimated rather unlikely, that* ***several aeroengines of a single aircraft suffer a bird strike simultaneously****. Especially, that both aeroengines of a twin-engined aircraft are concerned. The statistics of the US aviation authority however teaches us something different (Ref. 5.2.2.1-18). A spectacular case shows this picture.* ***Both aeroengines*** *of this airliner dropped out by bird impact shortly after the start. The pilot had to make an emergency landing, which he could manage in a large city at a river nearby. The airplane floated so long on the water, until the passengers could be rescued.*
*This is the more estonishing because, during the '****splashdown****' of a modern airliner with big fan engines at the wings ,a lateral touch at the water let expect rather catastrophic damages (volume 2, Ill. 10-12).*

Also turboprop engines can drop out by bird impact. They are already endangered from relatively small birds.

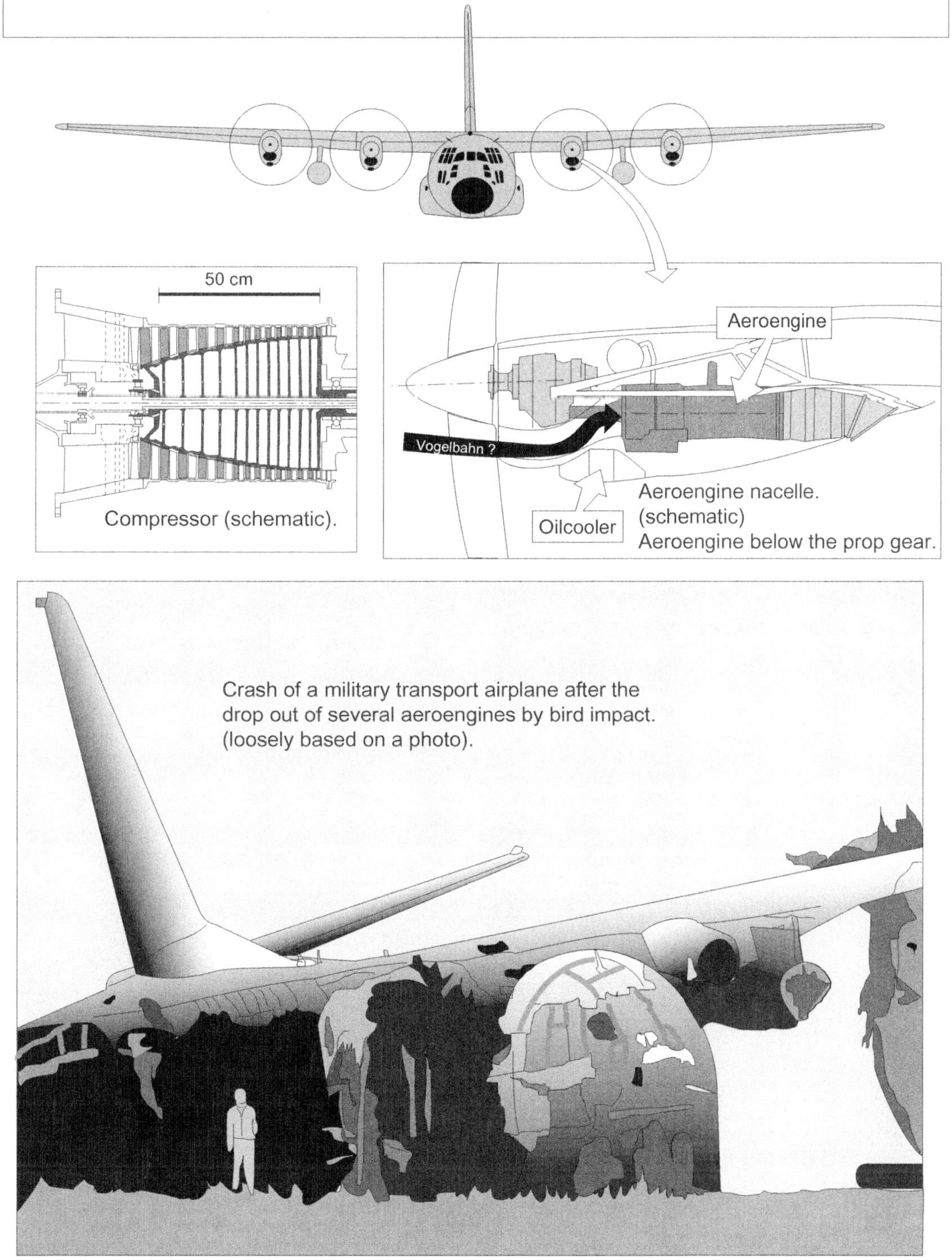

Ill. 5.2.2.1-6.1

Illustration description next page.

Ill. 5.2.2.1-6.1: *The displayed 4 engined* ***military transport aircraft*** *in the upper sketch crashed after a* ***flock of pewits*** *brought three of four turboprop engines to an outage (picture below, Ref. 5.2.2.1-18). The aircraft was in the landing approach and tried to avoid the flock of birds. It is not clear as far the evasive maneuvre or the loss of power from the aeroengines was cause of the accident.*
The documents on hand let not identify if the compressors have been damaged or if the opening of the oilcoolers intake, also is orientated to the front, are causative concerned.
So turboprop engines *are* ***not fully protected against birt impact*** *by the centrifuge effect of the propellers.*
It may be noted, that the turboprop engines are relatively small. The high-speed filigree compressor is as well FOD sensitive for the clogging by bird remains as also for mechanical damages. Additional aggravating is the mostly ***S- shaped bended intake channel****, which brakes the bird and so promotes a* ***low-speed-impact*** *(Ill. 5.2.2.2-7) in the compressor.*

Example 5.2.2.1-5 (Ill. 5.2.2.1-6.2, Ref. 5.2.2.1-2):

Excerpt: "....(the aircraft) which ingested lapwings in both engines on take-off. One crew member, of the eight persons on board, was injured, but the six occupants of a passing car were killed when the disabled aircraft slid across a road which passes the end of the runway"

Comments: The top diagram in Ill. 5.2.2.1-7 shows the aircraft after the accident. It is a relatively small civilian aircraft with two correspondingly small (and therefore sensitive to bird strikes) engines located on the rear fuselage. There have also been comparable damages in aircraft from other manufacturers. It is not clear whether or not this indicates an especially high sensitivity of this configuration. However, it is plausible that the small distance between the engines due to the relatively small fuselage diameter may increase the probability of both engines failing even due to small flocks of birds.

Flocks of small birds cause more catastrophic damage than single large birds.

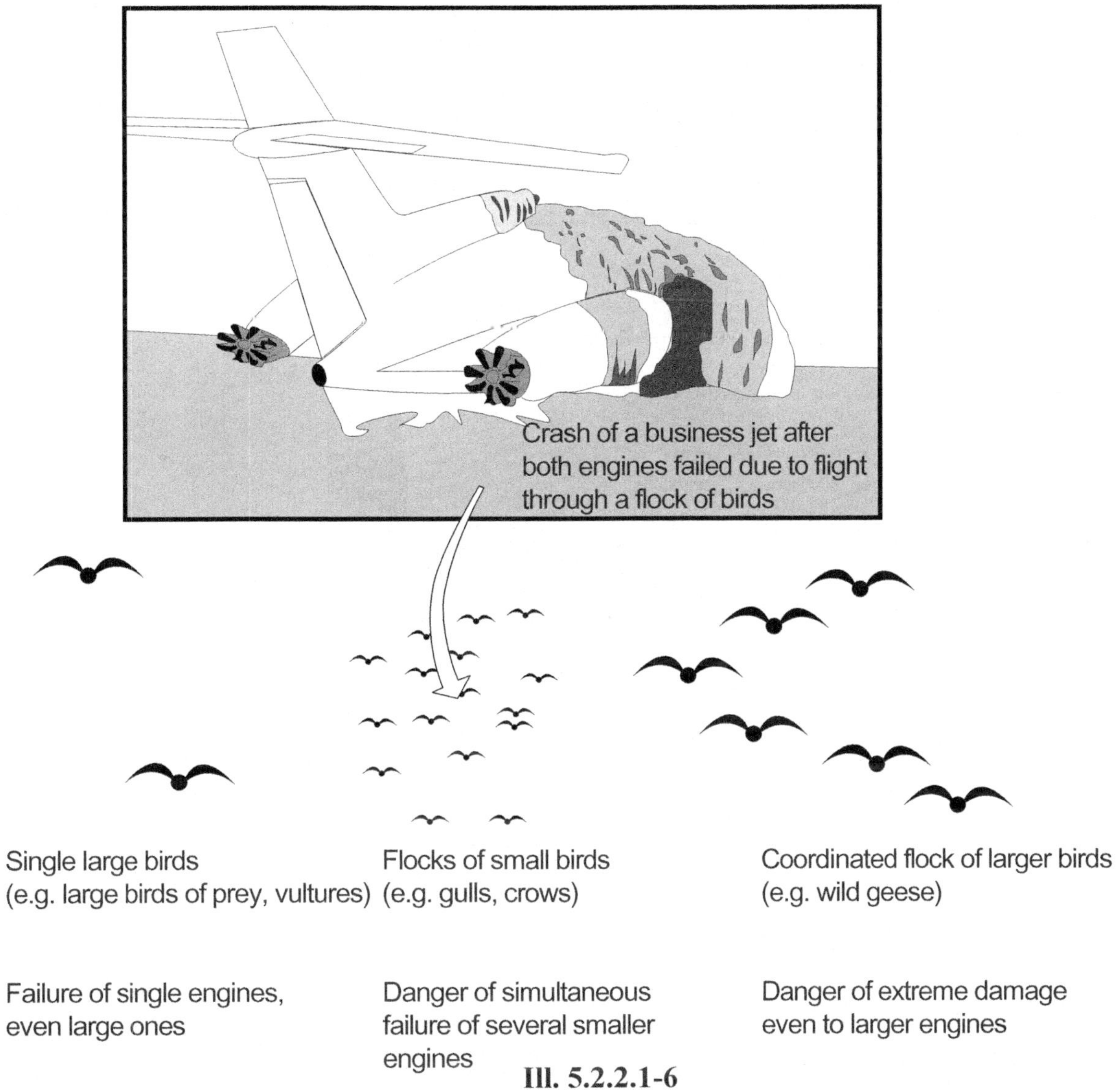

Ill. 5.2.2.1-6

Illustration 5.2.2.1-6 *(Ref. 5.2.2.1-2): The probability of* ***multiple engines failing*** *and possible total thrust loss depends largely on whether or not birds are encountered in* ***flocks****. There have been several reported cases in which a twin-jet aircraft crashed (top diagram) because both engines failed due to a flock of birds (Example 5.2.2.1-5). It is also important whether or not birds favor certain flight formations. These can cause* ***several birds to strike the same engine****. There is a documented case in which several wild geese caused catastrophic damage to a large fan engine (Example 5.2.2.1-6).*

An accident on the runway during start of the airplane occurred as also a bird flock upraised. Many birds hit the airplane so that the start had to be canceled. During deceleration of the aeroengines it came to the destruction of the right engine with a subsequent fire.

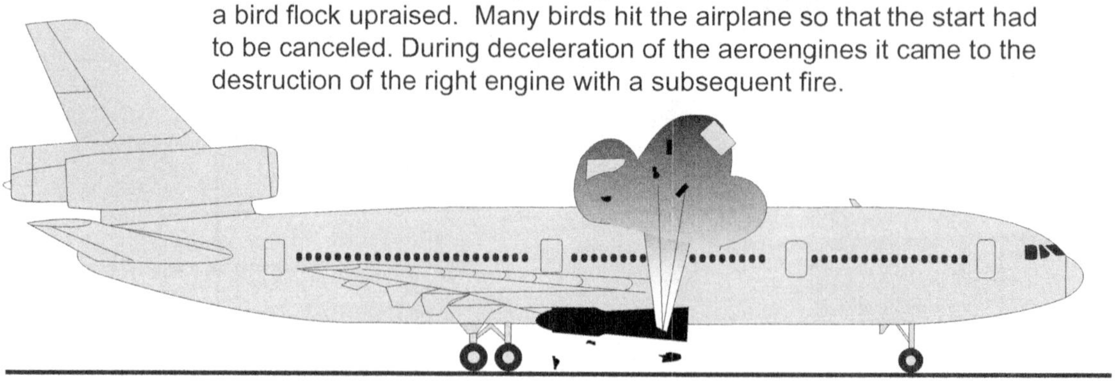

Distribution of the fragments at the runway from the aeroengine No 3.

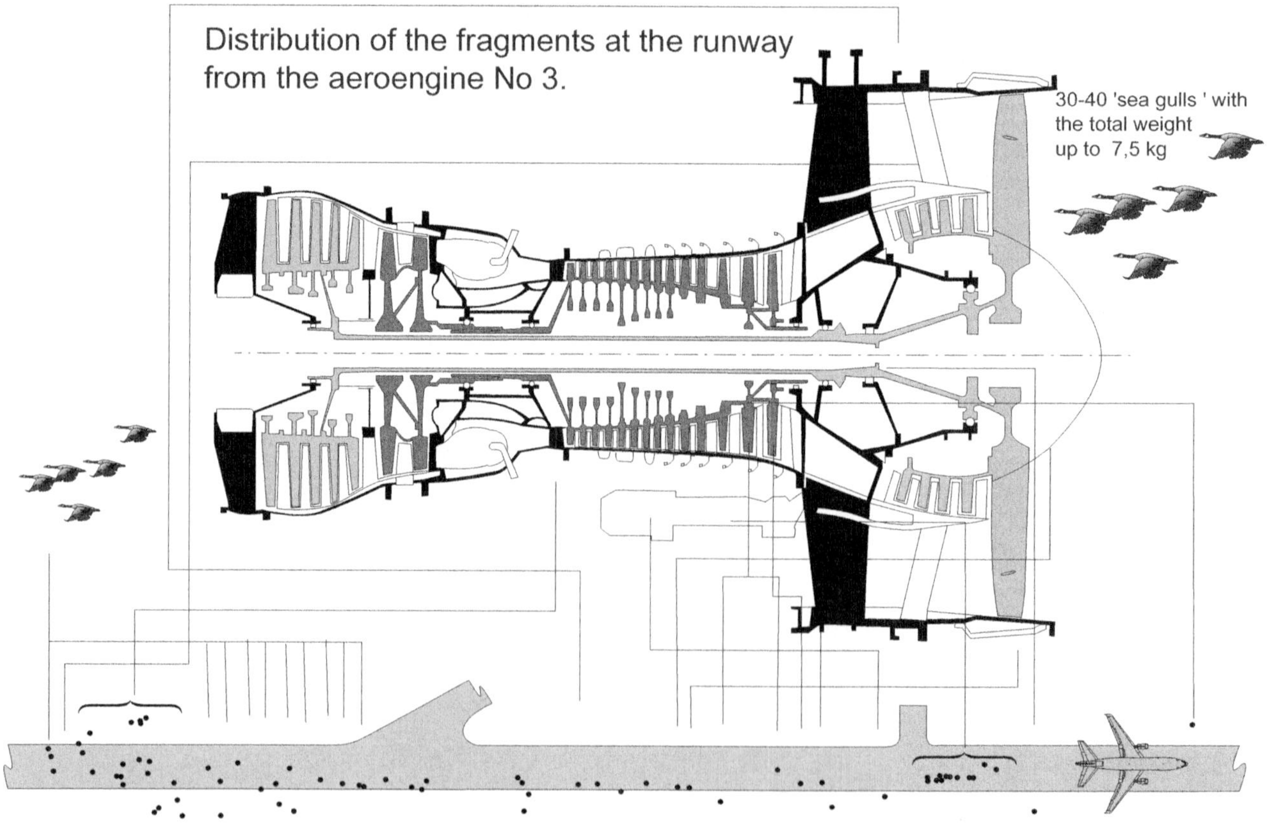

Example 5.2.2.1-6 (Ref. 5.2.2.1-11):

Excerpt 1 (Ref.5.2.2.1-10): „...after ingestion during takeoff of a number of turkey-sized seagulls...The No. 3 **engine on the right wing separated**, possibly as a result of unbalanced rotational forces, and the entire right **wing was engulfed in flames** by the time the pilot ... was able to brake to stop near the end of the runway....All 139 persons on board evacuated safely...

About **30-40 dead seagulls weighting up to an estimated 15 lb**. were found in the area of the runway were the strikes occurred....

Parts of the engine were strewn over half the length of the runway (see sketch below).

Excerpt 2 (Ref. 5.2.2.1-11): "...engine damage...was caused by a **massive multiple bird strike** of heavy birds **in excess of any FAA certification test requirements**....more than two birds and perhaps as many as **seven in the 5-6 lb.** weight class could have been ingested in the No. 3 engine...these large birds severely damaged the fan blades..."

Comments: This damage occurred in the mid-1970s and caused the certification and safety authorities (FAA, NTSB) to critically consider the problem of the ingestion of large birds into large engines with large bypass ratios (siehe auch volume 2 Ill. 9.4-7).. The results of these considerations are included in current certification and verification regulations. The birds involved in this case were probably **wild geese** rather than gulls (Ill. 5.2.2.1-6.5.1)..

For usually relative small turboprop engines already a small bird mass is enough to destroy the compressor.

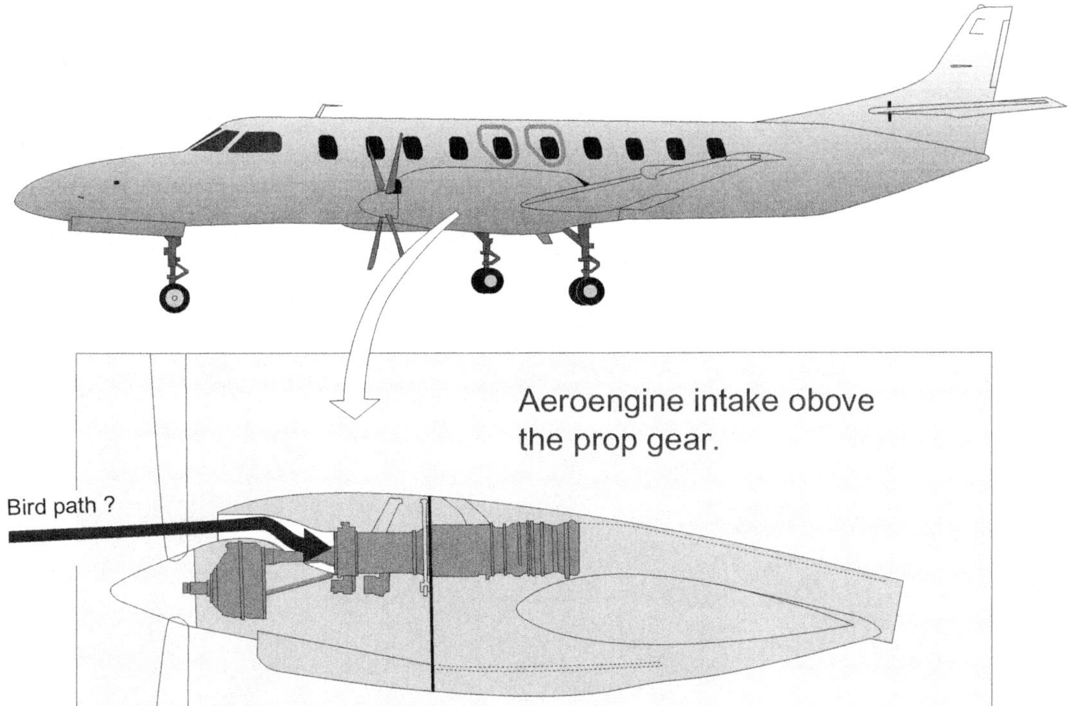

Ill. 5.2.2.1-6.3

Illustration 5.2.2.1-6.3: *The sketch shows a double engined civil aircraft at which the intakes of the* ***turboprop engines*** *are located at the upper side (compare intake below in Ill. 5.2.2.1-6.1). During start the dropout of one aeroengine occurred after the entrance of a 120 gram bird (Ref. 5.2.2.1-17). The compressor didn't show mechanical damages, although everywhere rests of the bird where found. Seemingly the remains have* ***markedly affected the quality of the compressor****.*

Heavy aeroengine failure after the entrance of a canada goose.

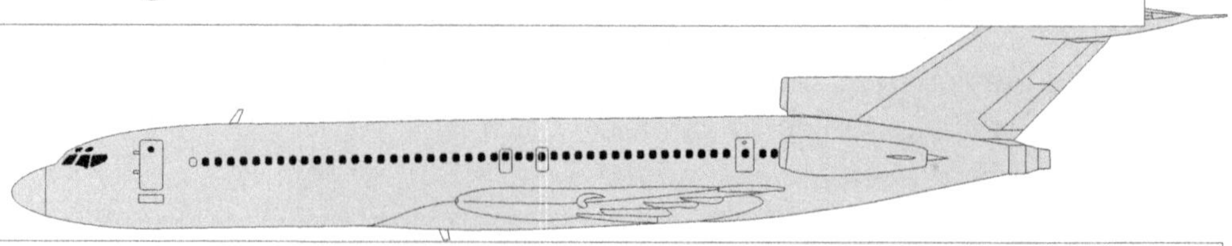

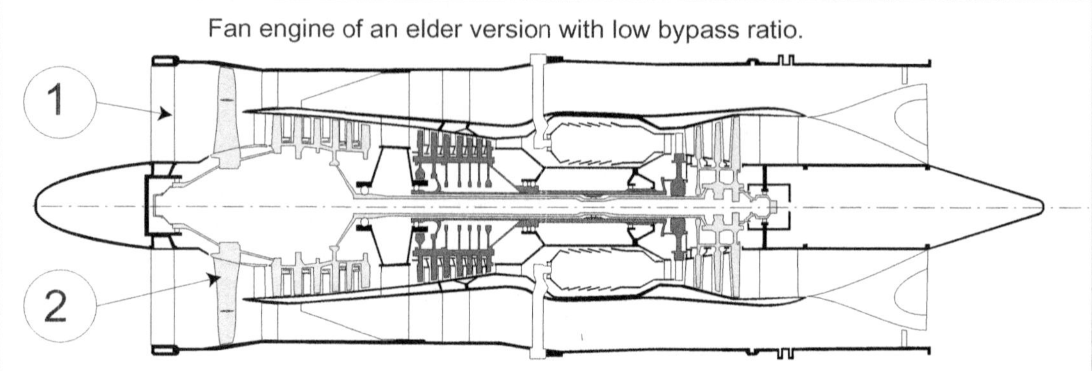

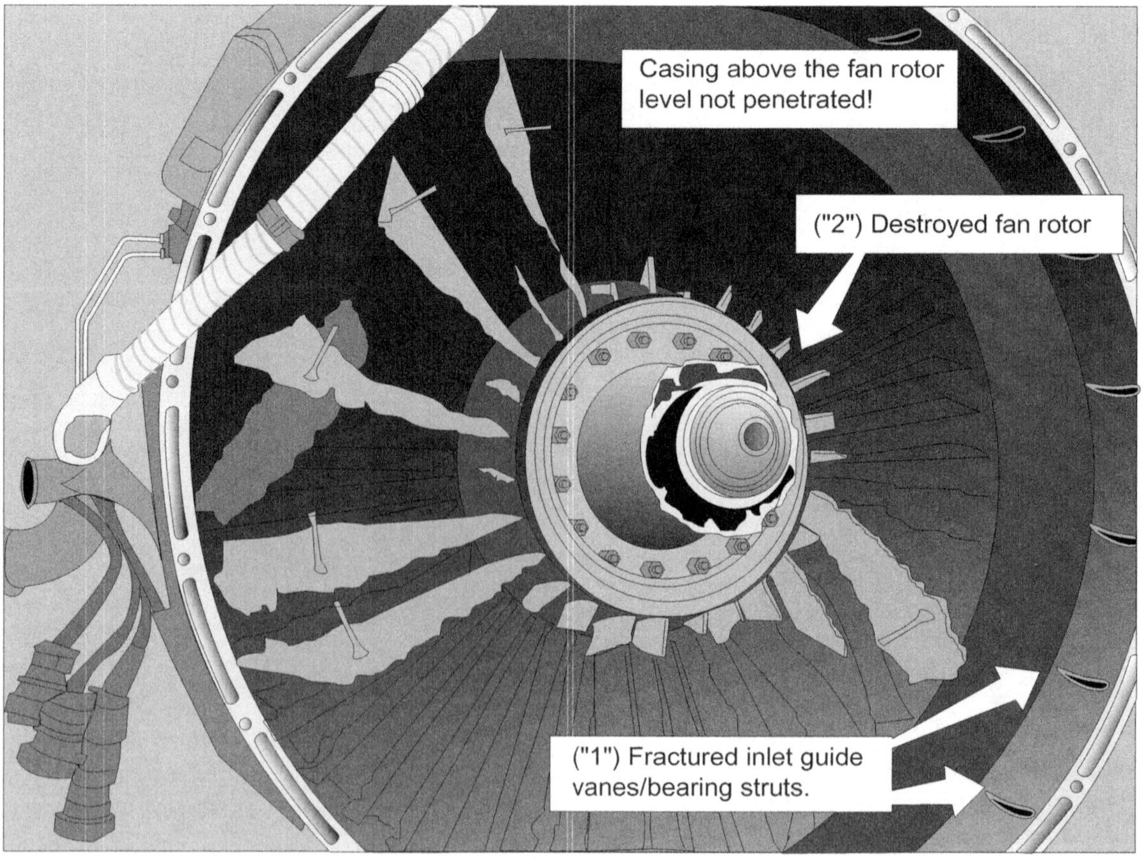

Remarkable findings of this failure picture. **Ill. 5.2.2.1-6.4.1**

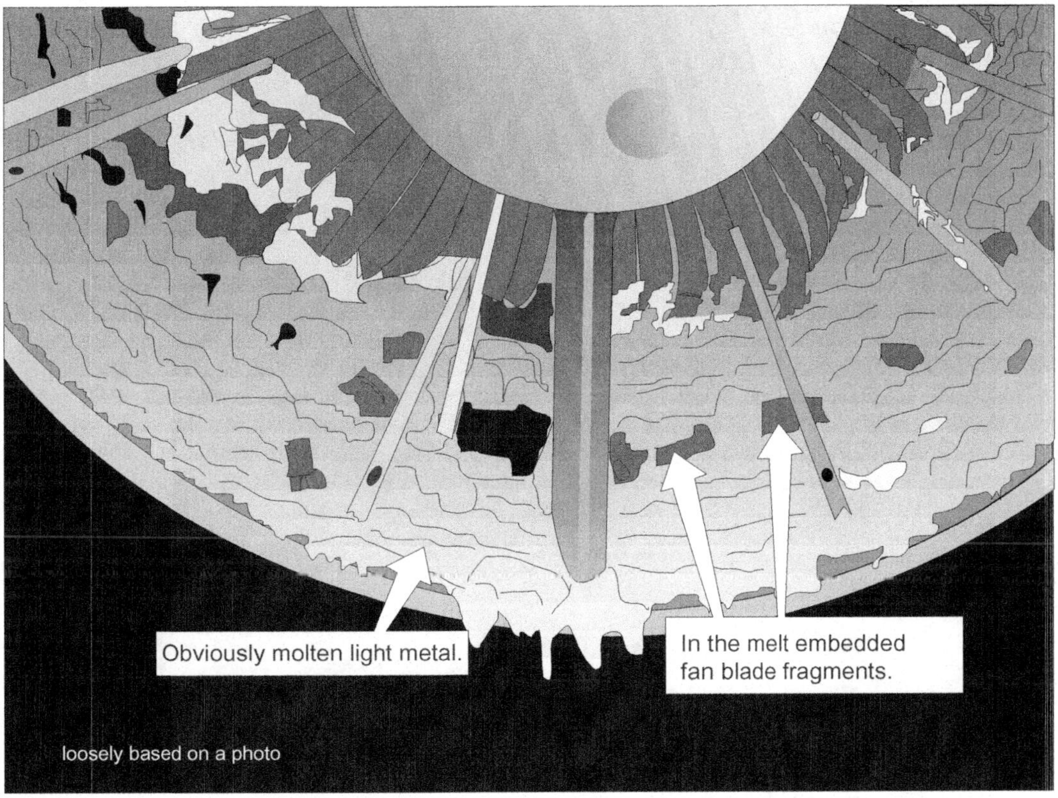

Ill. 5.2.2.1-6.4.2

***Illustrations 5.2.2.1-6.4.1** and **5.2.2.1-6.4.2** (Ref. 5.2.2.1-18)**:** A **canada goose** caused the displayed aeroengine failure. At the entrance of such a large bird extensive damages in the aeroengine can not be avoided. However the **casing wall** was not penetrated by fragments. With this, also such a failure **meets the specifications**. The failure pictures are obviously caused from two different incidents what, for example, can be seen in the region of the (fixed) nose cone. Concerned is in both cases an elder aeroengine type with **low bypass ratio**. The fan blades are, compared with modern fan engines (Ill. 5.2.2.1-6.5.1) relatively small and light. They have in the upper half of the blade supports (clapper). These can influence the bird impact behavior favourable (Ill. 5.2.2.1-12 and Ill. 5.2.2.3-10). The compressor rotor is **supported in front**. The bearing is supported by an entrance **guide vane apparatus** (struts to the bearing chamber „1"). This design has a markedly influence at the behaviour during a bird impact (Ill. 5.2.2.3-7) and the secondary failures to be expected.*

*In both cases obviously the struts in front have been destroyed. With this the low pressure rotor looses its support and it comes to an extreme rubbing process. This probably promotes the extensive damages of the fan blades („2"). Additionally a **fire of the escaping oil from the damaged bearing chamber** must be expected.*

*In the right picture obviously many **fan blades fractured above the clapper**. This is probably caused by an intense rubbing process after the drop out of the front bearing. Obviously the rubbing process and/or burning oil from the distroyed bearing chamber triggered a fire in the cowl lining. It camet to a **melting of light metal**. This may derive from the lining of the intake cowl, made from aluminium-honeycomb.*

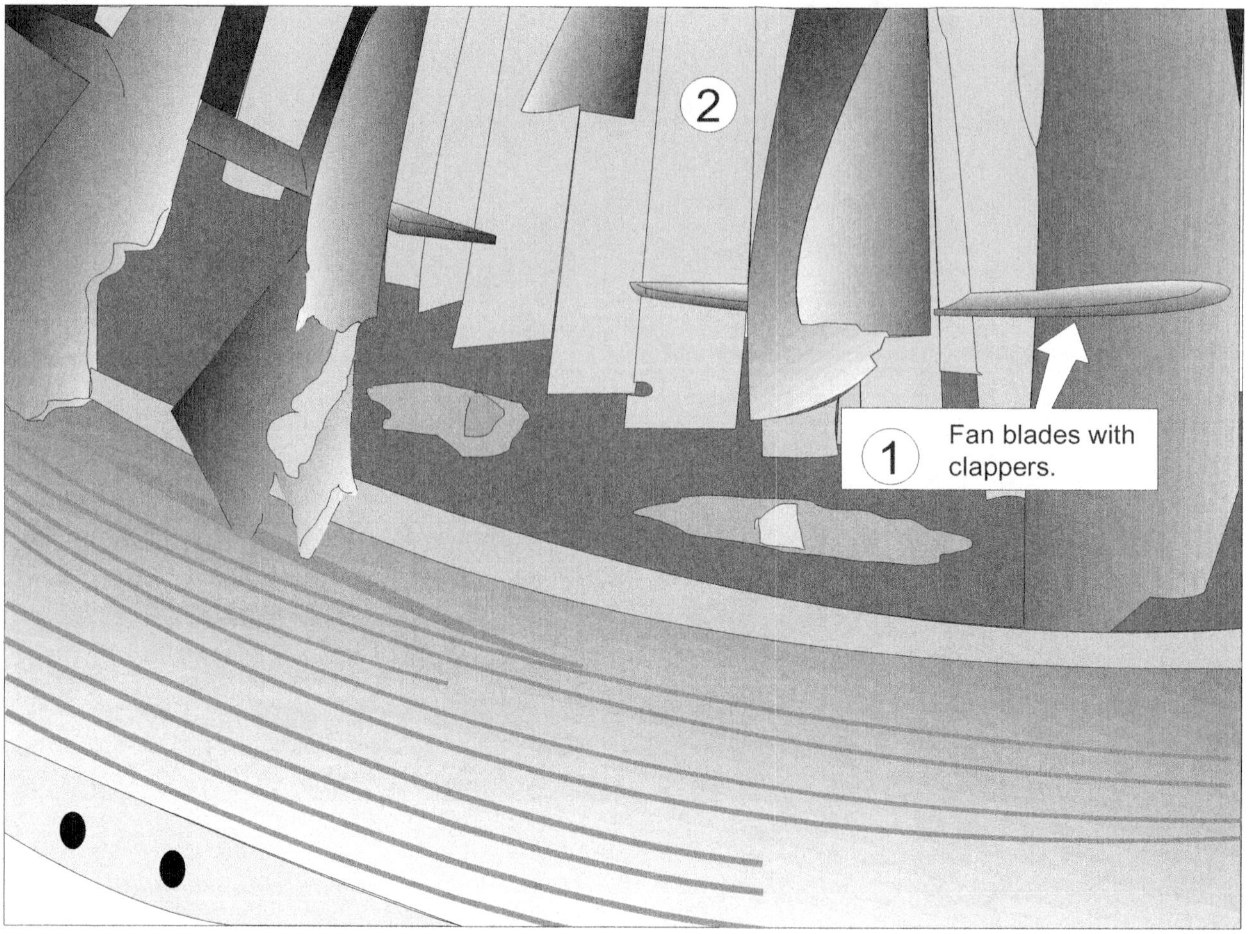

Ill. 5.2.2.1-6.5.1

Illustrations 5.2.2.1-6.5.1 *and* ***5.2.2.1-6.5.2*** *(Ref. 5.2.2.1-18): Those destructions in a modern fan engine are caused by a* ***canada goose****. During a direct hit, forces up to 40 tons must be expected (Ill. 5.2.2.1-2.3). With this serious aeroengine failures like the fracture of the* ***clapper stiffened fan blading,*** *can not be avoided. Important is, that the* ***fragments specification conform do not exit the aeroengine*** *(contained).*

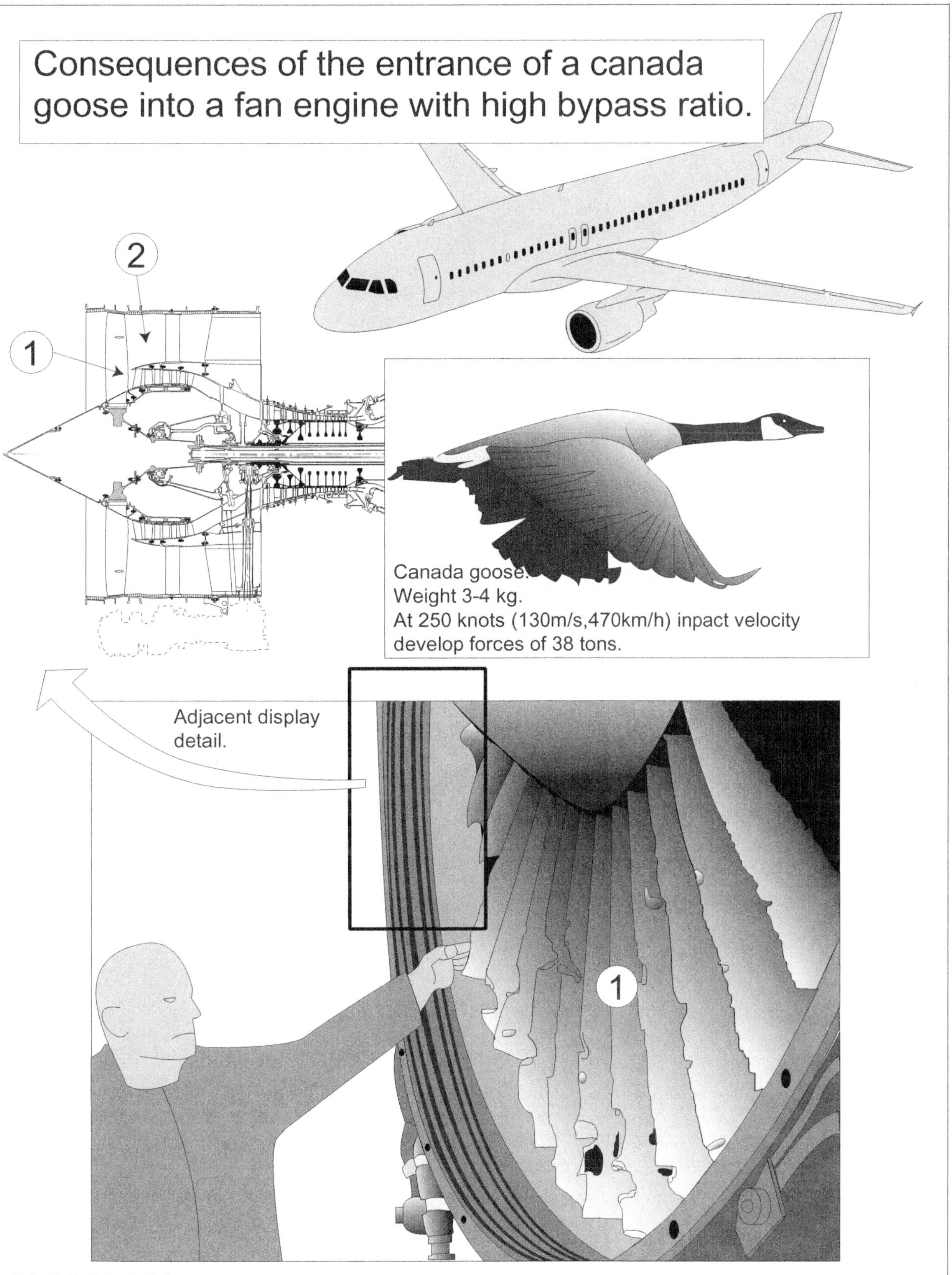
Consequences of the entrance of a canada goose into a fan engine with high bypass ratio.
2
1
Canada goose.
Weight 3-4 kg.
At 250 knots (130m/s,470km/h) inpact velocity develop forces of 38 tons.
Adjacent display detail.
1

Ill. 5.2.2.1-6.5.2

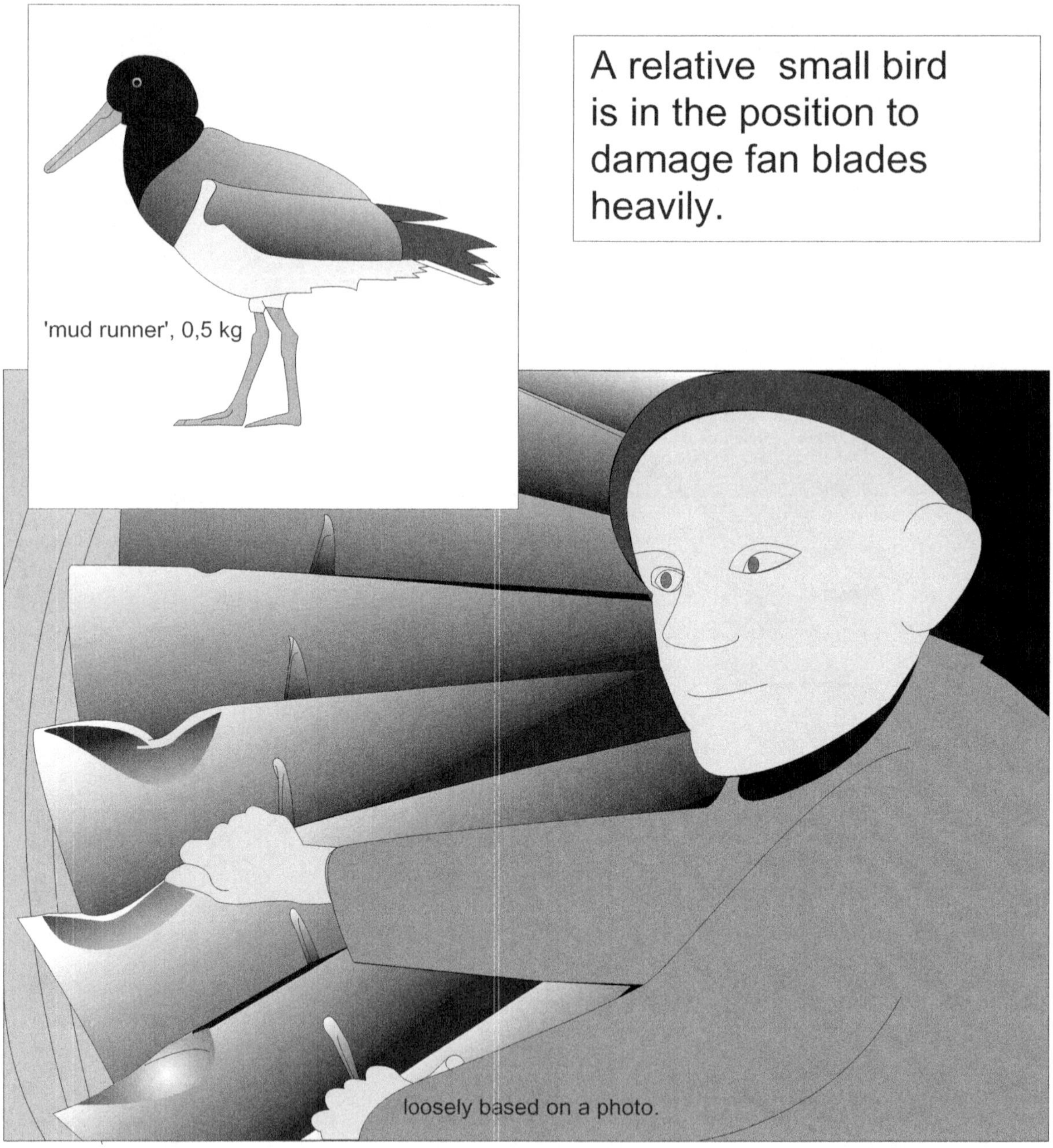

Ill. 5.2.2.1-6.6

Illustration 5.2.2.1-6.6 *(Ref. 5.2.2.1-18): This fanblade damage is caused by a relatively small bird of about 1/2 kg. Remarkably are the* ***plastic deformations of the blades*** *at the impact position. This adumbrates the* ***high pressures of the bird mass in the contact area****, which behaves as a sort of liquid (Ill. 5.2.2.2-1.1 and Ill. 5.2.2.2-1.1.1).*

Typical bird strike damage in the front engine area.

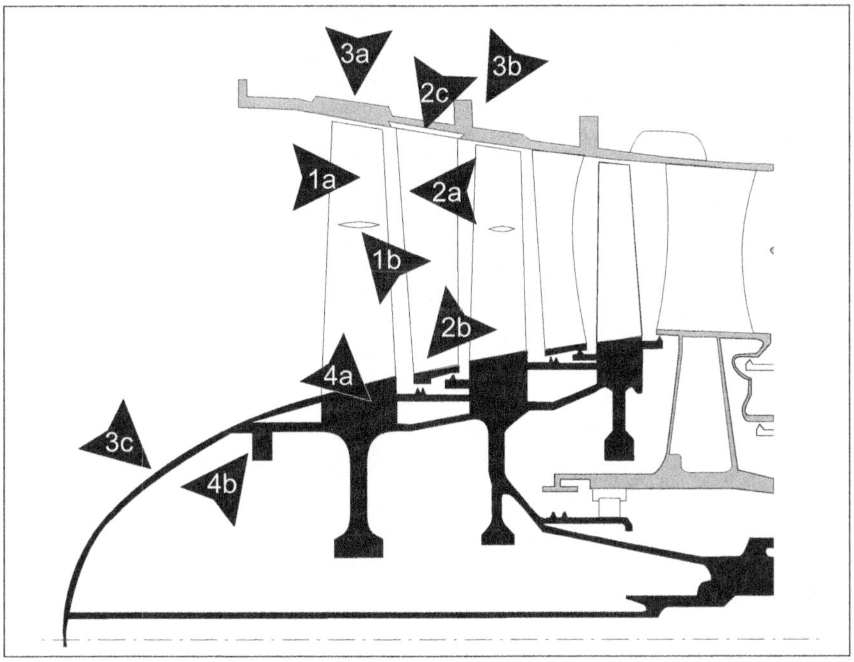

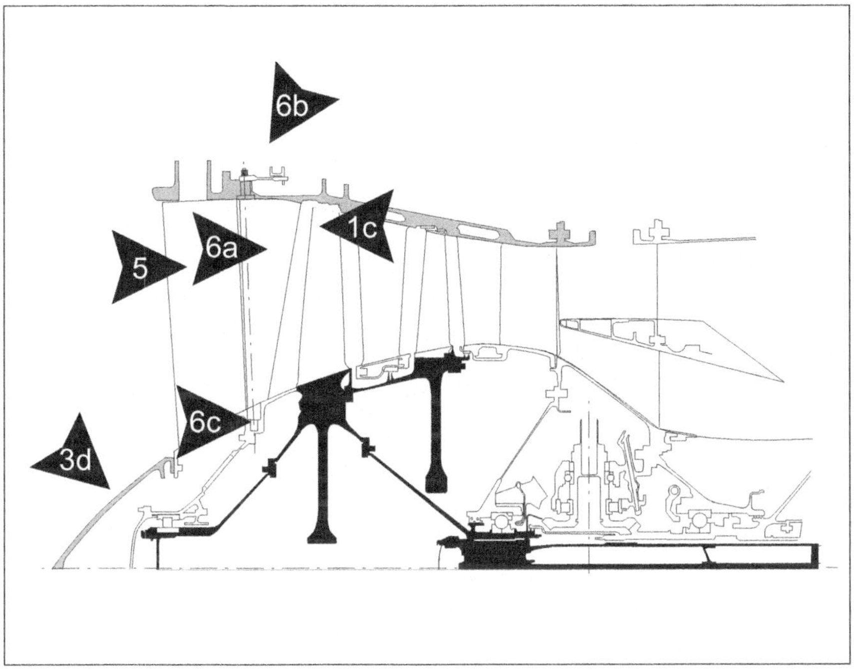

Ill. 5.2.2.1-7

External Factors
Bird Strikes: Damage

Illustration 5.2.2.1-7: *Bird strikes are especially dangerous for the following* ***typical parts*** *of multiple-shaft fan engines with small bypass ratios (frequently used in Fighter aircraft):*

Engines without inlet guide stars/struts *(top) and* ***no front bearings****, but with rotating nose cones (spinners):*

- Deformation or fracture of the fan ***rotor blades*** *(1a; Ill. 5.2.2.1-8)*

- Fracture of the fan rotor blades above the ***clapper*** *(1b; Ills. 5.2.2.1-9 and 5.2.2.1-12)*

- Deformation and fracture of ***guide vanes/stator*** *at the outer zones,* ***contact with the rotor blades*** *(2a)*

- Fracture of guide vanes at the ***inner shroud*** *(2b)*

- Fracture of guide vanes at the ***connection to the housing/casing*** *(2c)*

- Stressing and deformation of the ***housing/casing*** *after a rotor blade fracture (3a)*

- Fracture of ***threaded connections*** *on the housing/casing due to imbalances (3b)*

- Fracture (especially with fiber-reinforced synthetics; Example 5.2.2.1-2) or deformation (especially with metals) of the ***rotating nose cone (spinner****, 3c). In some cases overloading of the rotor's fixed bearing.*

- Overstressing of the axial ***blade*** *connections// fixing/****roots*** *(4a)*

- Fracture of the ***spinner connection****/fixing (4b)*

Engines with front bearings *(below), housing/* ***casing struts****, adjustable inlet* ***flaps*** *or* ***inlet guide vanes*** *(Ref. 5.2.2.1-14), and fixed inlet cones:*

- Deformation and fracture of the ***fan rotor blades*** *(1c)*

*- Fracture or deformation of the fixed/****stationary*** *(non rotating)* ***nose cone*** *(3d)*

*- Damage to the housing/****casing struts*** *(5) and possibly also the rotor bearings*

*- Damage (deformation, fracture) to the adjustable inlet flaps/****guide vanes*** *(Ill. 5.2.2.1-11) or* ***changes to the blade angle*** *(6a)*

- Damage to the ***adjusment mechansim*** *for the flaps or guide vanes (6b)*

- Damage to the inner ***flap bearings*** *(6c)*

Single engine aircraft operating at low altitudes are especially threatened by bird strikes.

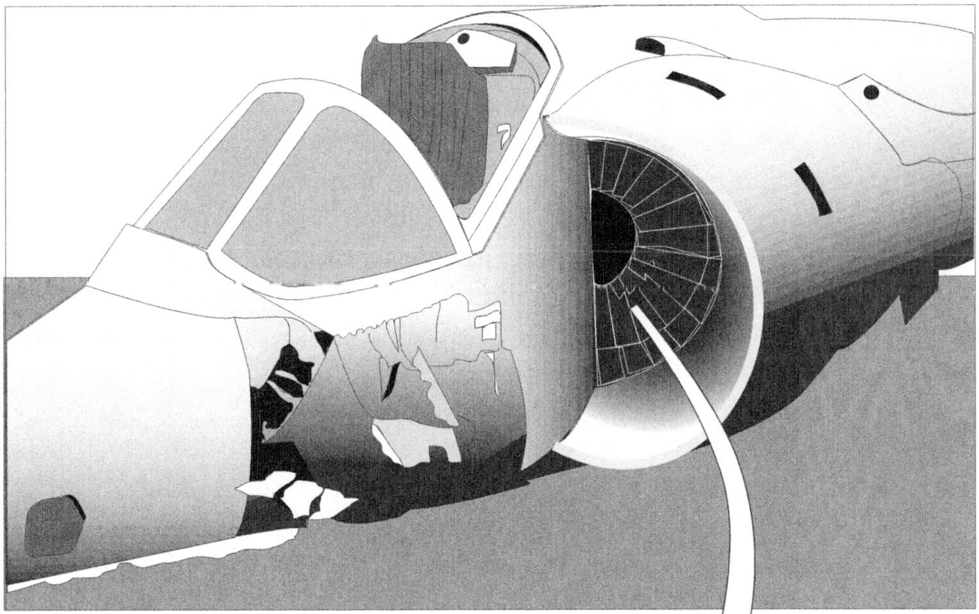

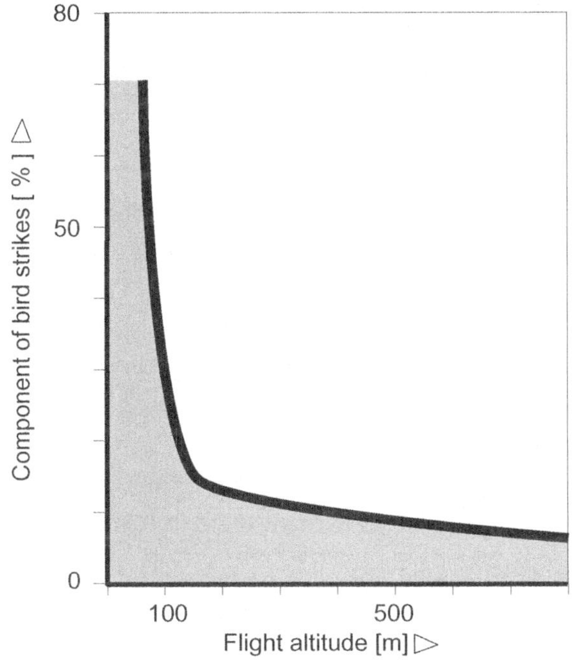

Influence of flight altitude on the probability of bird strikes (tendency)

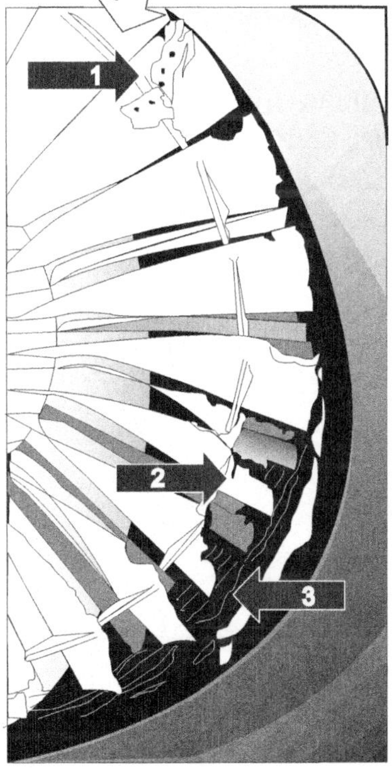

Ill. 5.2.2.1-8

Illustration 5.2.2.1-8 *(Ref. 5.2.2.1-12): This* ***VTOL*** *crashed after a* ***bird strike*** *(Ref. 5.2.2.1-12); this type of aircraft should be* ***especially sensitive to bird strikes*** *due to the relatively large engine inlets and extended operation at low altitudes. Parts (1) of the nacelle or engine inlet also appear to have entered the engine. Several fan blades have* ***fractured above the clapper*** *in typical fashion (2). This is due to the stiffening effect the clapper has on the entire blade ring (Ill. 5.2.2.3-10). The fan housing/casing (3) evidently contained the fragments.*

Damage types and damage development

The typical damage symptoms of bird strikes are **deformation**, crack initiation, and **fractures** of the engine components. This is caused either directly by foreign object impact or indirectly through developments such as **bent compressor rotor blades** damaging the neighboring stator vanes (Ill. 5.2.2.2-5) or becoming jammed in the latter (Ill. 5.2.2.2-6), causing one or both to break off. **Catastrophic blade failure** can be expected if the flexure causes the entire cross-section of the blade to be plastically deformed near the root (Ill.5.2.2.2-2). The corresponding maximum loads are several times greater than the loads at which the outer fibers of the profile (usually the edges and the blade spine) begin to be plastically deformed due to bending stress greater than the yield strength.

Mechanical impact stress on parts made from **brittle materials** (Ill. 5.2.2.2-4) results in visible effects from reflexion and overlaying of **impact waves**. This causes typical multiple fractures in locations where they would not occur under static loads. The extent of the damage caused by the impact itself may be limited at first, but continued engine operation can cause catastrophic development after a few seconds or minutes. Typical examples of this type of damage include dynamic **fatigue fractures on compressor bearings** after **rotor imbalances** and vibrations, and also dynamic fatigue fractures in deformed or damaged blading. These dynamic fatigue fractures frequently occur in plastically bent rotor blades, since the centrifugal force creates high mean stresses by unwinding the blade and causes additional flow disruption (flow stall at the deformed blades; Ill. 5.2.2.1-15) and **notch effects** (impact notches, cracks) which result in high dynamic stress. Bending of the blade along the circumference is less disruptive than a permanent twist. Permanent twist has a strong influence on the flow and can cause a **stall at the damage blades**. In subsonic compressors the **torsion angle change** must be less than about +- 10° of the set value, in transsonic compressors the maximum deviation is +- 2° of the desired angle.

If the above damage causes compressor surges, then the blading is subjected to additional dynamic loads. This is accompanied by the danger of the hot parts overheating due to a lack of cooling air. It has also been observed that before birds enter the compressor, they can break off parts of the cowling or inlet duct, and these fragments can also cause extensive FOD in the compressor.

Aside from the compressor parts described in Ill. 5.2.2.1-7, the following engine parts are especially sensitive to bird strikes:

Special damage due to bird strikes in the compressor area

- In compressors with **titanium blading**, if the engine continues to run after a blade fracture a **titanium fire** may ignite. This creates the danger of fire burning through the housing and escaping from the engine, ultimately causing the aircraft to crash.

- Bearings, especially **fixed bearings**, are stressed by powerful axial and radial forces caused by imbalances after a bird strike. This can result in bearing track fatigue, plastic deformation of the bearing tracks, or even **spontaneous force fractures**. Typical points of origin for a force fracture in an outer bearing ring are the corners of grooves or studs designed to keep the ring from rotating.

- **Connections** (bolts, nuts, serration) between the compressor disks can fracture due to large **deceleration moments** (deceleration of the rotor when the bird passes through), in extreme cases causing the compressor rotor to separate.

- **Blockage** of air intake **vents**, ducts, and **valves** (for cabin air, de-icing air, hot part cooling air, etc.)

Damage to the combustion chamber area due to bird strikes:

- **Blockage** of the combustion chamber vents and bores for the **cooling air film** and combustion air, resulting in danger of **uneven temperature distribution** and **local overheating** in the combustion chamber (can burn through the combustion chamber wall and combustion chamber housing, resulting in an engine fire), and unallowably poor temperature distribution in the gas flow, which can cause the downwind turbine blading to fail.

Damage in the turbine area caused by bird strikes:

- **Overheating** of turbine **guide vanes** (nozzles) due to poor temperature distribution in the gas flow

- **Blockage of cooling air** bores in turbine rotor blades and guide vanes, resulting in overheating with burning-through of walls and crack initiation

- Overheating of the turbine due to a lack of cooling air and reduced excess air during compressor surges; additional overheating caused by the regulator increasing the fuel flow in order to balance out the decrease in engine performance/efficiency

- Mechanical **damage through fragments or metallic melts** (for example, after a titanium fire).

External Factors
Bird Strikes: Damage

Fragments of the liner from the intake duct can make the situation worse.

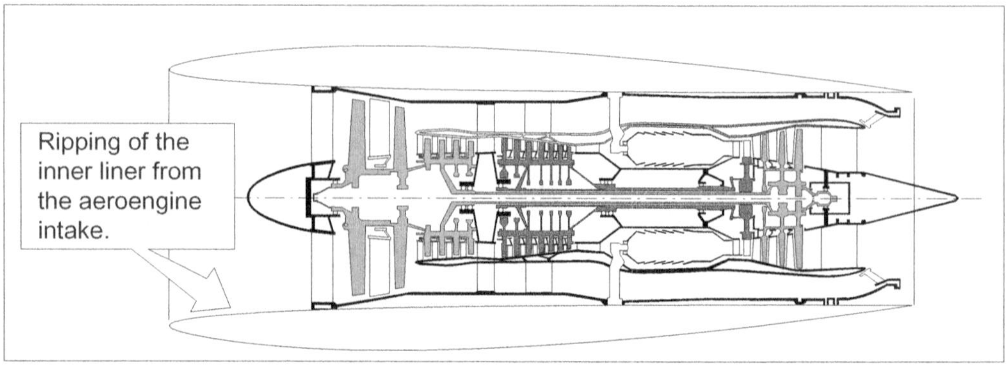

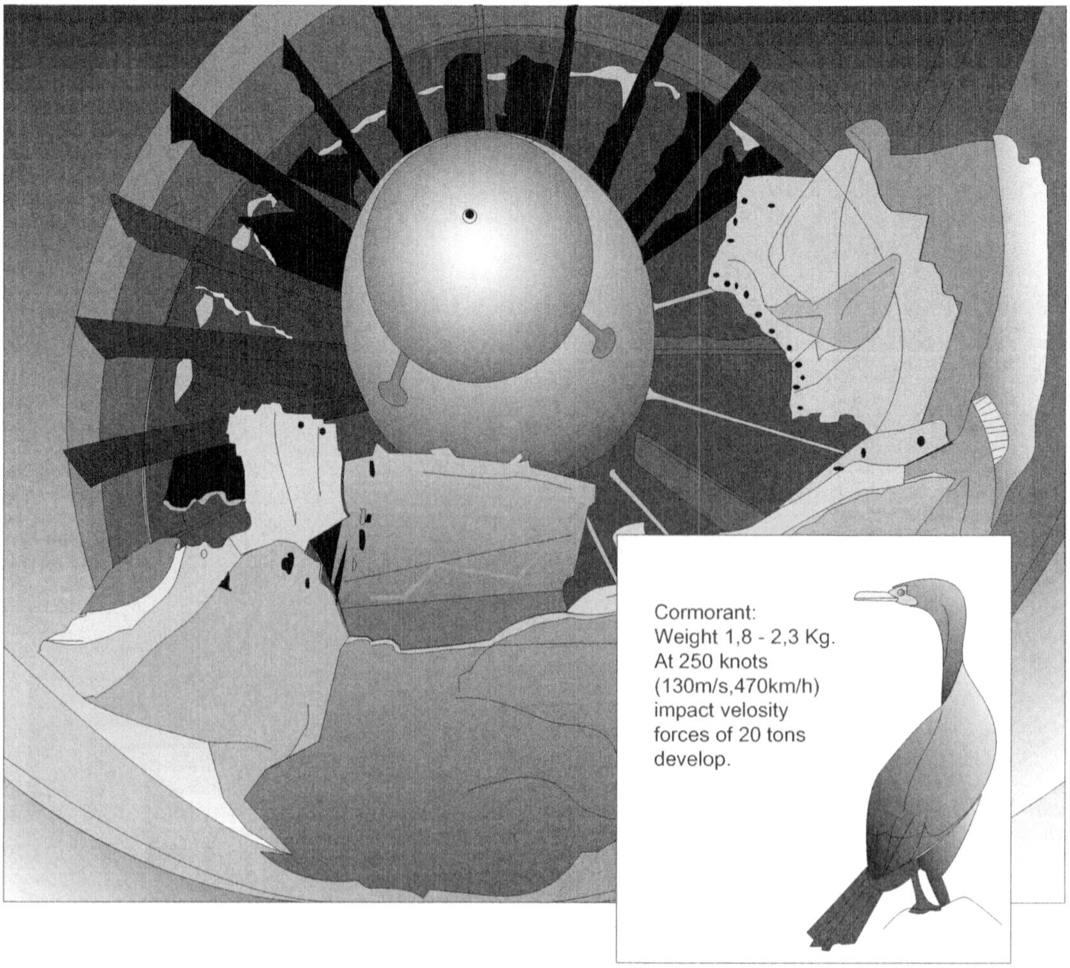

Ill. 5.2.2.1-9.1

Illustration 5.2.2.1-9.1 *(Ref. 5.2.2.1-18): In this case seems the dimension of the damage are also connected with* ***fragments of the liner from the air intake duct****. It can be supposed, that even the* ***disturbance of the airflow by the rests of the liner*** *prevented a further operation because of* ***compressor surging****.*

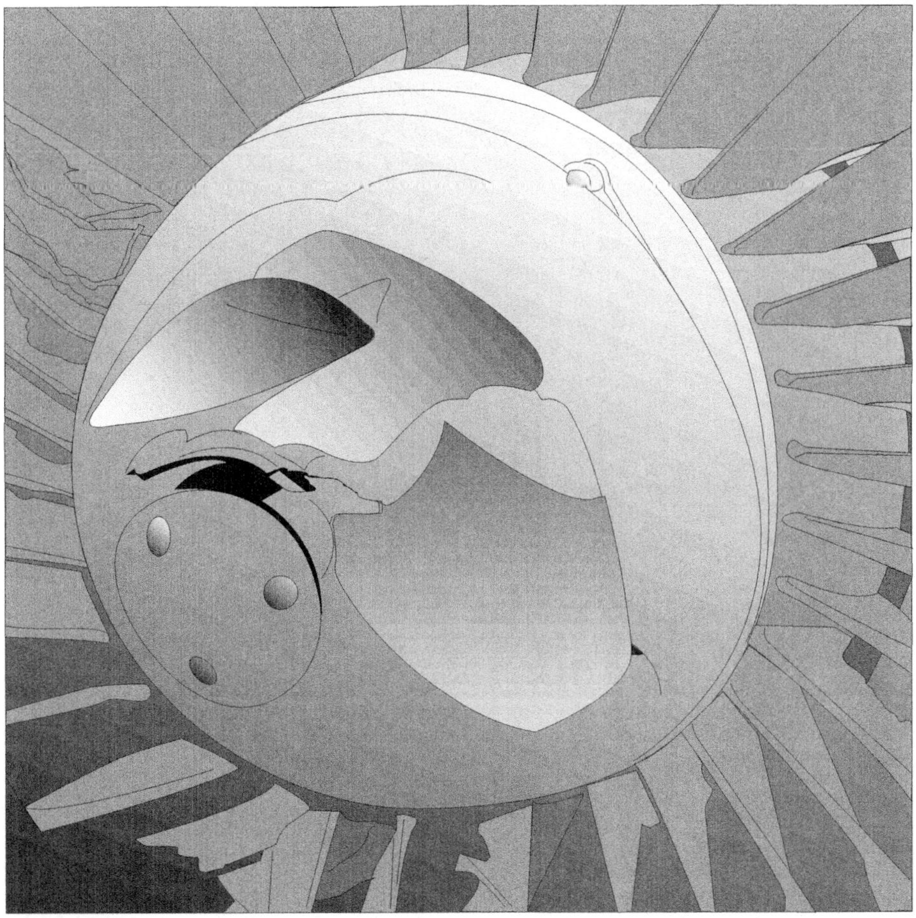

Ill. 5.2.2.1-9.1.1

Illustration 5.2.2.1-9.1.1*: Damage at the* ***rotating nose cone*** *(spinner) of a fighter engine by bird impact. The component is made from high strength* ***titanium alloy*** *sheet. A bird with the* ***size of a duck*** *hit the spinner at* ***high flight speed*** *near the ground and deformed it heavily. Inspite of extreme* ***unbalances and crack formation*** *obviously the centrifugation of fragments did not occur.*

"Low Speed Impact" occurs at fighter aeroengines by braking of a bird at the wall of the inlet duct. With this a overhung supported fanrotor is especially endangered.

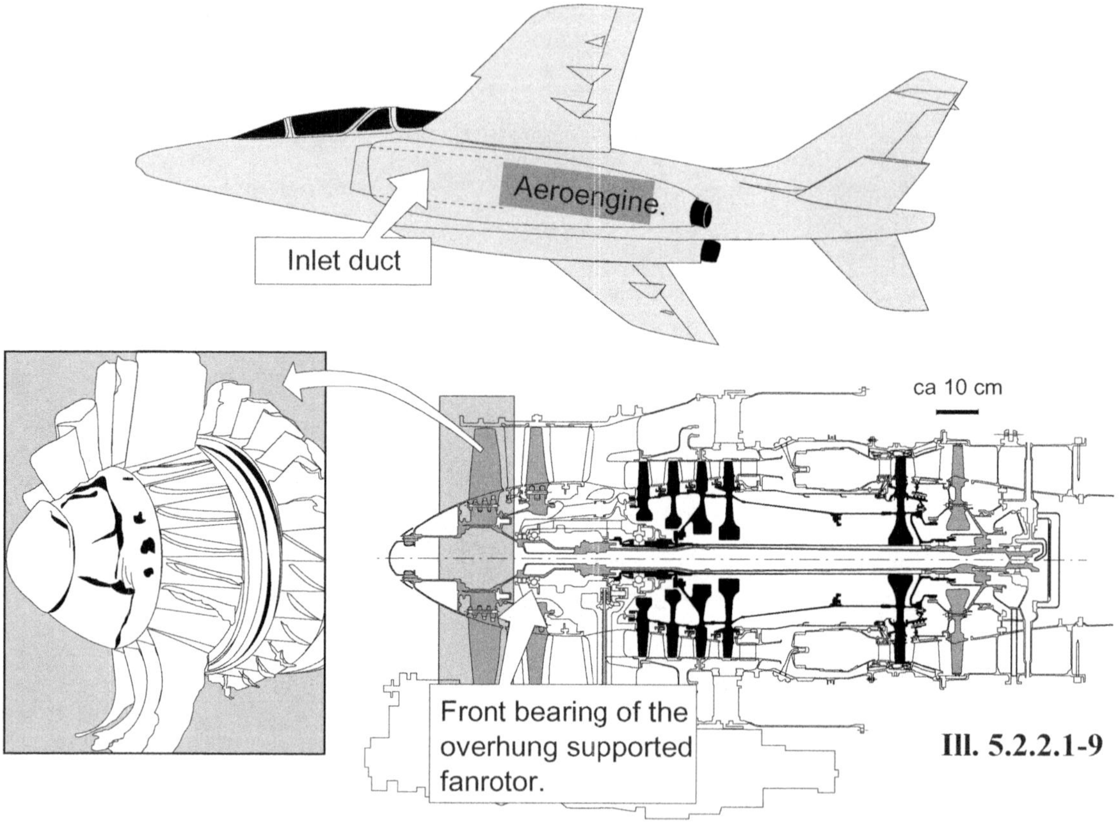

Ill. 5.2.2.1-9

Illustration 5.2.2.1-9 *(Ref. 5.2.2.1-13): This relatively small fan engine of a aircraft used ground attack and training has* ***wide chord fan blades****. The chord length of these blades gives them enough stiffness to make flow-disrupting clappers unnecessary. The lower diagram shows the fan rotor (Ill. 2.2.3-7) after a bird strike test with a 1.8 kg bird traveling at 210 m/s. This arrangement is sensitive for a „****low speed impact****" like in this case (Ill. 5.2.2.2-7). Such a low speed impact can be also expected at higher flight velocities if the bird* ***hits the wall of the inlet duct and is there braked****. With a bird of this size it is practically impossible to prevent fractures even in solid blades of this size. The relatively brittle fracture of several neighboring blades immediately above the foot platform is interesting. The fact that the preceding blades in the direction of rotation are heavily bent indicates a damage mechanism which started by one blade fracturing due to the bird impact. The fractured blade became jammed in the housing, and the following blades "ran over" it, either fracturing or bending in the process (volume 2, Ill. 8.1-15.1). Another interesting characteristic is the serious deformation of the metallic spinner, which indicates that at least some parts of the bird struck here first.*

Typical damage symptoms on rotor blades after a bird strike.

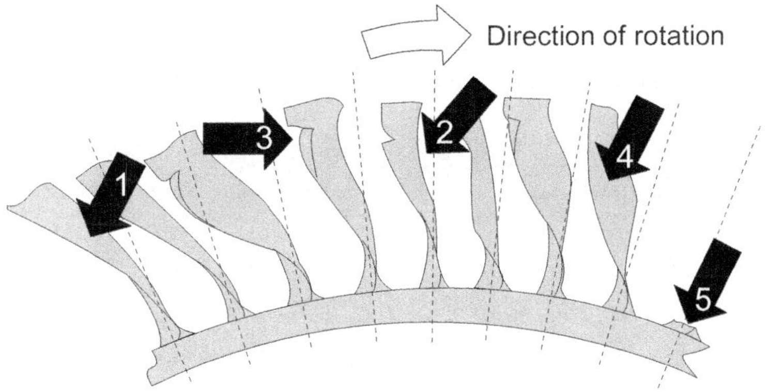

Housing deformations after a bird strike can be a cause of extreme rotor rubbing.

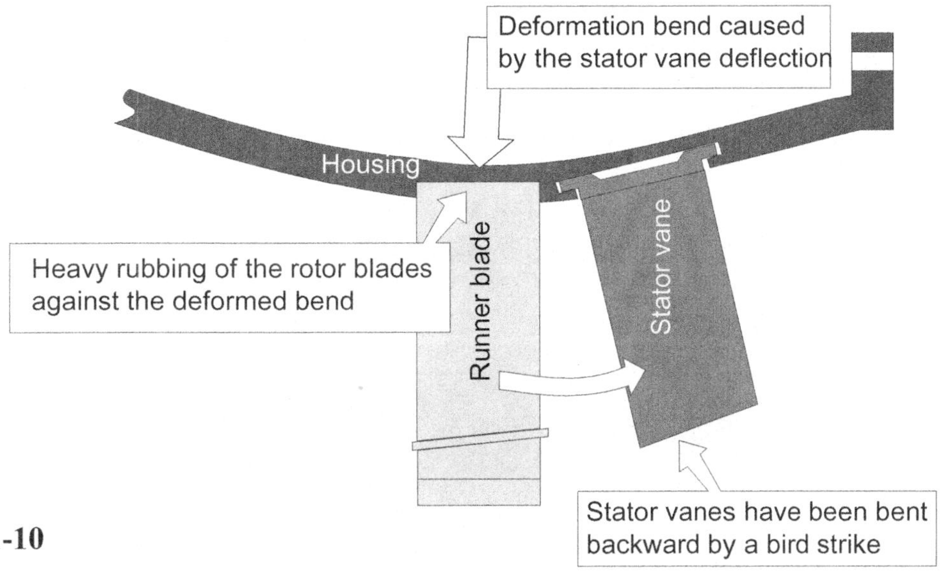

Ill. 5.2.2.1-10

Illustration 5.2.2.1-10: *This diagram is a schematic depiction of a compressor rotor stage that has* ***damage symptoms*** *typical of those caused by* ***bird strikes*** *(compare with Ill. 5.2.2.2-2):*

- ***Plastic deformation of a blade*** *that was struck at the tip "1"*

- *Blade that was struck slightly farther down and* ***plastically bent*** *in a typical S-shape "2"*

- *Blade* ***crack*** *caused by local overstress "3"*

- *Blade that has been* ***torqued by high loads*** *on the leading edge "4"*

- *Blade that as* ***fractured at the root*** *"5"*

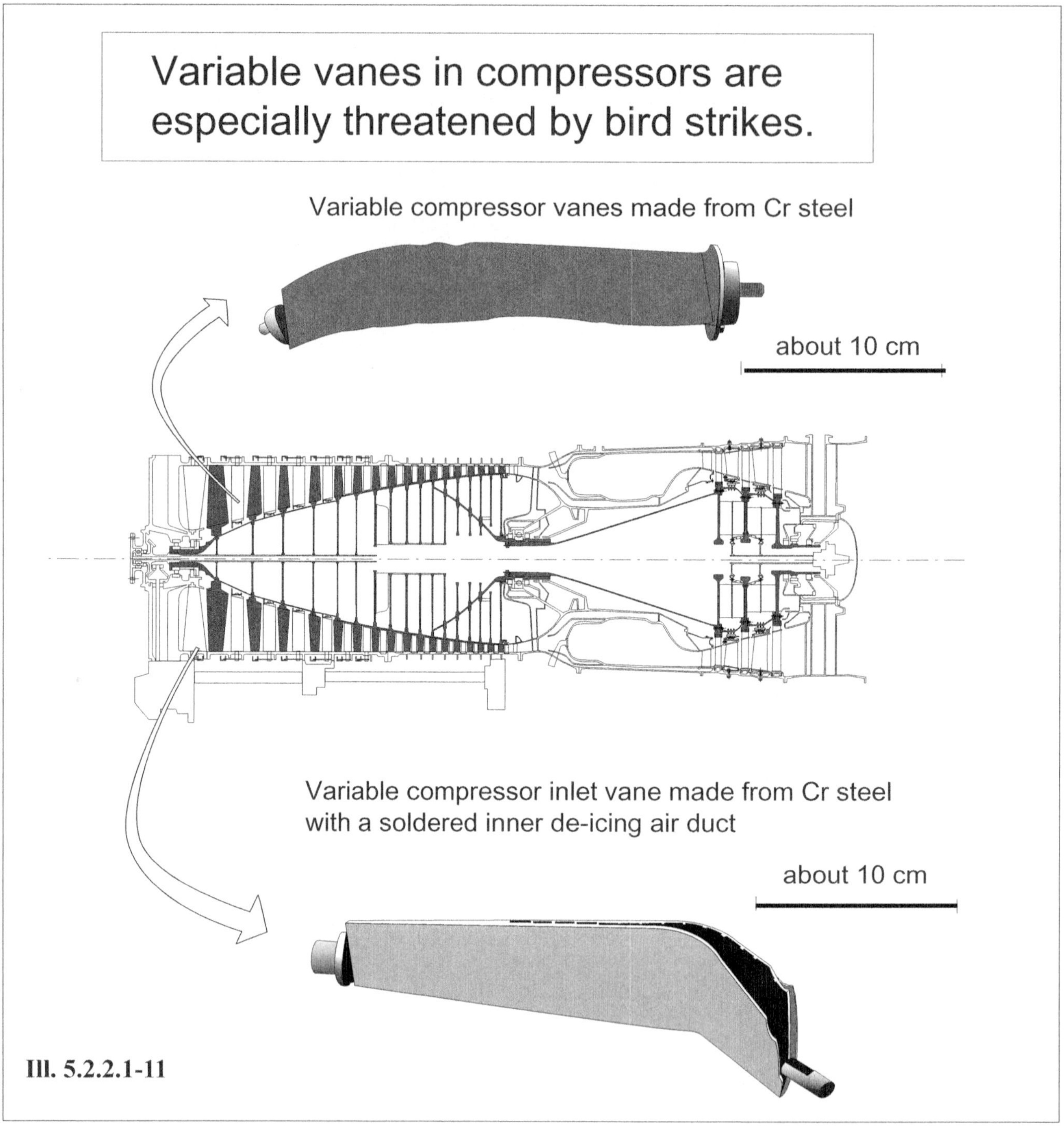

Ill. 5.2.2.1-11

Illustration 5.2.2.1-11: *This older single-cycle military engine was found to be prone to damage to the* ***adjustable guide vanes*** *of the forward stages through ingestion of relatively small birds such as gulls (Examples 5.2.2.1-4, Ref. 5.2.2.1-14). These blades are often hollow and designed to carry the de-icing air. As shown in the bottom diagram, overstress such as that caused by bird strikes causes the blades to crumple and/or burst open. The damage to the adjustable guide assembly can disrupt the compressor flow seriously, resulting in stalls and engine failure.*

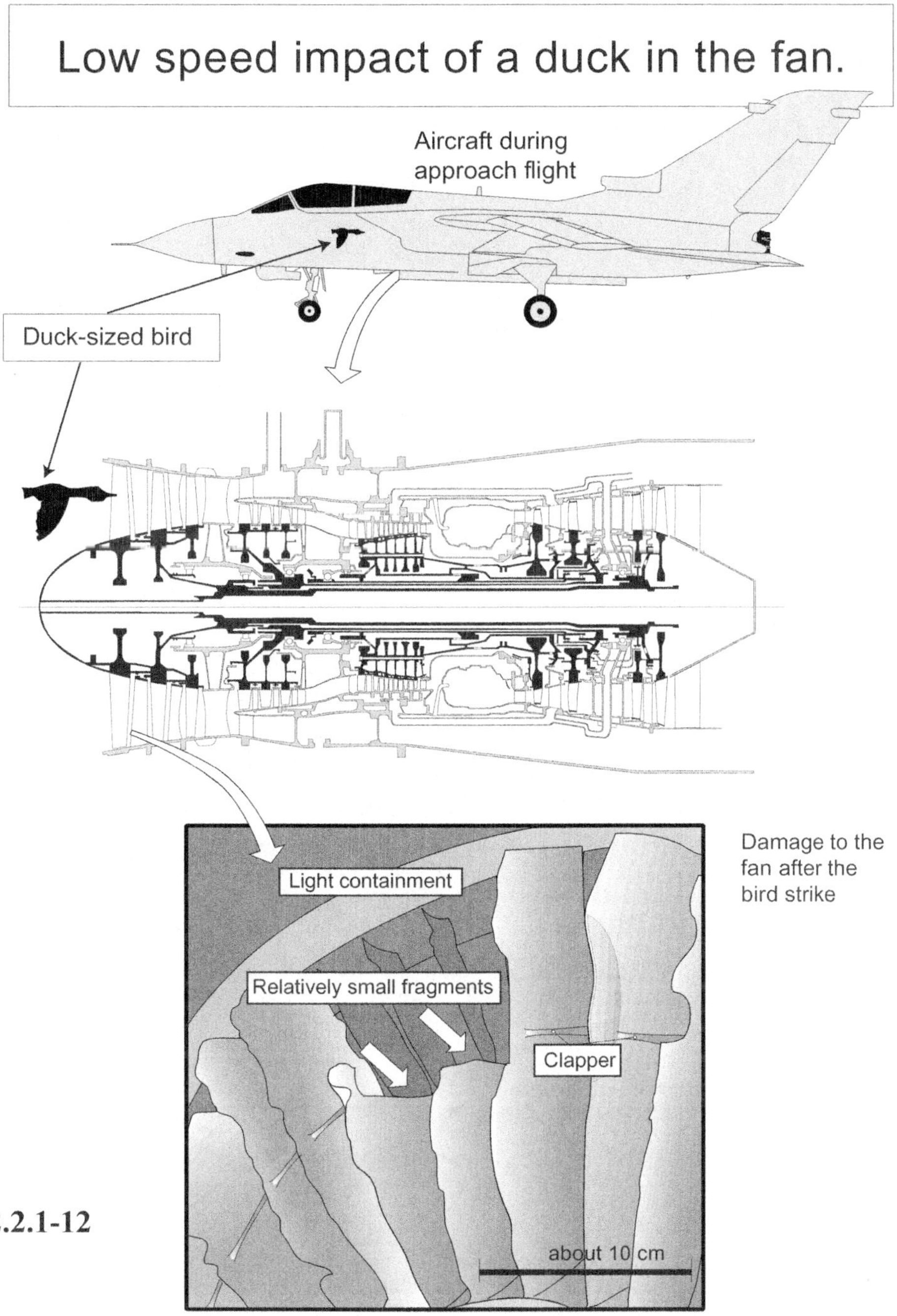

Ill. 5.2.2.1-12

Illustration 5.2.2.1-12: *Damage from the* ***low speed impact*** *of a duck during landing approach flight; the fan rotor blades of the first stage fractured above the clapper. All fragments were contained. This example shows the stiffening* ***effect the clapper*** *has on the blade ring (also see Ill. 5.2.2.3-10). The narrow fan blades of this older engine type created relatively low-energy fragments that did not threaten the housing. However, the heavy, clapperless, wide-chord fan blades (Ill. 5.2.2.1-9) found in modern engines require special blade design to ensure acceptable behavior during bird strikes. The large fragments resulting from a blade failure must be contained by a sufficiently elaborate, i.e. heavy, containment.*

In an engine without variable compressor inlet guide vanes a "High Speed Impact" endangers specially the guide vanes behind the first rotor stage.

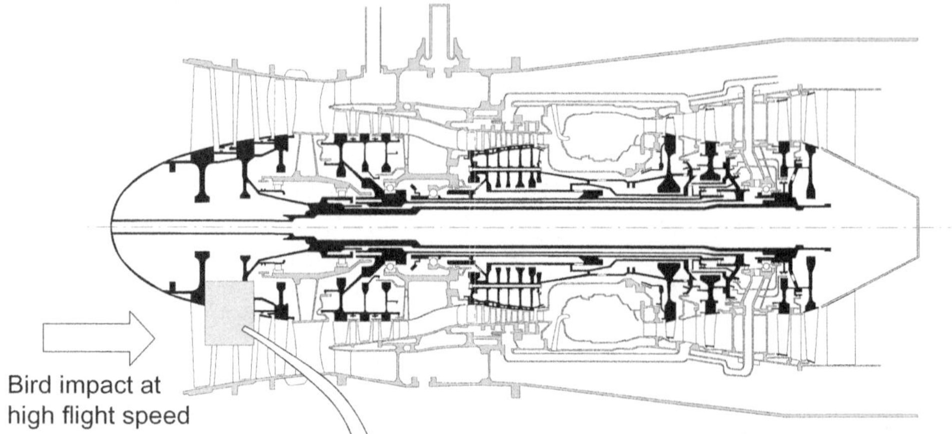

Ill. 5.2.2.1-13

Typical damage by a bird impact at fan exid guide vanes.

Illustration 5.2.2.1-13: *This diagram shows a* ***guide vane*** *behind the first fan rotor stage that* ***sheared off at the hub****. At high flight speed a bird traveled through the rotor blade largely without striking anything, due to the aerodynamic design of the blading; in extreme cases birds can pass whole between two blades. This stresses the stator with maximum kinetic energy. This will destroy part of the relatively filigreed stator and can lead to extensive consequential damages to the rotor.*

The bird itself is not the only FOD-threat to the engine.

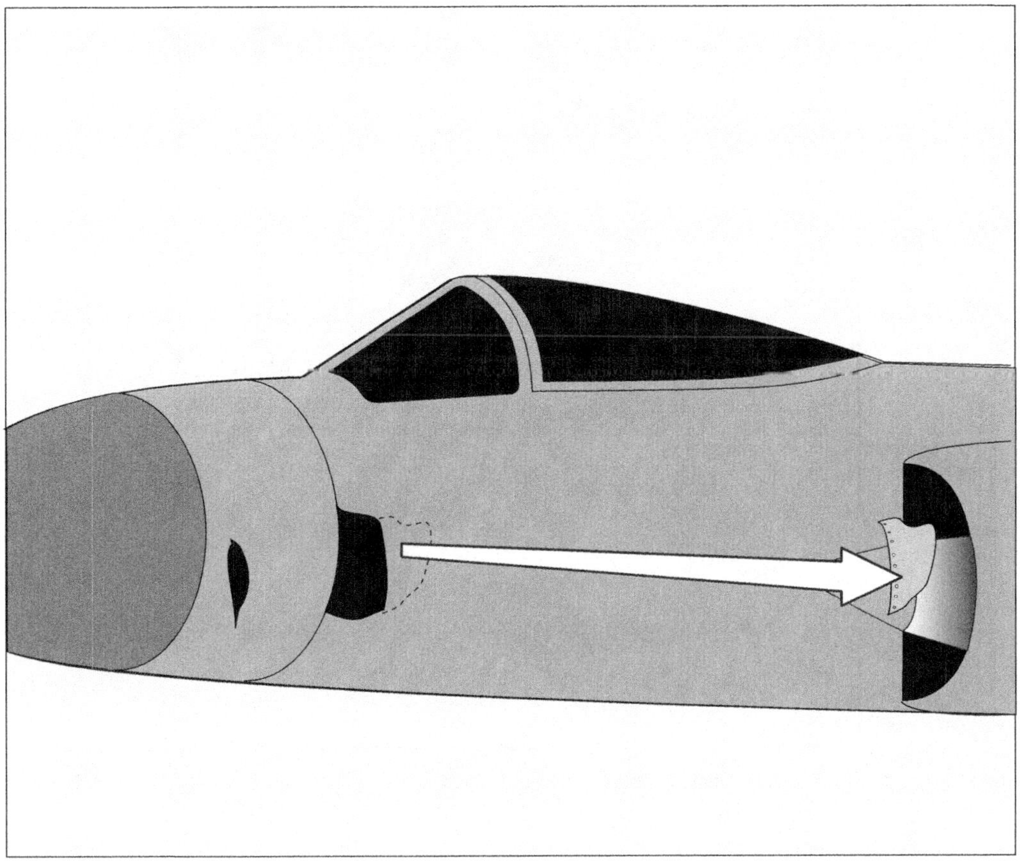

Ill. 5.2.2.1-14

Illustration 5.2.2.1-14 *(Ref. 5.2.2.1-15): In fighter planes, the danger that a bird will strike the engine directly is compounded by the possibility that a bird will* ***strike the fuselage*** *and then enter the engine. In this case, a bird will "spray" in a manner similar to a jet of water. The pieces of the bird are far less threatening to the engine than a whole bird. This effect is frequently taken into account by military certification regulations with regard to verification of engine behavior in case of large bird strikes. In the tests a bird is guided into the engine by* ***deflector plates****. This procedure is recommended especially with aircraft with engine inlet ducts that are shaped in a way that makes direct bird strikes on the engine unlikely. However, there is an increased risk if a bird strike creates* ***fragments*** *that can also enter the engine (parts of the radome, air inlet, covering, etc.). In the depicted case a single-jet fighter aircraft was struck by a seagull, causing parts fragments of the fuselage cover to break off and enter the engine.*

External Factors
Bird Strikes: Damage

Blades that have been deformed by bird strikes can be incited to vibrate and fracture.

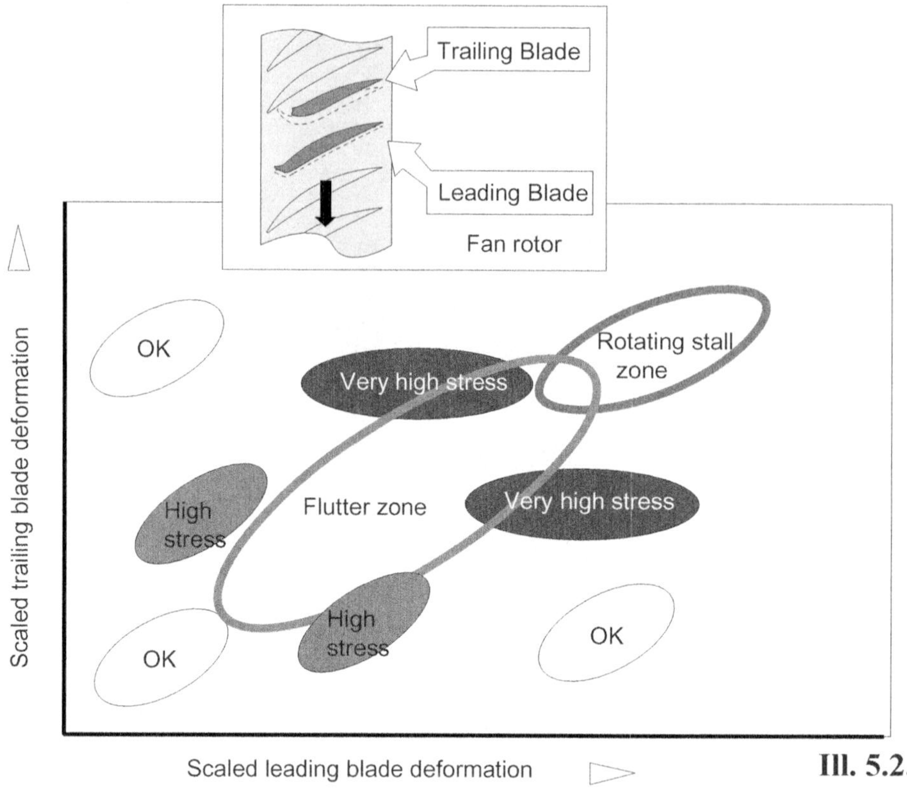

Ill. 5.2.2.1-15

Ill. 5.2.2.1-15 *(Ref. 5.2.2.1-16): In most cases in* ***large fan engines, two rotor blades will sustain damage****. The first blade against the direction of rotation, or "leading blade", and the one following it, or "trailing blade" (diagram). Depending on how serious the damage to these blades is, there is a risk of dynamic overstress through* ***flutter*** *or a* ***rotating stall****. The influence of the rotating speed is not shown in this diagram.*

Chapter 5.2.2.1 References

5.2.2.1-1 R.P. Paxson, J.D. Vance, "A Bird Strike Handbook for Base Level Managers Thesis", Report AD-A147 928 Air Force Institute of Technology, Wright-Patterson Air Force Base, Ohio, Nov. 1984.

5.2.2.1-2 J.Thrope, "Analysis of Bird Strikes reported by European Airlines 1972-1975", CAA Paper 77008, Civil Aviation Authority, London October 1977, ISBN 0 86039 055 1.

5.2.2.1-3 T.L. Alge, J.T. Moehring,"Modern Transport Engine Experience with Environmental Ingestion Effects", AGARD Conference Proceedings 558, "Propulsion and Energetics Panel Symposium", Rotterdam, The Netherlands, 25-28 April 1994, Chapter 9, pages 3 and 6.

5.2.2.1-4 NTSB Identification NYC961IA022, Index for Nov 1995.

5.2.2.1-5 NTSB Identification ANC931A188, Index for Sept 1993.

5.2.2.1-6 NTSB Identification ATL96FA101, Index for July 1996.

5.2.2.1-7 G. Fischbach, "916 Deutsche F-104 Starfighter, ihre Bau- und Lebensgeschichten", page 344.

5.2.2.1-8 G. Fischbach, "916 Deutsche F-104 Starfighter, ihre Bau- und Lebensgeschichten", page 189.

5.2.2.1-9 G. Fischbach, "916 Deutsche F-104 Starfighter, ihre Bau- und Lebensgeschichten", page 346.

5.2.2.1-10.1 "DC-10 Strikes Birds, Burns At JFK Airport", periodical "Aviation Week & Space Technology", November 15, 1975, page 24

5.2.2.1-10.2 M.L. Jaffee, "GE Pushing Effort on CF6 Engine Fix", periodical "Aviation Week & Space Technology", March 15, 1976, pages 197 and 199.

5.2.2.1-11 M.L. Jaffee, "NTSB, GE in CF6 Engine Conflict", periodical "Aviation Week & Space Technology", April 5, 1976, page 22.

5.2.2.1-12 H. Blokpoel, "Bird Hazard to Aircraft" Problems and Prevention of Bird/Aircraft Collisions, Verlag Clarke, Irwin& Company Limited, Canada., Diagram 2-9

5.2.2.1-13 J.M. Foueillassar, A.R. von der Muhl, "Petites Turbomachines: Experiences sur la rupture des Disques", AGARD Conference Proceedings No. 248, "Stresses, Vibrations, Structural Integration and Engine Integrity (Including Aeroelasticity and Flutter)", Chapter 16, Diagram 7.

5.2.2.1-14 E. Heckmann, "F404- das Jägertriebwerk unserer Generation", periodical "Wehrtechnik", 7/1986, pages 58 and 60.

5.2.2.1-15 J. Hild, "Vogelschlag und Flugsicherheit", periodical "Aerokurier", 10/1989, page 286.

5.2.2.1-16 M.Imregun, M.Vahdi, "Aeroelasticity analysis of a bird-damaged fan assembly using a large numerical moden", "Imperial College of Science, Technology and Medicine", London UK, The Aeronautical Journal, Paper No. 2487, December 1999, pages 569-577.

5.2.2.1-16 M.Imregun, M.Vahdi, „Aeroelasticity analysis of a bird-damaged fan assembly using a large numerical moden", „Imperial College of Science, Technology and Medicine", London UK, The Aeronautical Journal, Paper No. 2487, December 1999, page 569 up to 577.

5.2.2.1-17 NTSB Flugunfallbericht (NTSB/AAR-80-15),Flugunfall June 12, 1980.

5.2.2.1-18 A.Moffat, „Bird Strikes - Is the risk increasing?", Mount Cook Airlines, Paper der „Australian Society of Air Safety" 2006.

5.2.2.1-19 C.A. Huertas-Ortecho, „Robust Bird-Strike Modelling Using LS-DYNA", University of Puerto Rico, 2006, page 1-236.

5.2.2.1-20 J.Frischbier, „Vogelschlag in Flugtriebwerken - eine impulsartige Fluid-Struktur-Wechselwirkung in der Triebwerksauslegung", www.mtu.de/de/technologies/engineering_news/entwicklung. page 1-23.

5.2.2.1-21 J.Frischbier, „Multiple Stage Turbofan Bird Ingestion Analysis with ALE and SPH Methods", Herausgegeben von „American Institute of Aeronautics and Astronautics", 2005, page 1-9.

5.2.2.1-22 NTSB Flugunfallbericht, NTSB-AAR-76-19, Flugunfall vom 12. November 1975.

5.2.2.2 Bird Strike Damage Mechanisms

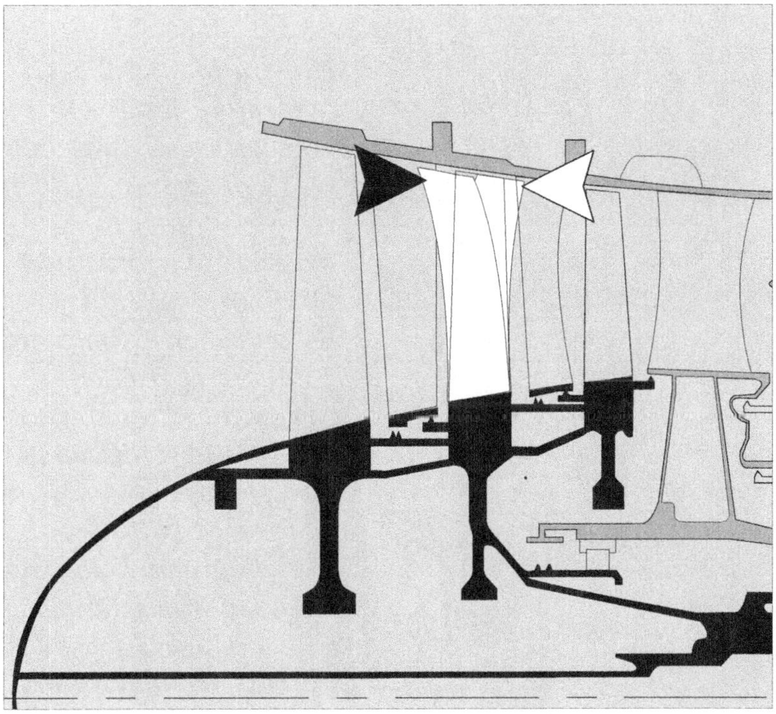

This chapter explains the extremely complex **sequence of events** that occur **during a bird strike**. This is intended to enable the reader to better understand damage symptoms and more accurately estimate risks.

This chapter begins by considering **fundamental processes** that occur when an **object strikes a beam** in bending that has been fastened on one side in order to simulate a rotor blade (Ill. 5.2.2.2-1).

Ill. 5.2.2.2-1 shows that the blade is not subject merely to bending loads, but also experiences a dynamic load process in the form of a decaying **vibration**. The movement of the blade tip can make it come into contact with the surrounding blade rows (Ills. 5.2.2.2-4 and 5.2.2.2-5) and cause extensive damage.

The flight speed of the affected aircraft, i.e. the impact speed of the bird against the blade screen, can be responsible for damage to various compressor components (Ill. 5.2.2.2-6). For example, low flight speeds can put especially high stress on the rotor blading (Ills. 5.2.2.2-6 and 5.2.2.2-7).

It has become possible to **simulate** the damage process with **computer** programs ever more realistically. This was and is achieved through extensive tests that make possible an iterative process that brings the calculated results ever closer to the actual physical processes. However, experience has shown that even today surprising damage mechanisms occur that are beyond current mathematical analysis Therefore, it is understandable that certification and acceptance of new engine types requires **practical verification** of satisfactory bird strike behavior.

These tests are usually conducted by using a pressurized-air cannon to shoot birds into the engine while it runs at a specified performance level. Ensuring sufficiently realistic conditions requires the use of real birds that are normally put to sleep shortly before the test, as this results in the most natural possible mass distribution.

Illustration 5.2.2.2-1.1 *(Ref. 5.2.2.2-4): For* ***laboratory tests*** *with gas pressure guns (air, helium) usually* ***alternative matters*** *instead of real birds are used. Different mixtures of natural rubber, 'plasticine' or gelatin with other materials (e.g., wood wool) however show results which are not sufficient realistic for* ***proofs*** *in the aeroengine. A mixture from 85 % water and 15% air reproduce the pressure flow most realistic (Ill. 5.2.2.2-1.1.1). Anyway in acceptance/****certification tests*** *of an aeroengine type, shortly before the test, euthanised* ***birds*** *must be used.*

The diagrams below apply for measurement values at sheets during a perpendiculary impact with a cylindric shaped water-air bird alternative matter. The ***shock pressure*** *(hugoniot pressure) is almost 10 times higher than the* ***flow off pressure****. The impact velocity increases the pressure exponential. This show measurement values in agreement with the calculated curve. For a better evaluation the practitioner should keep in mind, that the MPa indications correlate the ten times numerical value in bar. So impact pressures up to a multitude of 1000 bar and flow off pressures in the magnitude of several 100 bar. With materials like* ***fiber reinforced plastics*** *(FRPs) of a nose cone (example 5.2.2.1-2 and volume 3 Ill. 14-32), of the linings from the air intake ducts (Ill. 5.2.2.1-9.1), aeroengine nacelles as well as light metals of the walls from aircraft fuselages are even from an angular impact (Ill. 5.2.2.1-14) by far overloaded (Ref. 5.2.2.2-6). Its failing can produce dangerous* ***secondary failures of a bird impact*** *(Ill. 5.2.2.1-14).*

Also typical failure modes at compressor blades, especially fan blades (picture above left), can be amazing good simulated with computer based

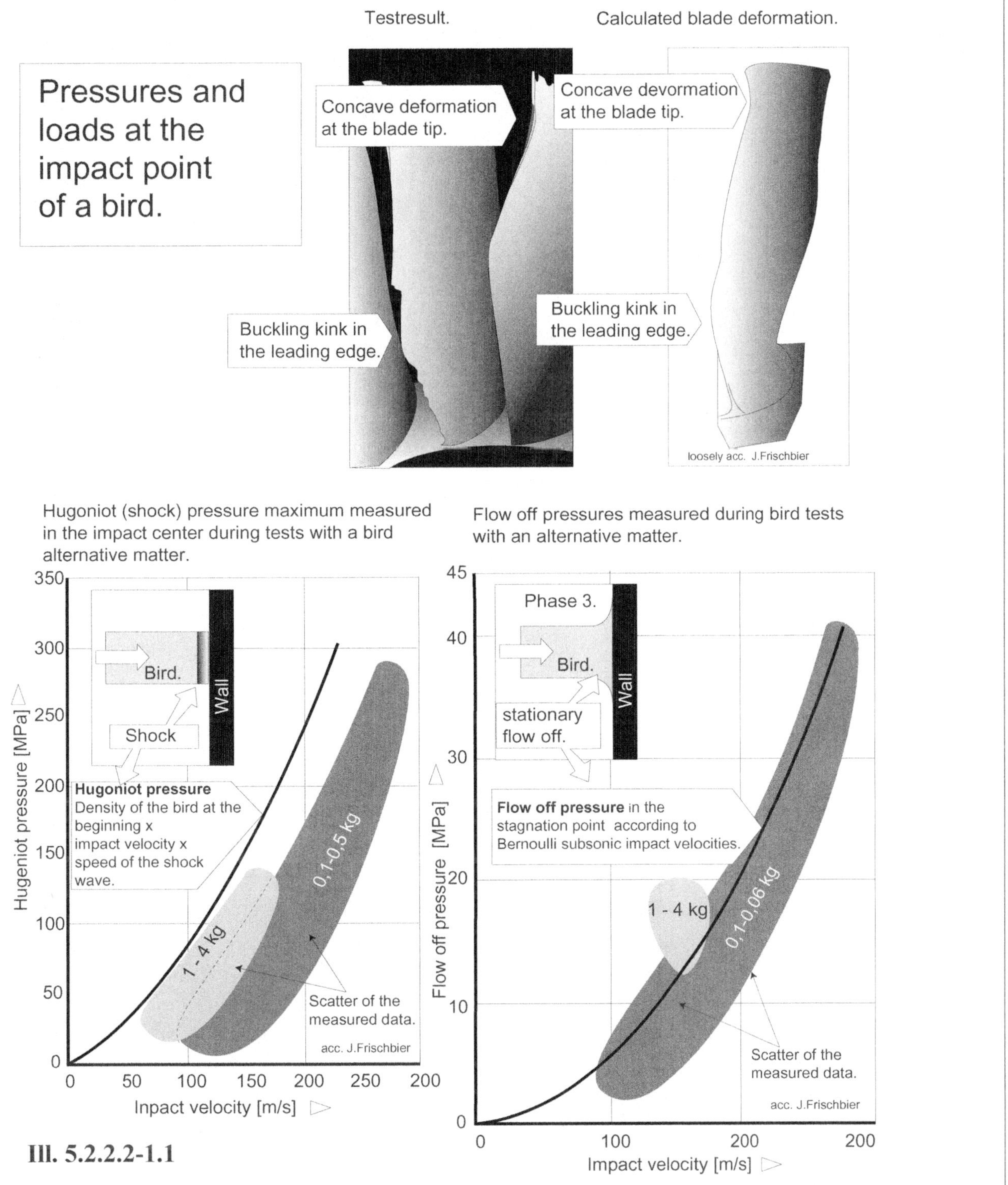

Ill. 5.2.2.2-1.1

calculations, which consider the impact effects (sketch above right). This demonstrates impressive the realistic approach. ***Crack formation*** *at the impact area must be seen in connection with an* ***embrittlement effect*** *of the blade material during* ***high speed deformation*** *(Ill. 5.2.2.2-1.2).*

External Factors
Bird Strike: Damage Mechanism

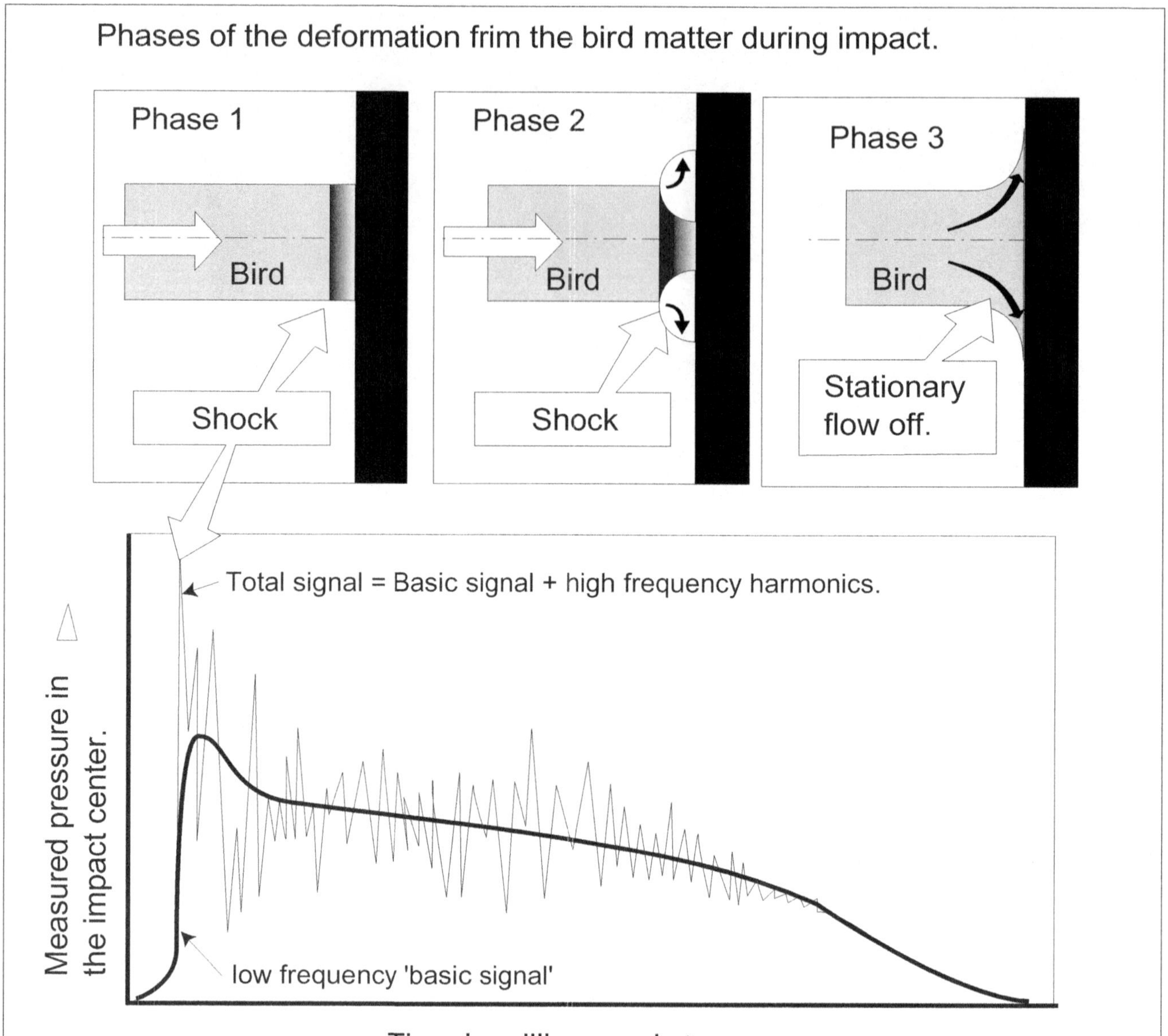

Ill. 5.2.2.2-1.1.1

Illustration 5.2.2.2-1.1.1 *(Ref. 5.2.2.2-4): Already in the seventies extremely high pressures during the impactof a bird at a rigid wall have been measured. Concerned was a perpendicular impact (sketches above). A* ***typischcal pressure progression*** *above the time is displayed in the diagram. Inside the bird respectively alternative matter during* ***beginning of the shock (phase 1)*** *back running* ***schock waves*** *produce the* ***pressure maximum*** *(shock pressure, hugoniot pressure). Actually extreme* ***high frequency pressure flictuations*** *occur by the shock wave and its reflection. Then the bird matter at the edge is accelerated outward parallel to the surface* ***(phase 2****). The result is a pressure drop inside, which leads to the rip of the bird matter. So it comes to the* ***flow off (phase 3)*** *correspondent a liquid. In the stagnation point the typical* ***pressure progression*** *develops, like it is measured in the* ***impact center*** *of a bird (diagram).*
The ***high frequency pressure oscillation*** *is real and obviously caused by a reflection of the shock waves.*

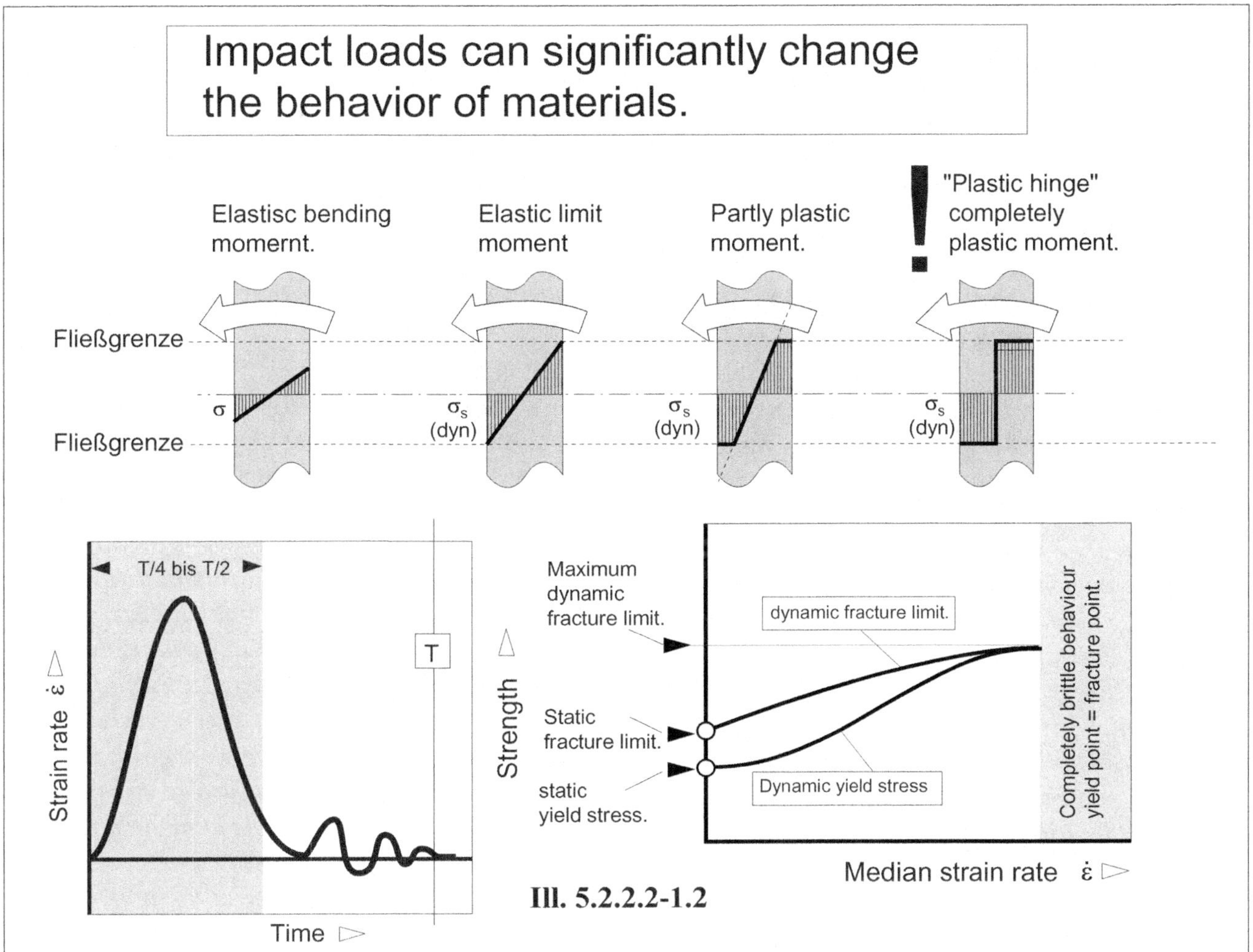

Ill. 5.2.2.2-1.2

llustration 5.2.2.2-1.2 *(Ref. 5.2.2.2-1): The* ***progress of plastic deformation*** *(upper sketch, Ill. 5.2.2.2-2) after a bird strikes a blade made from ductile, homogeneous metallic material can be schematically explained using the example of the beam in bending fastened at one end (top diagram). Of course, the blade shapes of a compressor are often very different from these idealized conditions. However, the insights gained here can be useful for understanding the damage mechanisms and blade behavior besides the* ***impact behaviour of the bird matter****.*

For a beam in bending with an even, prismatic cross-section which has been fastened securely on one end, the corresponding moment from a force "F" coming from the side and acting on the beam at distance "l" from the point where the beam is fastened is $M_E = F.l = \sigma_F \cdot h \cdot b^2/6$ *(h = height, b = width of the rectangular face), if the outer fibers are stressed by tension at the yield strength* σ_F*, and the cross-section has elastic-linear tension distribution. If the tension increases until the entire face cross-section begins to yield (i.e. is stressed above the yield strength), the corresponding moment is greater. In extreme cases no elastic core remains, and the moment is* $M_p = F \cdot l = \sigma_F \cdot h \cdot b^2/4$.

*This case, in which the cross-section is fully plastic, is referred to as a "plastic hinge" (****plastic collapse****). At the plastic hinge, i.e. the cross-section that the plastic hinge has passed through, the deformation resistance is higher than in the elastically deformed area. Therefore, the* ***plastic hinge*** *runs into this area.*

External Factors
Bird Strike: Damage Mechanism

In about half the time of the dynamic reaction to the bird impact, extreme deflection with high strain speeds takes place. The vibrations this incites decay rapidly (middle diagram). The high strain speeds can have a powerful, material-specific influence on the behavior of the part.
The greater the strain speed, the closer the yield strength comes to the breaking strength (bottom diagram). ***For example at high strength titanium alloys the yield strength increases at yield rates 199-1000 1/sec at 40%*** *(Ref. 5.2.2.2-4). It approachs equivalent the ultimate strength. This behavior results in embrittlement; i.e. the material fractures brittly due to overstress, even though it shows plastic behavior in a normal tensile test. behavior. Because of this, impact of a foreign object at high speed but with relatively little energy can fracture surprisingly thick wall sections. This effect is not very pronounced in titanium alloys, nickel alloys, and high-alloy steels, which are commonly used in compressor blading. Despite this, bird strikes can cause damage symptoms that could be plausibly explained by* ***high-speed embrittlement****.*

Illustration 5.2.2.2-2 *(Ref. 5.2.2.2-1): This picture shows a preception at the processes in the whole blade which is caused by the* ***mass inertia****.*
As model serves a ***one-sided mounted beam****. If the tip of a beam fastened at one end is struck by a sufficiently large force of deceleration (a mass with high speed and a corresponding amount energy), it forms a* ***plastic hinge*** *that moves towards the fastened end of the beam with decreasing speed (see the curve of plastic hinge progress). At the same time a curve is created as a rotation around the current location of the plastic hinge. When the plastic hinge reaches the point where the beam is fastened, it rotates as though it were stiff above this point. It could be shown that* ***masses which are small*** *in relation to the mass of the beam, but* ***moving very rapidly****, cause a* ***distinct curve in the beam tip****, but less* ***rotation at the point of fastening*** *(middle diagram). The beam behaves differently when struck by a* ***slow-moving, relatively large mass****. This results primarily in curvature at the point of fastening, but only minor permanent bending of the total length of the beam.*
If the impact does not occur at the upper part of the blade, then it will usually result in ***different damage symptoms*** *(Ill. 5.2.2.3-5). It causes a local bend, while the mass forces direct the* ***upper and lower parts of the blade against the direction of deformation****.*
The ***reaction forces on the blade root*** *(bottom diagram) are important if this is a weak point. For example, fiber-reinforced synthetic blades with a large jump in strength at the connecting point. This might be either an element that forms the hub contour between the blades (Example GE90) or the blade itself.*
If the FOD mass strikes below the ***blade leaf's center of gravity****, it creates a large reaction force at the root in the direction of the impact. If the blade is struck at the blade leaf's center of gravity, the reaction forces at the root should be small. If the mass strikes the upper blade leaf, then the reaction forces at the blade root will be directed against the direction of impact.*

External Factors
Bird Strike: Damage Mechanism

A strike-bending load is a complex process. The resulting deformations are greatly dependent on the load parameters.

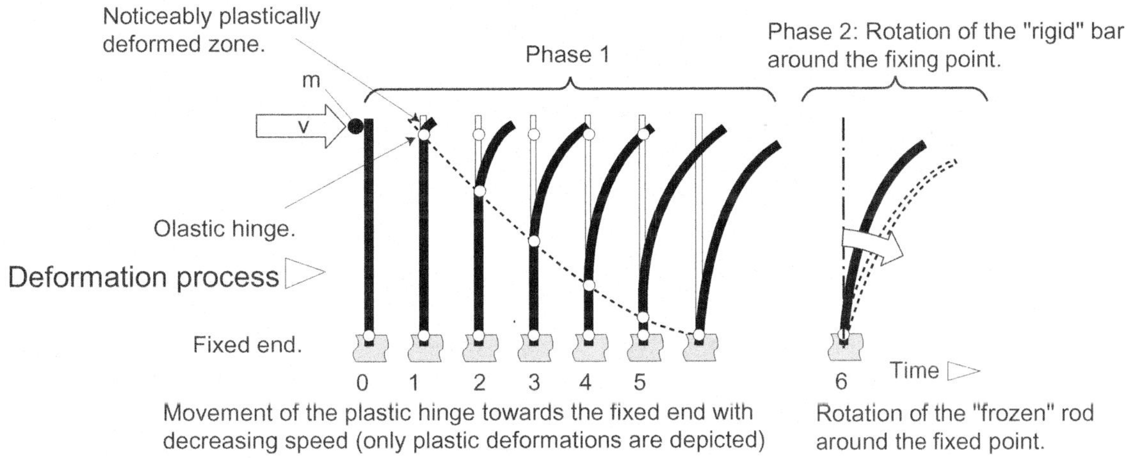

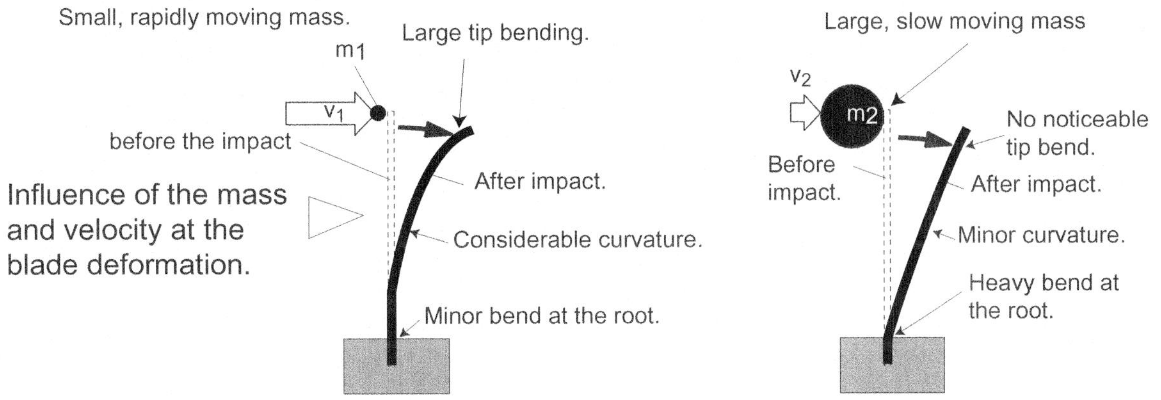

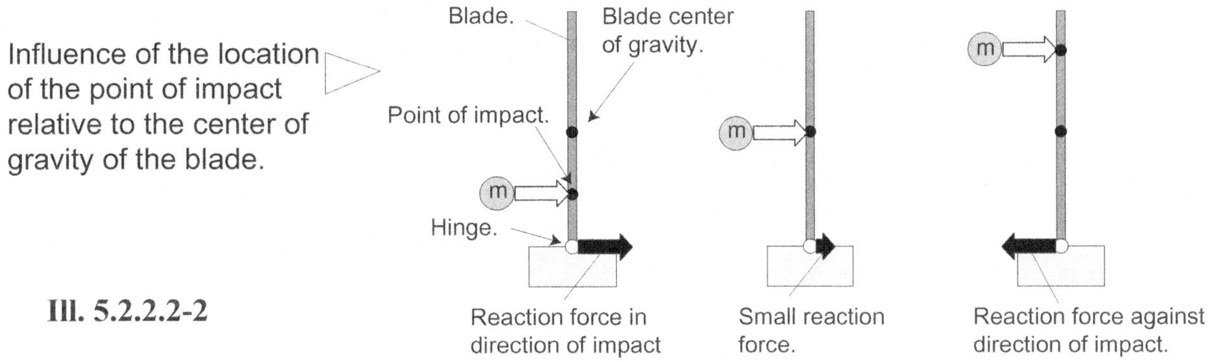

Ill. 5.2.2.2-2

Illustration 5.2.2.2-3: *The* ***impact process on a rotating blade*** *row is extremely complex and influenced by a large number of temporally changeable parameters.*
An ingested bird can strike the blade row at various speeds, depending on the flight conditions. If it is ingested at a high speed close to the air intake speed, it will follow the airflow until it is chopped into slices by the intake edges of the rotor blades, simultaneously accelerated axially, and guided into the following stator. The thickness of the slices is a function of the distance between blades and the speed at the impact radius of the blade, i.e. the RPM and the speed of the bird "V_E" in relation to the aircraft. The mass of these foreign object slices can be calculated by the size of the bird and its specific weight. If one assumes that the specific weight of the bird is roughly the same as water and therefore roughly 1000 times that of the air flow (for which the blading strength is designed), then it seems as though the blading would be overloaded.
Even modeling clay has proven to be insufficiently realistic as a substitute for birds. One reason for this may be the plastic behavior of modeling clay, which is highly dependent on the deformation rate and behaves brittly at high deformation rates.
*The impact of this mass creates a bending force that acts perpendicularly to the blade chord with a moment arm that corresponds to the distance between the point of impact and the blade root and an accelerating force resulting from the curve of the profile and related deflection of this mass, which acts at an angle in relation to the profile (also see Ref. 5.2.2.3-2). The impact forces due to foreign object strikes depend not only on the mass of the foreign object but also largely on the angle of impact; i.e. on the "incisive projection screen", the redirection of the foreign object, and the elasticities of the foreign object and the blade (**impact digit**! Ills. 5.2.2.2-4 and 5.2.2.2-5).*
The slower the birds enters a blade row, the thinner it is sliced, since more blades will be involved (Ill. 5.2.2.2-7). On the other hand, decreasing bird impact speeds are accompanied by increasing impact angles and axial force components.
When a large bird strikes a blade row, the relatively large distance between blades at the tips compared with the spacing at the hub makes the blade tip area do considerably more slicing and transport work than the larger number of engaged blades closer to the hub would have to do.

External Factors
Bird Strike: Damage Mechanism

Bird strikes are complex processes with many parameters that influence the loads on the blades.

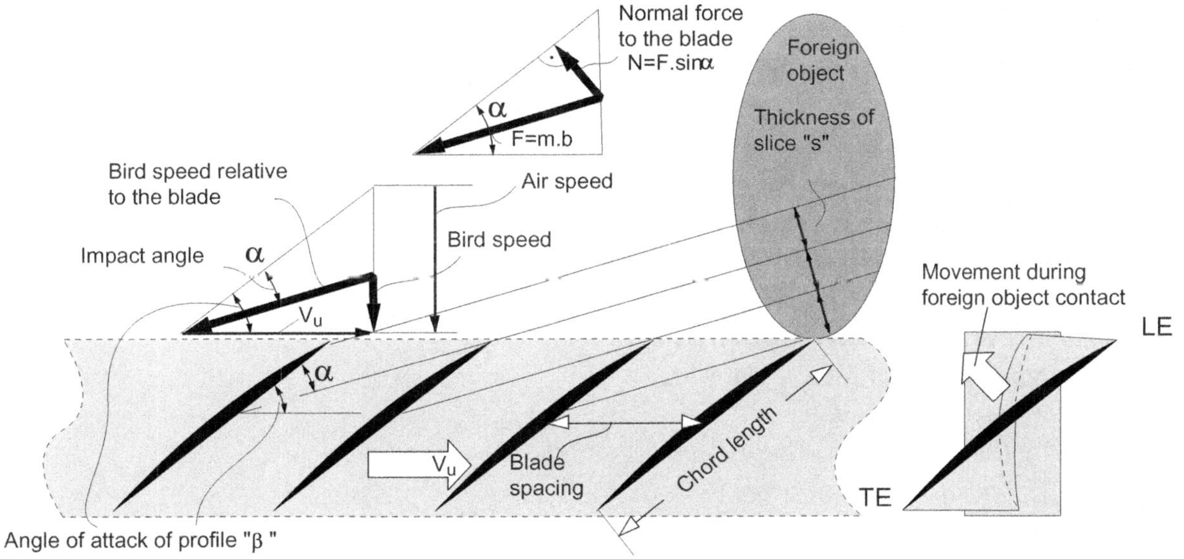

Values that vary over the blade length:

- Attack angle "β"
- Circumference speed "v_u"
- Flow duct width
- Blade spacing
- Profile (thickness ratio, leading edge radius)
- Chord length
- Stiffnesses
- Mass to be accelerated

Influencing parameters:

Blade:
- Density
- Materials
- Length
- Profile
- Spacing
- Twist
- Design (e.g. clapper)

Foreign objects:
- Mass
- Density
- Intake speed
- Toughness

Ill. 5.2.2.2-3

Illustration 5.2.2.2-4: *The top diagram schematically depicts a fluid model of the process that occurs when a* ***bird enters a rotating blade row****. It clearly shows that the leading blade edge is the first area to be stressed, putting* ***torsional and bending forces*** *on the blade that act diagonally forward corresponding to the angle of the blade. The farther the bird travels through the blade row, the more the force travels along the face of the blade to the trailing edge, changing the torsion moment.*

Compressor rotor blades struck by soft foreign objects are usually torqued, bent forward, and finally ***spring back*** *(up to several centimeters, depending on the blade length). The first deflection of the rotor blade due to the blade leaf bending forward bending around the* I_{min}*-axis is due to the angle of the blade face: "the blade row twists into the foreign object."*

After the soft body flows off the trailing edge the blade leaf springs back (bottom diagram).

Brittle-hard foreign objects *such as ice* ***burst or spray on impact****, resulting in a completely different energy transfer between the foreign object and the blade. The foreign object does not flow off the blade, which results in lower changing torsion stress (see bottom left diagram).*

The overlaying of torsion and bending deformations of the blade face reciprocally influences the progress of the impact forces and the ***resultant vibrations****.*

Compared with the twisting/untwisting forces on the rotor blade, the internal forces which bend back and square the blade are relatively small. This is especially true with modern compressor designs with typical ***sharp-edged inlet radii*** *(trailung edge) and small (or located far back) profile thicknesses. This should contribute to the observed characteristic that modern wide chord fan blades are sometimes surprisingly sensitive to bird strikes (Ill. 5.2.2.1-9).*

With hard-elastic foreign objects, the high internal forces cause the backward bending to be especially pronounced.

When blades untwist, especially ***blades with wide chords,*** *can lengthen radially and the edges may rub heavily against the housing. This can overstress the blades and/or the housing.*

External Factors
Bird Strike: Damage Mechanism

The blade loads change during a bird strike.

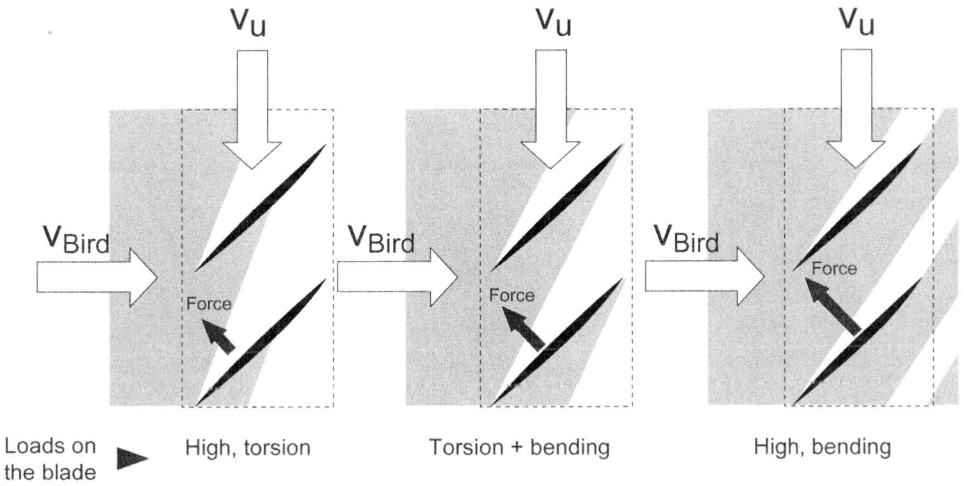

Schematic depiction of the impact force distribution when a soft foreign object passes through blading (pressure distribution according to the fluid model).

Ill. 5.2.2.2-4

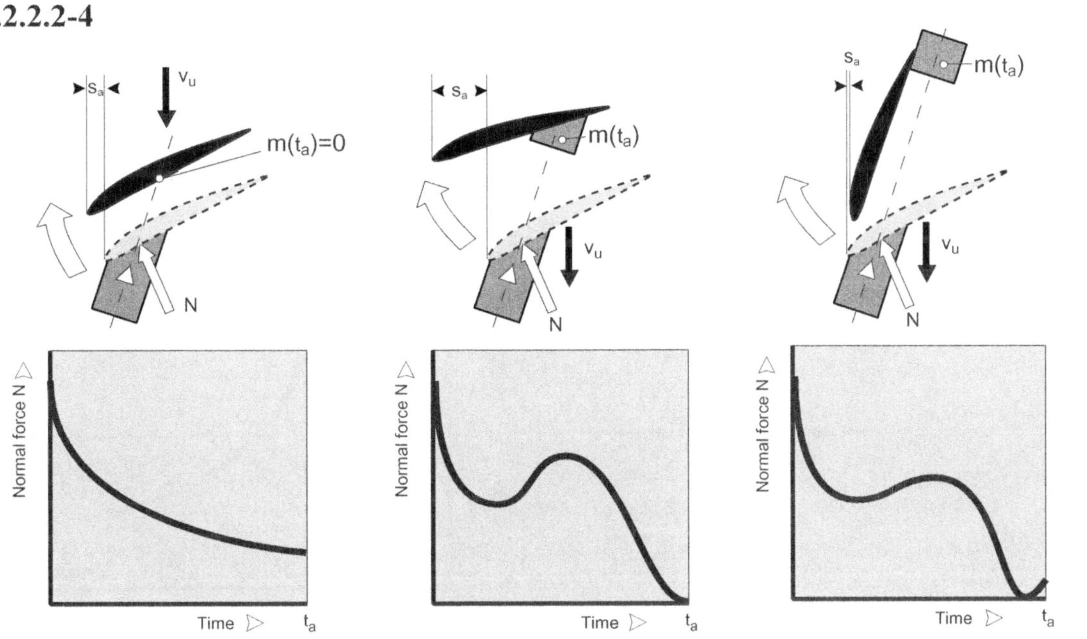

Brittle, hard particle (such as ice). The particle is "used up" (shatters) during the impact.

Tough small particle. N approaches "0", since the particle accepts the blade speed.

Tough large particle. The particle flows off across the twisted trailing edge.

Mechanisms of blade damage in the compressor during a bird strike

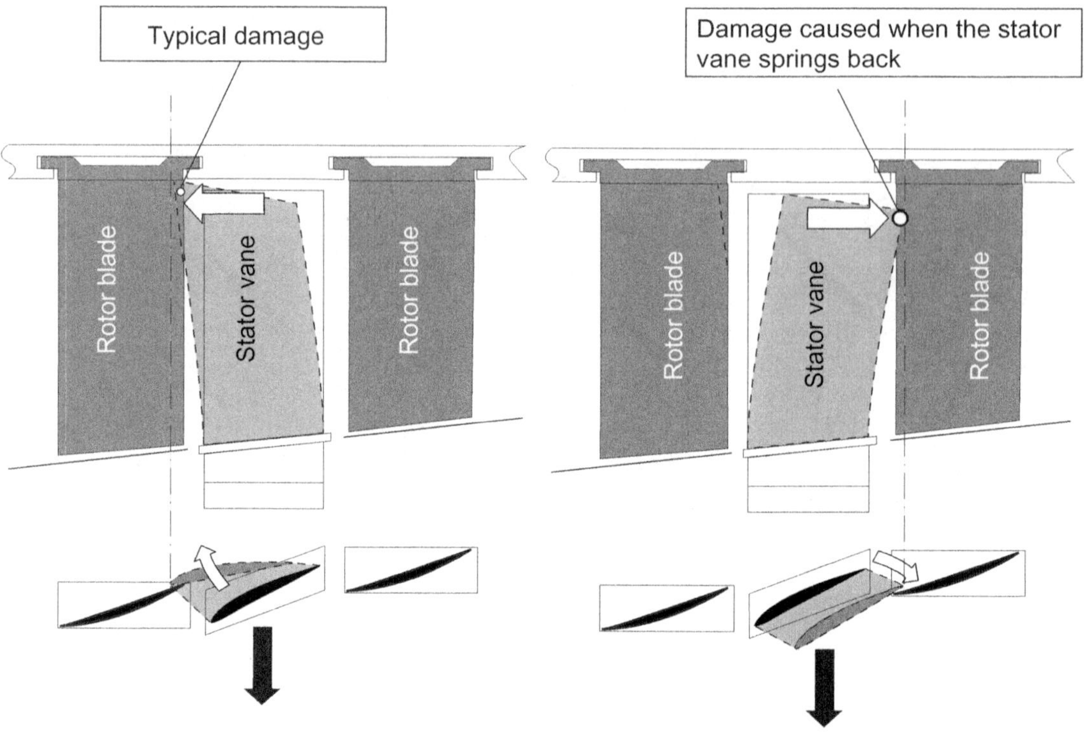

Ill. 5.2.2.2-5

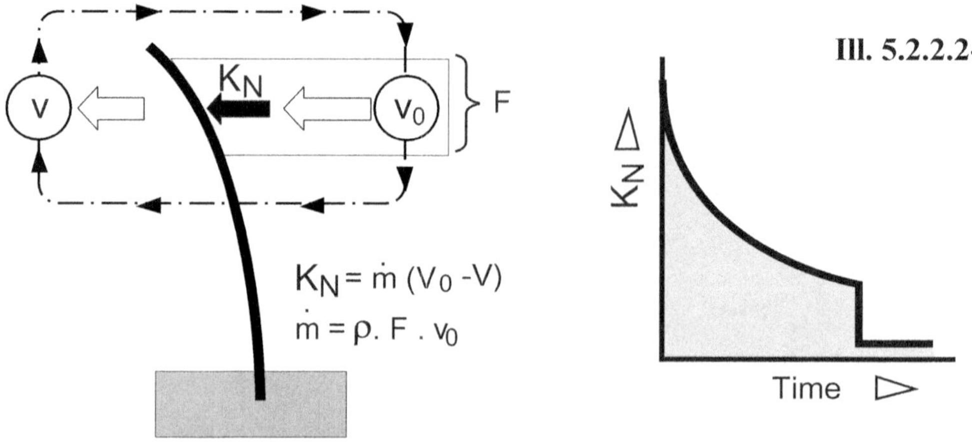

Temporal progression of forces during a bird strike (schematic)

Illustration 5.2.2.2-5: *Th e foreign object impact on the rotor blade first* ***bends the latter diagonally forward*** *due to the powerful forces required to transport it. The tip may be* ***deflected up to several centimeters****, depending on the size and elasticity of the blade.*

The deflection movement of the affected blade (top diagram), and therefore the progress of the impact forces (bottom diagram and Ill. 5.2.2.2-4), depend on the ***properties of the foreign object****, including:*

-brittle-hard*: ice, etc.*

-hard-elastic*: steel, etc.*

-soft-viscous*: birds, etc.*

Extensive research was necessary to find a ***bird substitute*** *that demonstrates impact behavior sufficiently similar an actual* ***bird strike****. It was found that water-filled plastic sacks are not a realistic substitute, since the damages they caused were considerably greater than those caused by actual birds. This is due to a lack of natural air pockets that dampen the development and spreading of the pressure waves during a bird strike, lessening the local damage (in fiber-reinforced synthetics, etc.) and impulse transfer. The amount of energy transferred during impact depends on the* **stossziffer**. *This consists of the relationship of the post-impact speed difference to the pre-impact speed difference between the involved bodies (blade and bird). The* **stossziffer** *indicates the elastic or plastic behavior of the bodies during impact and lies somewhere between 0 (completely plastic behavior) and 1(completely elastic behavior). The* **stossziffer** *is also dependent on speed.*

Illustration 5.2.2.2-6: *When a* ***bird strikes a rotor blade****, the angle of the blade and the force required to move the bird cause the blade to* ***bend diagonally forward*** *(Ills. 5.2.2.2-4 and -5). It then* ***springs back*** *elastically (middle diagram). These* ***oscillations*** *can cause the blade to bridge the gap between it and the neighboring stators and touch them (bottom diagram). If the* ***blade touches the front stator*** *the curvature of the blade may cause it to catch and/or be* ***pulled into the stator****, increasing the extent of offset. Therefore, this process is self-increasing.*
On the other hand, if the ***blade touches the stator located behind*** *the affected rotor stage, the blade is more likely to be thrown forward again.*
In both cases there is a danger of at least ***several stator vanes being weakened by notches*** *at the circumference near the connection to the root platform. Even if these stator vanes do not spontaneously fracture, they may fracture due to rapid fatigue.*

A bird strike can cause compressor rotor blades to come into contact with the stator vanes on either side.

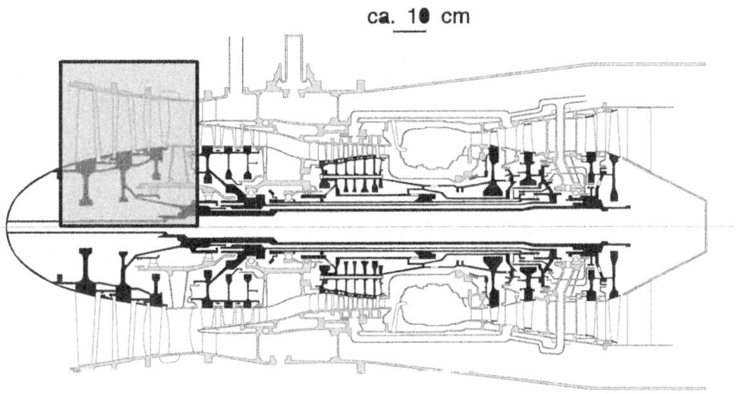

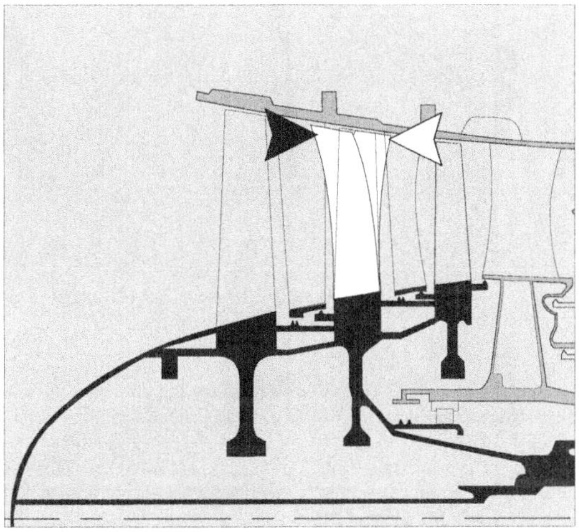

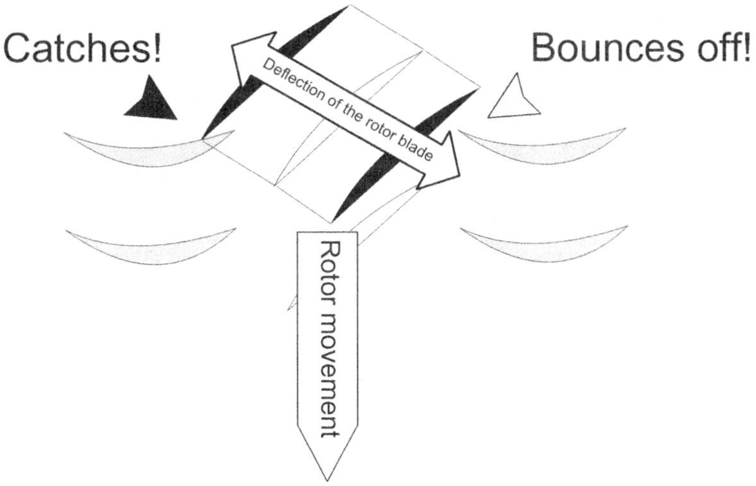

Ill. 5.2.2.2-6

External Factors
Bird Strike: Damage Mechanism

The flight speed largely determines the stressing of certain parts of the compressor.

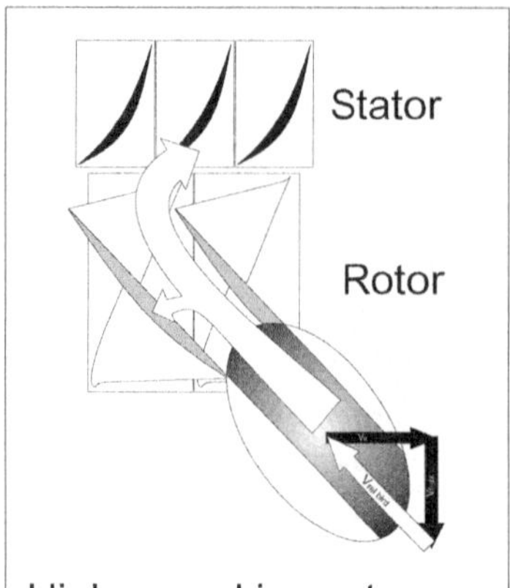

High-speed impact

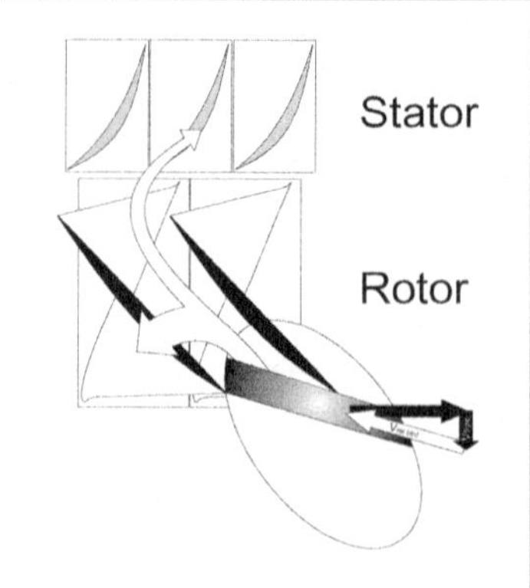

Low-speed impact

Compressor stator after a high-speed bird strike

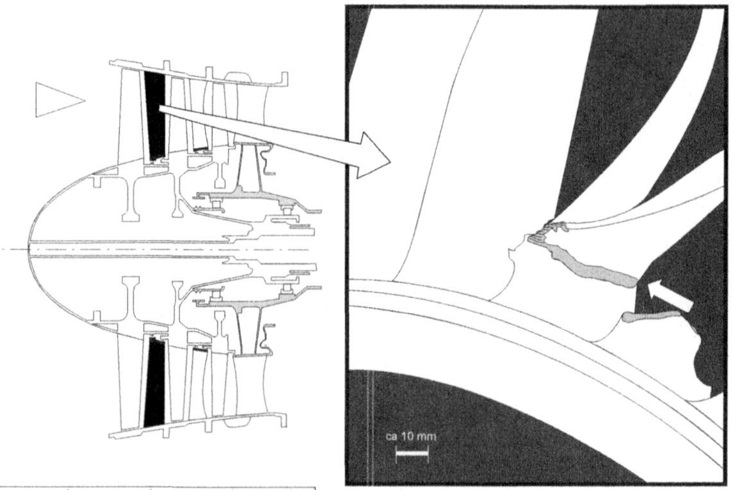

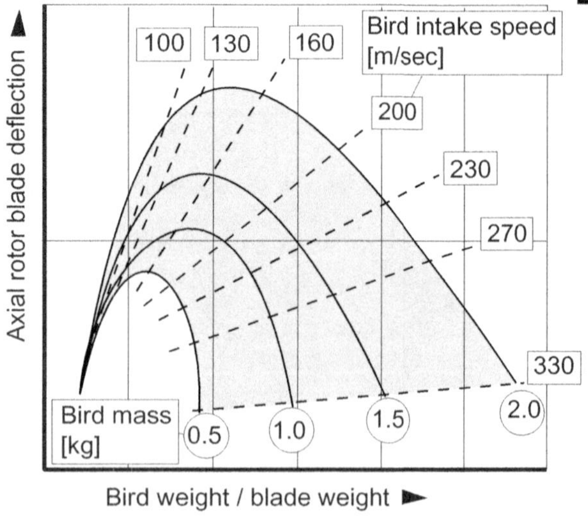

Typical deflection of a compressor rotor blade dependent on the mass of the bird and the flight speed

Ill. 5.2.2.2-7

Illustration 5.2.2.2-7 *(Ref. 5.2.2.2-2): If a* ***bird strikes a rotating blade*** *row at* ***high flight speeds****, and therefore high relative speed, it is referred to as a* ***high speed impact*** *(top left diagram). This situation occurs, for example, when birds strike fighter aircraft on low-altitude missions traveling at high speeds. In this case the inlet conditions are virtually the same as the aerodynamic configuration (inlet angle and speed correspond to the air flow). Thus the bird is* ***cut by the leading edge of the rotor blade****. In extreme cases with proper size ratios, the bird can pass whole between two blades. This puts minimal stress on the rotor blading, but the stator is struck by the entire bird and subject to extreme stress.*

At low flight speeds (top right diagram), ***low speed impact*** *puts especially high bending stresses the rotor blading (see Ill. 5.2.2.2-3). However, the following stator blades are only struck by parts of the bird.*

Stator blades/vanes*, especially those fastened on one end, are bent backward along the circumference due to the intake energy of the foreign object and the reaction speed it received from the preceding rotor stages.*

This results in powerful ***shear stress*** *on the blade fastening, which usually results in* ***cracking or fracturing*** *of the stator blade leaf. This can in turn lead to extensive consequential damages.*

If the ***inner shroud*** *is heavily bent or cracked, it can cause extensive damage to the rotor blading in front of and/or behind it. Because this will damage* ***all of the rotor blades at the circumference****, it can result in unexpectedly large damages that greatly exceed the stress levels on components such as housings which are normal for bird strikes. This can represent an uncertainty in acceptance tests.*

The bottom diagram is based on calculations and shows how a ***compressor rotor is deflected*** *by various bird strikes. It is clear that the maximum deflection, i.e. stressing of the rotor blade at relatively low intake speeds, is caused by low speed impact.*

In ***designs with inlet guide blades*** *(compare with Ill. 5.2.2.3-7), i.e. housing/****casing struts in the inlet****, the bird may be decelerated in this area. This can create conditions for a pronounced low speed impact with increased stressing of the following rotor blades, even at high flight speeds.*

External Factors
Bird Strike: Damage Mechanism

Low flight speeds can be especially dangerous for the rotor blading.

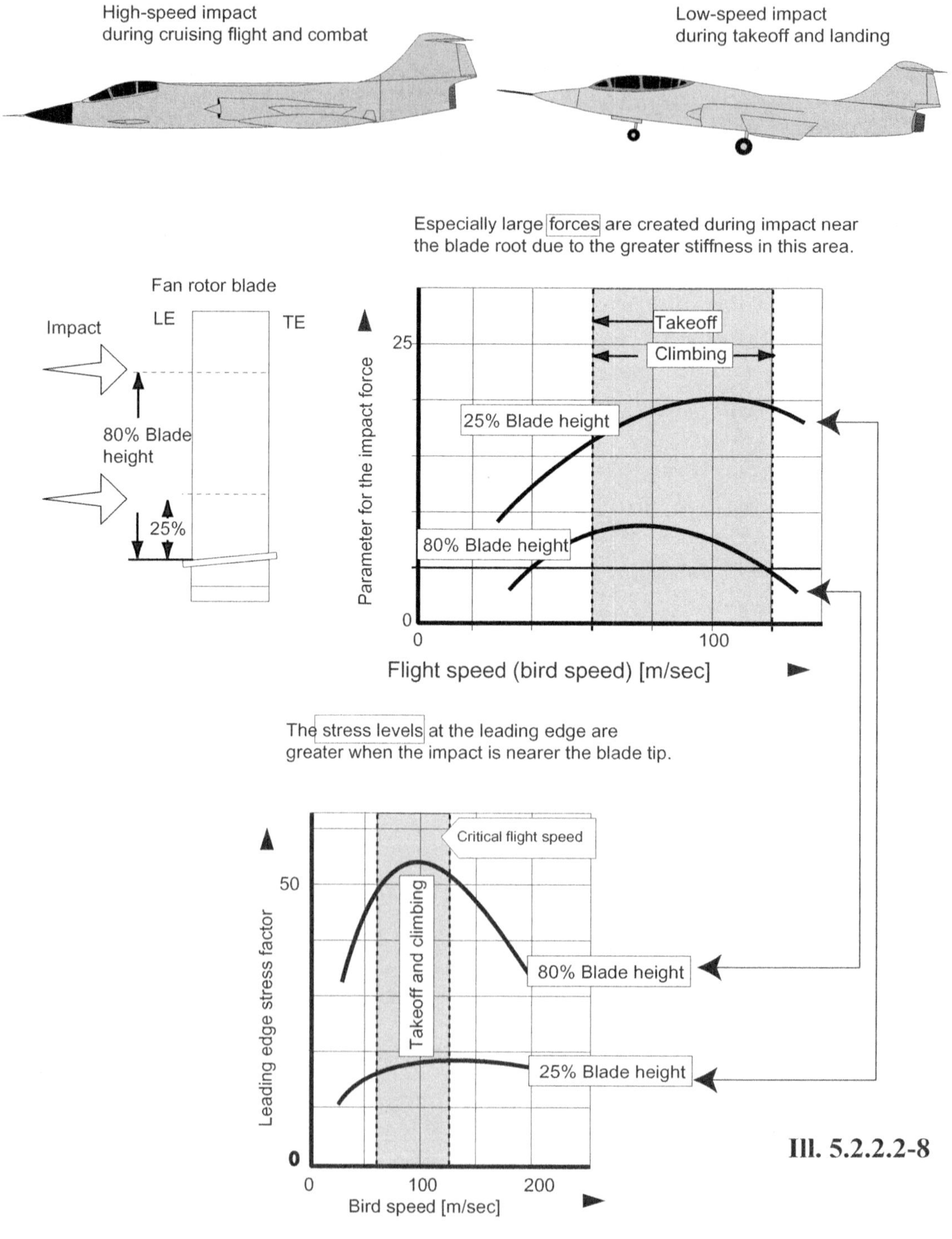

Ill. 5.2.2.2-8

***Illustration 5.2.2.2-8** (Ref. 5.2.2.2-2): It was shown in chapter 5.2.2.1 that aircraft are especially prone to **bird strikes during takeoff and landing** approach flight, i.e. at low altitudes. Also, under these conditions an additional factor makes the engines very sensitive to damage: at low intake speeds and high RPM (climbing after takeoff or direct landing approach flight, flaps fully extended), the bird is **cut into thinner slices** per blade, since more blades strike it due to the slow entry speed. On the other hand, the **angle of impact increases** along with decreasing bird entry speed (see Ill.
5.2.2.2-3). This increases the **ratio of axial force** components on the blade, meaning that a greater impulse is transferred to the blade. This effect increases with the difference in **bird speed relative to the designed air intake speed**, since the aerodynamic configuration was designed in accordance with the latter. At full RPM and low flight speeds (takeoff and climbing), maximum blade stress is reached. This critical flight speed is marked in the diagrams with a gray background.*

*The **impact forces** are greater near the hub than at the blade tip due to the greater **stiffness of the blade** and the **smaller angle of attack** (middle diagram and Ill. 5.2.2.2-5). The effective lever arm is smaller near the root, and the section modulus of the blade profile is greater. Due to the relatively large space between blades at the tips, the number of blades that engage the bird (especially large birds) is less than it is near the hub. This means that the tips are stressed by a larger bird mass per blade. Because of this, the overall **blade strains and deformations** are greatest when impact occurs at the tips (bottom diagram). While the **impact forces** at 25% of blade height are greater than at 80% blade height, the **leading edge stress**, as a parameter for blade failure, is greatest at an impact at 80% blade height.*

Chapter 5.2.2.2 References

5.2.2.2-1 W. Johnson, A.G. Mamalis, "Gegenüberstellung statischer und dynamischer Schadens- und Deformationserscheinungen", Fortschr.-Ber.VDI-Z. Row 5 Nr. 32. VDI Publishing, D-4000 Düsseldorf, 1977.

5.2.2.2-2 R.S. Cox, Rolls-Royce Limited, Bristol Engine Division, "Bird Ingestion Problems Relating to Gas Turbine Engines", September 1969.

5.2.2.2-3 C.A. Huertas-Ortecho, „Robust Bird-Strike Modelling Using LS-DYNA", University of Puerto Rico, 2006, page 1-236.

5.2.2.2-4 J.Frischbier, „Vogelschlag in Flugtriebwerken - eine impulsartige Fluid-Struktur-Wechselwirkung in der Triebwerksauslegung", www.mtu.de/de/technologies/engineering_news/entwicklung. Page 1-23.

5.2.2.2-5 J.Frischbier, „Multiple Stage Turbofan Bird Ingestion Analysis with ALE and SPH Methods", Herausgegeben von „American Institute of Aeronautics and Astronautics", 2005, Page 1-9.

5.2.2.2-6 P.Starke, G.Lemmen, H.Drechsler, „Validierung von Verfahren für die numerische Simulation von Vogelschlag", DYNAmore GmbH, LS-DYNA Anwenderforum, Bamberg, 2005, Page J-1-47 up to J-1-52.

5.2.2.3 Constructive and Technological Influences on Bird Strike Behavior

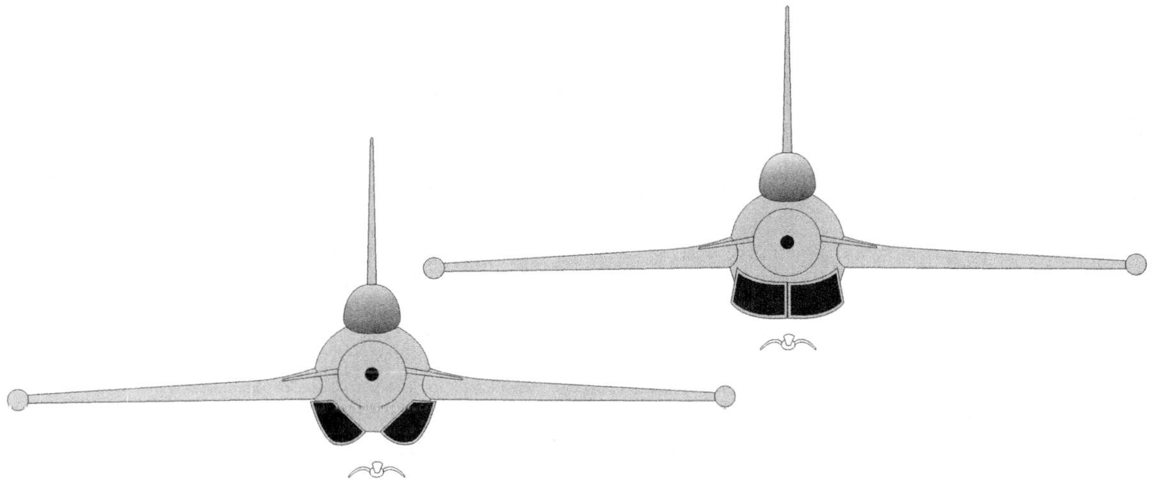

The behavior of an engine during a bird strike is largely dependent on **constructive and technological factors** such as the use of suitable materials specific to each engine part. As mentioned earlier, the **engine location** is also important. This is partly because it influences the **static probability of a bird strike occurring** (Ill. 5.2.2.1-5), and partly because it affects the bird itself during a bird strike.

For example, statistics show that engines **on the wings** are more frequently subject to bird strikes than those on the **vertical tail**. Engines **mounted tightly on the rear fuselage** are the least likely to be struck by birds. The values are fairly scattered, but a rough estimate of the respective hit rates would be 1:0,5:0,3. According to the damage rates, rear fuselage engines seem to be even better protected. Unfortunately there are no comparable available data for the various inlet configurations of military aircraft.

Further influences include the fuselage and wings acting as deflectors or feeders, and the flow conditions around these areas and the engines.

Inlet cross-section and engine type:

Although the bird strike rate increases with the engine inlet cross-section, the increase is not linear as might be expected, but is rather clearly at a declining rate (Ill. 5.2.2.1-5). Additionally, scattered data makes conclusive statements regarding this very difficult. It is assumed that

- this is related to **differences in the flow conditions** around the inlets of conventional and large fan engines.

- that large fan engines allow small birds to **pass through the bypass duct** without leaving any traces.

The damage rates (bird strikes with registered damage) show a clear sensitivity of the **large fan engines**. It is interesting that a large percentage of damages occur in the fan engines.

This may be due to the relatively **long fan blades**, which result in large **bending moments** (Ill. 5.2.2.3-9), and the wide chords that cause a great deal of stress to be put on the **leading edge** (torsion stress on the blade, Ill. 5.2.2.3-8). On the other hand, very small engines are naturally sensitive to bird strikes due to their filigreed structures and smaller open cross-sections.

However, independent of these damage susceptibilities, which are related to engine size, every **engine design** also has typical damage s**usceptibilities and related damage characteristics** and **failure types**.

Illustration 5.2.2.3-1: *Slightly curved inlets, annulus collector outer contours ("**swan-necks**"), and inlets and intermediate casings with many struts can have either a damage-decreasing or a damage-increasing influence.*
*In accordance with the fairly realistic image of a bird striking an engine at **high impact speeds** corresponding to the flight speed "**spraying**" in a manner similar to a jet of water, the smaller "droplets" (bird pieces) are easier for the blading to process due to their distribution and low weight, resulting in considerably **less damage** than when a bird enters the engine whole.*
*These conditions can be optimized through specific constructive measures, such as **inlet duct design**, making **direct hits on the engine impossible**. Naturally, this process also includes suitable design of the **wall strength and stiffness**.*
*This type of strategy can, for example, make possible the use of light **fiber materials in the blading** of fighter aircraft engines.*
*Even if the **impact angle** and/or the yielding of the impacted surface are sufficient to "spray" the bird, it must be assumed that the bird may be decelerated by **friction with the wall** and/or collect in areas such as on struts or in a swan-neck duct (as in Ill. 5.2.2.3-11), etc. This creates conditions for a **low speed impact** (Ill. 5.2.2.2-7) with a larger impact angle on the rotor blades.*
This shifts the conditions for the "critical flight speed" towards the normal-impulse-maximum (Ill. 5.2.2.2-8).
Internals in the inlet duct such as air guide vanes and flaps can be broken off by an ingested bird and cause additional damages.
*In certain Russian fighter aircraft types, **grill-like structures in the engine inlet** duct are used to catch ingested birds (Ill. 5.2.2.3-3).*
This certainly requires relatively solid and heavy constructions in order to avoid the risk of parts of the grill being broken off by a large bird. In addition, all internals in front of the engine present an increased danger of FOD through loose or fractured parts (Ill. 5.2.1.2-11). Influences on the inlet flow to the engine must also be considered.
*An important, virtually decisive influence on inlet design is its radar signature. Because the rotating fan blades could create a strong radar echo, a **grill or specially designed inlet guide vanes are placed in front** of it (Ref. 5.2.2.3-7).*
This type of internal (called radar blockers or) would most likely have a negative effect on the bird strike danger. They could hardly be made solidly enough to withstand a dangerous high speed impact without damage or fracturing. For this reason, curved inlets which cover the engines are advantageous.

Careful designing of the inlet duct can minimize the bird strike risk for an engine.

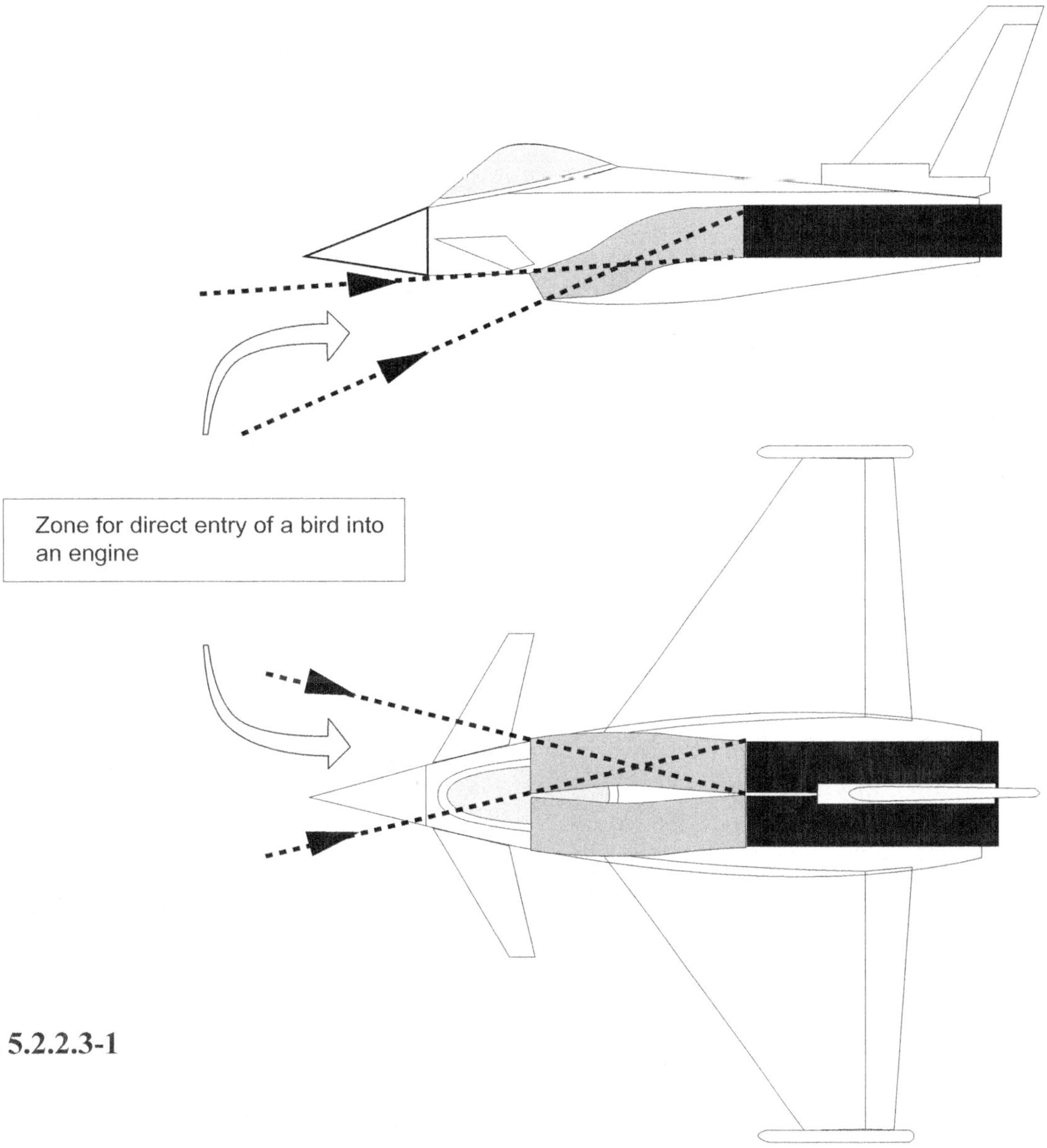

Ill. 5.2.2.3-1

The bird strike risk is influenced by the location of the engines. It is very unlikely, but has occurred before: one bird enters simutaneously both engines.

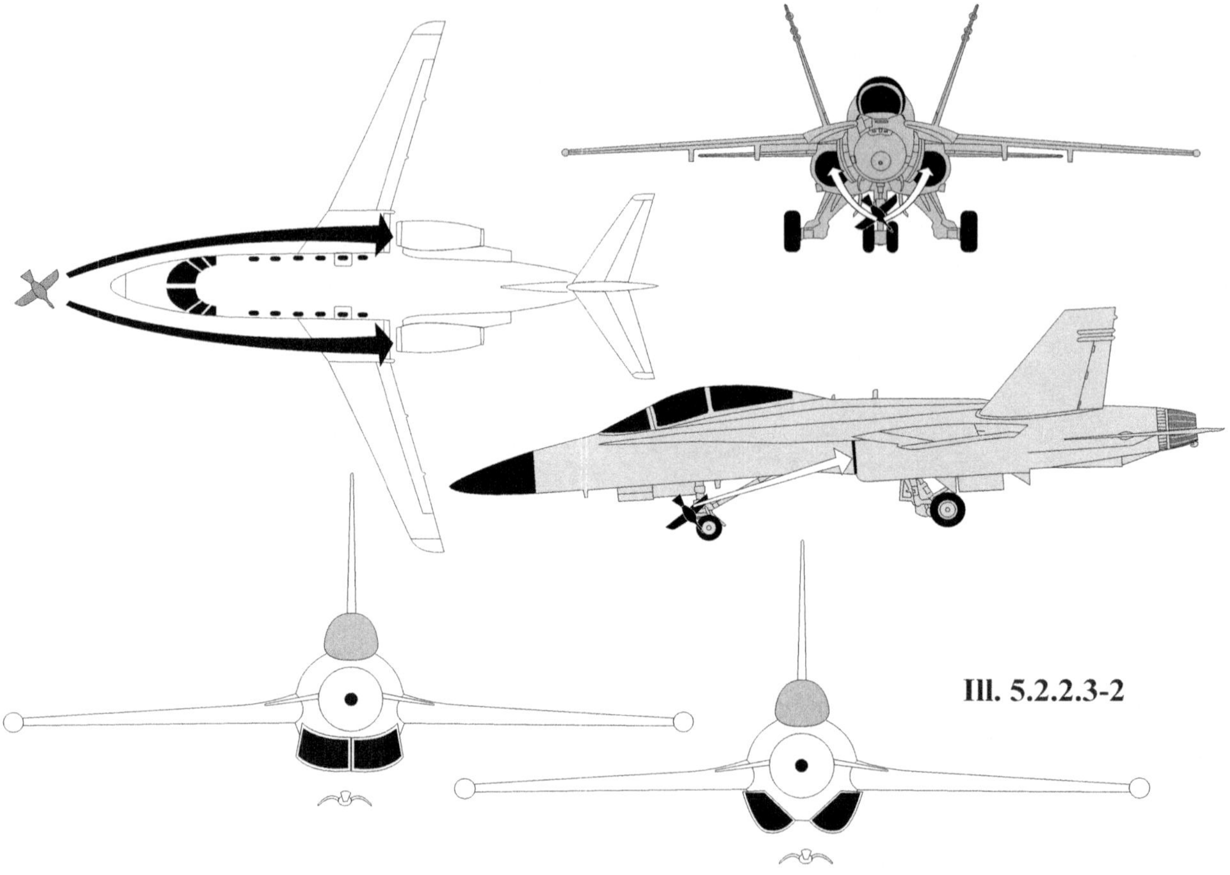

Ill. 5.2.2.3-2

Illustration 5.2.2.3-2: *There have been several reported cases in which* ***multiple engines*** *on an aircraft suffered virtually simultaneous bird strikes (see Examples 5.2.2.1-5 and 5.2.2.3-1). Experience has shown that this is entirely possible through several birds of a single flock. It seems practically impossible for a single bird to* ***strike the fuselage or landing gear*** *so symmetrically, that the pieces damage both engines and cause them to fail (top left diagram). However, experience has shown that this type of incident can and does occur, with both civilian and military aircraft.*
The affected aircraft are twin-jet types in which the ***engine inlets are located on the fuselage****.*
It is especially remarkable that when the bird strikes the tip of the fuselage, it does not spray apart and be deflected outward by the fuselage as expected, but rather splits into relatively large pieces that travel along the fuselage and enter the engine inlets (top left diagram).
The two bottom diagrams show different engine configurations in modern twin-jet fighter aircraft that should have considerably different bird strike risks (danger of a single bird damaging both engines). The probability of a larger bird damaging both engines simultaneously should be considerably greater for the configuration with

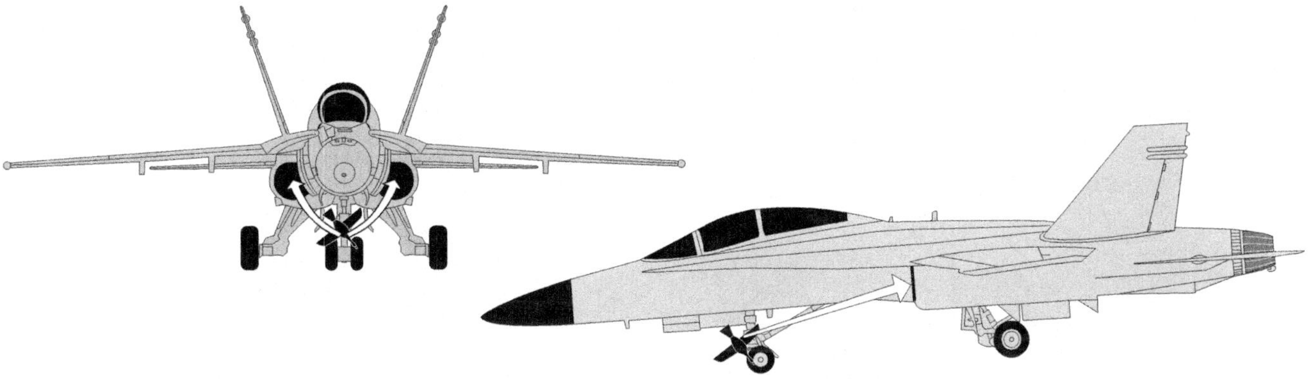

Example 5.2.2.3-1 (Ref.5.2.2.3-1):

Excerpt: "The seagull split.

One of those 'unusual' occurrences happened last July at an air show in Michigan. During a routine landing after an afternoon practice flight, one of the jets experienced a bird strike on the nose-gear landing light (Ill.5.2.2.3-2 right diagram).

The remains of the large bird split in two with parts being ingested into both engines, causing the total failure of one engine and partial failure to the other."

Comments: This incident involved a fighter aircraft type. It is remarkable that such compact pieces of the bird passed by the fuselage into the engine inlets located relatively high above the front landing gear. Evidently it cannot be assumed that, at relatively low flight speeds such as in this case during landing, the bird will spray on impact and no dangerously large pieces will remain.

both engine inlets close together under the fuselage (in order to minimize radar echo?) than for the configuration in the bottom diagram.

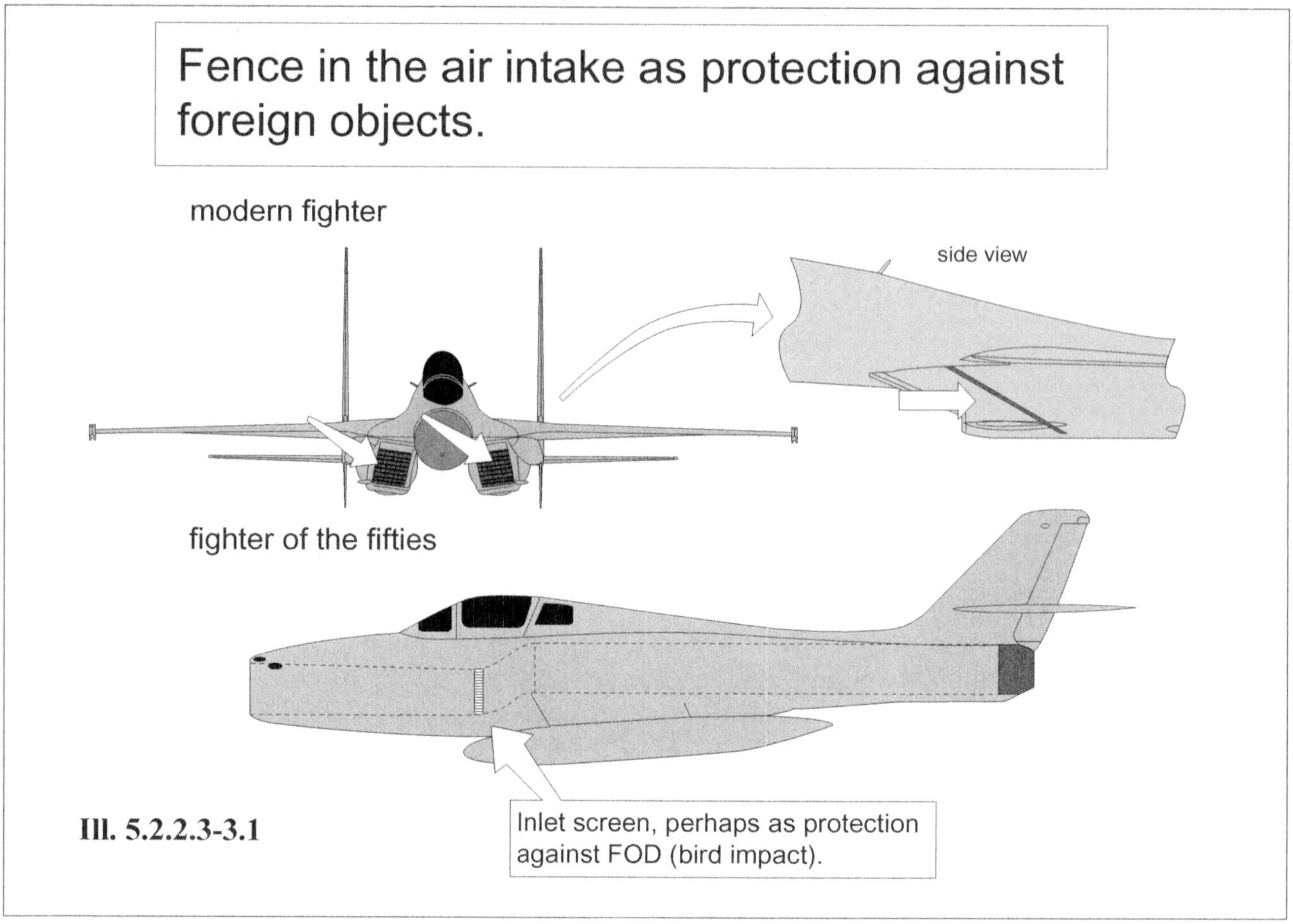

Ill. 5.2.2.3-3.1

Illustration 5.2.2-3.1 *(Ref. 5.2.2.3-5): This modern Russian fighter aircraft has **angled grills** (Ref. 5.2.2.3-3.2) **in the engine inlets** that are designed to protect against FOD. According to Ref. 5.2.2.3-6, the **grills are retractable**. It can be assumed, that these grills are only extended into the intake duct in especially FOD-susceptible operating conditions. If foreign objects include birds, then this is one of the rare cases in which this type of protective measure is used (compare with Ill. 5.2.1.2-9). A very early use of an **„inlet screen“** shows the scetch of an old fighter type at the bottom (Ref. 5.2.2.3-8).*

Diverse designs of FOD protection grills have been already tried, none was fully sufficient.

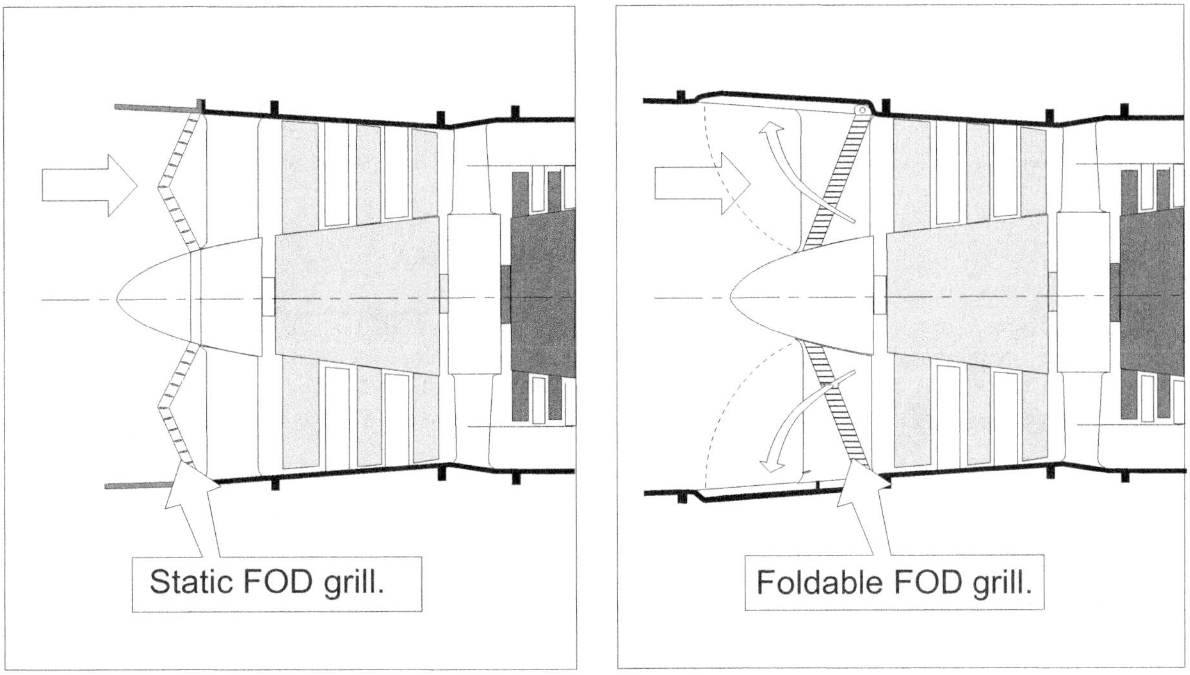

Ill. 5.2.2.3-3.2

Illustration 5.2.2.3-3.2 *(Ref. 5.2.2.3-9): This two variations of* ***FOD protection grills*** *are from early russian specialist literature. They serve not only the protection against bird impact but also against other foreign objects like bigger stones and hail. Obviously these have been realized in several types of fighters (Ill. 5.2.2.3-3.1). As drawback are termed:*

- *Increased* ***intake drag*** *equivalent 3-5 % thrust loss respectively increase of the SFC.*
- ***De-icing*** *of the grill (only the static type).*
- *Increased* ***weight****.*

The left sketch shows a static grill made from profiled metal sheet with slots of 2x2 mm up to 6,5x6,5 mm. It is located in the region of low flow velocity.

At the right a pivoting grill from several segments is displayed. It is mounted at the casing on the outside with pneumatic or hydraulic piston rods. If the main ***landing gear is actuated also the activation of the grill*** *takes place.*

With some engine concepts it is possible to achieve a bird strike-proof inlet into the core engine.

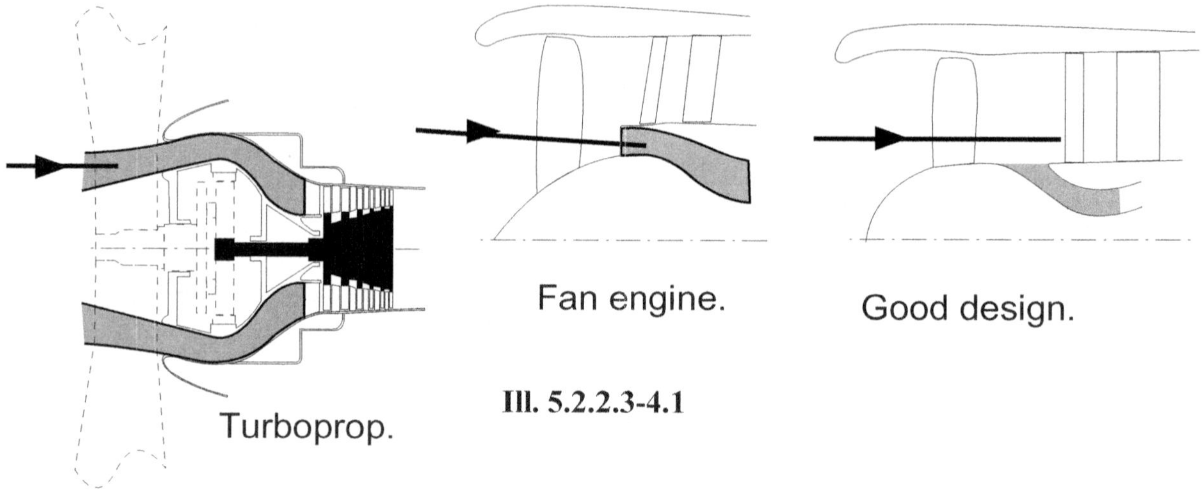

Ill. 5.2.2.3-4.1

llustration 5.2.2.3-4.1: *The shape and* ***location of the inlet*** *to the engine core can considerably reduce the probability of a bird, or at least large pieces of a bird, from entering the sensitive compressor.*

In ***propeller engines,*** *the* ***rotating nose cone*** *(spinner) and the propeller act to deflect foreign objects. A similar effect is consciously used in fan engines (middle diagram).*

With regard to FOD, the configuration in the right diagram ***should be the safest*** *(also see Ill. 5.2.1.3-9). However, it is necessary that the flow conditions at the engine inlet are acceptable for the operating behavior of the compressor and the longer design. This concept has been successfully used for years in an engine type for business jets. These engines use three* ***blisk*** *stages in the high-pressure compressor, and these have never had any FOD problems.*

Illustration 5.2.2.3-4.2 *(Ref. 5.2.2.3-10): Also at* ***turboprop engines*** *the danger of a* ***bird impact*** *exists (volume 2, Ill. 5.2.2.1-6.1). This picture shows the concept of a deflection (sketch in the middle) of the bird out of the* ***S-shaped intake duct*** *of an aeroengine with separate positioned prop gear (upper sketch). For this, obviously a* ***flap*** *is provided. Additionally there exists a* ***heated grill*** *against icing to minimize failures by FOD. The deflection of the bird shold especially prevent the* ***danger of a torsion fracture of the gas producer shaft*** *(sketch below) caused by braking of the rotor from this relatively small aeroengine. The exact layout of the whole design could not be seen from the available literature and was completed by the author correspondent to his understanding.*

External Factors

Bird Strikes: Construction

Concept of optional devices against bird strike damages at a smaller turboprop engine.

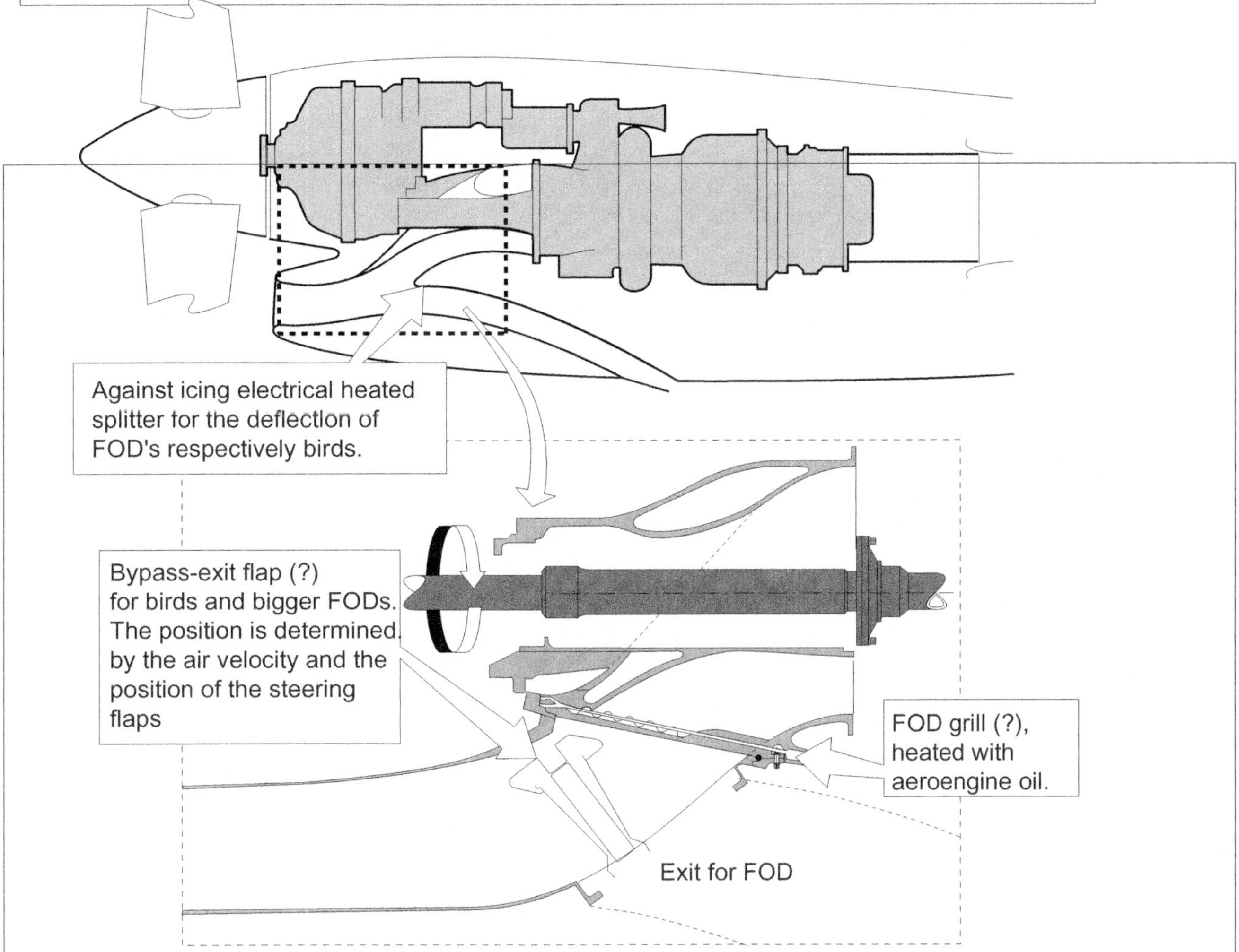

The possibility exists, that details ofthe sketch have benn misinterpreted because the literature on hand.

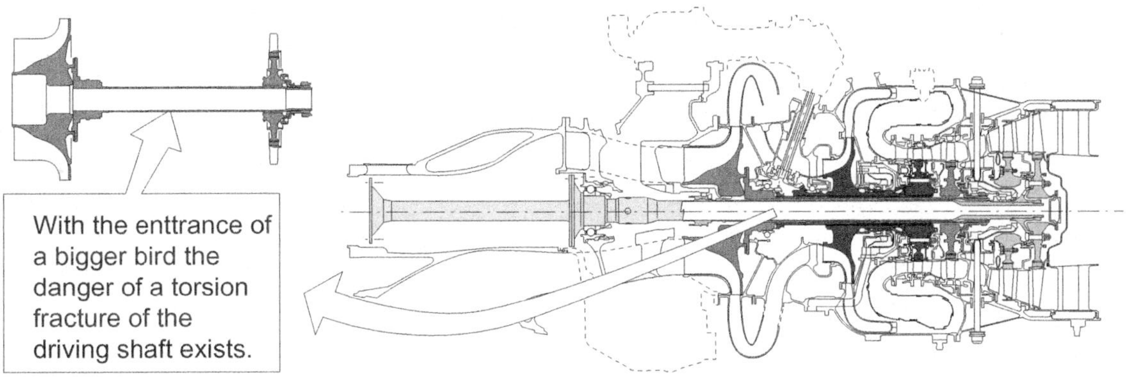

Ill. 5.2.2.3-4.2

Examples of tested protection grills of turboprop engines against bird impact.

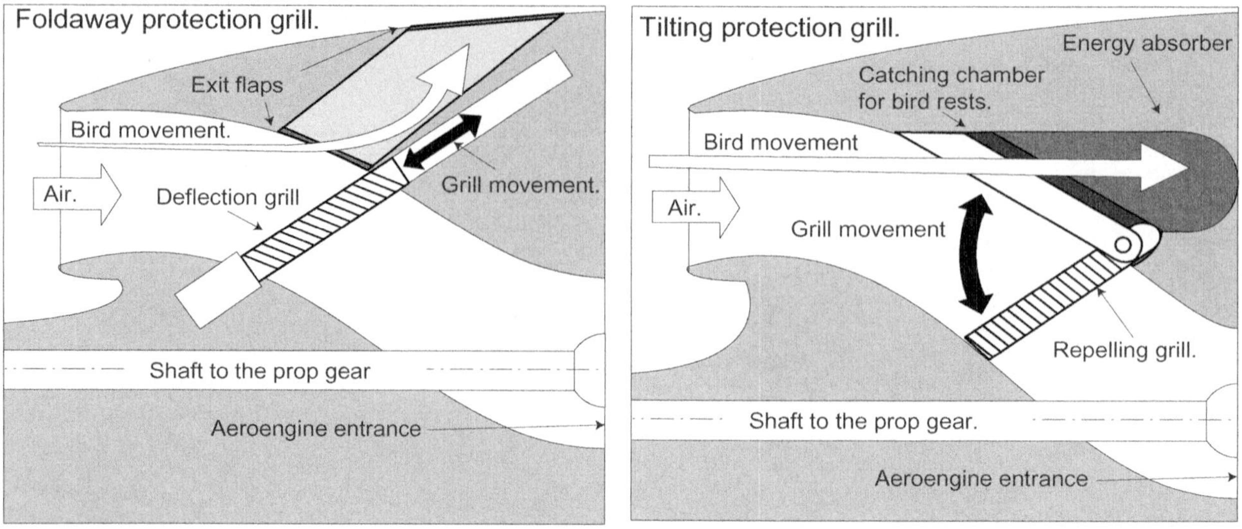

Ill. 5.2.2.3-4.3

Illustration 5.2.2.3-4.3 *(Ref. 5.2.2.3-11): Already at the beginning of the seventies last century,* ***protection grills*** *in the intake duct against bird impact have been tested. Also activities from the middle of the eighties deal with this problem (Ill. 5.2.2.3-4.2). Firstly the question arises: Why just for turboprop engines which yet seem to be protected by the rotation propeller? However as emerged cases show, this is obviously not true (Ill. 5.2.2.1-6.1 and Ill. 5.2.2.1-6.3). A special problem is the curved intake duct in which the bird mass, also after a splashing at the wall, can again accumulate in front of the aeroengine (Ill. 5.2.2.3-11, Ill. 5.2.2.3-12 and Ill. 5.2.2.2-7). Because of the braking by friction with the wall, this leads to a „****low speed impact****". From this the relatively filigree compressor rotor blades of the first stage from this relatively small aeroengine are especially endangered.*

As remedy there are concepts of grills in the intake duct.

The ***left version*** *concerns a* ***slidable system****. In the probable impact area of the wall of the duct, presstressed elastic exit flaps are mounted. During impact of big or small birds a spring triggers the opening. Then the birds can exit from the aeroengine nacelle through a channel.*

The ***scetched case at the right*** *shows a pivoting mechanism. It is driven by a motor. In the point of impact the duct wall has a '****retention chamber****' which obviously shall gather remains of birds. The* ***energy absorber*** *placed behind is thereby obviously deformed by the retention chamber. An energy absorbing filling shall prevent the overloading of the nacelle.*

In both cases to minimize the impact forces, the grill is oriented angular (30°) to the bird trajectory. Anyway, in an extreme case during the direct hit of a 2 kg bird with about 320 km/h a force of about 50 tons, must be absorbed without unacceptable deterioration (Ill. 5.2.2.1-3). This force will be doubled at 400 km/h.

In layout the steel lamella of the grit have been 2,25 mm thick, its distance was 15 mm.

Till now protection grills for aeroengines of airliners seem only to be ideas.

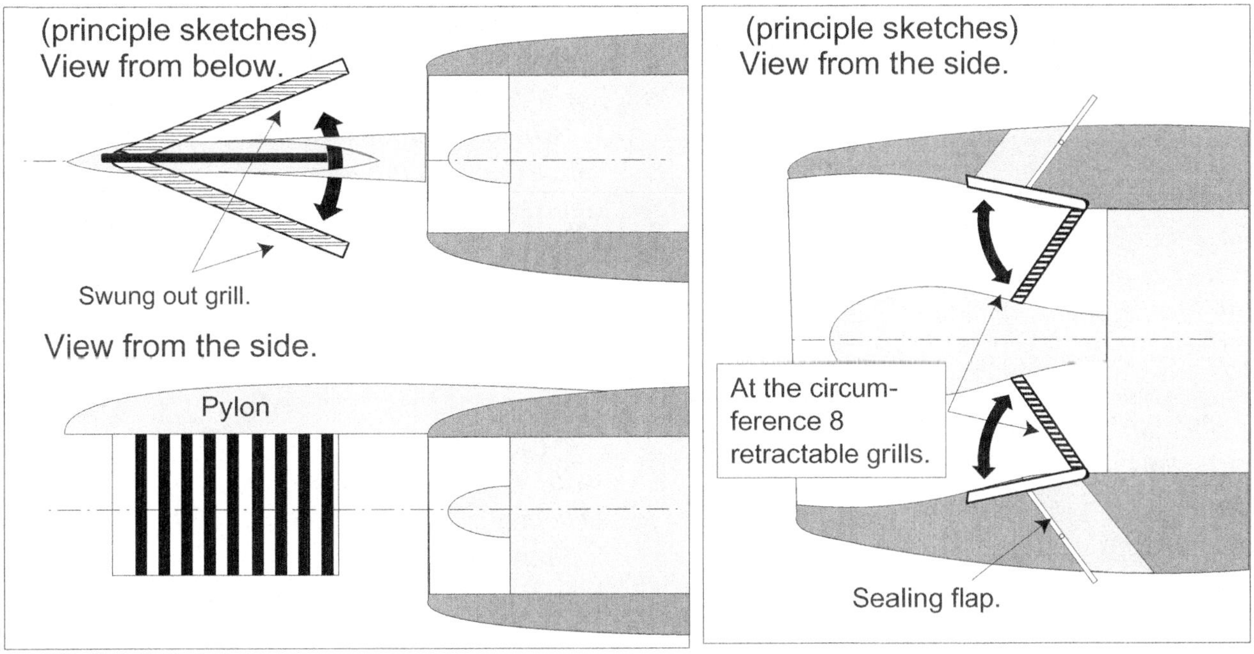

Ill. 5.2.2.3-4.4

Illustration 5.2.2.3-4.4 *(Ref. 5.2.2.3-11): Obviously concerned are not realized* ***studies*** *for the protection against bird impact. Anyway these are interresting as basis for a discussion and stimulation.*

This ***arrangement of the left*** *is as well designed for the mounting at the wing with a pylon as also for aeroengines at the rear aircraft fuselage. In this case the orientation of the device, which bears the pylon, is horizontal. The sketches show an elder aeroengine type with a* ***lower bypass ratio****. The bird is directed* ***around the aeroengine****. For modern aeroengines with* ***high bypass ratio*** *the tilted grill merely deviates* ***outward into the bypass region of the fan****. Main problem seems to be the* ***additional weight of the construction****. As problematic also the* ***icing*** *of the system is mentioned, which may require a de-icing device. Further also possibly efficiency loss and vibration exitations of the aeroengine by deteriorations of the flow (vortex formation) must be considered.*

The ***arrangement at the right*** *shows an into the intake duct hinged protection grit, similar Ill. 5.2.2.3-4.3. An additional weight in the range of 100 kg will be expected for aeroengines with lower bypass ratio. A version for aeroengines with high bypass ratio is not evident.*

***Illustrations 5.2.2.3-5:** The **reaction of a blade** to a to impact momentum is largely dependent on the blade material. This usually results in elastic and plastic deformations and is followed by vibrations at various resonance frequencies due to the blade springback (see Ill. 5.2.2.2-1). The impact time is an increment of 10^{-3} seconds. Experience has shown that steel blades perform considerably better under bending loads than do **titanium** ones under the same condition, and titanium blades perform better than blades made from **Al alloys** or **fiber composites** (for example, boron fiber-reinforced aluminum or carbon fiber-reinforced synthetics). While steel and titanium blades primarily show plastic deformation under bending loads, the same amount of energy makes the fiber material fracture brittly after purely elastic deformation (primarily due to the brittle fracture behavior of the reinforcing fibers; bottom diagram).*

*This behavior at first seems rather implausible, since the strength of currently used fiber-reinforced materials and titanium alloys is certainly comparable to that of steel. However, as the top diagrams show, the blade tip **deflection** after absorbing a certain amount of energy is greater, the lower the density of the blade material and the greater its elasticity (the smaller the E-modulus). This is understandable, since less deflective movement would be expected from a heavy, stiff steel blade than from a light blade under the same impact energy (Ref. 5.2.2.3-2). Due to this behavior, the same amount of energy deflects the light blade more and therefore subjects it to correspondingly higher bending loads, depending on the elasticity modulus. This makes the light blade more sensitive to bird strikes.*

A further drawback of large deflection is the possibility of contact with neighboring blade rows, which is combined with extensive consequential damages.

This behavior has also led to blades in smaller engines not being made from light, fiber-reinforced materials, especially fiber-reinforced synthetics. Only recently have fan rotor blades in large engines, which weigh more than 10 kg, started being made from fiber-reinforced synthetics. Blades this solid are sufficiently robust to handle impacts even from large birds.

*In order to use the advantages of fiber-reinforced synthetic blading in fighter aircraft engines, a material-specific safety philosophy should be considered, but it would not necessarily have to worsen the failure rate of the entire engine. On the contrary, when using **fiber-reinforced synthetics in rotor blades**, this can be accomplished by shaping the inlet duct so that birds cannot strike the engine directly, but first hit the duct wall and spray into small, harmless pieces which then enter the engine.*

*The use of **fiber-reinforced synthetics for guide vanes** in the fan is realistic if one considers, for example, twin-jet fighter aircraft and must only achieve the same reliability as with titanium blades. With the relatively thin and narrow guide vanes used in these engines, high speed impact of a somewhat larger bird (Ill. 5.2.2.2-7) will in all likelihood result in failure of the blade which was struck. In this case, even the fracture of a fiber-reinforced synthetic is acceptable, especially since fragments of this material will cause considerably less damage downflow than ductile, metallic fragments would.*

The lighter and more elastic a blade material is, the greater the deformations and danger of overstress during a bird strike.

Elastic deformation of blades made from various materials during a bird strike

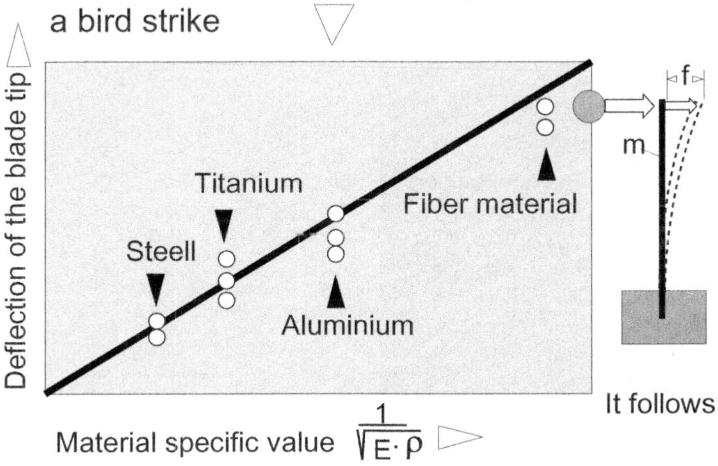

Kinetic energy of the rod after impact:

$A \sim m.v^2$

$m \sim \rho$

Impulse transfer during impact:

$v \sim 1/m$

$A \sim 1/\rho$

The kinetic energy A is stored in the bending rod as elastic deformation energy:

$A \sim f^2 \cdot E$

It follows for the blade tip deflection "f", that

$$f \sim \frac{1}{\sqrt{E.\,\rho}}$$

Typical bird strike damage symptoms of blades made from various materials

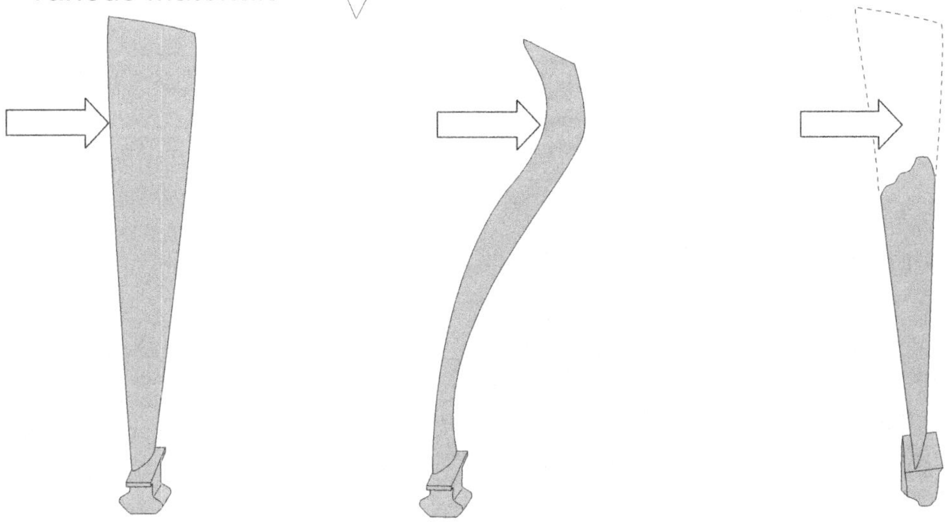

Cr steel. Titanium alloy. Fiber reinforced aluminium (B-Al).

Projectile mass: 14 g gelatine/wood shaving mixture, projectile velocity 260 m/s.

Ill. 5.2.2.3-5

The notch impact strength can be an indicator of sensitivity to bird strikes.

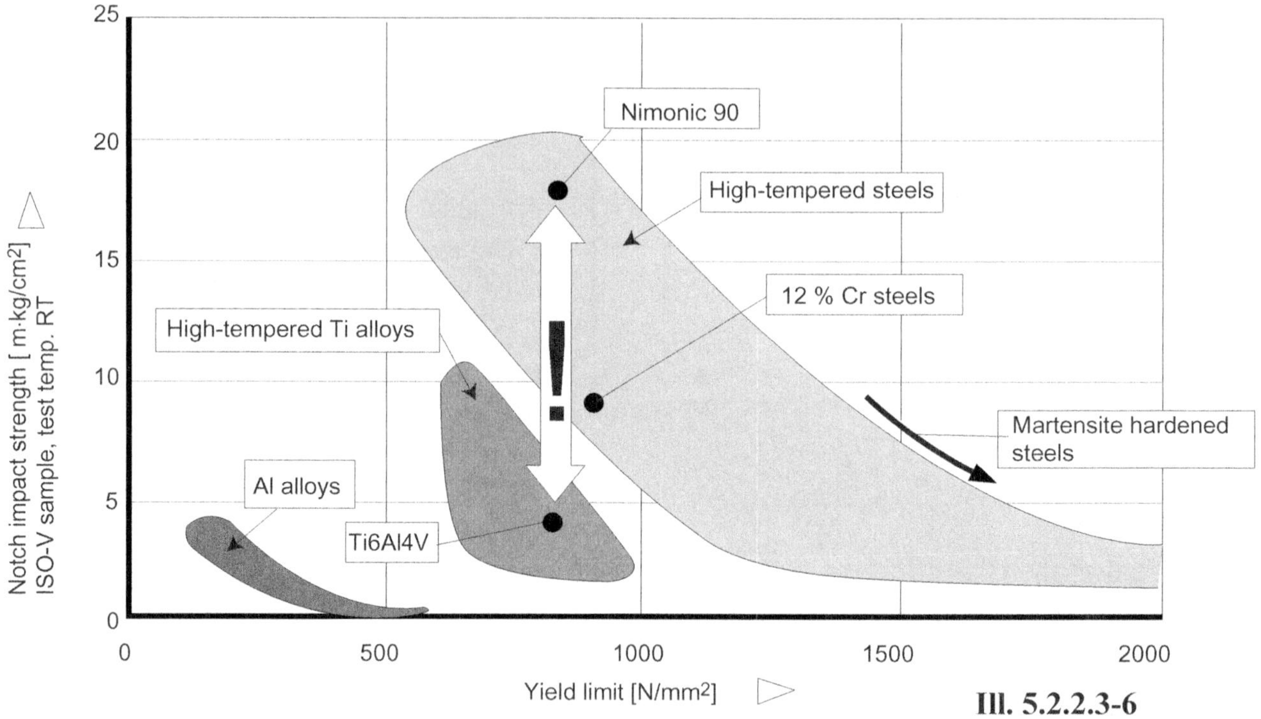

Ill. 5.2.2.3-6

Illustration 5.2.2.3-6: *The different compressor* ***blade materials*** *(alloys based on Al, Ni, Ti, heat treated steels, and fiber-reinforced materials) can all behave very differently during* ***foreign object strikes****, depending on the material characteristics. The modulus of elasticity and yield strength are of primary importance for elastic deformations. However, impact damage also depends on the* ***notch impact strength*** *and the density of the blade material (Ill. 5.2.2.3-5). The diagram shows that high-strength* ***titanium*** *alloys have a considerably lower notch impact strength than steels of similar strength. Al alloys have rather poor strength as well as poor notch impact strength, which was undoubtedly a factor (along with low dynamic strength and sensitivity to erosion and corrosion) in the decision no longer to use these materials for compressor blading in modern engines.*

Low notch impact strength can drastically increase the extent of damage if, for example, during a bird strike notches in the blading are created through contact with other engine parts or fragments, and this is followed by heavy stressing of the blading (through heavy rubbing, etc.). This type of notch can be the initiation point for an unexpected ***blade fracture****. Blades made from high-strength titanium alloys, especially, may show this undesirable behavior and thus deviate from their pre-calculated behavior.*

Fundamentally, all blades and disks should be made from material with the most possible impact

resistance. This is not only so that the directly struck blades do not fracture, but also in order to limit the consequential damages after a blade failure and ***prevent haircuts****, etc. (a haircut is the fracturing of all blades on a rotor stage).*
Interestingly, Ti-alloys in general have a considerably lower notch impact strength than steels or forged super alloys. The plastic deformation and possible stable and unstable crack initiation or fracture are influenced by the plastic deformability and the ***hardening behavior****. The ultimate strain is a characteristic value for the undamaged material under static force. The lower the notch sensitivity of the material, the less likely it is that dynamic stress during vibrations will cause crack growth. At the high stress speeds relevant here, the plastic strain behavior, the impact bending strength, and the notch impact bending strength all decrease. The material tends to fracture more brittly with increasing* ***deformation speed****. At the same time, the ultimate strength and the yield strength usually increase, the latter increasing more quickly before meeting with the ultimate strength. In this case the material is* ***completely brittle*** *(Ill. 5.2.2.2-1). Because of this material behavior, for fractions of a second the stress may be greater than the static yield strength (deformation speed effect). In structural steels yield strength may be exceeded by 2 or 3 times without any noticeable permanent deformation. With total damage to bladings this embrittlement can sometimes be observed on fracture surfaces.*

***Illustration 5.2.2.3-7:** In engines equipped with them, **inlet guide vanes** (front guide annulus, top diagram) are especially susceptible to damage. Unlike rotor blades, which experience a decrease in the normal impulse beyond the critical flight speed (Ill. 5.2.2.2-8), with guide blades the normal impulse continues to increase due to the fact that the impact angle does not approach zero. Therefore, with increasing **intake speeds**, the damage to be expected also increases. Other high-risk areas are **adjustable guide vanes**, especially inlet guide vanes, during landing, because the blade annulus is closed and may be struck by a foreign object at a large angle.*
***Inlet guide blades or bearing struts at the compressor inlet** can create low-speed impact conditions for the following rotor blades during high-speed flight by decelerating the bird mass (compare with Ill. 5.2.2.2-7)*
The high impact speeds can bend (Ill. 5.2.2.1-11) rotor blades or crack and/or fracture them at the transition to the housing (Ill. 5.2.2.2-7). Adjustable guide vanes can fracture at the adjustment bearings and/or all guide vanes may close due to the coupling of the adjustment mechanism, causing a sudden thrust loss in the engine. Even if only single adjustable guide blades suffer permanent angle changes, it can lead to serious operating problems in the compressor (stalls, blade vibrations).
Additionally, guide vanes are fundamentally subject to higher stresses during bird strikes than rotor blades:
*While, in normal cases, multiple rotor blades of a stage slice up birds and distribute the stress, **all of these bird pieces strike the same stator blade position**. For this reason, the design and fastening of stator blades on the housing must be undertaken with sufficient care.*

Experience has shown that,

*- in **light metal housings**,/casings especially, fastening blades with a T-**root is preferable to using dovetail** systems.*
*- fastening the guide vane tips with an **inner ring** (shroud), or even **grouping** them into guide vane sections, provides favorable results. These designs **stiffen the construction**, **distribute loads**, and prevent the blades from bending over and rubbing on the following rotor stage. If contact occurs, there is an additional danger of self-increasing deformation caused by the stator annulus being pulled into the rotor due to the characteristic blade angles (Ill. 5.2.2.2-6). It is also important that the **inner shroud is sufficiently strong** to prevent fractures, since fractures of the stiffening ring will usually result in catastrophic damages.*

Elastic and plastic deformations of the V-guide vane annuluses (Ill. 5.2.2.2-7) in the area of the bird strike, combined with decelerating effects cause foreign object movement in the compressor that varies greatly from the movement of the normal flow, increasing the impact angle and therefore also the stress on the following rotor blades (middle diagram; Ref. 5.2.2.3-2). On the other hand, a sufficiently strong and stiff guide annulus in front of the first rotor stage can considerably reduce rotor damage during bird strikes if it causes the bird to strike the rotor blades at a more favorable angle of impact.

External Factors

Bird Strikes: Construction

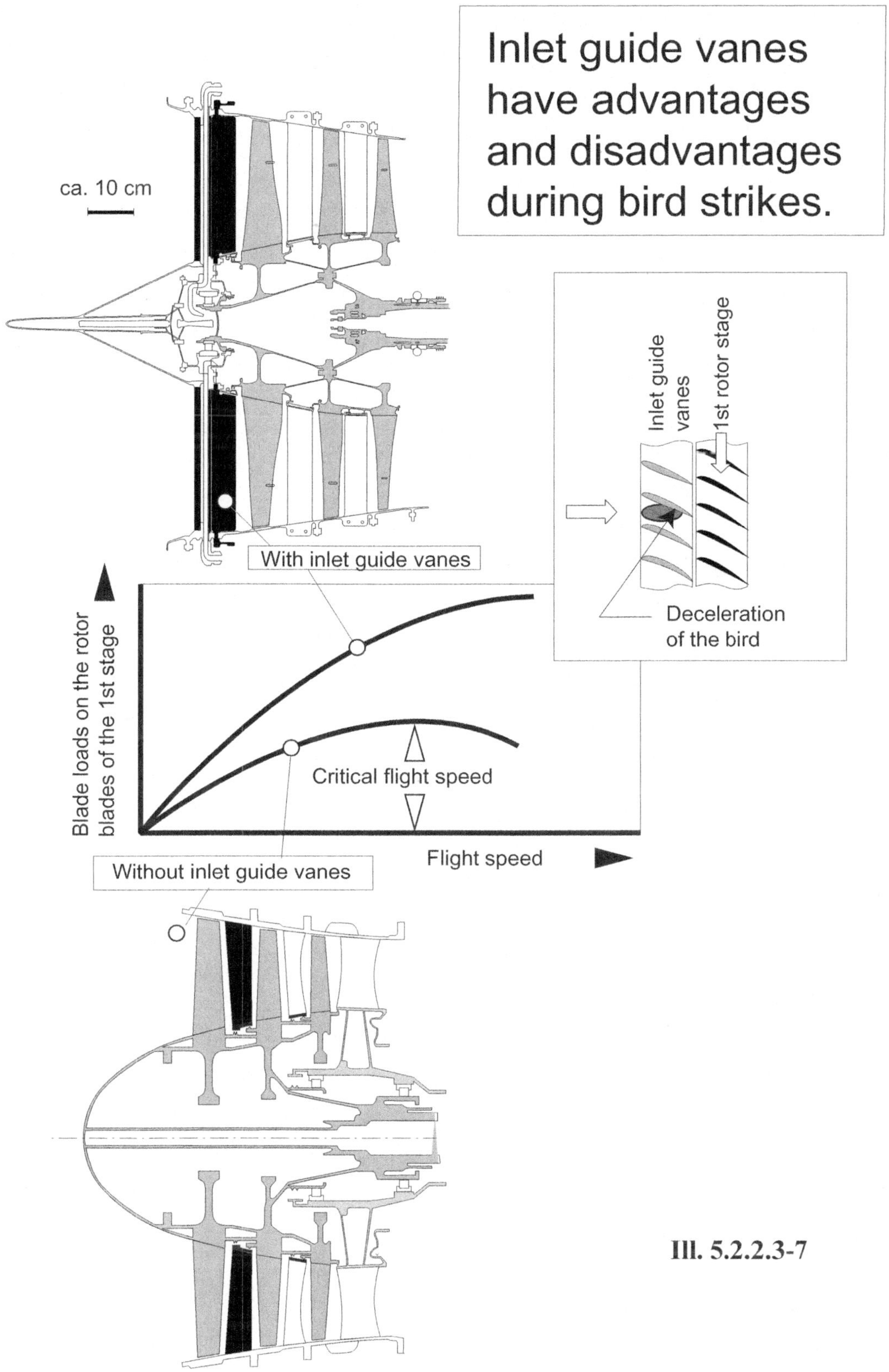

Ill. 5.2.2.3-7

External Factors
Bird Strikes: Construction

"Modern" blade shapes can also prove to have disadvantages during bird strikes.

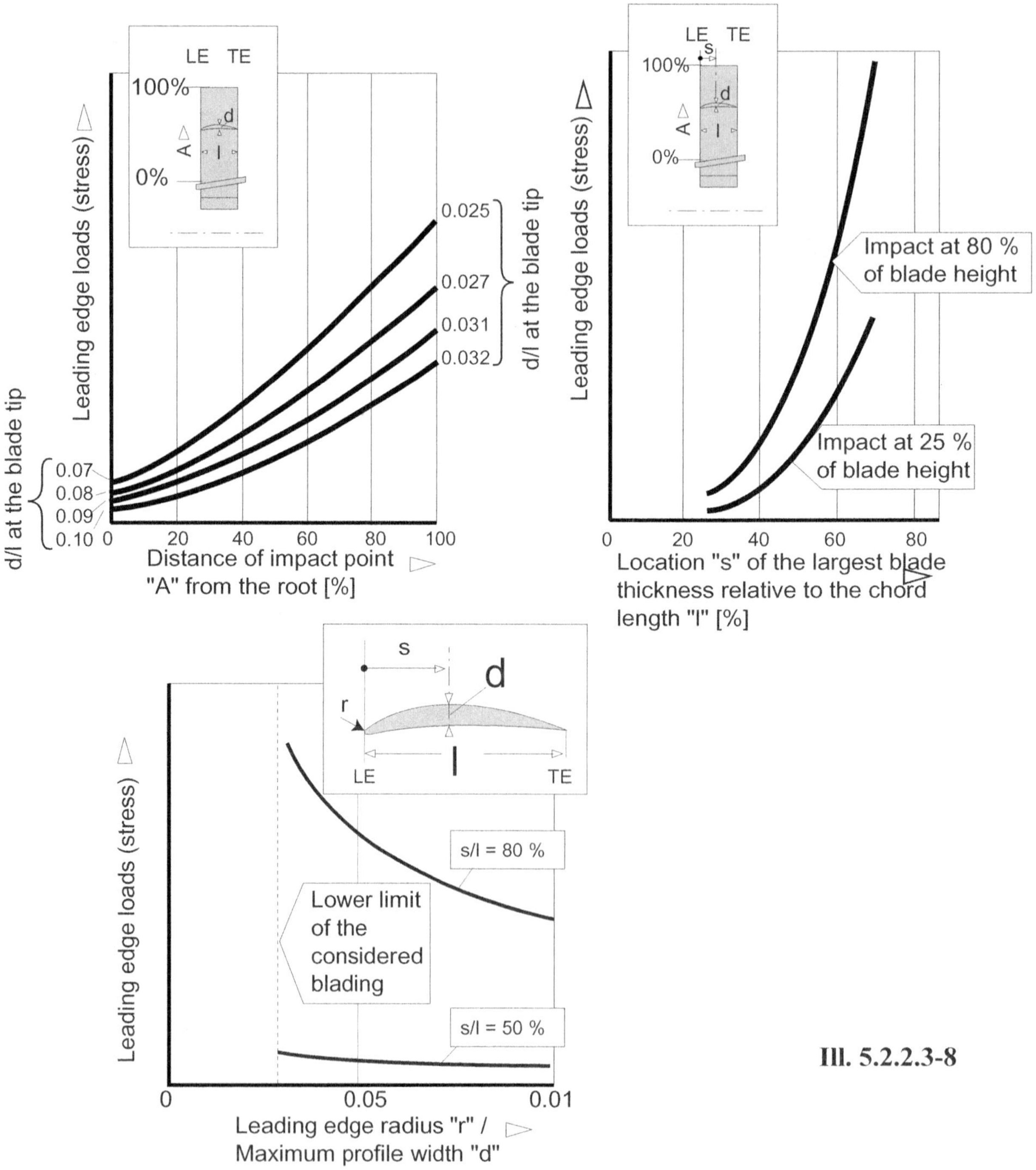

Ill. 5.2.2.3-8

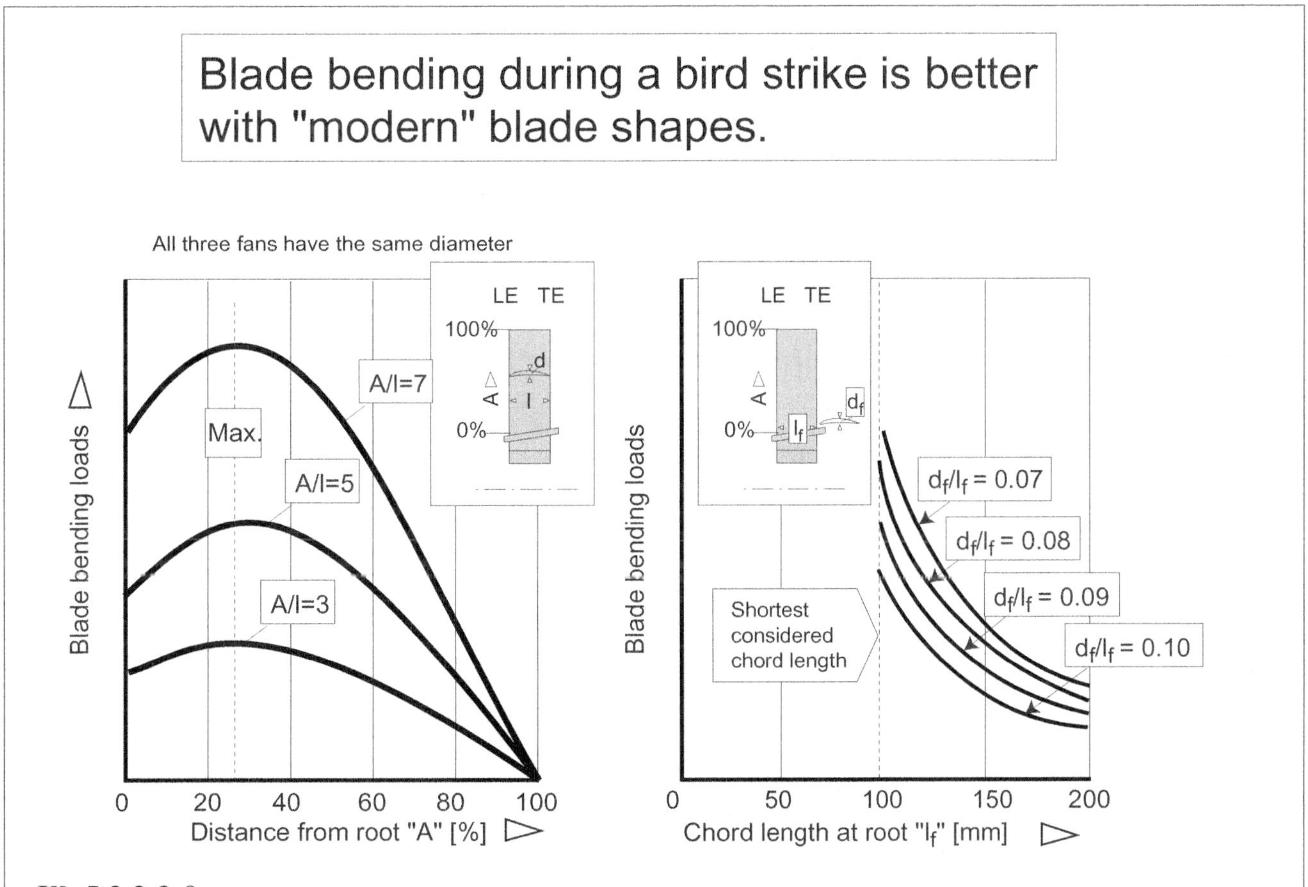

Ill. 5.2.2.3-9

Illustrations 5.2.2.3-8 *and* ***5.2.2.3-9*** *(Ref. 5.2.2.3-3): The stator and rotor blade zones most* ***sensitive to FOD and bird strike*** *damage are:*

- the leading edges of the blade, which are extremely thin for reasons of weight and aerodynamics (supersonic flow speed at the blade tips). The ***edges fail by bending or fracturing*** *(Ill. 5.2.2.1-9).*

- the area of the blade ***near the root*** *(most stressed area is usually a little* ***above the root platform****, in Ill. 5.2.2.3-9 a little above 20% of blade length). If the blade is already fairly wide at this point, then the most sensitive zone is not directly at the blade root. The blades are overstressed by bending loads, plastically deformed, and may fail due to dynamic fatigue in case of further operation.*

As would be expected, blades with thick profiles with a large inlet edge radius and the maximum thickness located far towards the leading edge sustain less damage during bird strikes than modern blades with slim, sharp-edged high speed profiles. With regard to the ***blade bending stress*** *(Ill.5.2.2.3-9), it decreases with wider chords and shorter blade lengths.*

However, the leading edge stress may be different (Ill. 5.2.2.3-8 top left and right). This stress is directed by the torsion loads when the foreign object strikes the leading edge area. In this case a ***wide chord*** *can be a disadvantage, as practical experience has shown (Ill. 5.2.2.1-10 bottom).*

Blades with clappers can have considerable advantages in their bird strike behavior when compared with wide chord blades.

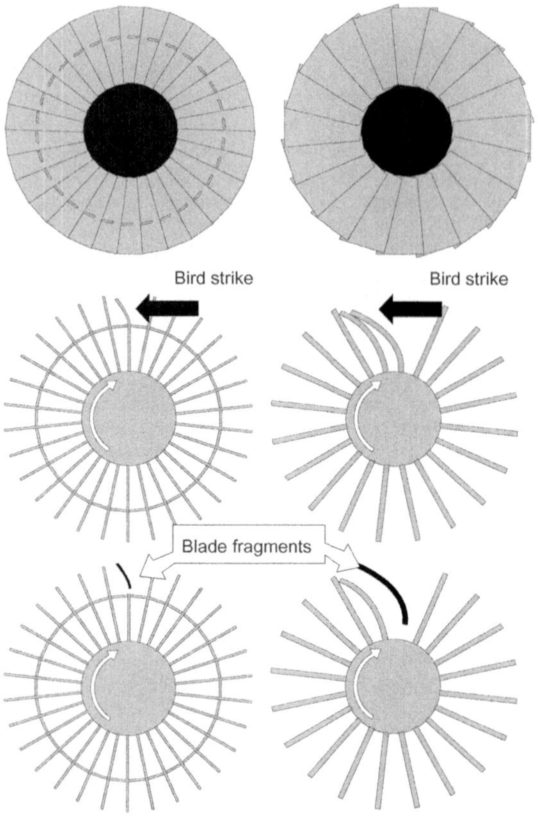

The disadvantages of wide chord fan blades during bird strikes:

- The blades do not brace each other
- The blade loads in the root area are relatively high
- This can result in large fragments which cause more serious consequential damages
- Serious imbalances require stronger bearings or special measures
- The containment must be stronger (increased housing weight)

Ill. 5.2.2.3-10

Illustration 5.2.2.3-10: *Practical experience (see Ill. 5.2.2.1-12) has shown that the relatively narrow* ***fan blades with clappers*** *that are used in older engine types have considerable advantages over* ***wide-chord fan blades without clappers****. If a large bird strikes the blade tip area, then the clappers act as braces and distribute the bending loads across several blades. In addition, the lever arm to the braced area is relatively small, which means that overloads result in the creation of relatively* ***small fragments****, which can pass through the bypass without threatening the core engine.*

Due to the small resulting fragments (and also the thinner and narrower blades), the resulting ***imbalances are easier to control****, making overstress of the main bearings less likely. In order to safely control the large imbalances caused by fractured wide-chord fan blades and prevent unallowable damages (such as overstressing of the engine suspension), it may be necessary to take elaborate measures to construct the bearings to behave elastically, at least in case of overstress.*

The weight of the casing in the fan area is largely determined by the containment, which must prevent the escape of blade fragments. The relatively small and low-energy fragments of blades with clappers require correspondingly ***lighter containments****, which is an advantage with regard to the total engine weight, especially in big fan engines with large bypass ratios.*

However, despite this advantage, modern fan engines exclusively use wide-chord blades for aerodynamic reasons.

Illustration 5.2.2.3-11: *Experience has shown that so-called* ***swan-neck ducts*** *in intermediate compressor casings can behave poorly in case of bird strikes (also see Ill. 5.2.2.3-12). These casings are used in* ***multi-shaft engines*** *and direct the compressor air flow of the core engine into the high-pressure compressor zone. Although birds are sliced into small pieces by the blades of the fan and/or low-pressure compressor before reaching the swan-neck duct, these small pieces*

Bird strike sensitivity caused by S-shaped flow ducts in compressor housings.

Ill. 5.2.2.3-11

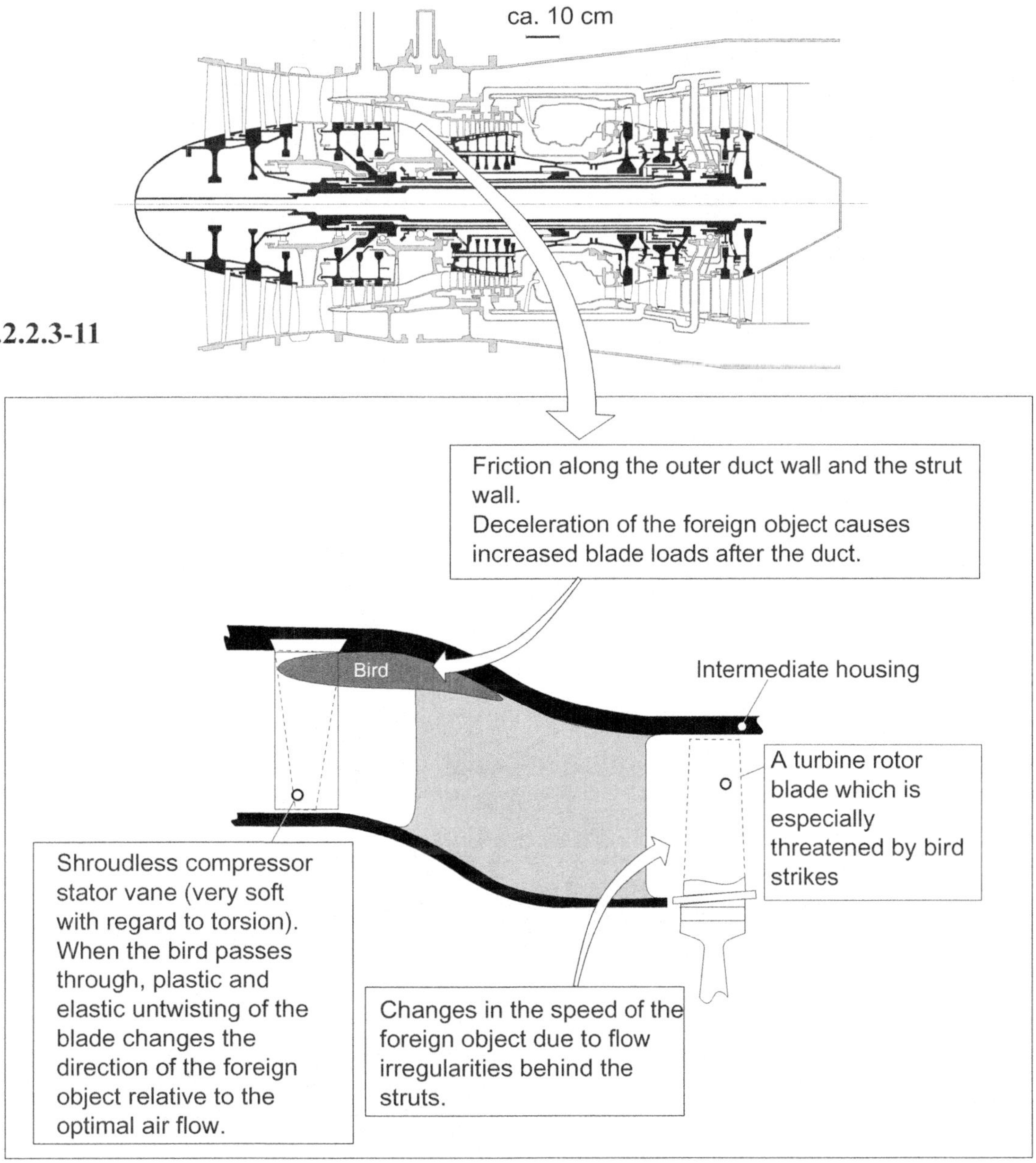

*can **collect on the struts** to the outer wall of the s-duct and enter the **high-pressure compressor** as a larger mass. Also, the mass travels slowly and is concentrated at the tip region of the rotor blades. This creates conditions for a **low-speed impact**, resulting in **maximum bending loads** (Ill. 5.2.2.2-7). This process can easily overstress filigreed rotor bladings. For this reason, it is recommended that at least the first rotor stage of the high-pressure compressor be made especially robust.*

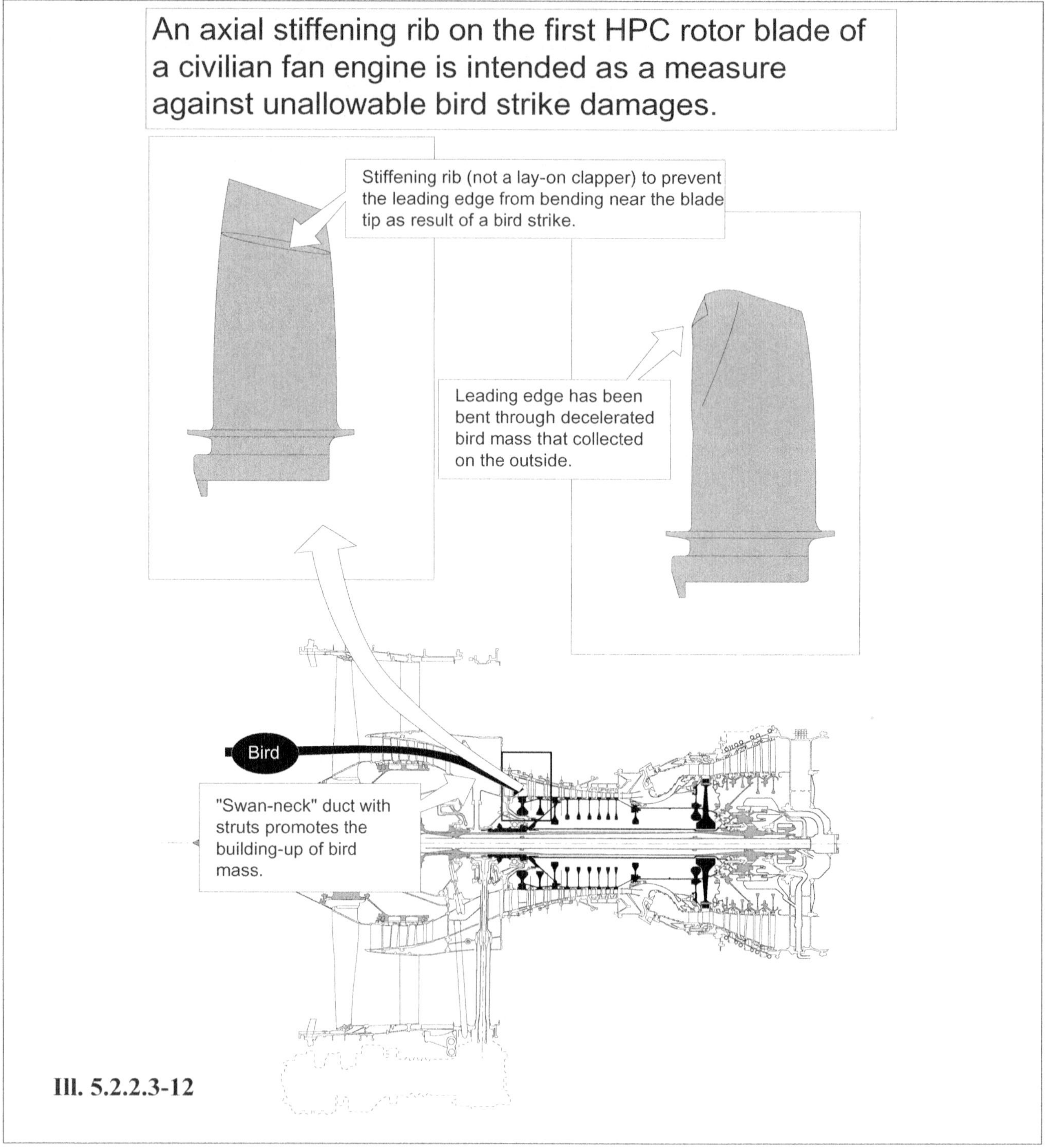

Ill. 5.2.2.3-12

Illustration 5.2.2.3-12: ***If pieces of a bird collect in a s-duct****, on casing struts, and/or on stator blades and are decelerated, it can cause extensive damage to the blades of the rotor stage immediately downflow (also see Ill. 5.2.2.3-11). In this case in a modern fan engine for use in passenger aircraft, the leading blade edges were so deformed near the tips that an axial stiffening fin became necessary. The seriousness of the problems can be seen in the acceptance of the flow disturbances caused by the poor aerodynamics of the fin.*

Chapter 5.2.2.3 References

5.2.2.3-1 D.Smith, ""Blue Angel's Maintenance Team", periodical "Aviation Maintenance", February 1999, page 35.

5.2.2.3-2 A.J. Tudor, "Bird Ingestion Research at Rolls Royce", proceedings of the symposium "The Mechanical Reliability of Turbo-Machinery Blading". April 1-3, 1968.

5.2.2.3-3 R.S.Cox, Rolls-Royce Limited, "Bird Ingestion Problems Relating to Gas Turbine Engines", proceeding paper Sept. 1969.

5.2.2.3-4 H.L.Hillebrand, "Kommt ein Vogel geflogen", periodical "Flug Revue" 2/1990, pages 95 to 98.

5.2.2.3-5 periodical "Flug Revue" August 1997, pages 48 and 49.

5.2.2.3-6 periodical "Air International" October 1990, pages 206 and 207.

5.2.2.3-7 D.A. Dulghum, "Stealth Engine Advances Revealed in JSF Designs", periodical "Aviation Week & Space Technology, March 19, 2001, page 90.

5.2.2.3-8 „Republic F-84, Startschwierigkeiten", Zeitschrift „Aero", 1957, Heft 61, page 1753.

5.2.2.3-9 Z.S. Palley, I.M. Korolev, E.V. Rovinsky, „Structure and Strength of Aircraft Gas-Turbine Engines", Übersetzung FTD-HT-23-903-68 aus dem Russischen von „Foreign Tecnology Division", 1968, page 16 and 17.

5.2.2.3-10 D.L.Cook,"Development of the PW100 Turboprop engines", SAE Paper 850909 des „General Aviation Aircraft Meeting and Exposition", Wichita, Kansas, April 16-19, 1985.

5.2.2.3-11 I.G.Horeff, "Development or a Prototype Turbine Engine Inlet Device for Protection Against Bird Ingestion", Proceeding of the World Conference on Bird Hazards to Aircraft, Kingston, Ontario, Canada, 2-5 September, 1969, page 403-411.

5.2.2.3-12 J.Frischbier, „Multiple Stage Turbofan Bird Ingestion Analysis with ALE and SPH Methods", Herausgegeben von „American Institute of Aeronautics and Astronautics", 2005, page 1-9.

5.2.2.4 Measures Against Bird Strike Damage

• Sufficiently early discussion with authorities and operators to **determine the certification requirements** for bird strike resistance before beginning construction. Otherwise there is a danger of the alternatives being limited considerably. This includes the **interpretation of allowable damages** and influences on the engine.

• Sufficiently early, **realistic verification of the engine behavior** required during bird strikes.

• Optimization of the **fan area** (spinner geometry, fan blades, splitter, Ill. 5.2.2.4-1, Ref. 5.2.2.3-4, also see Ill. 5.3.3-3)

The sensitivity to bird strikes can be reduced by optimizing the geometric conditions of the fan area.

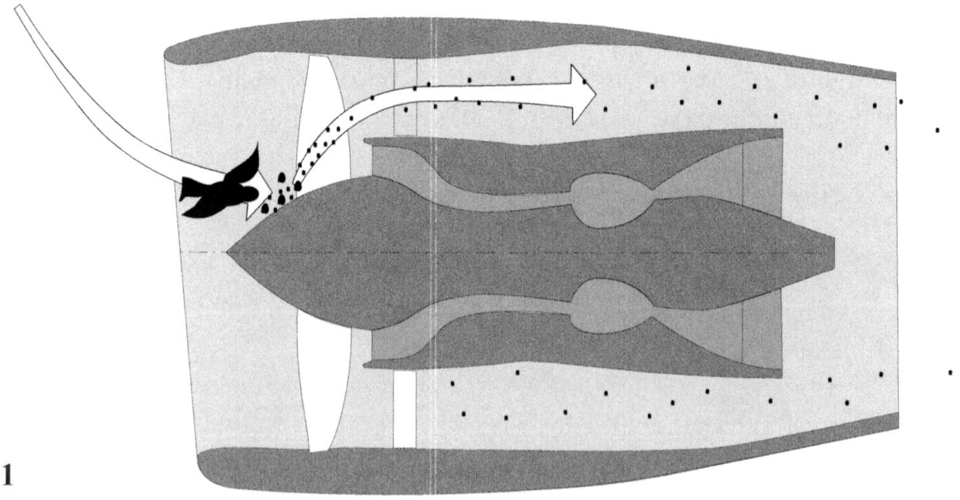

Ill. 5.2.2.4-1

• Constructive design of "**robust**" **blading**

- **Impact resistant materials** (high impact- and notch impact strength) for especially highly stressed engine blading. If this is not sufficiently safely possible (for example, in fan stators), one may consider materials such as fiber-reinforced synthetics, which may fracture, but will only cause minor consequential damages.

- Use of **thickest possible blade profiles** with the maximum thickness located towards the leading edge, **rounded leading and trailing edges**, and short **aspect ratio**; these demands do not correspond to the geometry of modern compressor blades with wide chords and supersonic profiles. Despite this, the demands posed by bird strikes should be given as much consideration as possible.

- **Wide spacing** between stages, tightly spaced blades.

- **Bracing** long, slender (fan) blades with **clappers**, if possible set on alternating sides to further improve the (slicing effect) bird strike resistance

- Resilient, heavily **damped rotor blade fastening**

- Stator blades/**vanes** that are **as stiff as possible** and fastened together in **rings or ring segments**

- Hollow (e.g. for de-icing air) guide vanes should be **designed to prevent buckling**, through use of stiffening spars and a non-bursting trailing edge.

- Guide vanes in housings made from weaker materials such as light metals (Al and Mg alloys) should be fastened with more robust **hammerheads rather than dovetails** (Ill. 5.2.2.2-5).

- Carefully designing inner **bearings** and bearing connections of adjustable **guide vanes.**

- **Predetermined break points** should prevent closing and blocking of the blades in case of extreme overloading of the guide blade adjustment levers.

- **Burst protection rings** above fan rotor blades.

- **Sufficient ring-reinforcements** on the compressor casings to prevent bursting.

- **Preventing large amounts of flammable dusts** from being created due to imbalances cause by bird strikes; no flammable rub-tolerant coatings and blade materials in potential contact areas.

- If there is a risk of high-speed direct hits, there should be **no** inlet stator annulus or housing/ casing **struts ahead of the first rotor stage**.

- **Avoiding swan-neck air ducts**, including in the rear compressor area. If a swan-neck duct is unavoidable, at the least the first rotor blading immediately following the duct must be especially robust in order to prevent overstressing due to collected bird pieces.

- Air extraction bores and slits (, for cabin air, de-icing air, cooling air, etc.) should be designed in such a way that they will **not become blocked**.

- Rotor **disk connections** and rotor shaft teeth should be designed to be **shear-proof** in case of polar weight moments of inertia.

- With bladings made from titanium alloys, the **ignition of titanium fires** as consequential damage after a bird strike should be minimized through **suitable pairing of materials** (rub coatings) on potential rub surfaces.

- **Rotor disks** should be **sufficiently solid** near the annulus to prevent the annulus breaking off due to the high loads caused by a bird strike.

- Design of the **nose cone**; there are two "philosophies":

- Sufficiently **resistant rotating** (spinner) **or fixed nose cone** (Ills. 5.2.2.1-9 and 5.2.2.3-7, Example 5.2.2.1-2). If the cone is thin and "soft", then the impact force of a bird strike is decreased and the energy is absorbed by deformation. This may have the drawback of resulting in unallowable imbalances, damaging bearing stresses, and in extreme cases, failure of the bearings or casings. In this case bearings should be selected that are **suited for high imbalances**.
There are potential advantages to cones made from fiber-reinforced synthetics. When parts made from these materials fail, the resulting fragments are far less dangerous than fragments of metallic parts.

- Use of a s**tiff cone**; while this construction is robust with regard to the creation and **toleration of imbalances**, it is relatively heavy and its stiffness results in high impact forces that are transmitted into the **bearings and the supporting structure**, making these more likely to become overloaded and cause extensive consequential damages. In this case it must especially be ensured that the rotor bearings are sufficiently strong against axial impact forces resulting from bird stikes in the hub area.

- Sufficient **strength of the fan area in case of blade failure** and/or large imbalances. This includes continuous housing/**casing flanges** with **additional external tracks**, **oversized screwed flange** joints possibly with increased expandability (tension screws), and a strong housing/casing **against radial deformation**.

• Including the **inlet duct in the design concept for the bird strike resistance** of the engine. For example, the shape of the inlet duct should ensure that no direct hits of the bird on core engine are possible.

• Making use of devices such as **shutters or fences in the inlet duct** (Ill. 5.2.2.3-3) which stop and/or slice up the bird in a sufficiently safe manner, but do not endanger the engine in case of a failure/ fracture.

5.3 Erosion

5.3.1 Erosion Mechanisms

5.3.2 Erosion Damage

5.3.3 Measures Against Erosion Damage

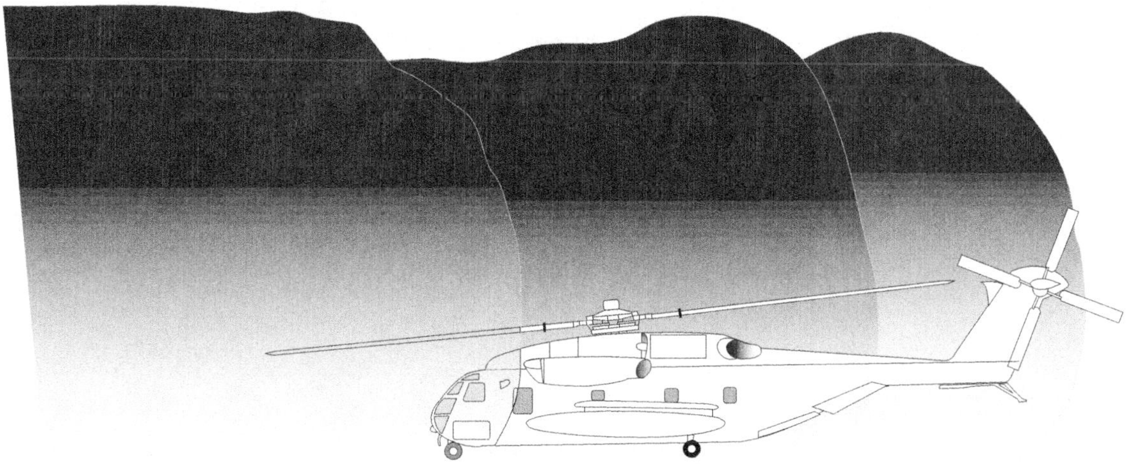

Erosion is a type of tribological stress in which a **solid body in flowing liquid or gas wears** engine parts in an abrasive, shattering, and tribochemical process. All of these **erosion mechanisms** can occur in turbine engines. This chapter focuses mainly on **wear caused by particles** in the gas flow.

Aside from erosive particles in the air and **gas flows**, particles in the **fuel** can also damage the engines. Even **fuel droplets** themselves can have an erosive effect when they strike a surface (Ills. 5.3.1-1 and 5.3.1-11.1). **Vapor lock** in the fuel flow can cause **cavitation** (Ill. 5.3.1-11.2). Overheated fuels can create microscopic coke particles that damage fuel system components such as fuel nozzles and can make them fail.

With regard to their ability to ingest particles along with the inlet air flow, engines are essentially large vacuum cleaners. However, other erosive **particles may originate in the engine itself** (Ill. 5.3.1-2) and cause erosion damage. These particles can damage engine parts in many different ways, especially through their abrasive qualities. Wherever there is an unfiltered air flow, erosion damage can occur. One of the components most heavily stressed by erosion is the compressor with its **blading** (Ill. 5.3.2-2).

Ingested particles that can erode the blading are primarily **dust**, **rain drops** (see Chapter 5.1.1), and **hailstones** (Chapter 5.1.2). Among these, **dust erosion** damage is the most frequent. It occurs in all stages, decreasing towards the rear of the engine. Rain and hail damage is only to be expected in the front stages, especially in fan engines with large bypass ratios, which usually do not have long air ducts ahead of the fan. Intense erosion conditions are present with engines on cargo aircraft when using thrust reversers, which throw up dust from the runway. A critical erosion problem in fan blades is likely to occur especially when **fiber-reinforced materials** with a synthetic matrix are in serial production. In this case, suitable protective coatings are a prerequisite for successful long-term serial use.

Engines in aircraft that frequently operate at low altitudes, such as **helicopters**, are at especially high risk. Large amounts of dust may be thrown up and affect the engines of **VTOL aircraft** during takeoff and landing. Heavy dust and rain erosion can also be expected in the engines of military aircraft that are frequently operated at high speeds near the ground.

As mentioned above, the erosive effect of ingested particles decreases in the rear compressor stages. This can be explained by the **bursting of the particles** when they strike the blades of the front stages. Therefore, no significant direct erosion of those dust particles is to be expected in the turbine area. However, this is not the case when the erosive particles are created in the rear part of the engine. These particles are usually **spalled ceramic thermal barriers** and **coke particles** from the combustion chamber. A special sort of 'erosion' is **triggered by dust melts** (in **desert** environment or **vulcano ash**) which **stuck and react at the hot parts** and accelerate the wear by spalling markedly (Ill. 5.3.2-12.3 and Ill. 5.3.2-15).

5.3.1 Erosion Mechanisms

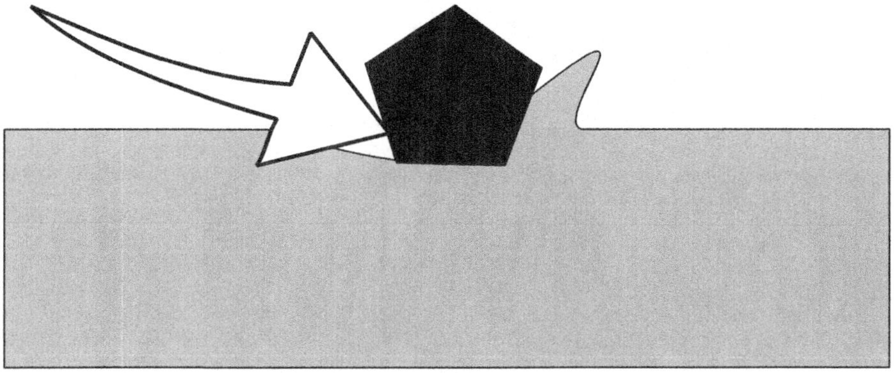

The **erosion mechanisms** in an engine can be influenced both microscopically and macroscopically in a characteristic manner by the locally **effective parameters**. For example, depending on the angle of attack of erosive particles, erosive wear can occur primarily through cutting or splintering mechanisms (Ill. 5.3.1-3).

The damage symptoms understandably depend on the many parameters of the erosion process. These include the **particle size** and **type**, **impact speed**, **impact angle**, **number** of particles, and the **duration** of the particle influence. In addition, the extent of the damage is also dependent on the **properties of the erosion-stressed part**, such as **material** properties (strength, ductility, oxidation behavior, etc.), **surface topography** (size and geometry of roughness), and **geometry** (such as the leading edge radius of blades), to name a few important factors.

It is easy to see that an erosion process is a combination of extremely complex single processes that must be understood.

The dominant erosion mechanism in engines is **abrasion** through **impacting particles**. This abrasion mechanism can occur in various ways, depending on the particle energy and the properties of the part surface. With ductile materials micro-level chip removal can be expected. Hard, brittle materials can suffer micro-level fatigue processes resulting nicks and disruptions, and brittle materials and coatings can spontaneously splinter out.

Other types of wear in which the dominant wear mechanism is surface disruption, such as **rain erosion** (**droplet impact**) or **cavitation** (erosion in liquids due to vapor bubble implosion), are briefly covered here (Ill. 5.3.1-1). Another type of erosion that has not been recognized as being damaging until now should also be mentioned (referred to here as "**fine erosion**"). It is usually observed at the micro-level on the back sides of blades and lays bare the microstructure (especially in titanium alloys) in a manner similar to etching (sputter effect of the air molecules?).

With erosion processes in hot parts, additional mechanisms are involved that are based on the **oxidation of blank metal surfaces** (erosion surfaces) or on **reactions with the base material** or the **oxide layers** (such as sulfur particles in gypsum dust). These mechanisms can significantly increase the wear rate and damage caused by otherwise relatively weak abrasive erosion processes.

Materials can behave differently when subjected to different erosion mechansims.

Comparison of the erosion strengths of various materials against sand and rain erosion under equal erosion energy
(Re: W.Kayser)

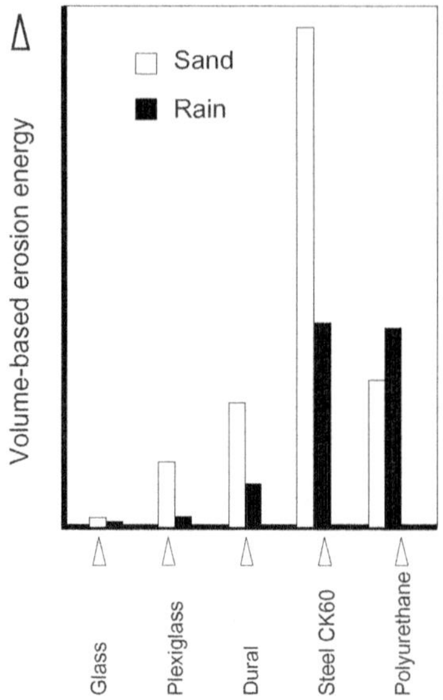

Relationship of the erosion rates of rain erosion and cavitation
(Re: H.Rieger)

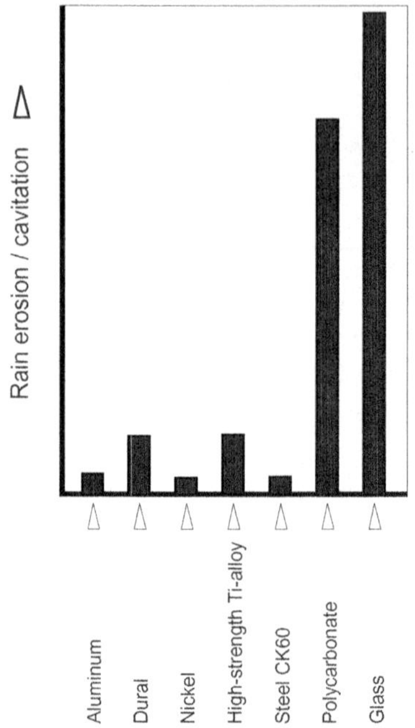

Ill. 5.3.1-1

Wear type	Tribologic loads		Mechanisms	Examples of erosion-stressed engine parts
Blasting wear - Glide plane wear - Angled blast wear - Colliding blast wear Sand and dust erosion Mineral wear Abrasive wear	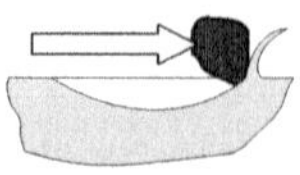	Cutting process Splintering of the surface zone Fatigue crack initiation	Abrasion	Compressor blades
Droplet impact (Rain erosion)	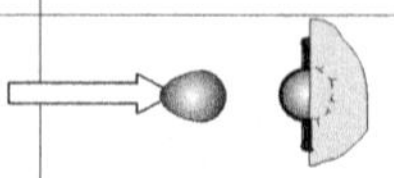	Shock waves: Deformation and fatigue with crack initiation	Surface disruption, tribo-chemical reactions	Fan blades
Cavitation Flow cavitation Vibration cavitation	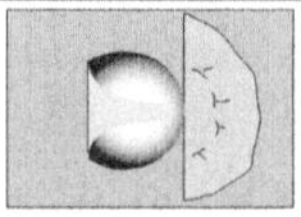	Imploding vapor bubble creates shock waves: deformations and fatigue with crack initiation	Surface disruption, tribo-chemical reactions	Regulator parts Friction bearings Injection nozzles Fuel pumps
Erosion through a gas stream	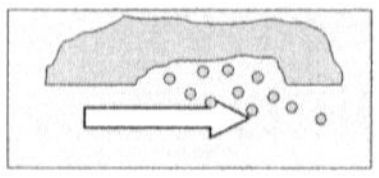	Crumbling of brittle and porous coatings due to gas forces in the flow	Abrasion, surface disruption, tribo-chemical reactions	Spray coatings in housings and labyrinths
It is possible for several wear types to combine				

***Illustration 5.3.1-1:** If **erosion** is understood to **mean surface wear** caused by the mechanical action of a media, then there are many different mechanisms which fit this definition (bottom diagram).*
*Erosion through solid particles is generally referred to as **peening wear**. Other terms that are frequently found especially in older literature, but are not in DIN 50320, include **mineral wear, sand and dust erosion, and abrasive wear**. The material being eroded through the impact of hard particles is experiences slice/furrow-like chip removal. With hard materials, and especially hard coatings on softer, the wear occurs primarily as splintering and/or fatigue crack formation, as well as the breaking out of particles. This type of erosion process occurs primarily on the aerodynamic surfaces of the compressor and turbine bladings. Abrasive wear does not occur solely through the air flow, but can also be caused by hard particles carried in liquids. A typical example is coke particles in an overheated fuel flow that can cause extensive erosion damage to the injection system.*
***Droplet impact** always occurs as an erosive load when droplets of liquid strike a surface at high speeds. If the liquid is rain, then it is referred to as **rain erosion**. This process creates shockwaves and deformations in the material which, in case this process is frequently repeated, lead to fatigue and breaking-out of the material. This type of wear can occur in rotor blades and fan outlet guide vanes.*
***Rain erosion can be a serious problem for compressor blades (fan blades) made from fiber-reinforced plastics/synthetics** (Ref. 5.3.1-19).*
*With **cavitation**, the erosion process takes place in a **flowing liquid**. If, in accordance with Bernolli's law, pronounced underpressure zones occur in a liquid that result in the formation of vapor bubbles, the latter can implode if the flow carries them into an area with higher pressure. The damage model is a "liquid stinger" that is created in the imploding vapor bubble and strikes the part surface at high speed, creating a damage mechanism similar to droplet impact (Ills. 5.3.1-11.1 and 5.3.1-11.2). This type of wear, which is primarily based on fatigue of the surface, is also referred to as "**cavitation erosion**". In watery media, the cavitation process is often increased by simultaneous corrosion (**cavitation corrosion**).*
*Cavitation damages have been recorded on **components of the fuel system** such as injection nozzles and regulators as well as on the pump gears of turbo pumps.*
*Cavitation has also been blamed for damages to **gear pumps and transmission gears** in high altitude flights (easier vapor bubble formation due to the low air pressure).*
*Rain can cause considerably heavier erosion damage than cavitation (top right diagram). The extreme sensitivity of **polycarbonate** and **glass** to droplet impact is especially conspicuous.*
*In the same way that strong wind throws up sand, an extremely intense **gas flow**, especially in the rear compressor area, can cause **particles to break free from soft coatings** such as the **abradable coatings** in compressor casings (e.g. porous nickel-graphite spray coatings). This process can be self-increasing, since these particles, in turn, have an erosive effect.*

*The top left diagram shows that materials can **behave considerably differently** during erosion through solid particles than they do during erosion through drops of liquid (Refs. 5.3.1-1 and 5.3.1-2). For example, it can be seen that soft, elastic, very ductile polyurethane has a high erosion resistance against sand and rain (compared with plexiglas or aluminum, etc.), which makes its use as a **protective coating against erosion** on fan parts (rotating nose cone, fiber-reinforced synthetic outlet guide vanes, etc.) understandable. The great resistance to rain erosion relative to the resistance to sand erosion cannot be observed in any other (harder) materials.*

Illustration 5.3.1-2: *Many different* ***particles*** *can have an* ***erosive effect*** *in engines. These particles may be ingested with the inlet air, be created indirectly by erosion, or form inside the engine in some other way.*

Erosion particles in the inlet air "E1": *These are usually dust from the surrounding atmosphere. There is a correspondingly large variation in the size, consistency, and composition of the dust particle.*

Wear products from housing/casing coatings "E2" and "E3": *In the fan and compressor area, tight tip clearance is maintained through the use of rubbing systems. In order to prevent the blade tips from being unallowably damaged or worn, relatively soft abradable coatings are used on the housing side. However, the macroscopically soft behavior during rubbing does not mean that all of the coating particles are also soft. The coating often consists of hard particles bonded into a soft matrix, and the abradable behavior occurs by these particles breaking out at the micro-level. Ingested particles that blow over these coatings in concentrated amounts can erode these coatings, and the erosion products can further increase the total process in the following stages.*

Wear products from the blading "E4": *With compressor blades with hard coatings (e.g. TiN erosion protection), if the kinetic energy of the particles is too great, the thin coatings may splinter. In turn, the extremely hard coating particles can have an abrasive effect. If blades with tip armor are used, then particles of the armor can break out and increase the erosion stress.*

Wear products from rotor spacer rings "E5": *In order to minimize the tip clearance between the stator vanes and the rotor, hard ceramic rub-tolerant coatings are used on the spacers (Ill. 5.3.2-10). If these coatings spall, the resulting particles can even damage the blading near the rotor.*

Soot and coke particles from the combustion chamber "E6": *Coke and soot feel so soft, that dangerous erosive effects are not usually expected. However, experience has shown that these particles can cause unallowably serious wear in the hot part area in a relatively short time (Example 5.3.1-1, Ill. 5.3.2-12). In the combustion chamber, coke can collect on the surfaces of the walls and the injection system and break off as erosive particles. The formation of soot during the combustion process in the gas flow can also lead to heavy erosion of the turbine blading. Erosion processes on hot parts create fresh metal surfaces and alternatingly accelerate oxidation.*

Particles of combustion chamber coatings "E7": *The walls of modern combustion chambers are increasingly being coated on the inside with* ***ceramic thermal insulating layers*** *and* ***thermal barriers*** *in order to minimize the required cooling air. These coatings can spall or their surfaces may crumble out, causing the abrasive particles to threaten the turbine blading, especially.*

Particles of the rub coatings of the outer seal ring of the HPT "E8": *In order to minimize the tip clearance at the high-pressure turbine rotor blades, the housings are outfitted with seal segments that are coated with a hard ceramic rub coating (usually zirconium oxide) that the blades can "grind" into. Abrasive particles of these coatings can be created in a short time due to rubbing or over longer periods through "crumbling out", causing damage to the rear hot parts.*

Labyrinth wear products "E9": *In labyrinths, rub coatings can create particles either through rubbing or erosion. Even the armor on labyrinth*

Typical erosive particles and where they originate in an engine.

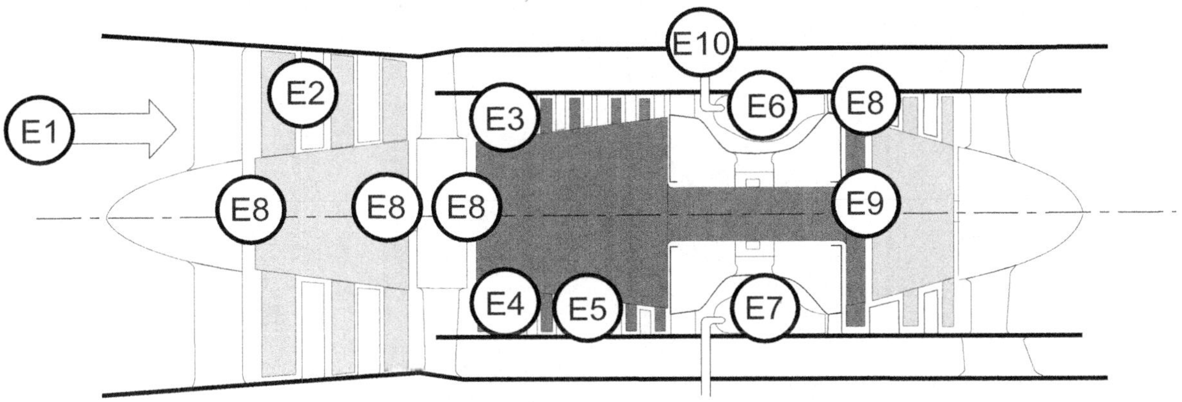

E1 Erosive particles in the intake air

E2 Wear products from housing coatings in the fan/LPC (abradable coatings)

E3 Wear products from housing coatings in the HPC (abradable coatings)

E4 Wear products from spacers

E5 Wear products and coating particles from the blading

E6 Soot and coke particles from the combustion chamber

E7 Particles from combustion chamber coatings (thermal barriers)

E8 Particles of the outer seal ring of the HPT (thermal barriers)

E9 Labyrinth wear products

E10 Coke particles in the fuel

Ill. 5.3.1-2

tips can spall and allow the base material to be worn. These particles can have erosive effects in the labyrinth itself or in other areas of the engine (Ill. 5.3.1-8)

Coke particles in fuel "E10": *If fuel temperatures are too high, the* ***fuel may decompose****, creating fine, erosive* ***carbon particles****. These particles can damage fuel nozzles and change the angle of the fuel stream so it burns through the combustion chamber and the combustion chamber liner.*

External Factors
Erosion: Mechanisms

Understanding the acting mechanisms during an erosion process makes designing measures against damage easier.

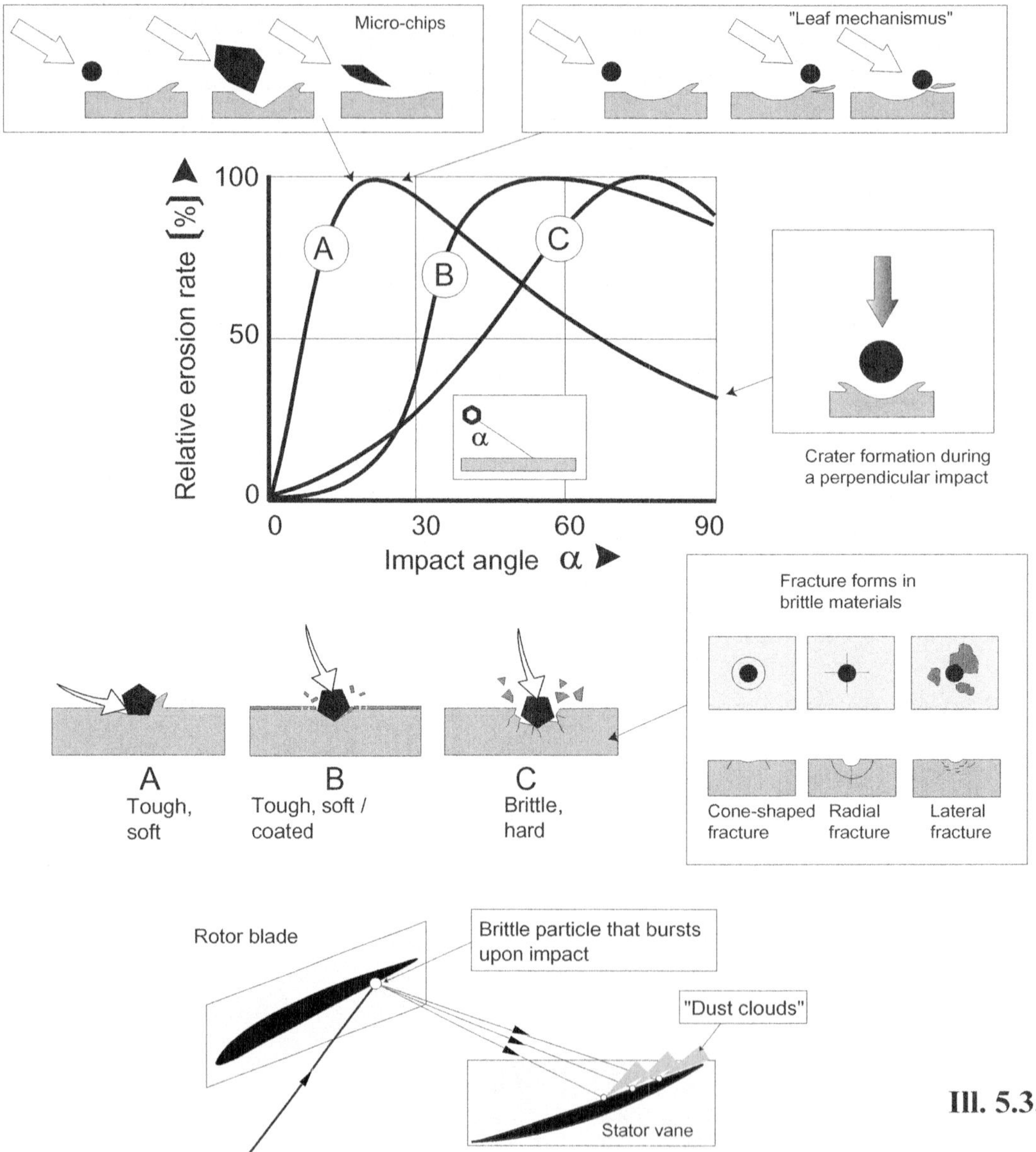

Ill. 5.3.1-3

In the rear compressor stages, the impact energy of the individual particles considerably decreases due to comminution.

Illustration 5.3.1-3: *The* ***angle of attack*** *is especially important for the wear effect of* ***peening wear****. The extent of damage depends largely on the angle of attack and the ductile or brittle behavior of the material (top diagram).* ***Brittle materials such as glass*** *(curve C) exhibit a wear sensitivity that increases almost linearly with the angle of attack. Due to the large energy transmitted (shockwaves), these materials fail by shattering if the erosion particles strike them at a steep angle. The wear process includes micro-crack initiation (middle detail). At flatter angles and high hardness, these materials are highly resistant to slicing mechanisms.*

Ductile materials (curve A) such as ***titanium alloys****,* ***martensite heat-treated steels****, and* ***aluminum alloys*** *exhibit a steep increase in wear until 25°, but then the wear rate drops almost continuously until the angle of attack is roughly 90°. This is understandable from the* ***chipping wear mechanism*** *(top diagrams) at the micro-level: when the angle of attack is flat, the cutting process is especially effective on the soft materials, while particles striking the surface at steep angles bounce off without cutting. This characteristic is used, for example, when lining blasting cubicles with* ***rubber coatings****. For this reason, soft erosion protection coatings, such as PU coatings on the rotating nose cone, are especially suitable for surfaces with relatively steep angles of attack.*

Curve "B" is a material system with a hard, ***brittle coating*** *(e.g. TiN, see* ***Ref. 5.3.1-3****) on a relatively soft base material (titanium alloy, etc.). In this case a sufficiently large energy transfer between the particle and surface will shatter and* ***break through the surface****. This causes increase erosion wear. Because the energy transfer in extremely hard surface coatings (no cutting) increases with the angle of attack, the destructive process mentioned above occurs relatively suddenly above a discrete threshold value (sudden rise in the curve).*

Because the erosion process is influenced by the ***energy transfer****, it can be concluded that curve "B" is also determined by the* ***size of the erosion particles*** *and their* ***impact speed****.*

If mineral particles (larger than 0.010 mm , Ill. 5.3.1-9) strike a surface, they may burst. The low energy of the smaller resulting particles (due to their small mass) means that ***thin, hard coatings can be used*** *as erosion protection* ***in the rear area of compressors****. These coatings will not be destroyed by the low-energy particle fragments.*

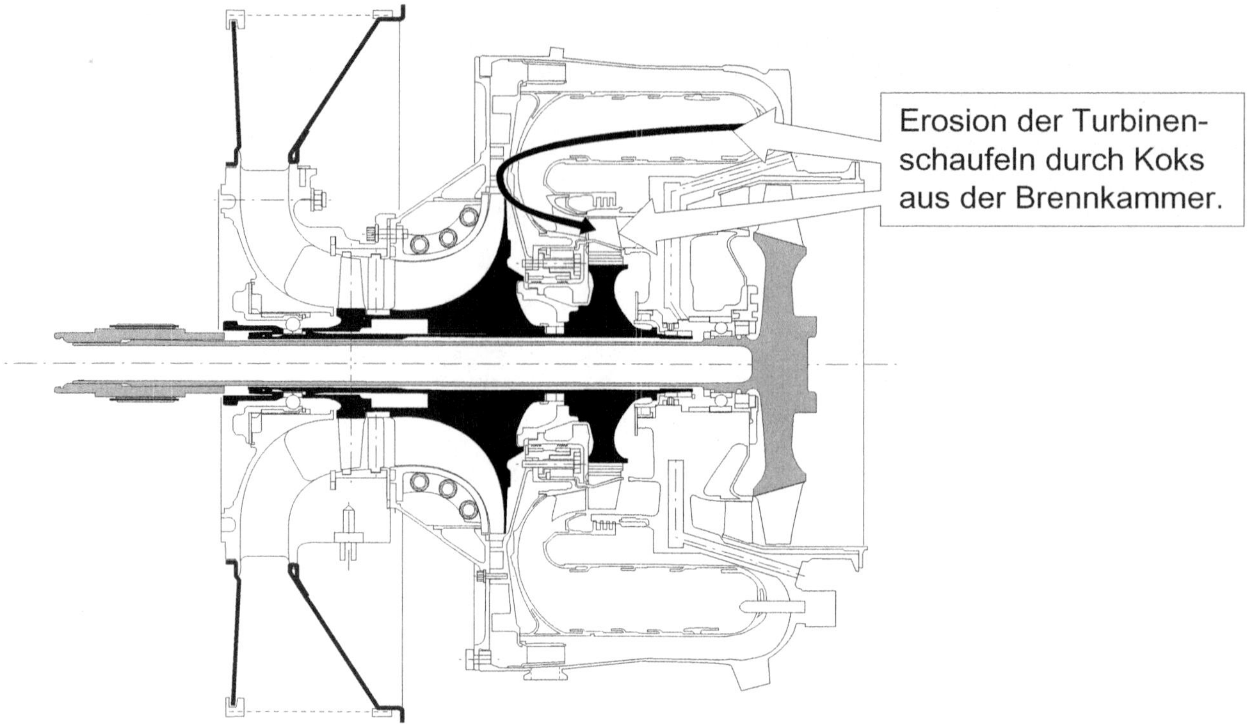

Example 5.3.1-1 (Ref. 5.3.1-16, Ill. 5.3.2-12.2):

Excerpt: "...To further reduce the... unscheduled removal rate,...engineers are examining at least two additional improvements- a **low coke combustor** and...

Officials expect the new **combustion liner** to eliminate blade erosion caused by carbon buildup, increasing blade longevity. The new liner also will cut visible smoke and increase liner life, since the effusion cooled liner eliminates hot spots and improves the distribution of temperatures in the component. Company engineers have not yet fully quantified the improvement in component lives, but they believe blade live will be doubled or tripled."

Comments: This concerns a helicopter engine with a reverse-flow combustion chamber. The considerable decrease in the life span of the rotor blading on the first turbine stage caused by the coke deposits is interesting (volume 3 Ill. 11.2.2.2-6.2).

The erosion behavior during abrasion understandably depends largely on the hardness of the particles and part surface.

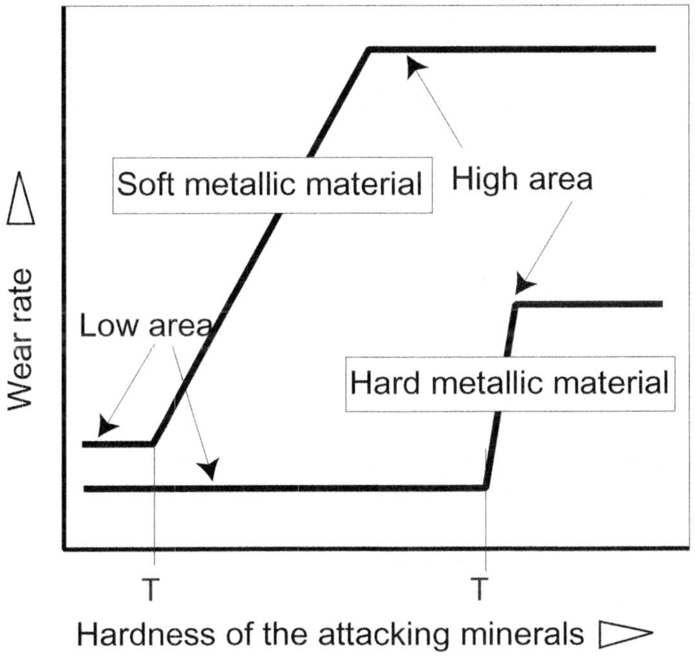

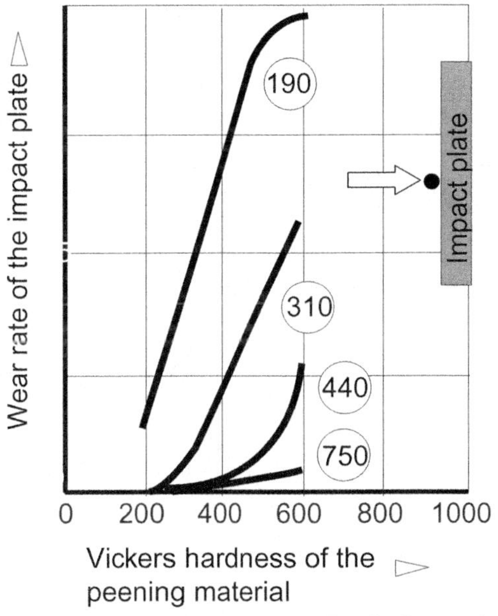

Ill. 5.3.1-4

***Illustration 5.3.1-4:** The **strength and hardness** of the **particle material** and wear-stressed metallic engine part, as well as the relationship of the hardness of the part surface to the hardness of the particles, have a decisive influence on the wear rate (top diagram; schematic). Up to a certain hardness of the impacting particles, relatively little erosion wear occurs (low area), depending on the hardness of the engine part. When this hardness is exceeded at point **T**, the **erosion rate** increases rapidly (high area). This behavior explains why apparently small differences in the hardness of the attacking minerals and/or the erosion-stressed material can lead to large differences in the erosion behavior. However, in the middle parts of the high and low areas, even relatively large differences in hardness may only have a minor influence on the wear rates.*

The erosion rate then remains relatively constant at this level.

*Relatively small differences in the chemical composition of the ingested dusts (such as proportion of SiO_2 -, Al_2O_3) strongly influence the wear rate through the proportion of hard particles. This behavior must be considered when **selecting erosion particles for tests**.*

*With soft ductile materials such as **rubber or PU coatings** the described effect does not apply, since the wear rate may even decrease with increasing particle hardness.*

*In the right diagram the described **erosion behavior of steel plates** with various hardnesses (numbers in the circles are the Vickers hardness of the test plate) at right-angle (pure) impact of a metallic peening material. It is easy to recognize the shifting of the steep increases, depending on the hardness of the test plate.*

Abrasive erosion is material-specifically dependent on the angle.

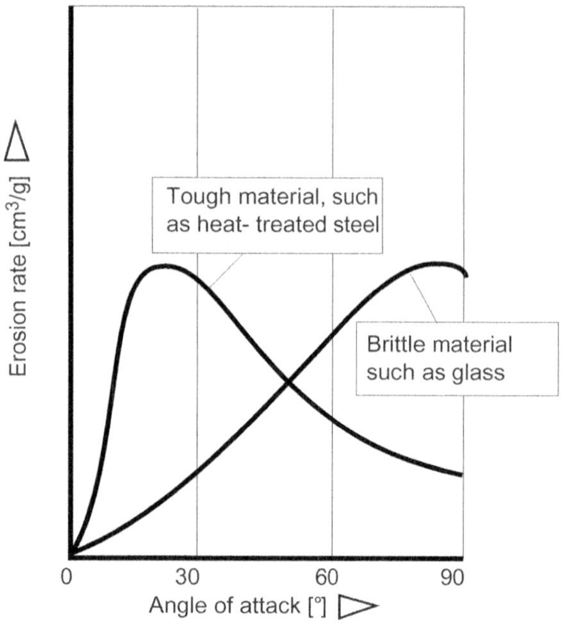

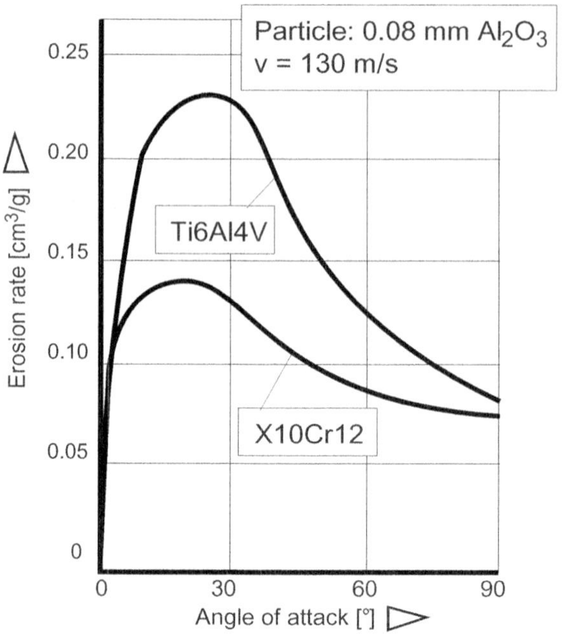

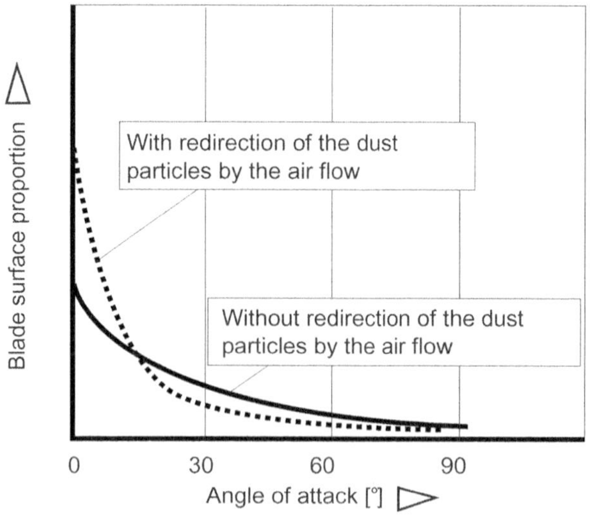

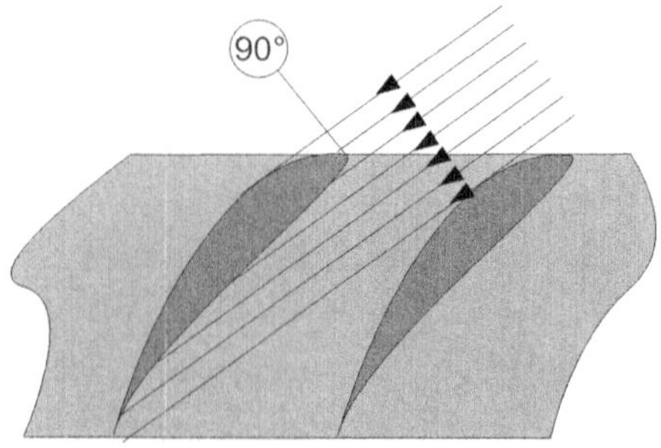

Schematic depiction of angles of attack on a stator vane profile of particles which have not been redirected by the air flow

Ill. 5.3.1-5

Illustration 5.3.1-5: *(Ill. 5.3.1-3) Depending on the* ***angle of attack*** *of the particles, the erosion behavior of an engine part made from soft and ductile materials is clearly and characteristically different from the wear behavior of brittle and hard materials. Typical "soft" blade materials in the compressor (top left diagram) are heat-treated steels (typically 13% Cr steels in older engines) and titanium alloys.* ***Titanium materials*** *are almost twice as sensitive to abrasive erosion as are steels. The erosion maximum seems to be slightly shifted towards greater angles of impact. Considering that the blade profiles of modern compressors have considerably sharper edges than the steel blades in earlier engines, then it must be assumed that the erosion sensitivity has also increased accordingly.*

Because the impact angles of particles on the blade profile vary greatly (bottom diagram), the erosion behavior of blades with the same shape but different materials can be expected to be different. Blades made from ***fiber-reinforced synthetics****/plastics or with lacquer coatings, especially, should have considerably different erosion behavior than titanium alloys and coatings of brittle hard materials (e.g.* ***TiN****) on titanium alloy bases. The variation in behavior must be considered when developing and verifying the operating properties of bladings.*

Of course, the particle-carrying air/gas flow also has an important influence on the ***particle trajectories*** *and therefore also the impact angles. It is usually assumed that particles smaller than 0.010 mm will largely follow the air flow. For this reason, the leading blade edge should exhibit large impact angles in a small area, while the blade surfaces will be struck at very flat angles (dotted curve in bottom left diagram). However, larger dust particles are less and less likely to be deflected by the air flow, making them more likely to damage the trailing edge (bottom right diagram). In the rear compressor area, in which only small, already shattered particles occur,* ***hard, thin coatings can successfully protect*** *against erosion.*

According to the depicted erosion model, it can be expected that the trailing edge of the blade and the suction side of the blade are not subject to erosion stress. However, this is not always true. There have been many cases in which especially the ***trailing edges of stator vanes*** *show considerable damage from particle impacts or the suction side has significant erosion damage. This can be explained through ricocheting particles, such as sand grains that bounce off of neighboring blades and/or the blades of the following stage and are thrown forward (Ill. 5.3.1-9).*

Erosion behavior of a modern (titanium alloy) and an older (Cr steel) compressor blade material.

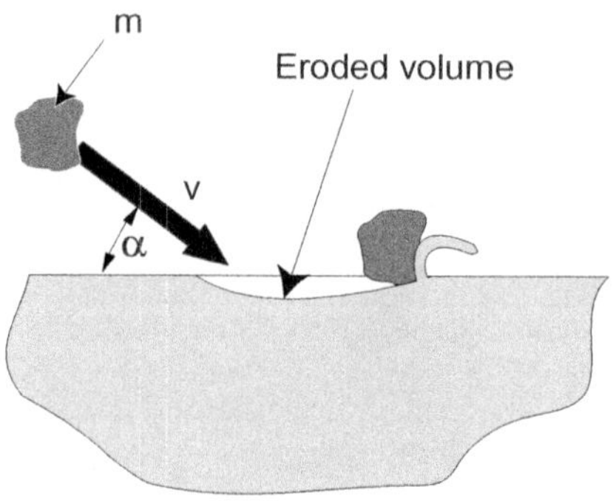

Term definitions:

Erosion rate $E = C.f(\alpha).\ m.\ v^2/\ \sigma$ ductile material
Erosion rate $E = C.f(\alpha).\ m.\ v^{3-7}/\ \sigma$ brittle material

C = constant of the erosion system
$f(\alpha)$ = function of the impact angle
m = mass of the dust particle
v = velocity of the dust particle
σ = lowest yield limit of the wear-stressed material operating temperature

Ill. 5.3.1-6

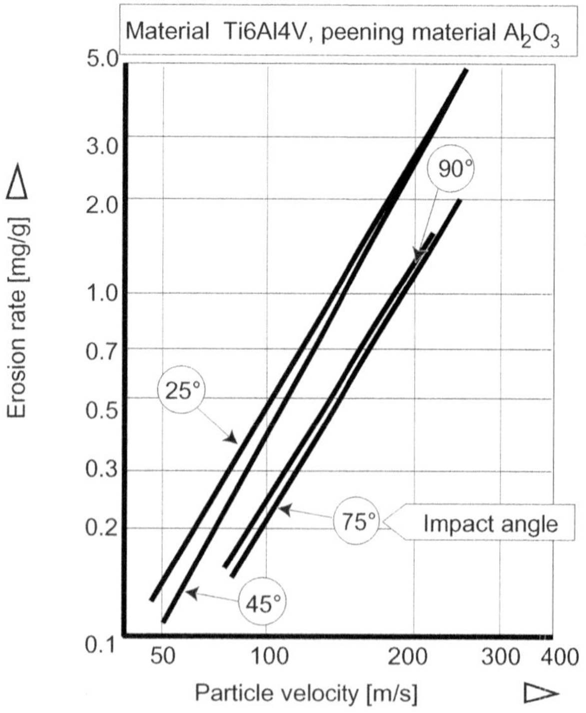

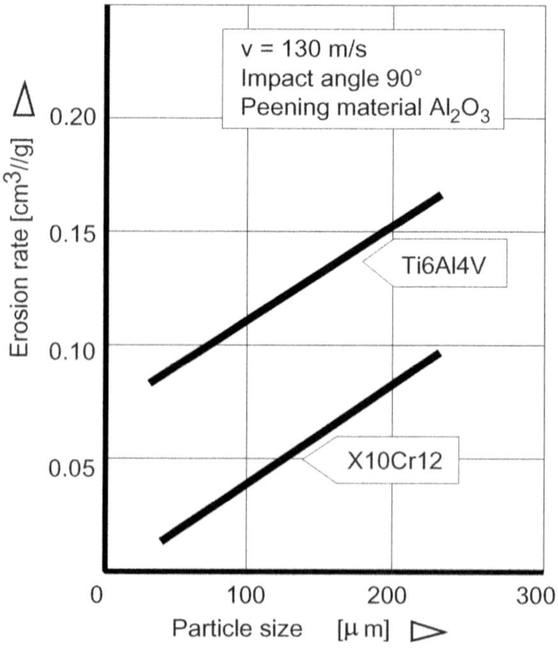

Illustration 5.3.1-6 *(Ref. 5.3.1-4): The* ***particle size*** *is important for the wear/****erosion rate*** *and the damage to the blades for several reasons (such as notch effect, dynamic strength, geometric changes to the edges):*

- At the same impact speed, the ***effective energy*** *of a heavier is correspondingly greater*
- The particle experiences ***less deflection*** *from the flow and has different impact angles than smaller particles (Ill. 5.3.1-5).*
- The relative ***speed perpendicular to the blade surface*** *(circumferential speed of the blade), i.e. the* ***impact speed****, should be higher than with smaller particles.*
-The ***impact craters*** *are considerably larger and* ***change the blade profile*** *(burr formation, craters), especially at the leading edge (Ill. 5.3.2-2).*

The impact speed of the particles on compressor parts is roughly that of the gas- and circumferential speeds, i.e. between 100 m/s and over 300 m/s.

According to the bottom right diagram and the relationship described above, at the same impact speed and a 90° impact angle, the erosion rate increases exponentially with the particle size (particle material Al_2O_3).
The ***particle speed*** *goes into the erosion wear exponentially (see top diagram and bottom left diagram). This dependency can be far greater in brittle materials than it is in ductile materials. The increase of the wear rate with increasing particle impact speed is independent of the impact angle. However, the absolute erosion rates clearly show the angle-dependency described in Ills. 5.3.1-3 and 5.3.1-5.*

Several materials show unexpected erosion behavior.

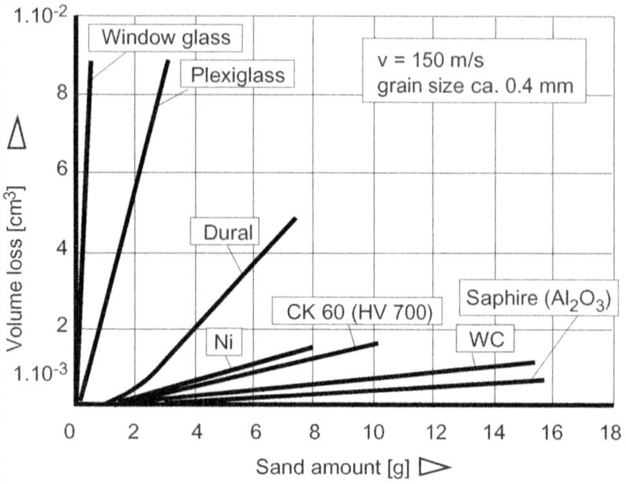

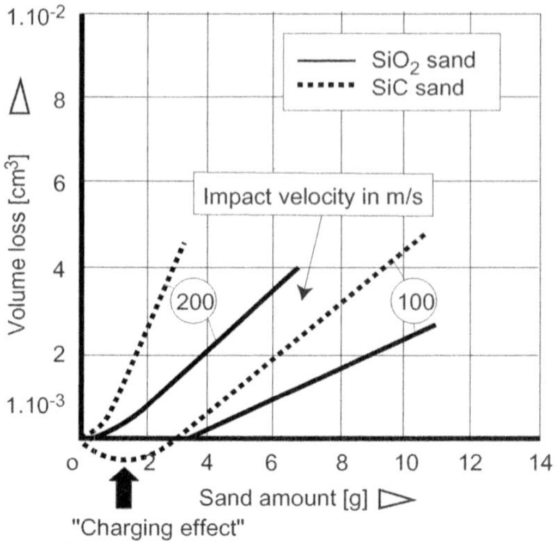

Ill. 5.3.1-7

Illustration 5.3.1-7 *(Ref. 5.3.1-1): Materials with apparently very similar* ***material parameters*** *(strength, hardness, ductility, etc.), can have very different erosion behaviors. For example, under the erosion conditions given in the top left diagram, the performance of window* ***glass is extremely poor*** *compared with that of sapphire, and also considerably worse than plexiglas ® (acryl glass) erosion. Therefore, it is understandable that enamel and* ***glass coatings*** *(enamel)* ***on compressors were unsuccessful****. On the other hand, pure nickel performs remarkably well under erosion stress, which explains why it is used as* ***erosion protection*** *on the edges of fan rotor blades made from* ***fiber-reinforced plastics****.*

*At the beginning of an erosion process at certain impact speeds, a slight weight increase in the erosion-stressed part can be observed (right diagram). This "****loading effect****" is due to erosion particles sticking in the part and seems to be less pronounced at higher impact speeds.*

Illustration 5.3.1-8: Erosion processes in and around labyrinths *are not as well known, but can certainly be damaging.*

Erosive labyrinth wear products from fin armor or abradable coatings can be blown out of the labyrinth (C2) and become caught in neighboring housing areas (A). This ***whirling dust film*** *can wear its way through metal/steel cross-sections that are several millimeters thick.*

Labyrinth wear products or dust-laden barrier air can become ***trapped between the labyrinth fins*** *and be whirled around (C1). This can seriously erode the relatively soft abradable coatings across from the labyrinth fins (B).*

Erosion is an important damage mechanism in labyrinths.

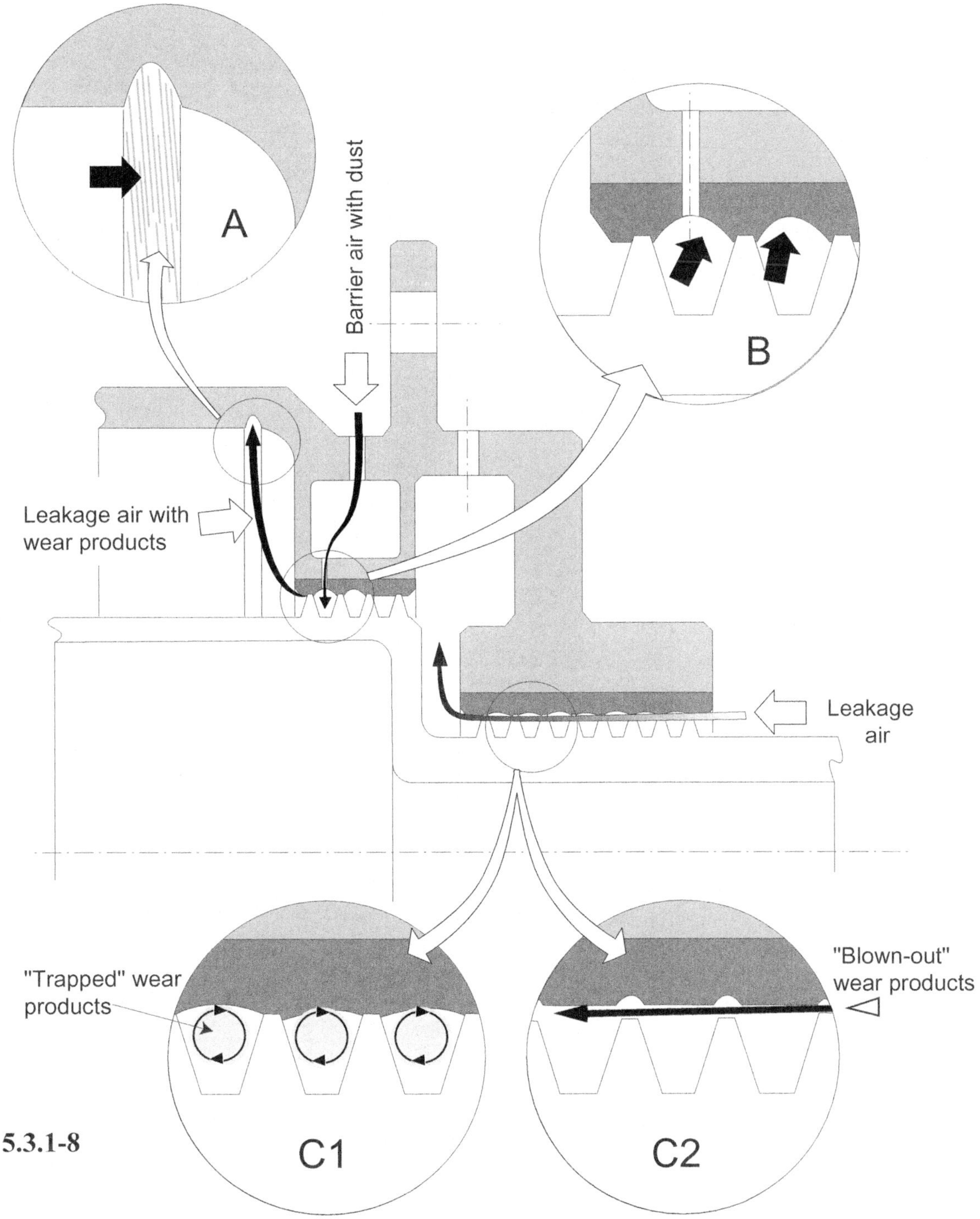

Ill. 5.3.1-8

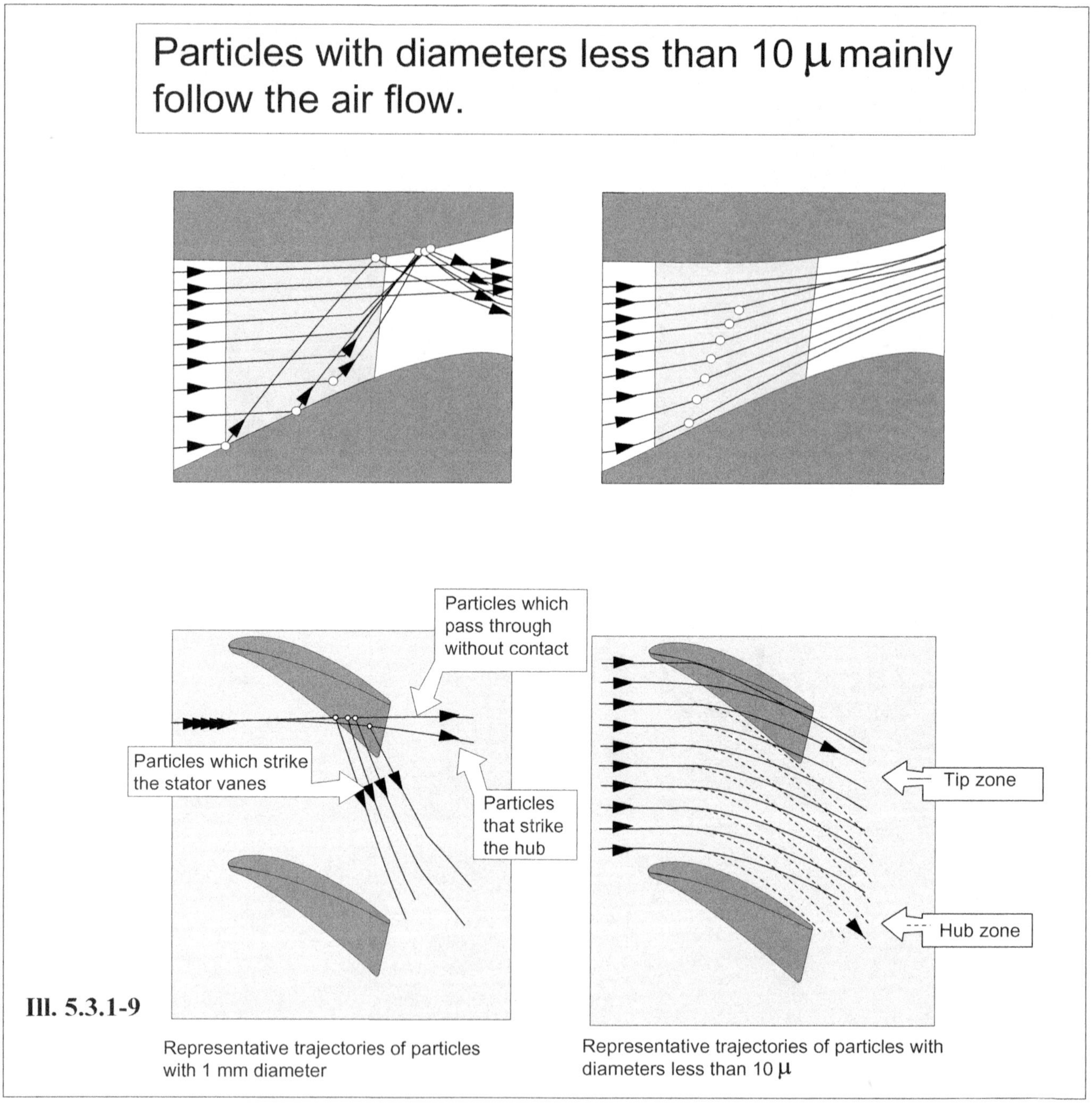

Ill. 5.3.1-9

Illustration 5.3.1-9 *(Ref. 5.3.1-14): The tests with mineral particles in the air flow have shown that particles smaller than 0.010 mm largely follow the* ***air flow*** *and therefore put relatively little erosion stress on the blading. However, larger* ***particles can ricochet*** *and be* ***reflected*** *around the air duct. For this reason erosion damages can, specific to the engine type, occur in unexpected areas such as on the rear edges of stator vanes. Typical signs of these ricochets are* ***impact marks on trailing edges****, which indicate movement of the particles against the direction of the air flow.*

It should also be considered that larger mineral particles burst upon impact and the erosion stress in the rear compressor area is correspondingly lower.

The erosion sensitivity of different compressor concepts is different.

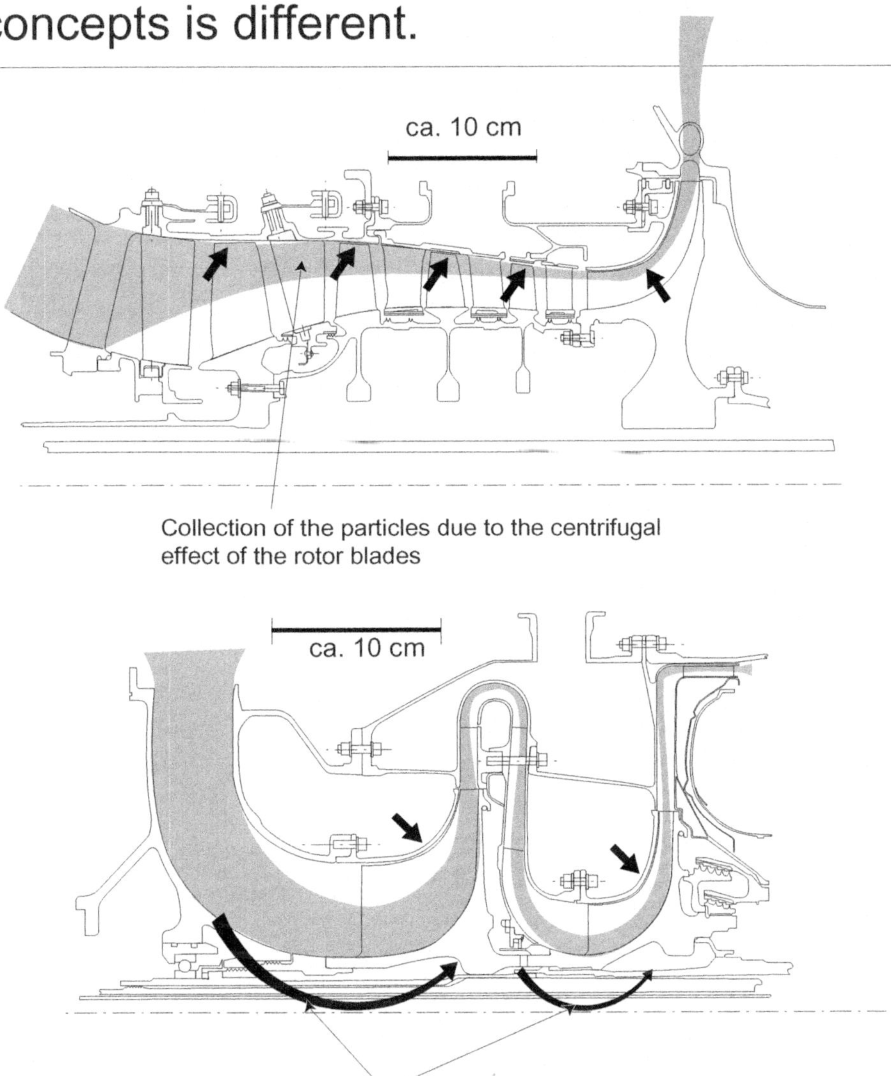

Ill. 5.3.1-10

Illustration 5.3.1-10 *(Ref. 5.3.1-5, 5.3.1-6, 5.3.1-7): Depending on the configuration of a compressor,* ***characteristic zones*** *in the gas flow are especially erosion sensitive. The top diagram shows an axial-radial compressor (Ref. 5.3.1-18) of the type common in smaller helicopter engines. The dust is centrifuged outward and puts erosion stress especially on the tips of the rotor blades and the casing wall: This increases the corresponding tip clearances (arrows). The* ***dust is concentrated outside on the air flow*** *and also erodes the outer area of the radial compressor blading (Ill. 5.3.2-8).*

The bottom diagram shows a two-stage radial compressor. Here the centrifugal forces from the air flow deflection are greater than the outwardly acting centrifugal forces, causing the ***dust to concentrate in the hub area****. This causes the blades in the radial compressor to be especially erosion-stressed in the less sensitive hub area. The erosion behavior of this type of compressor can be expected to be more robust.*

Wear/Erosion Under the Influence of Liquids: Droplet Strikes, Rain Erosion

Rain erosion (Ill. 5.3.1-1) is a type of damage that has been known for a long time in **helicopter rotor blades**. Even in the 1970s extensive investigations regarding this were conducted. Additionally, **rain erosion on cockpit glass** of military aircraft designed for high speed low-altitude flight was also researched. The available literature therefore primarily concerns the typical materials and **erosion parameters** from these operating fields.

Erosive damaging rain can be expected in the front compressor changes or fan when there is no long air **inlet duct**. Rain erosion should be considered in cases where **fiber-reinforced plastics** (especially carbon fiber and aramdid fibers, see Ref. 5.3.1-19) are used in the compressor blading. Conditions for rain erosion are more likely to occur with engines of aircraft that operate at low altitudes, such as helicopters.

In soft materials, the wear mechanism is primarily plastic deformation, constriction, and breaking out of small surface areas (Ill. 5.3.1-11.1). Hard materials may break out due to hardening and crack initiation (exhaustion of the deformability) or fatigue. Weight loss is often noticeable only after a certain incubation period. Rain erosion causes material removal. However, the geometric changes in, for example, a blade leaf, is not necessarily the most serious damage. The micro-crack initiation and/or pitting material removal can cause the **fatigue strength** of the part to decrease considerably.

The following example is intended to make clear the expected material removal due to rain erosion. If rain drops of 1.2 mm diameter (drop density of 0.01 drops per cm^3, temperature 25 °C) strike a surface at a right angle at nearly the speed of sound, then the amount of aluminum removed would be about 0.01 mm/s, but with titanium alloys it would be far less (0.01mm/10^1-10^2h). Typical rain densities have been measured to be between 0.001 to 0.1 drops per cm^3 (see Chapter 5.1.1, Ill. 5.1.1.-6).

Drop impact damage must not always be related to rain. For example, if droplets of liquid from a water injection into the compressor strike the blading or from a water injection into combustion chamber area (e.g. in industrial turbines) strike the turbine blading, it can cause erosion damage to coatings. In the temperature range between 10°C and 50°C, rain erosion is virtually unaffected by the temperature. Above 50°C the erosion effect decreases.

Cavitation

Cavitation (5.3.1-11.2) is a type of damage related to droplet impact. In this case, as well, the material is stressed by the impact of a liquid. However, with cavitation a "liquid spike" forms in an **imploding vapor bubble** in a liquid flow. Cavitation damage is especially found in components of fuel systems.

The damage mechanism of cavitation.

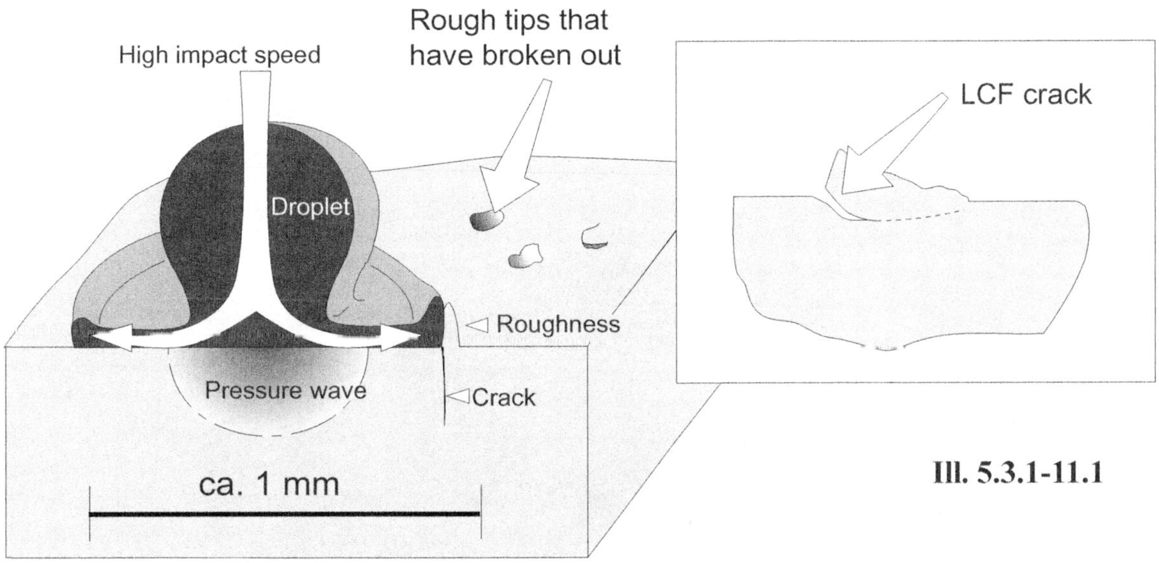

Ill. 5.3.1-11.1

Illustration 5.3.1-11.1 *(Ref. 5.3.1-8): A* ***droplet of liquid that strikes a solid surface*** *at high speed create shockwaves that cause large local elastic and/or plastic deformations, depending on the material. This can* ***remove protecting oxide*** *coatings and result in* ***increase corrosion*** *attacks. Many impacts one after the other with elastic deformations can cause dynamic fatigue at the micro-level. This leads to small break-outs of material that accumulate into macroscopic wear. In case of plastic deformations, craters can form in the impact, and in these craters overexpansion can initiate concentric micro-cracks that lead to material breaking out when struck by further droplets. If micro-breakouts have occurred, then droplets can create "liquid spikes" similar to those during cavitation (Ill. 5.3.1-11.2) in the hollow space, accelerating the erosion process. On the surface the liquid is distributed concentrically to the droplet. This liquid disk can build up on rough areas of the surface and cause micro-cracks there, as well.*

The ***dynamic cracks*** *(LCF) cause pitting-like material breakouts up to 0.1 mm deep on the surfaces of* ***titanium parts****. This damage form was observed on part surfaces on which coating remnants were removed with a* ***high-pressure water jet*** *after oxide peening (creates relatively high roughness).*

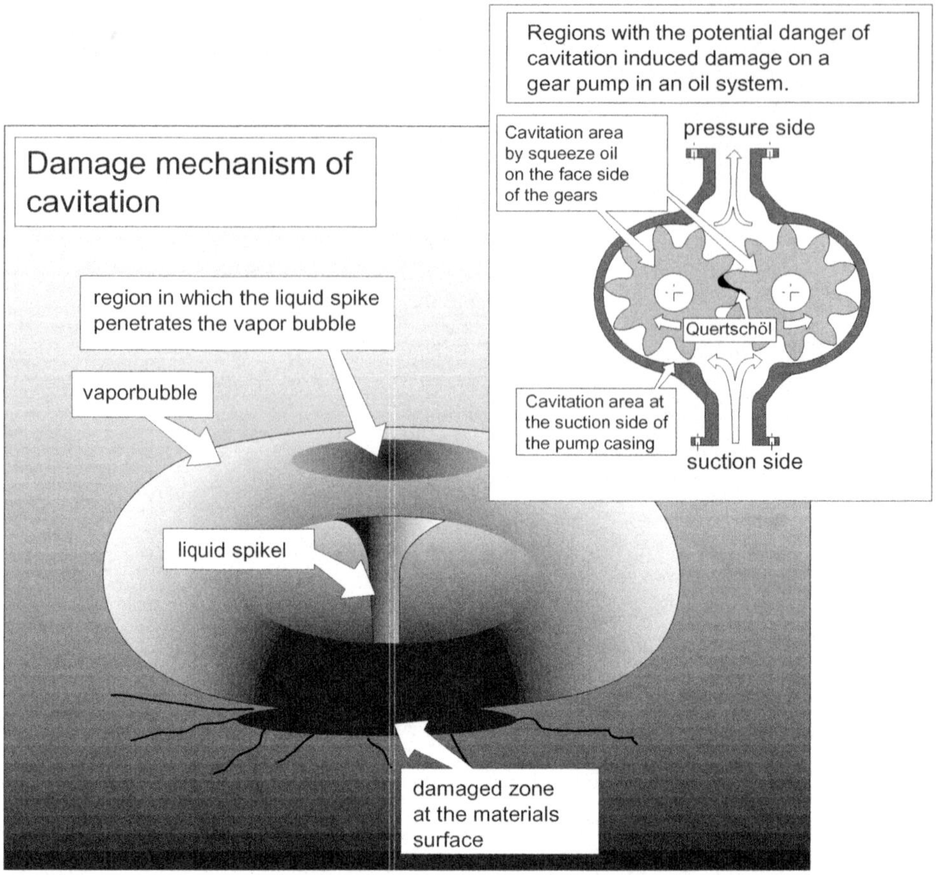

Ill. 5.3.1-11.2

Illustration 5.3.1-11.2 *(Ref. 5.3.1-15): This diagram is based on a high-speed photograph and shows the implosion of a vapor bubble in a liquid during supersonic vibrations. One can clearly recognize the liquid spike that forms during the implosion process and damages the material surface in a manner similar to the damage mechanism of droplet impact (Ill. 5.3.1-11.1). If there are many of these impulses after one another, they can destroy protective coatings and additionally increase corrosion. This damage/wear type is referred to as* ***cavitation corrosion****. Continued stressing results in local deformation and fatigue of the surface.*

Damage-causing ***vapor bubbles*** *form in underpressure zones like those that can occur in fast-flowing or vibrating liquids (Bernoulli's law).*

Thereby it is not necessary that only vapor bubbles of the flowing liquid itself are concerned. Also ***dissolved gases*** *and/or* ***earlier evaprateing liquids*** *like water (e.g., condensate that during unsuitable storage was introduced into the oil) can during elevated temperature and/or pressure drop similar to a bottle of fizzy water trigger the formation of vapor bubbles and so cavitation.*

It is interresting that with ***increasing roughness****, different ro rain erosion (Ill. 5.3.1-14) the erosion attack gets slower (Ref. 5.3.1-25).*

The sketch above right shows areas of ***gear pumps*** *in an oil system, where damage by cavitation was oserved (sketch above right, Ill. 5.3.2-1, Ref 5.3.1-21; volume 5 Ill. 23.2.1-18). Especially affected are the* ***suction side*** *of scavenge pumps and the front side of the gears as a result of the expansion of the squeeze oil.*

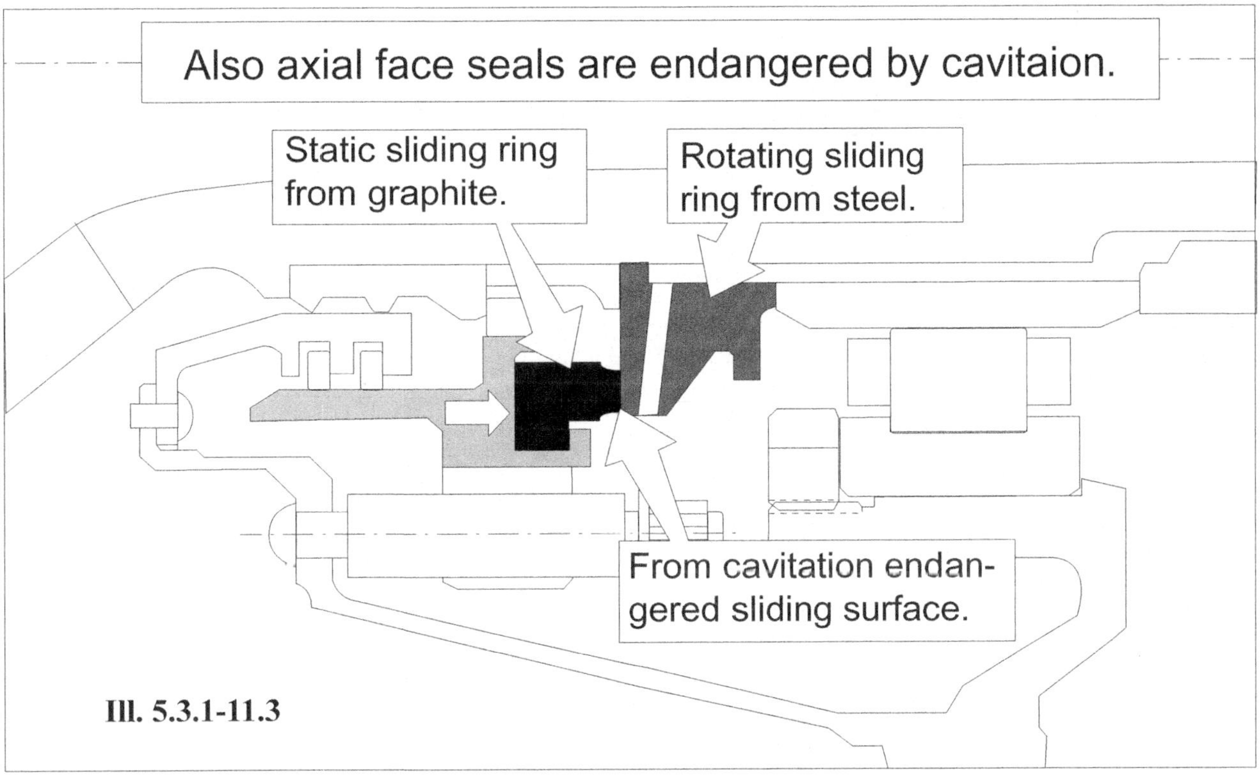

Ill. 5.3.1-11.3

Ill. 5.3.1-11.3 (Ref. 5.3.1-22): Aslso ***axial face seals*** *(volume 5 Ill. 23.4.2.2-1), as they are especially used in the region of bearing chambers, are endangered by* ***cavitation*** *caused failures (volume 5 Ill. 23.4.2.2-2 and Ill. 23.4.2.2-3).*

Ill. 5.3.1-11.4 (Ref. 5.3.1-24): In the region of the ***contact surfaces from oil dampened anti friction bearings*** *also cavitation was observed (volume 5 Ill. 23.1.1-2). The typical high frequency vibrations of the antifriction bearings (outer ring) during unsufficient oil pressure and/ or high oil temperature can promote the formation of vapor bubbles.*

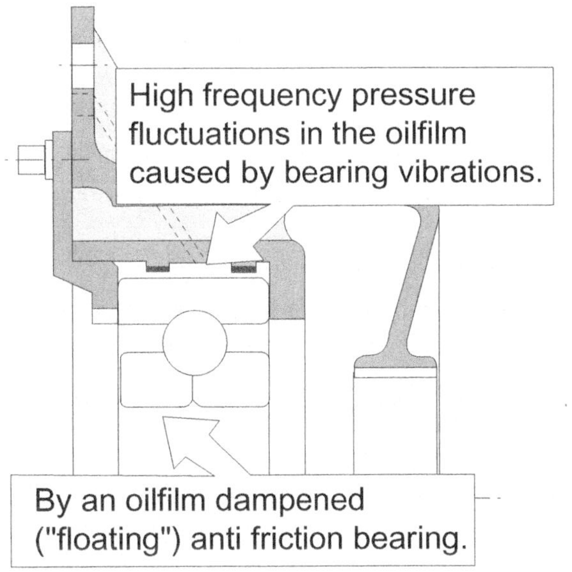

Ill. 5.3.1-11.4

Erosion parameters

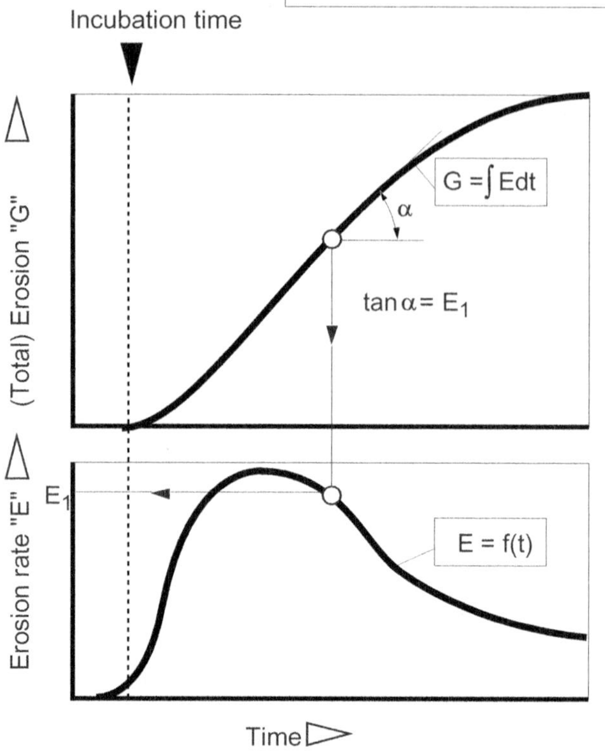

Instead of time, a time-dependent erosion parameter can be used to indicate the erosion rate:

- total mass of the acting erosion particles
- total volume of the acting erosion particles
- total erosion energy

The erosion damage can be given in:

- eroded total volume
- eroded total mass
- wear depth
- strength loss (such as fatigue strength)

The total erosion "G" (erosion sum, cumulative erosion) is the total erosion during the erosive loading

The erosion rate "E" is the momentary erosion (erosion speed), i.e. the erosion per unit of time or relative to a corresponding erosion parameter.

Ill. 5.3.1-12

Illustration 5.3.1-12: *If the* ***erosion rate*** *"E" refers to the volume of material eroded during a specific time from a certain material, divided by the volume of the impacting liquid, then a typical pronounced erosion behavior during rain erosion can be determined:*
With a certain ***incubation period*** *(corresponding to an incubation volume of droplets), the erosion rate increases quickly at first, reaches a maximum, and then decreases. After a longer period the erosion rate is virtually constant. The incubation time depends both on the material and on the kinetic energy of the droplets during impact. During the incubation period, micro-crack-initiating fatigue processes occur, which cause the surface material to break off or break out after the incubation period. The steep erosion increase at first is accompanied by a clear increase in roughness.*

Rain drops can damage compressor materials when they strike them at high speeds.

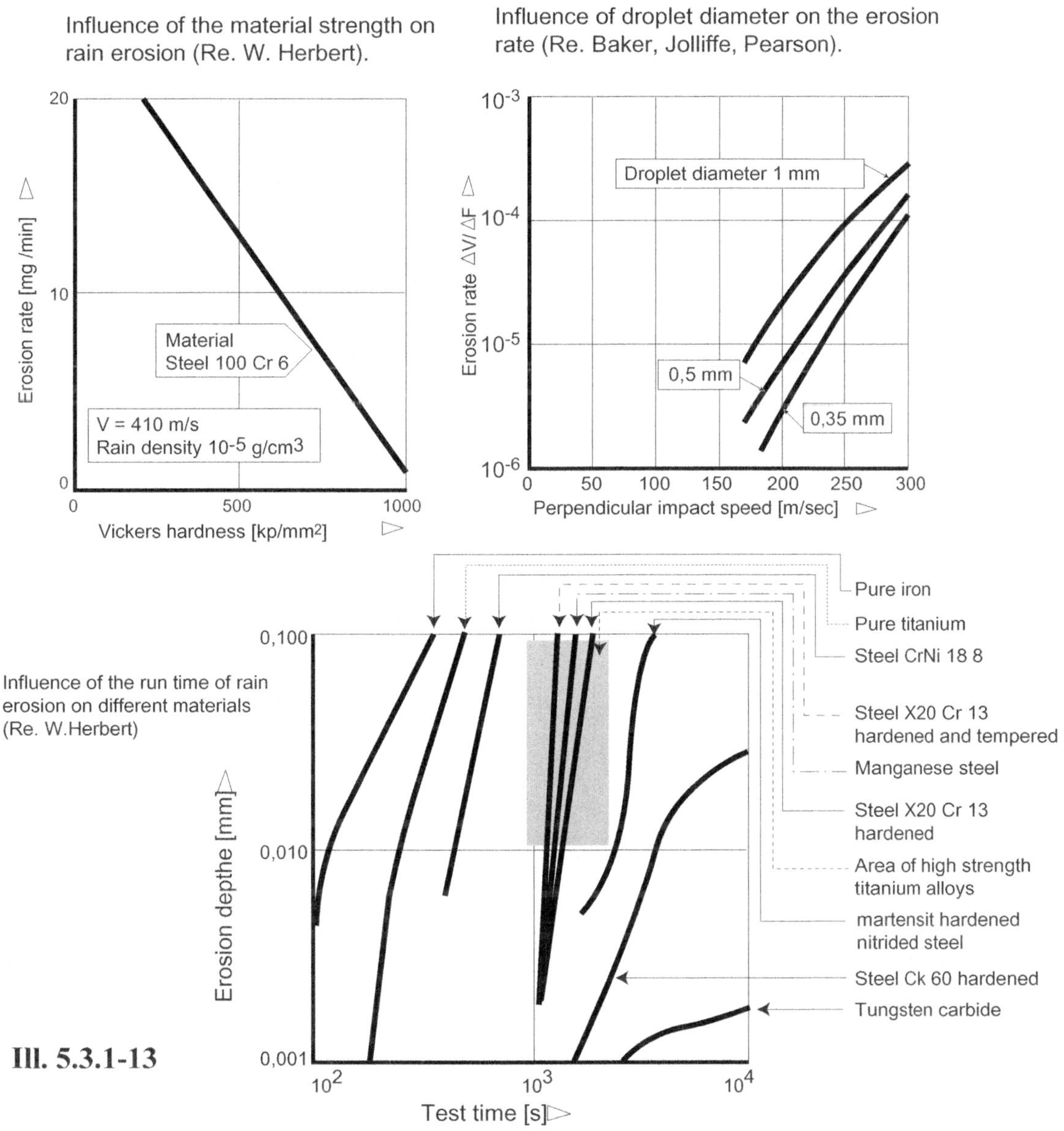

Ill. 5.3.1-13

Illustration 5.3.1-13 *(Refs. 5.3.1-9 and 5.3.1-10):*

Influence of materials on rain erosion:

In metals with pronounced strain hardening (increase of the yield strength during plastic deformation), the mechanical stress during erosion can increase the erosion resistance. In general the erosion resistance of metals increases with hardness (top left diagram), whereas for minerals and synthetics this is often not the case. Aside from the macro-hardness, the micro-hardness is also important, i.e. whether or not there are soft or hard phases. In areas with large variations in hardness, such as along grain boundaries (carbide fortification, etc.), increased *erosion attacks occur. Because **fine-grained metals** have more grain boundaries than coarse-grained metals, the former are **more sensitive to rain erosion**. Depending on the kinetic energy of the rain drops, the order of the erosion resistance of different materials may change. For example, soft PU lacquers perform very well with droplets at low impact speeds, i.e. with kinetic energy that can be absorbed elastically. However, if the impact speed is above a certain limit, the coating is quickly destroyed. It is has been conclusively determined that **materials behave very differently during rain erosion than they do during sand erosion**, and the performance of a material during one of these does not indicate its performance during the other.*

Influence of droplet size on rain erosion:

The influence of droplet size on erosion behavior can vary greatly between materials. There have even been observations of contradictory dependencies: the erosion rate of pure aluminum decreases with the droplet diameter while it increases with droplet diameter in PU surfaces. This can be explained with the overlaying of several different influences such as material properties, number of impulses, and surface topography.
*The most common **rain drop diameter** is between roughly 1 mm and 3 mm (see Ill. 5.1.1-6), which corresponds to a mass of about 1 mg to 10 mg. Both the volume and shape of droplets influence the erosion rate. The critical droplet speed (v), above which noticeable erosion begins to occur after longer periods, is dependent on the droplet size (top right diagram). The influence of the droplet size is considerable near the critical speed and decreases at higher impact speeds (Ill. 5.3.1-12). These threshold conditions, above which considerable erosion occurs, are given approximately for metals by the relationship $v_c^2 \cdot d = 10^7$ (v_c in m/s, d in 0.001 mm).*

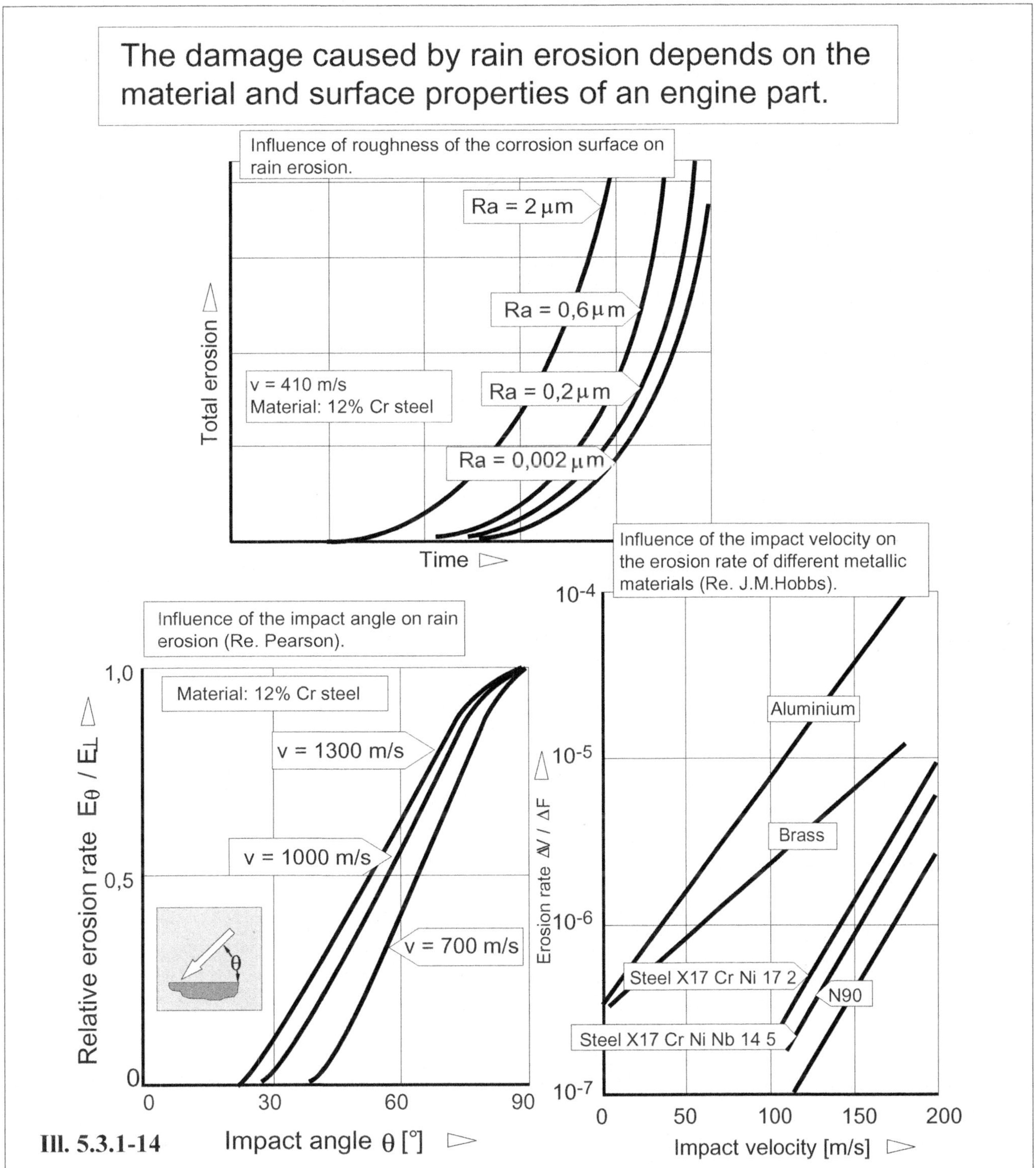

Ill. 5.3.1-14

***Illustration 5.3.1-14** (Ref. 5.3.1-9, 5.3.1-10, 5.3.1-11):*

Influence of the droplet impact speed on rain erosion:

The impact speed v (m/s) determines the kinetic energy released by the droplet and therefore its erosive effect (bottom right diagram). If v_c is the critical droplet speed (Ill. 5.3.1-13), below which no noticeable erosion occurs even over long

periods of time, then the following relationship applies to the erosion rate:
$E = a\,(v\text{-}v_c)^n$. "a", "n" and "v_c" depend on the material and erosion parameters such as he droplet diameter and shape. For example, with steels, $E = a\,(v\text{-}125)^{2\ \text{to}\ 2{,}16}$.

Influence of the impact angle on rain erosion:

The erosion rate increases rapidly with the impact angle (bottom left diagram). This shows that (with smooth surfaces) the normal component of the impact speed is decisive. The tangential component only plays a minor role in droplet strikes on smooth surfaces. However, the rougher the surface, the greater the influence of the tangential component. For this reason, only the inlet edges of compressor blades are noticeably affected by rain erosion. The advantage of blades with sharp inlet edges is that they slice up the drops and prevent stressing pressure waves from forming.

Influence of surface roughness on rain erosion:

The top diagram shows that with decreasing roughness, the wear due to rain erosion decreases considerably. High erosion resistance in the shape of considerably longer incubation times can therefore be expected primarily from smooth surfaces. As soon as the incubation period has passed, the roughness of the erosion surface increases and the erosion attack accelerates correspondingly.
Also for **cavitation** *(cavitation erosion) a higher starting roughness increases the attack. However during progrssion of the deterioration with the jaggedness a slower weosion speed can be observed (Ref. 5.3.1-25).*

Chapter 5.3.1 References

5.3.1-1 W. Kayser, "Erosion durch Festkörper", proceedings of the "2. Forschungskonferenz Regenerosion", 16.-18. August 1967, Meersburg Germany, pages 278-290.

5.3.1-2 H.Rieger, "Vergleichende Untersuchungen zur Werkstoffzerstörung beim Tropfenschlag und bei der Kavitation", proceedings of the "2. Forschungskonferenz Regenerosion", 16.-18. August 1967, Meersburg Germany, pages 260-277.

5.3.1-3 D.R. Nagy, V.R. Parameswaran, J.D. Mac Leod, J.P. Immarigeon, "Protective Coatings for Compressor Gas Path Components", proceedings AGARD-CP558 of the conference "Erosion, Corrosion and Foreign Object Damage Effects in Gas Turbines", Rotterdam, The Netherlands, 25-28 April 1994, chapter 27.

5.3.1-4 R.Ball, W. Tabakoff, "An experimental Investigation of the erosive Characteristic of 410 Stainless Steel and 6-Al-4V Titanium", Report No. 73-40, Department Aerospace Engineering- University of Cinncinati, US-Army Research Office Nr. DA-ARO-D-124-G-154, 1973.

5.3.1-5 D.L. Mann, G.D. Warnes, "Future Directions in Helicopter Engine Protection System Configuration", proceedings AGARD-CP558 of the conference "Erosion, Corrosion and Foreign Object Damage Effects in Gas Turbines", Rotterdam, The Netherlands, 25-28 April 1994, chapter 4, page 4-6

5.3.1-6 S.C. Tan, R.L. Elder, P.K. Harris, "Particle Trajectories in Gas Turbine Engines", proceedings AGARD-CP558 of the conference "Erosion, Corrosion and Foreign Object Damage Effects in Gas Turbines", Rotterdam, The Netherlands, 25-28 April 1994, chapter 14, page14-12.

5.3.1-7 C.G. Horton, H.Vignau, G. Leroy, "The Calculation of Erosion in a Gas Turbine Compressor Rotor", proceedings AGARD-CP558 of the conference "Erosion, Corrosion and Foreign Object Damage Effects in Gas Turbines", Rotterdam, The Netherlands, 25-28 April 1994, chapter 15, page 15-14.

5.3.1-8 "Failure Analysis and Prevention", Metals Handbook Ninth Edition, Volume 11, American Society for Metals, Metals Park, Ohio 44073, page 164.

5.3.1-9 W. Herbst, "Tropfenschlagverhalten von Eisen und Vergütungsstählen", Proceedings der "2. Forschungskonferenz Regenerosion", 16.-18. August 1967, Meersburg, Germany.

5.3.1-10 Baker, Jolliffe, Pearson, contribution to the proceedings of the "2. Forschungskonferenz Regenerosion", 16.-18. August 1967, Meersburg, Germany.

5.3.1-11 J.M. Hobbs, Beitrag in den Proceedings der "2. Forschungskonferenz Regenerosion", 16.-18.

August 1967, Meersburg, Germany.

5.3.1-12 "V2500 designed for maximum core protection", diagram, periodical "Aircraft Engineering", May 1993, page 3.

5.3.1-13 "Delta Weighs Changes to CFM56-2", periodical "Aviation Week & Space Technology", December 20, 1982, page 31.

5.3.1-14 A.Hamel, "Solid Particle Dynamic Behaviour Through Twisted Blade Rows", proceedings of the "2. Forschungskonferenz Regenerosion", 16.-18. August 1967, Meersburg, Germany.

5.3.1-15 J. Hancock, "Ultrasonic Cleaning", ASM Handbook Volume 5 Surface Engineering.

5.3.1-16 S.W.Kandebo, "Alliedsignal Commits to LT101 Improvements", periodical "Aviation Week & Space Technology", March 11, 1996, page 70.

5.3.1-18 J. Pugnale, "Dual Cebtrifugal Compressor: The Helicopter Solution to Sand and Ice Ingestion", "American Helicopter Society", Washington DC, 44th Annual Forum Proceedings, June 16-18, 1988, pages 673-681.

5.3.1-19 J.M.S. Keen, "Development of the Rolls-Royce RB.211 turbofan for airline operation", proceeding paper 700292 of the ASE "National Transportation Meeting" New York,N.Y., April 20-23, 1970.

5.3.1-20 D.I.G. Jones, C.M. Cannon, „Control of Gas Turbine Stator Blade Vibrations by Means of Enamel Coatings", Zeitschrift „Journal of Aircraft", Vol. 12, No. 4, April 1975, page 226 -230. (2497)

5.3.1-21 Z.S. Palley, I.M. Korolev, E.V. Rovinsky, „Structure and Strength of Aircraft Gas-Turbine Engines", Übersetzung FTD-HT-23-903-68 aus dem Russischen von „Foreign Tecnology Division", 1968, page 372.

5.3.1-22 B.S. Nau, „Film Cavitation Observationw in Face Seals", Proceeding der „Fourth International Conference on Fluid Sealing", Philadelphia, Pa., 1969, page 190-198.

5.3.1-23 K. Steffens, „Technik der Luftfahrtantriebe", Vorlesung an der TU- Aachen, März 2003.

5.3.1-24 H.R.Carr, G.J.Ives, P.Jenkins, „A Joint Study on the Computerisation of In-field Aero Engine Vibration Diagnosis", Proceedings AGARD-CP448 der Konferenz „Engine Condition Monitoring-Technology and Experience", Quebec, Canada, 30May-3.June 1988, Kapitel 31, page 31-5.

5.3.1-25 „Kavitation von Stellventilen", Technische Information V74 / Schulung der Fa. SAMSON AG page 3-58.

.

5.3.2 Erosion Damage

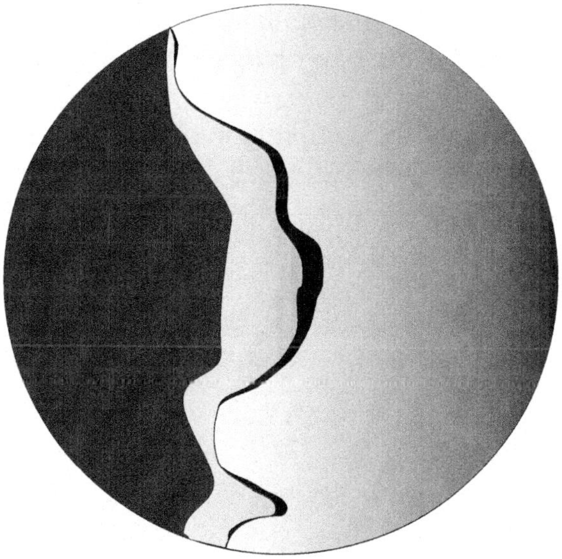

Erosion damage can occur on many different engine components (Ill. 5.3.2-1) and take many different forms. Typical effects are:

- **Worsening of the blading** (profile changes, increased roughness, increased tip clearance) and therefore also the operating behavior of the compressor leads to temperature increases with correspondingly high turbine temperatures.
- **Erosion in labyrinths** can also cause performance decreases, but it can also change the bearing thrust or cause oil to escape from the bearing chambers.
- **Erosion wear products** can **block the cooling air ducts** of the hot parts and raise their temperature, considerably reducing the life span of the hot parts (especially turbine rotor blades).
- **Dynamic fatigue fractures** on the **rotor blading i**n the compressor and turbine due to cross-sectional weakening and notches, in some cases combined with flow stalls due to changed profiles and increased roughness.
- **Overheating** damage in the **combustion chamber** due to erosion damage to the injection system caused by coke resulting from overly high fuel temperatures.

- **Oxidation** damage to **turbine blades** and stator vanes (Ill. 5.3.2-12) resulting from erosion damage to a protective coating (caused by soot particles, ceramic particles from thermal insulation layers, etc.).

In the compressor inlet dust is usually still distributed evenly across the cross-section. The farther the dust particles are inside an axial compressor (Ill. 5.3.1-10), the more they are centrifuged outward, resulting in the dust and therefore also the erosion damage being concentrated more and more at the blade tips (Ill. 5.3.2-9). A large amount of dust may pass through the bypass duct and not enter the core engine.

In cases where air is extracted from the compressor to cool the hot parts or cabin air, **extraction takes place on the rotor side** whenever possible, since the air in this area is relatively clean.

Some of the parts of an engine that are affected by erosion.

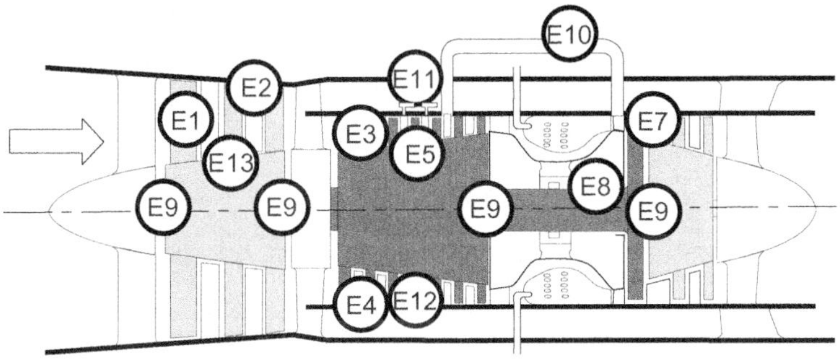

- E1 Fan and LPC blading
- E2 Housing coatings (abradable coatings) and liners.
- E3 Blading HPC
- E4 Housing coatings HPC
- E5 Spacer coatings behind 'S-channel'
- E6 Injection system
- E7 HPT and LPT blading
- E8 Bearing casings, gas ducts
- E9 Labyrinths, labyrinth coatings
- E10 Air extraction/bleed ducts, air ducts
- E11 Variable vane actuators
- E12 Blade root contact surfaces
- E13 Interstage seals in the LPC

Ill. 5.3.2-1

Illustration 5.3.2-1: *Many different engine components can be directly or* ***indirectly damaged through erosive particles****:*

"E1" fan and LPC blading: *Roughening through material removal and/or deposits (fouling),* ***changes in blade profiles****, increased* ***tip clearance****,* ***blade fractures*** *caused by dynamic fatigue fractures originating in erosion notches.*

"E2" housing/caing coatings in the fan and LPC: *The erosion material removal causes increased tip clearances to the rotor as well as* ***steps and roughness in the path of the gas flow****.*

"E3" HPC blading *(Ill. 5.3.2-4): see "E1".*

"E4" HPC housing coatings *(Ill. 5.3.2-4): see "E2".*

"E5" spacer coatings *(Ill. 5.3.2-10): Erosion of the relatively hard spacer coatings increases the tip clearances between the rotor and compressor stator blades with no inner shroud. Additionally,* *the roughness of this area of the hub contour can be increased.*

"E6" injection system: *Erosion of the injection nozzles through coke formation in* ***overheated fuel****. This can result in overheating in the combustion chamber area, even including flames escaping from the combustion chamber housing/ casing.*

"E7" turbine blading *(Ill. 5.3.2-11): Erosion and increased* ***oxidation*** *especially at the leading blade edges caused by dust, coke from the combustion chamber, and spalled ceramic thermal barriers.*
Blocking of cooling air ducts *and cooling air film bores, which can cause local overheating.*
Chemical reactions and erosion material removal in ceramic thermal barriers.
If dust collects in a fir-tree joint, it can unallowably alter the dynamic behavior of the blades (Ref. 5.3.2-1).

"E8" bearing damage *(Ill. 5.3.1-11): Erosion in bearing casings and gas ducts caused by circumferentially-traveling trapped particles of ingested dust and labyrinth material wear products. Weakening of wall cross-sections, even causing the walls to fail.*

"E9" labyrinths *(Ill. 5.3.1-11): Erosion in* ***labyrinths****, especially on softer abradable coatings.*

"E10" air extraction ducts and air ducts *(Ill. 5.3.2-14): Erosion of walls, especially in the outer area of* ***pipe bends****.*

"E11" Variable vane adjustment: *Dust in the* ***guide vane bearings*** *can cause jamming and wear.*

"E12" blade root contact surfaces: *In rotor blades and guide vanes, an abrasive process can* ***increase fretting wear*** *due to dust that has penetrated into these areas.*
Changes in the friction relationships can alter the dynamic behavior of the blading (Ref. 5.3.2-1).

„E13" Elastomeres*: Concerned are* ***rub in coatings*** *from filled* ***silicone rubber****. This is used in the front region of the compressor (fan, booster, LPC) at* ***interstage seals****. This coating usually is located on the* ***inner rings of the stators*** *opposite the labyrinth tips on the rotor. There is the suspicion, that traces of wear on these coatings are in connection with ice formation. The exact mechanism seams not yet clear. However it is known from overhaul processes, that* ***particles from water ice can be used as blasting abrasive*** *to remove soft contaminations (Ref. 5.3.2-25). In this case however the provider calls it a non abrasive process. An abrasion effect however may merely depend from the process parameters. This is also true if the ice contains abrasive particles (e.g., dust) as contamination.*

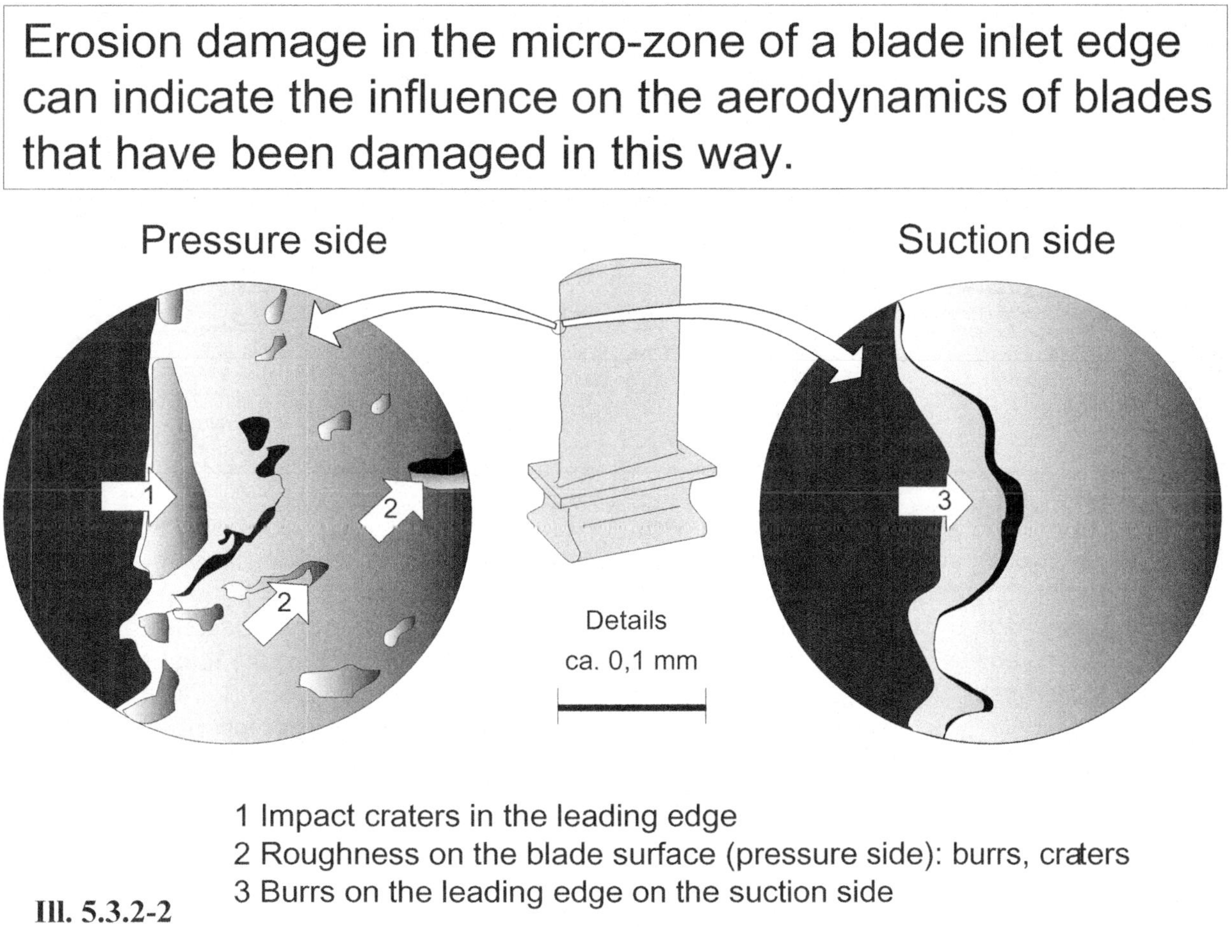

Ill. 5.3.2-2

Illustration 5.3.2-2 : Changes to the blade profile *at the micro-level, especially the leading edge, can affect the aerodynamic behavior of the blade considerably. The impact of sand grains causes notches and burrs, as well as increased roughness primarily on the pressure side of the blade. On the suction side, especially, burrs promote local flow stalls (bottom diagram).The consequences include a decrease in compressor efficiency and also increased dynamic loads on the individual blades. Combined with the notch effect of the impact notches, this* ***increases the risk of fatigue fractures****. Clear signs of high dynamic loads (extrusions, see Ill. 5.4.2.1-2) in erosion notches were verified through use of scanning electron microscopes (SEM).*

Typical behavior of a shaft-power engine under erosive loads.

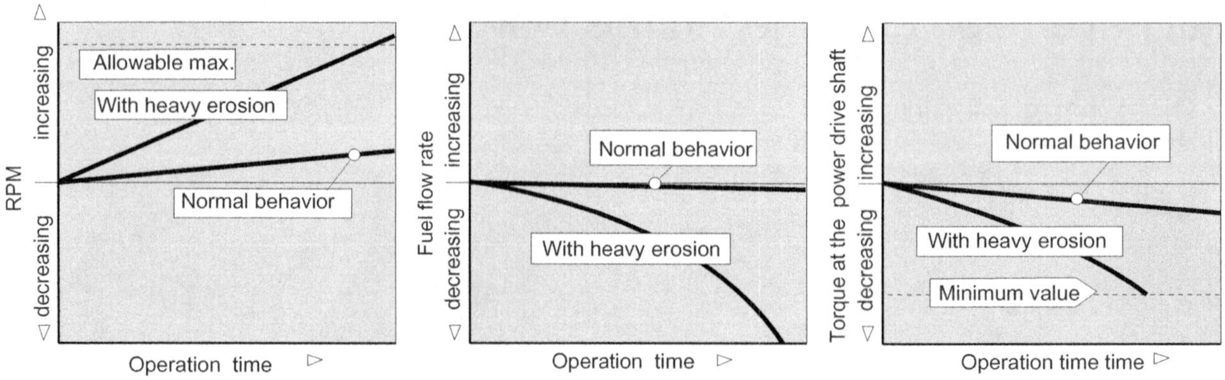

Because compressor efficiency of the eroded blading decreases, the compressor can no longer take on as much power. As a result, RPM increases.

In order to keep the compressor RPM and/or the turbine inlet temperature from rising unallowably, the fuel throughflow must be reduced.

In shaft engines, the torque decreases at the power drive shaft because the fuel flow rate must be reduced (to maintain the compressor RPM and turbine temperature).

Ill. 5.3.2-3

Illustration 5.3.2-3 *(Ref. 5.3.2-2): The* ***erosion-induced changes to the compressor blades****, such as increased roughness and/or changes to the blade profile at the micro- and macro-levels, result in a considerable* ***worsening of the compressor efficiency****. This, in turn, affects the operating behavior of the engine and the damage frequency and life spans of other components, especially the hot parts. In order to maintain the same engine performance, the efficiency decrease must be compensated by increasing the gas temperatures. This* ***raises the temperatures of the hot parts*** *(combustion chamber and turbine). In addition, the decreased play to the surge limit can cause it to be crossed, leading to extreme short-term overheating in the turbine with extensive consequential damages.*

Erosion attack can dangerously change the natural frequency of blades, so that resonance occurs.

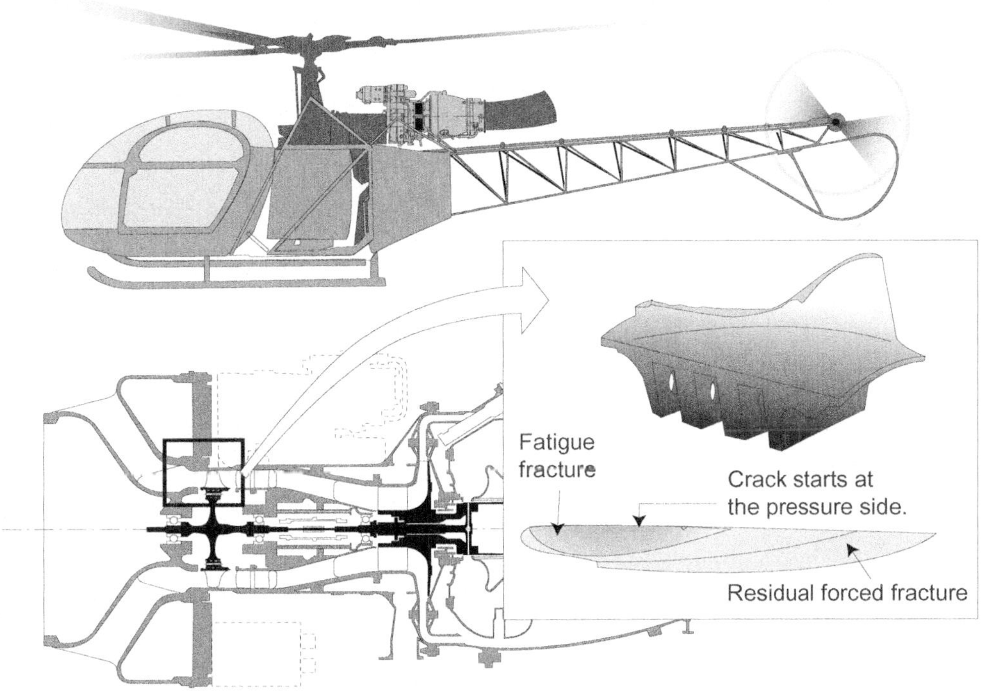

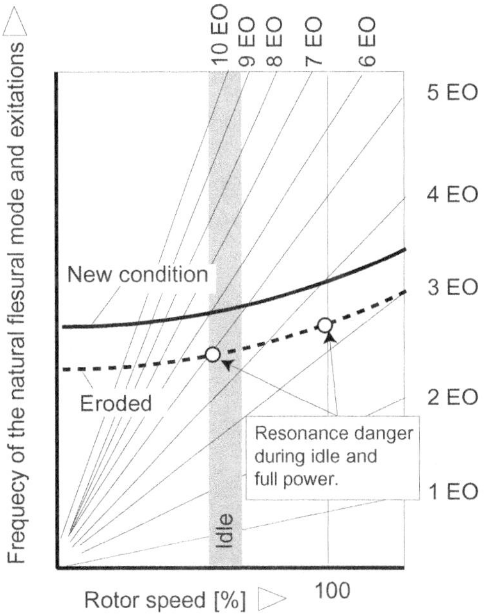

Example 5.3.2-1.1 (Ref. 5.3.2-19):

In a 400 kW class helicopter engine with a side inlet, after longer operating times there were cases of dynamic fatigue fractures of individual compressor rotor blades made from the titanium alloy 6Al 4V, resulting in extensive secondary damages. The single-engine helicopters were operated in an environment with **dust conditions,** that were at least equal to those in a north African desert. Due to costs and efficiency demands, the engines were **not equipped with filters or sand separators**. The blades, which had been run for long periods, showed **serious erosion and notches near the root platform.** This is related to the side inlets and the corresponding path of the particles in the inlet flow (Ill. 5.3.2-7, right diagram). After a full overhaul interval, the blades were eroded down to 50-70 % of the normal profile thickness. This **changed the resonant frequency of the blades so much, that it resulted in resonance and dynamic fatigue fractures.**

Typical erosion on blades in a compressor.

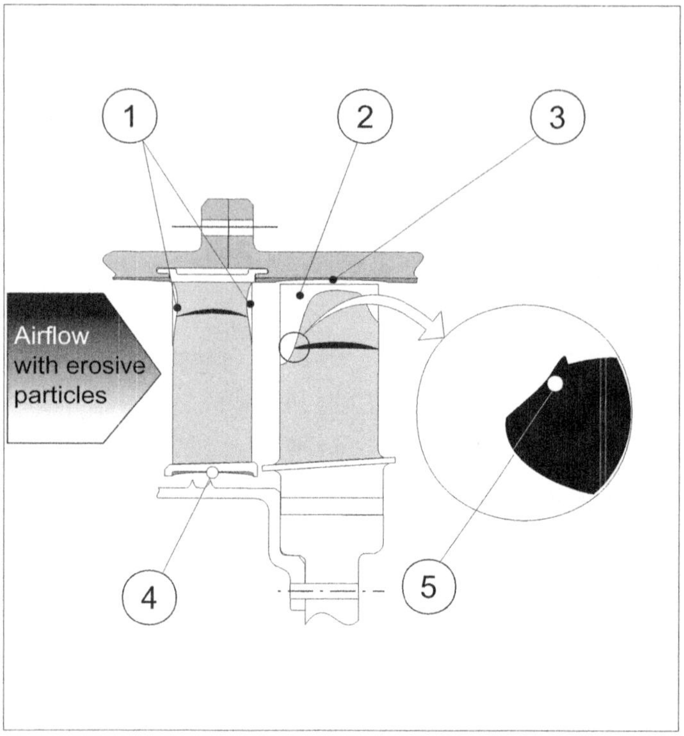

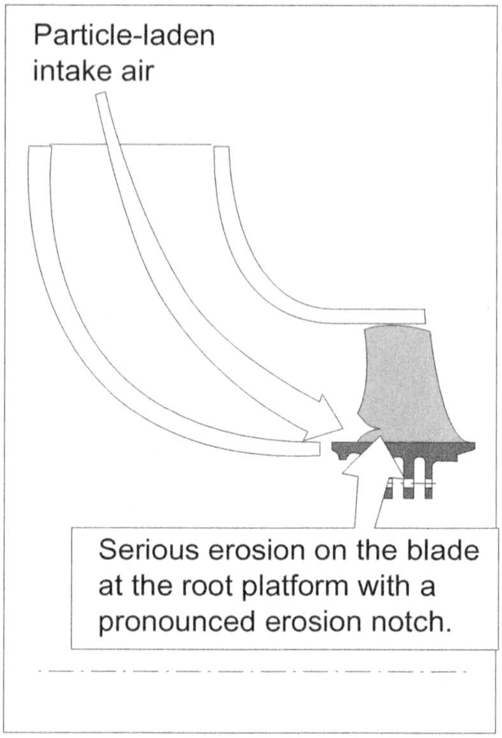

Ill. 5.3.2-4

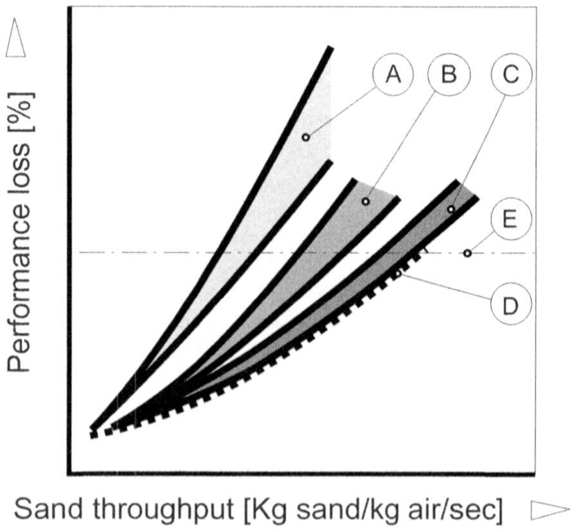

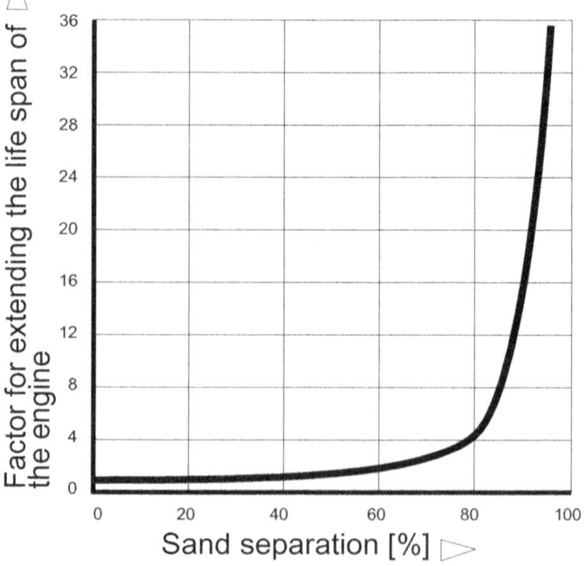

Illustration 5.3.2-4: *The life span of a* ***helicopter engine*** *depends largely on the presence of a* ***dust filter*** *in front of the engine inlet and the separation effectiveness of the filter (bottom right diagram, Ref. 5.3.2-18). It can be seen that only filter systems with high separation rates permit desirable engine life spans.*

The top left diagram is a schematic depiction of the typical erosion symptoms in a rear compressor stage. The ***centrifugal effect*** *of the blading in the air flow concentrates the particles near the casing wall. This makes the erosion effects more pronounced in radially farther outward areas. This especially weakens the root areas of stator blades ("1"), promoting dynamic fatigue fractures in case of flexure. However, the clearance between the rotor and blades without inner shrouds is not affected, minimizing the efficiency decrease in this part of the compressor. The rotor blades, on the other hand, are worn primarily at the tips ("2"), where the* ***clearance increases*** *are spread over a larger circumference, affecting the compressor efficiency more than the stator blade erosion.*

In addition, increased erosion of the relatively soft ***abradable coatings*** *on the* ***inside of the casing*** *("3") further increases the clearance at the rotor blade tips.*

With stator blades with inner shrouds and soft abradable labyrinth coatings facing the rotor ("4"), there is a danger of clearance increases due to erosion, although the erosion stress is lower than near the rotor blade tips (lower dust concentration near the hub).

The erosion of the blade profile, with roughening of the leading edge and burr formation on the suction side ("5"), promotes local flow stalls and therefore also compressor stalls.

The erosion behavior (erosion sensitivity) varies considerably between different compressor designs (see diagram bottom left, Ref. 5.3.2-2). Axial compressors ("A") show the greatest sensitivity. ***Combined axial-radial compressors*** *("B"), such as are common in helicopter engines (Ill. 5.3.1-10), perform better.* ***Side-mounted inlets*** *can result in the blades being damaged especially at the highly dynamically stressed* ***transition to the root platform*** *(sketch above right, example 5.3.2-1.1). Machines with pure* ***radial compressors*** *show the most robust behavior ("C"). This positive performance can be further improved through erosion protection on disks and diffusers ("D"). The horizontal line "E" shows the operating limits of the engine according to regulation AS/AV E-85 930.*

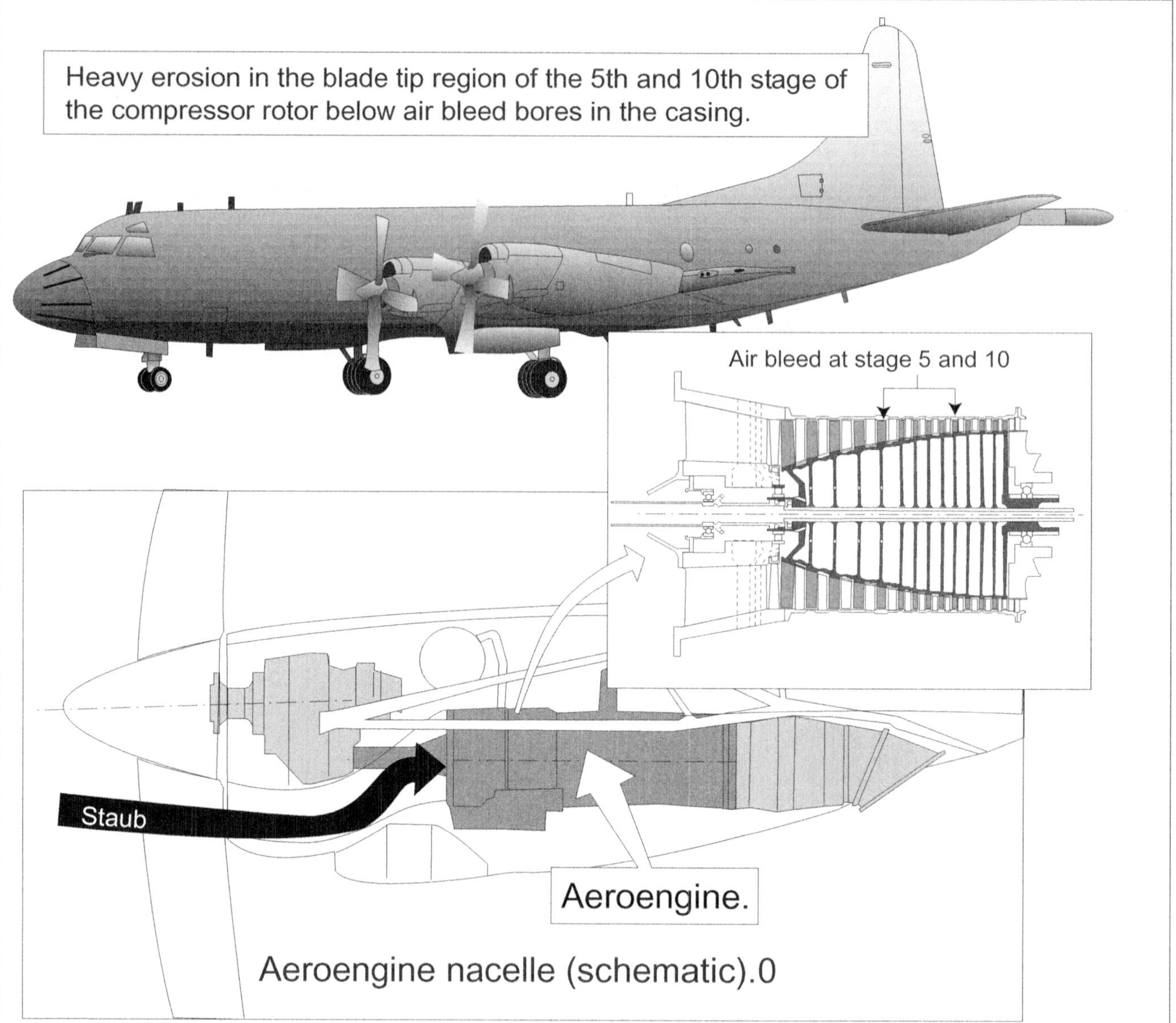

Example 5.3.2-1.2 (Ref. 5.3.2-3):

Excerpt: "severe trailing edge erosion of an (helicopter engine) 5th and 10th stage compressor blades has been observed in Field Service Evaluations and sand ingestion tests. The erosion typically occurs within **3 millimeters of the blade tip** and has been attributed to overtip flow produced by the **bleed holes** present over the **5th and 10th stage rotors**.... (measurements of the air flow showed). The flow was found to be unaffected by the bleed holes beyond 3-4 mm from the blade tip.

Comment: An increased erosion at the blade tips of the rotor blades below the air extractions respectively **openings for air bleeds** are observed on-again (Ill. 5.3.2-7). Here obviously the dust acts especially deteriorating, concentrated by the centrifugal force, supported by the disturbed flow.

Rollover of a leading edge due to heavy particle action in the direction of the profile.

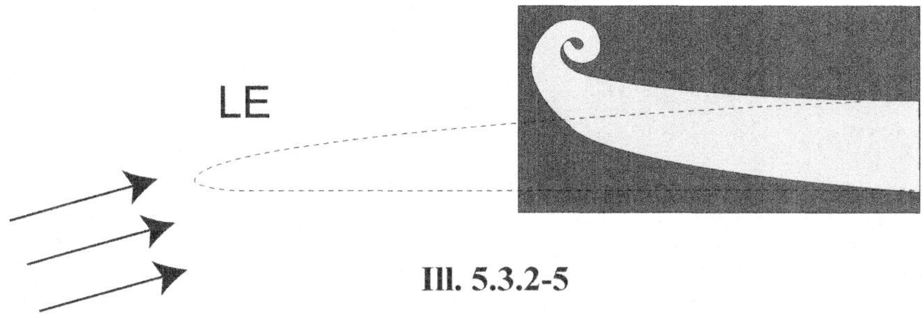

Ill. 5.3.2-5

Illustration 5.3.2-5 *(Ref. 5.3.2-4): "Rollover" as an extreme profile change. The leading edge of a helicopter engine compressor rotor blade has been rolled up by sand erosion.*

Heavy compressor surge as consequence of coating erosion with widening of the tip gap in the casings of the high pressure compressor.

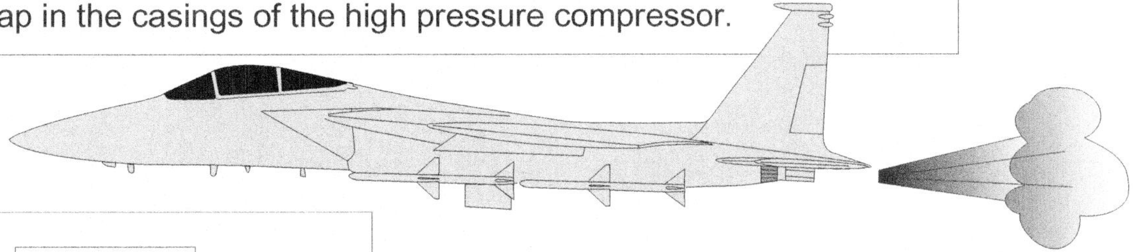

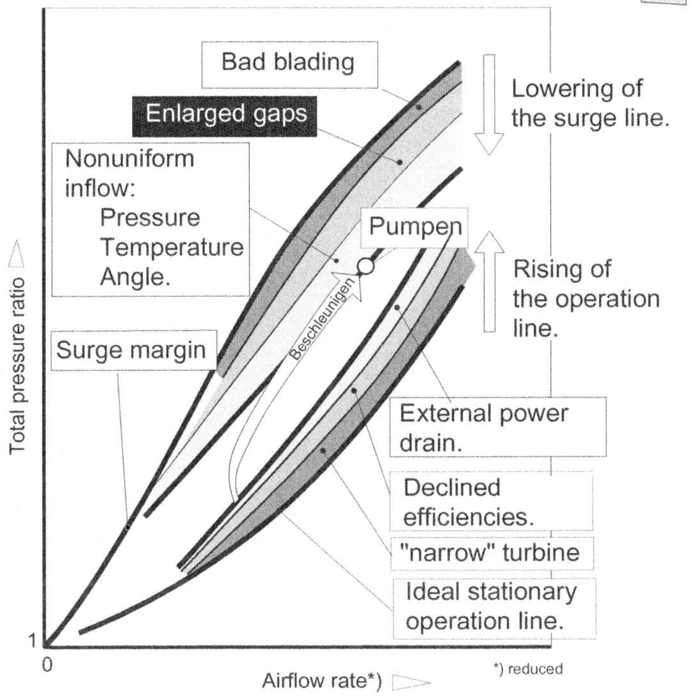

Example 5.3.2-1.3 (Ref. 5.3.2-7):

Excerpt: "To eliminate compressor **stalls caused by deterioration of unit's abradables**, (the engine manufacturer) recommended repositioning the compressor's sixth stage vanes and rescheduling the unit's variable geometry through a software change in the engine's electronic control unit."

Comments: The affected military engine type had massive problems with compressor stalls (surges?) during operation, **because the tip clearances in the high-pressure compressor had become too large** due to erosion (example 5.3.2-3). This evidently contributed to inquiries as to whether the involved aircraft type would be better outfitted with engines from a competing company.

External Factors
Erosion: Damage

Tough blade coatings that have been damaged through erosion and/or foreign objects can delaminate and cause massive flow disruptions.

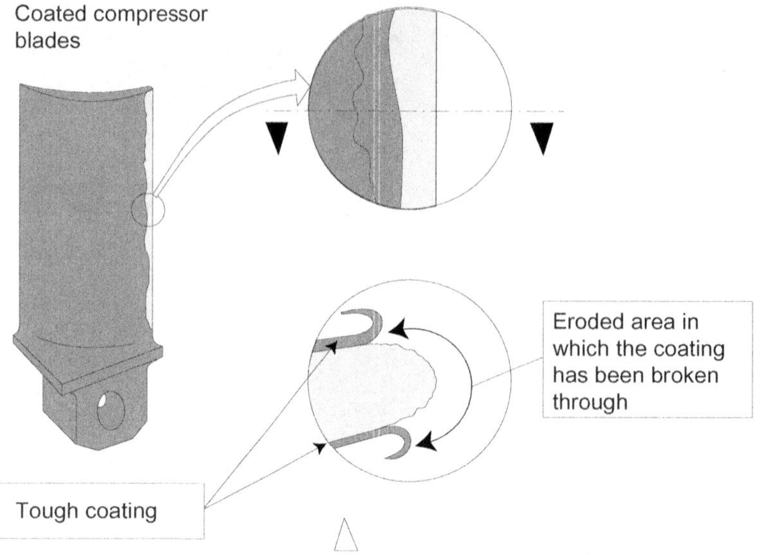

A tough organic coating that has rolled up after damage due to erosion

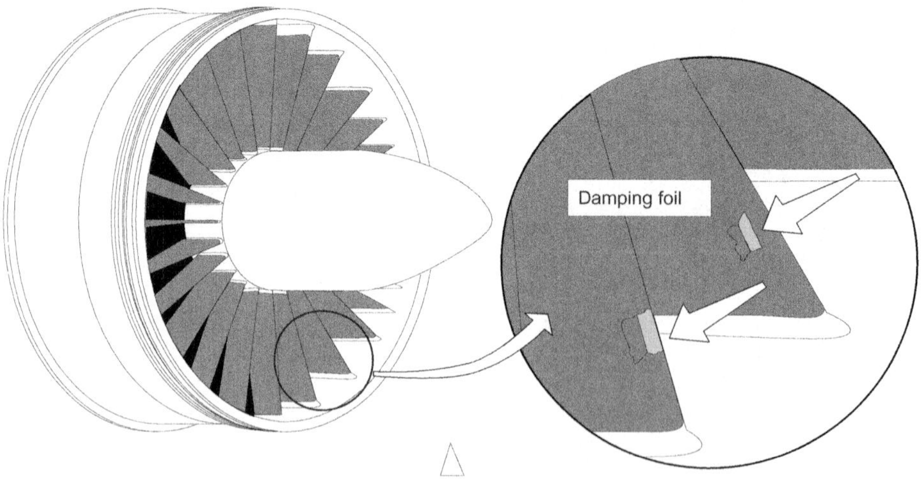

Ill. 5.3.2-6

Delaminating metallic damping foil after damage due to FOD or erosion

Illustration 5.3.2-6: Coatings *on engine parts, especially blades, can locally erode or be damaged by foreign objects so that the flow, combined with blade vibrations, cause the coating to* ***delaminate****. This is especially the case with ductile coatings such as organic erosion protection coatings (top diagram) or metallic foils such as* ***damping foils*** *(bottom diagrams), in which the bond with the base material is considerably weaker than the coating strength. This kind of delamination or* ***rolling-up*** *of the coating causes considerably more flow interference than superficial material removal or flaking. The delamination of larger coating areas also poses a threat of damage or* ***blockages in the rear engine*** *area (for example, the cooling ducts of hot parts).*

Flow irregularities, such as around air vents, can considerably change the erosion pattern.

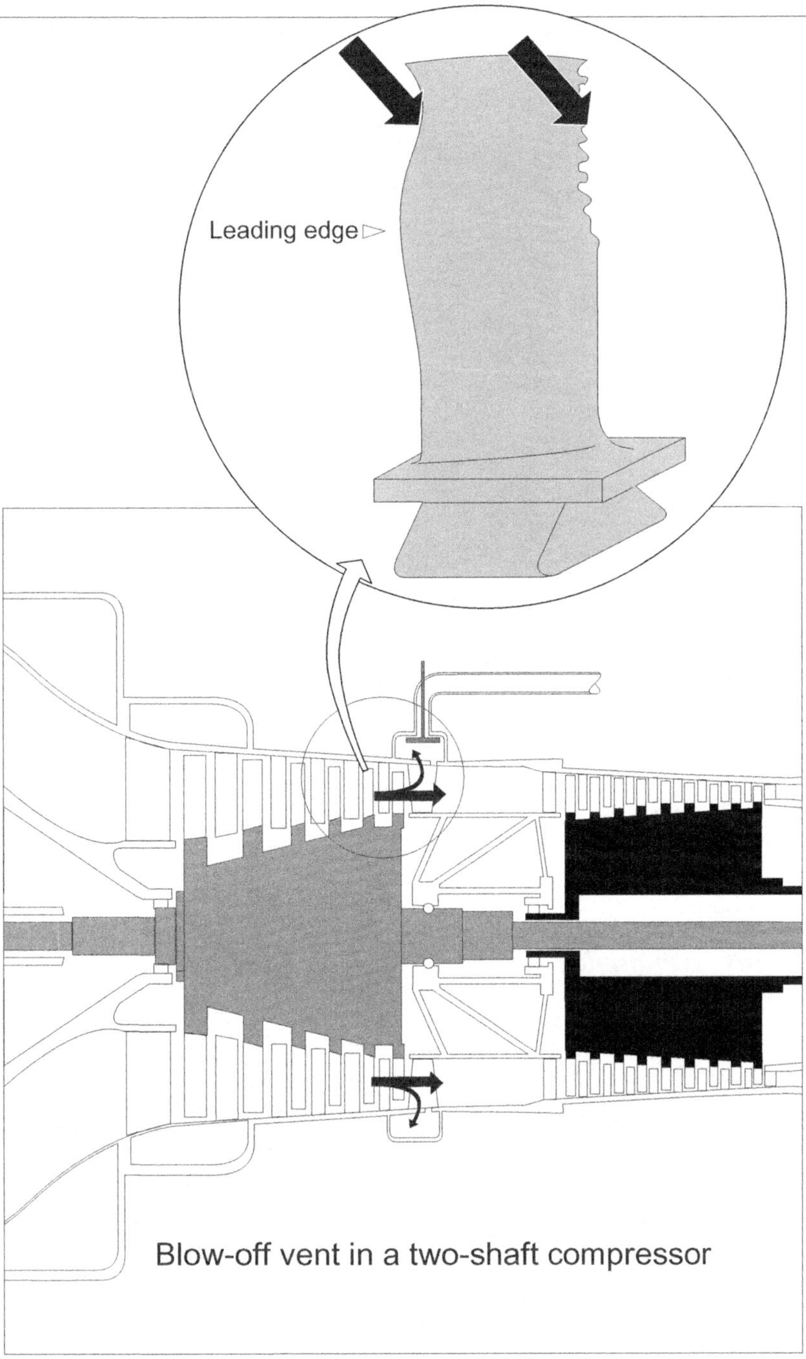

Ill. 5.3.2-7

Illustration 5.3.2-7 *(Refs. 5.3.2-3, 5.3.2-2, and 5.3.2-17): In* ***blade tips****, especially those located in turbulent flow zones such as around areas where air is extracted through* ***bleed valves*** *(Example 5.3.2-1), a typical erosion symptom known as "****hooking****" can appear in the rotor blade tip area. This causes increased wear of the trailing edge below the corner of the blade.*

Erosion patterns on a radial compressor disk

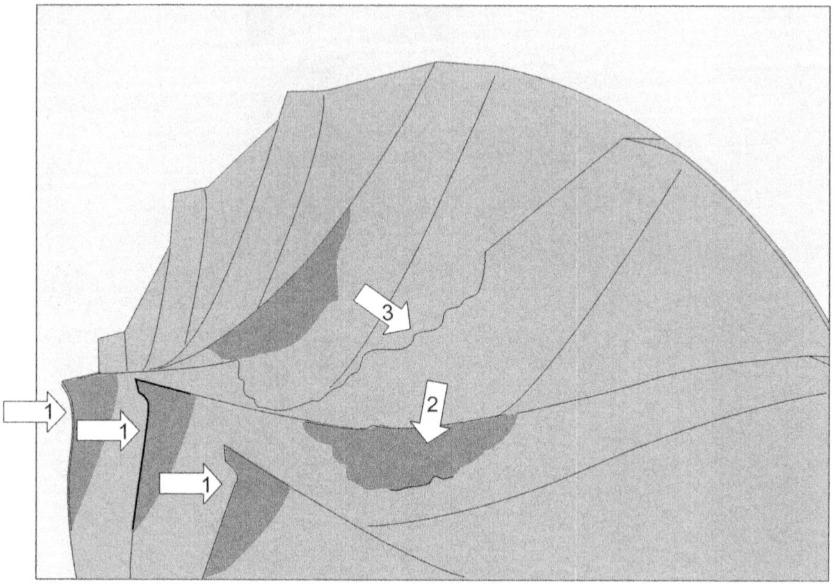

Ill. 5.3.2-8

Illustration 5.3.2-8 *(Ref. 5.3.1-5):* ***Radial compressors*** *are usually particularly erosion resistant (Ill. 5.3.2-4). Depending on the upstream configurations, such as* ***radial or axial inlets*** *and axial or radial compressors, the erosion symptoms on the blading can vary greatly (Ill. 5.3.1-9). In this case a multiple-stage axial compressor is located upstream. During an erosion test with coarse dust a sudden power decrease of 15% occurred. This was followed by the mechanical failure of several blades.*

Illustration 5.3.2-9: *This diagram illustrates the technical relationships in the extreme erosion stress mentioned in Example 5.3.2-2, which can evidently occur during a* ***sand storm****. The engines involved were shaft power engines of the 3000 kW class. The blading of the axial compressor in this engine type is made from a high-strength titanium alloy and is relatively filigreed. The involved helicopter is not known to have been equipped with any particle separating devices such as filters. Therefore it can be assumed that the compressor blading experienced conside-*

Even today, in a sandstorm, Mother Nature will get the better of technology.

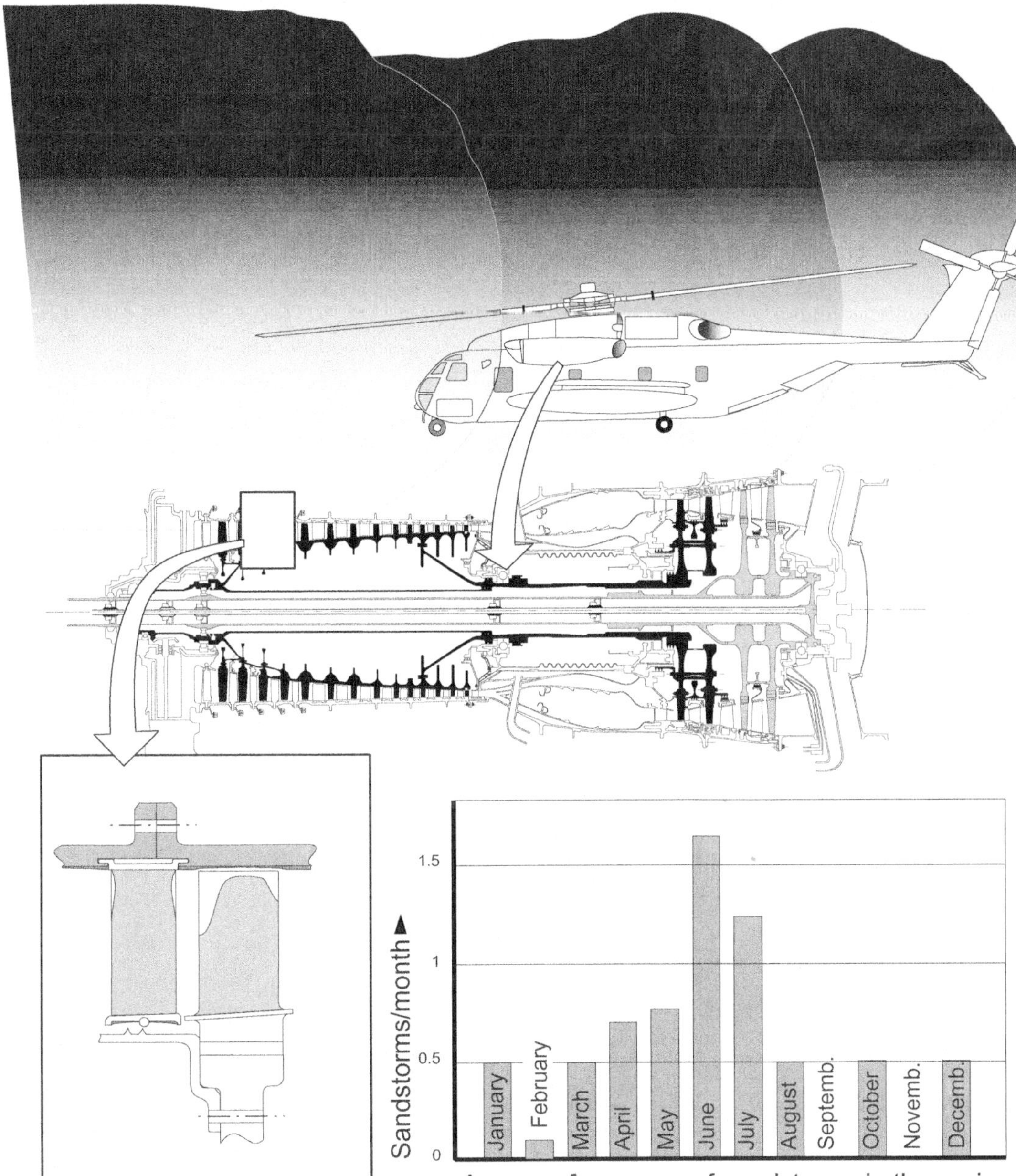

Ill. 5.3.2-9

Average frequency of sandstorms in the region of the Arabian Peninsula.

*rable **profile changes** of the blades (similar diagram at bottom left) due to the **sand erosion**. In this case an unallowable influencing of the engine's operating behavior and power output was concluded to have occurred.*

Erosion occurs not only on the blading, but also on the spacers and in the housing.

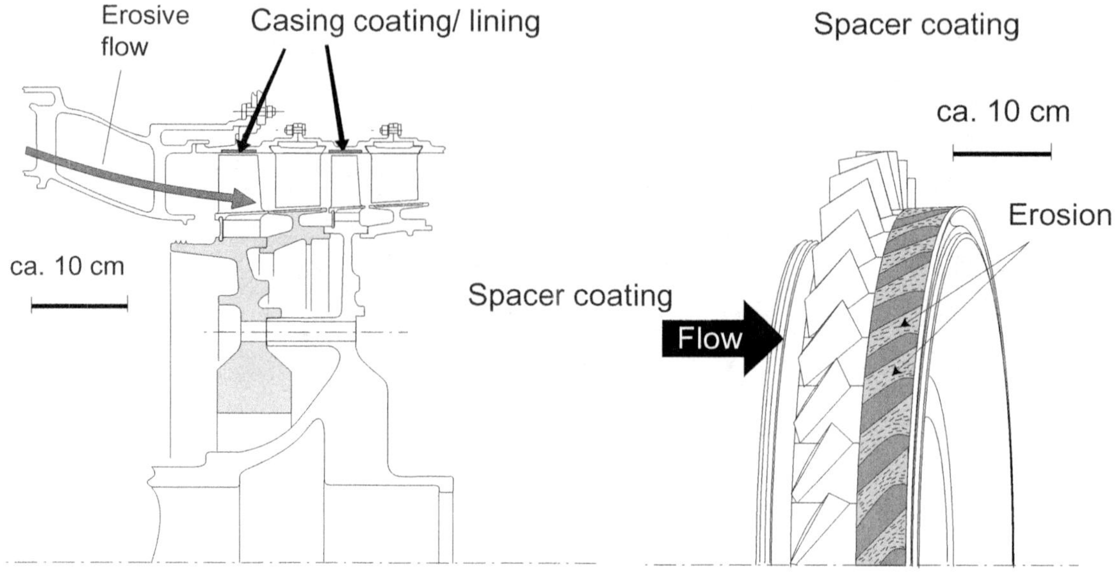

Cross-section from a high-pressure compressor

Ill. 5.3.2-10

Example 5.3.2-2 (Ill. 5.3.2-9):

Excerpt 1 (Ref. 5.3.2-5): ..."a table... indicated, by location and month, the frequency of suspended **dust occurrences.** Helicopter pilots, however, were surprised when they encountered the dust, were unprepared to accurately assess its impact on their flight, and stated that they were not advised of the phenomenon."

Excerpt 2 (Ref. 5.3.2-6): "The remaining (helicopters) were also having problems after running into two totally unexpected and vicious sandstorms. The one flown by the force commander... and No2 chopper made forced landings to ride out the worst of the storms. Luckily, after over 20 minutes on the ground, the storms passed and both crews resumed their journeys.... No5 (helicopter) suffered a major mechanical failure and, unable to carry on with the mission, was forced to return..."

Comments for Example 5.3.2-2:

Both excerpts refer to the hostage rescue operation in Tehran on April 24, 1980. The helicopters which were used were a type **without filters or separators in front of the engines**. Evidently the sand storm danger was estimated to be relatively minor. According to Ref. 5.3.2-6, the crew was forced to land due to poor visibility.

It is also interesting how difficult even short-term prediction or location of a sandstorm is. The bottom right diagram in Ill. 5.3.2-9 (Ref. 5.3.2-8) shows that in April, the sand storm frequency on the neighboring Arabian Peninsula is roughly one in every 60 days. Evidently this relatively probability was sufficient to select a helicopter type with no particle separator. This decision may have been supported by demands for **long flight range**, which would not allow the increased **fuel consumption** caused by the **inlet resistance** of a filter.

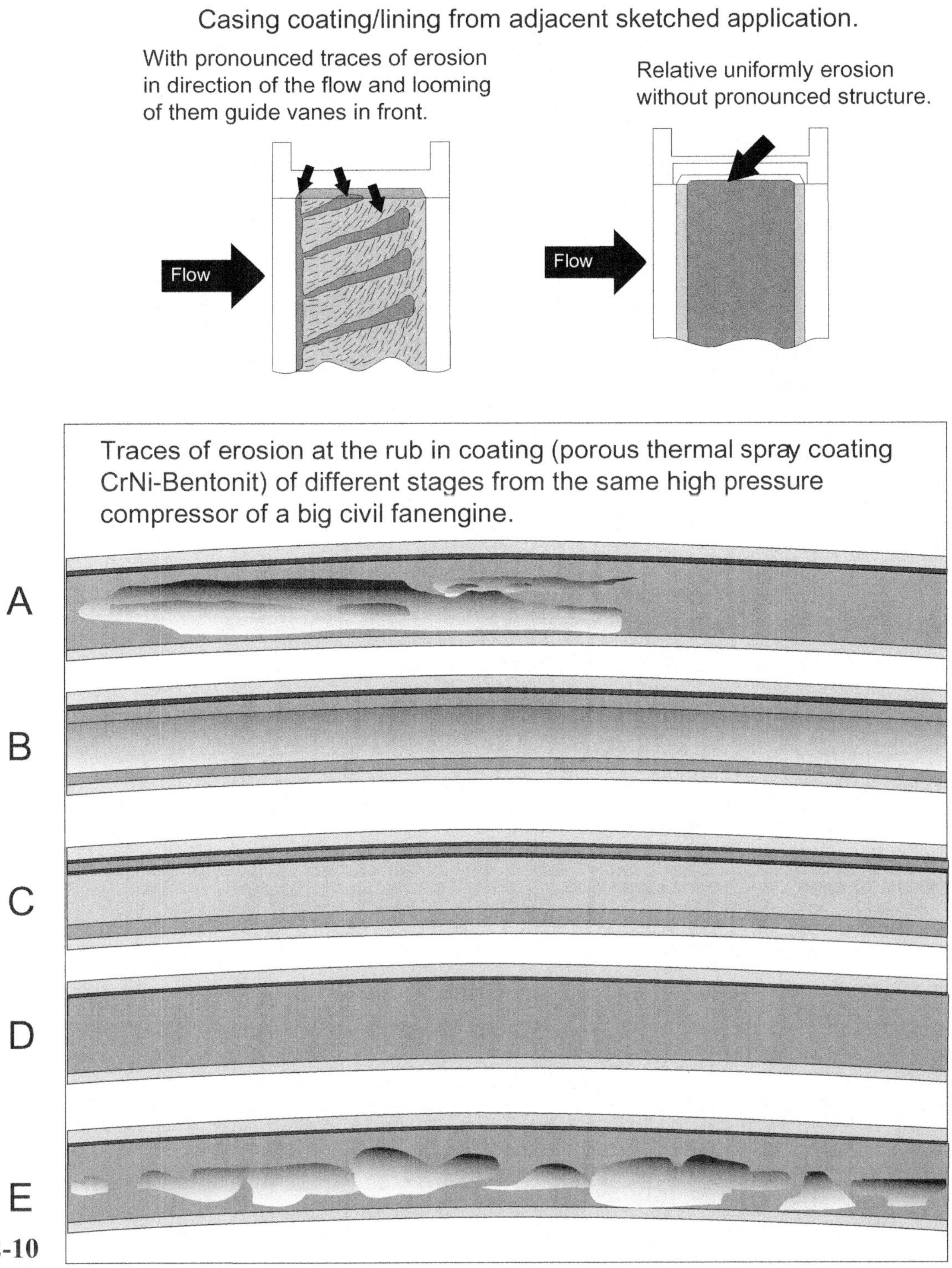

Ill. 5.3.2-10

***Illustration 5.3.2-10:** Engine-specific **singularities in the gas flow direction can affect the erosion symptoms** and the especially erosion-stressed components in characteristic ways. In engines that have a **S-shaped housing/casing ducts (swan neck duct**, left side left sketch) between the high-pressure and low-pressure compressors, particles are evidently deflected towards the hub. This can be understood by considering the particle deflection in a radial* *compressor (see Ill. 5.3.1-10). The first* **compressor spacer** *(left side left sketch) after the swan neck duct is especially erosion stressed. This stress should decrease for the following spacers due to the repeated outward centrifuging of the particles as they travel farther through the compressor. Experience seems to confirm this. The erosion symptoms of the spacers usually show the spacing of the upstream rotor blades in a more or less pronounced manner. The blades maintain their position to the spacer. Experience shows that the particles collect on the pressure sides of the blades.*

The sketches on the right side are a schematic depiction of typical erosion symptoms on soft housing coatings (such as Ni/graphite spray coatings). The darker zones are the more heavily eroded areas. Similar to the case with rotor spacers, the **erosion pattern** *corresponds to the upstream blades right side upper sketch left), although in this case the blades are the stator vanes, which do not move relative to the housing coating. Evidently the particles can also travel along the stator vane profiles and become concentrated. This typical damage symptom is usually not very pronounced when the rotor blades run-in heavily (bottom right side).*

How erosion of the rub in coatings in compressor casings can tangent the safety.

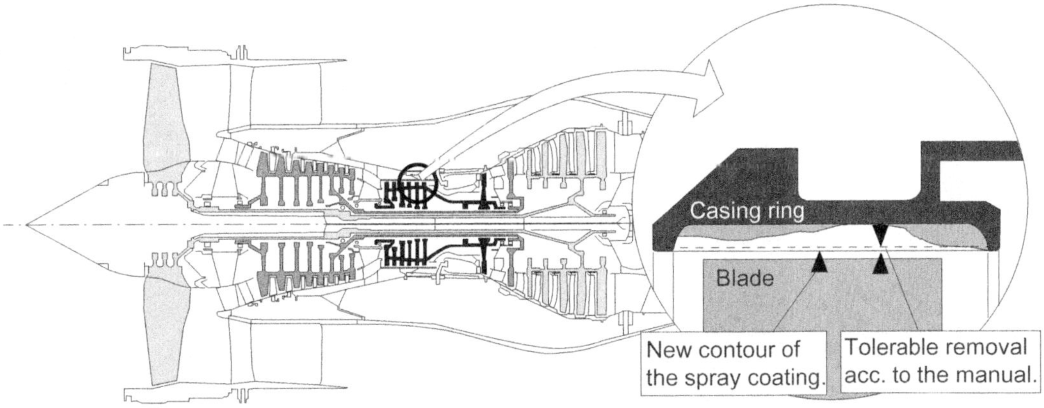

Example 5.3.2-3 (Ref. 5.3.2-22):

In the moment of lift off at the left aeroengine a loud bang occurred. After this in about one minute further 50 bangs followed. Concerned are **surge shocks** (flow stalls in the compressor, volume 3 Ill. 11.2.1.1-1). In the same moment the low pressure rotation speed of this three shaft aeroengine dropped. Contrary the rotation speed of the medium pressure rotor and the high pressure rotor increased. After dumping the fuel the aircraft landed again about 1 hour later at the departure airport.

Already about 3 weeks before the **engine health monitoring** of the aeroengine (= EHM, volume 5 Ill. 25.1-5.1) showed an **increase of the turbine gas temperature** (TGT, Ill. 5.3.2-3 and volume 2 Ill. 7.0-2). However the following **borescope inspection** is only limited possible. It produced no alarming indication.

An investigation of the aeroengine by the OEM, together with the responsible aviation authority, determined markedly **erosion of the rub in coatings** (porous NiCrAl-Bentonit) from the casing rings of the high pressure compressor (HDV, detail scetch in Ill. 5.3.2-10). An inspection of further aeroengines showed two more cases with a similar result. The **enlarged tip gaps** of the rear HPC stages caused an **efficiency drop** and triggered the surge (volume 3 Ill. 11.2.1.1-8). They also explained the rise of the gastemperature.

In dispute remained the cause for the unusual heavy erosion attack. The aviation authority called the deviations in the **quality** of the rub in coatings (thermal spray coatings) to account. These promote an increased oxidation and with this the erosion during long time operation. The OEM rather believed at an unusual amount of ingested dust.

As remedy a visual **inspektion** of the high pressure compressor, for coating erosion was ordered, if a step like increase of the gastemperature is registrated by the EHM-system.

Surge problems in a modern fighter engine following rub-coating wear.

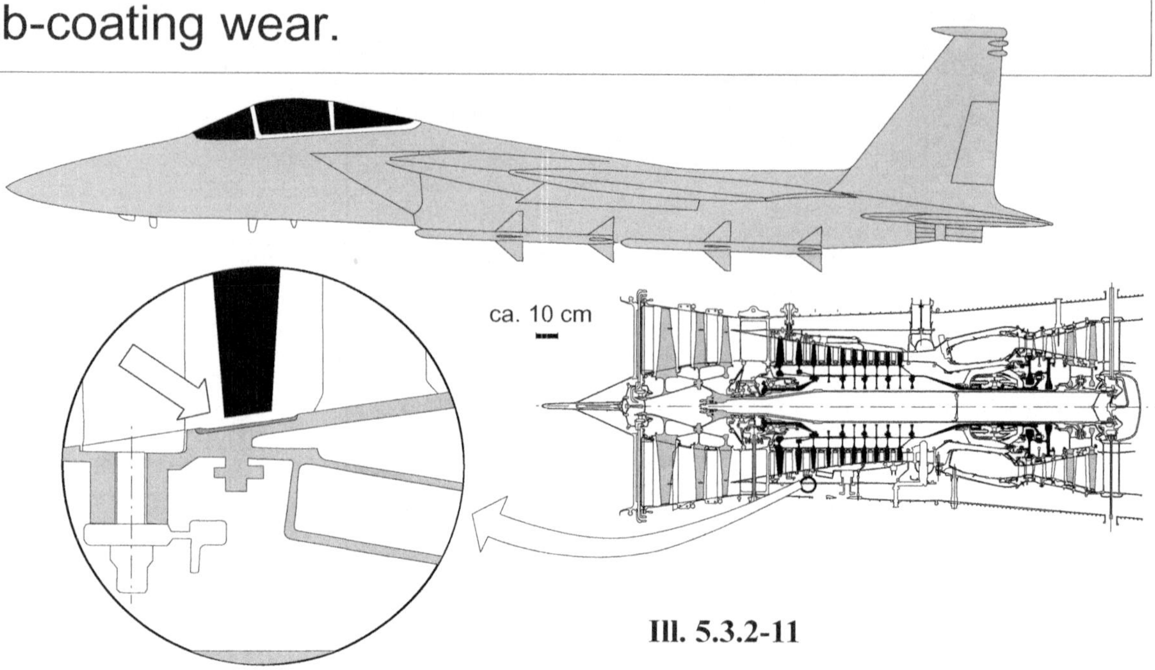

Ill. 5.3.2-11

Illustration 5.3.2-11 *(example 5.3.2-4, Ref. 5.3.2-7): The erosion of the usually soft* ***abradable coatings in compressor housings/casings*** *is a common problem in modern engines. The high power concentration at the best efficiency levels and good operating behavior demands, among other characteristics, minimization of the compressor rotor blade tip clearance. This can only be achieved by having the blades run-in when the clearance gap is breached. In this case, the blade tips must not be unallowably damaged. This requires soft coatings. Evidently even a multi-stage fan in a bypass engine is not sufficient for centrifuging all the ingested particles and preventing the core engine from experiencing any erosion stress. It can be assumed that, along with the particle concentration which increases near the housing, housings that become conically smaller along with the gas flow promote the erosion of housing coatings as in the depicted case (middle and bottom diagrams).*

Serious erosion can also occur on turbine blades.

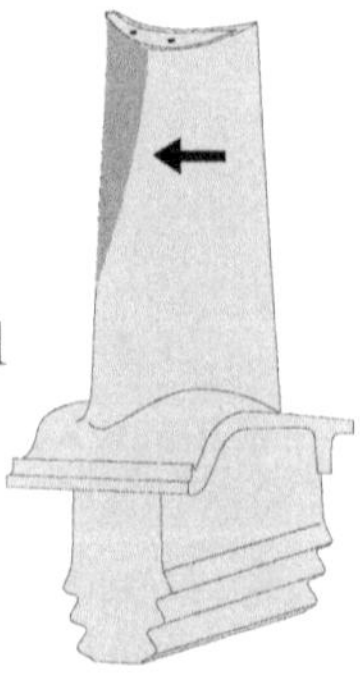

Ill. 5.3.2-12.1

Illustration 5.3.2-12.1: Particles from the combustion chamber *and other engine zones, located upstream in the engine, can cause serious erosion on the pressure side of turbine stator blades and the suction side of turbine rotor blades (also see Ills. 5.2.1.1-11, Ill. 5.3.2-12 and 5.2.1.1-12.2). Typical particles include soot from the combustion chamber (example 5.3.1-1) and particles from crumbling* ***ceramic thermal barriers****, rub coatings of the seal segments, or*

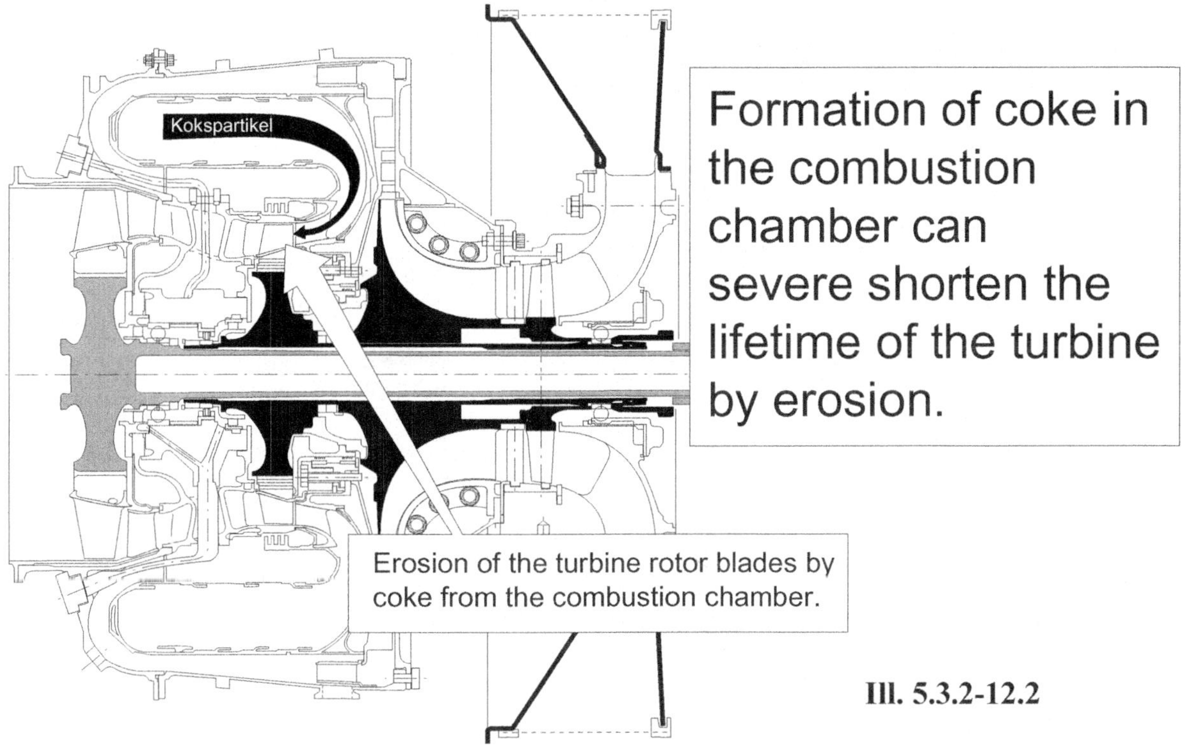

Ill. 5.3.2-12.2

hard armor coatings *on blade tips and labyrinths.*
Even damage to thin diffusion coatings (typical thickness less than 0.1 mm) can result in oxidation of the base material, promoting damaging erosion which can considerably reduce the life spans of the parts.

Illustration 5.3.2-12.2 *(example 5.3.1-1): To minimize the life shortening* ***erosion by coke from the combustion chamber*** *of this helicopter aeroengine, a version of the combusion chamber with less coke formation (low coke combustor) was introduced. This combustion chamber has a new designed* ***effusion cooled inner wall****. So there was the hope to double or triple the life time of the blades. At the same time also the visible* ***smoke production should be minimized****. The effusion cooling enables a* ***more even temperature distribution without hot spots****. With this ,the lifetime of the inner wall may be markedly extended.*

Illustration 5.3.2-12.3: *The material removal from a hot part surface can be highly accelerated by a* ***combination of oxidation and erosion****. Brittle oxidation layers are especially erosion sensitive at steep impact angles (Ill. 5.3.1-5), such as those expected on leading edges of blades. The erosion material removal from the* ***protective oxide coatings*** *exposes reactive fresh metal surfaces which oxidize correspondingly heavily. A similar scenario (Ref. 5.3.2-17) occurs when* ***dust particles are melted*** *in the combustion chamber and then strike and* ***stick*** *(glassing, Ref. 5.3.2-4) to the relatively cold cooled surfaces of the hot parts, especially the blades.*
The melted dust can cramp and/or react with the relatively coarse oxide layers, creating a powerful bond. Thus especially the ***flow through cross section at stators (nozzle) of the high pressure turbine*** *can be clogged, so that during a power increase the aeroengine can easy get into a surge (Ref. 5.3.2-4).*
When the engine is shut down, these layers harden into brittle coatings with crack initiation

Accelerated erosion on the turbine blading caused by a reaction with dust melts.

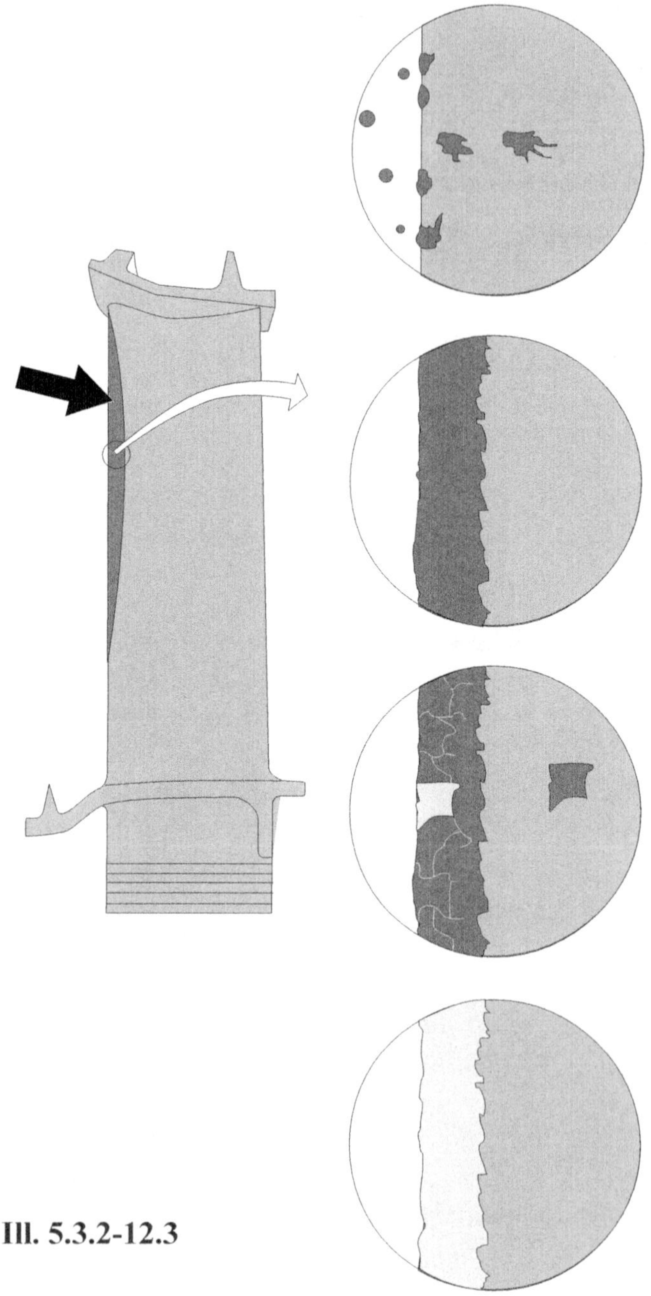

Ill. 5.3.2-12.3

and break off partly or completely along with the protective oxide layers (Ref. 5.3.2-20)..
When engine operation is ***resumed, the fresh metallic surfaces*** *oxide especially heavily, and the process is repeated when dust again enters the engine.*

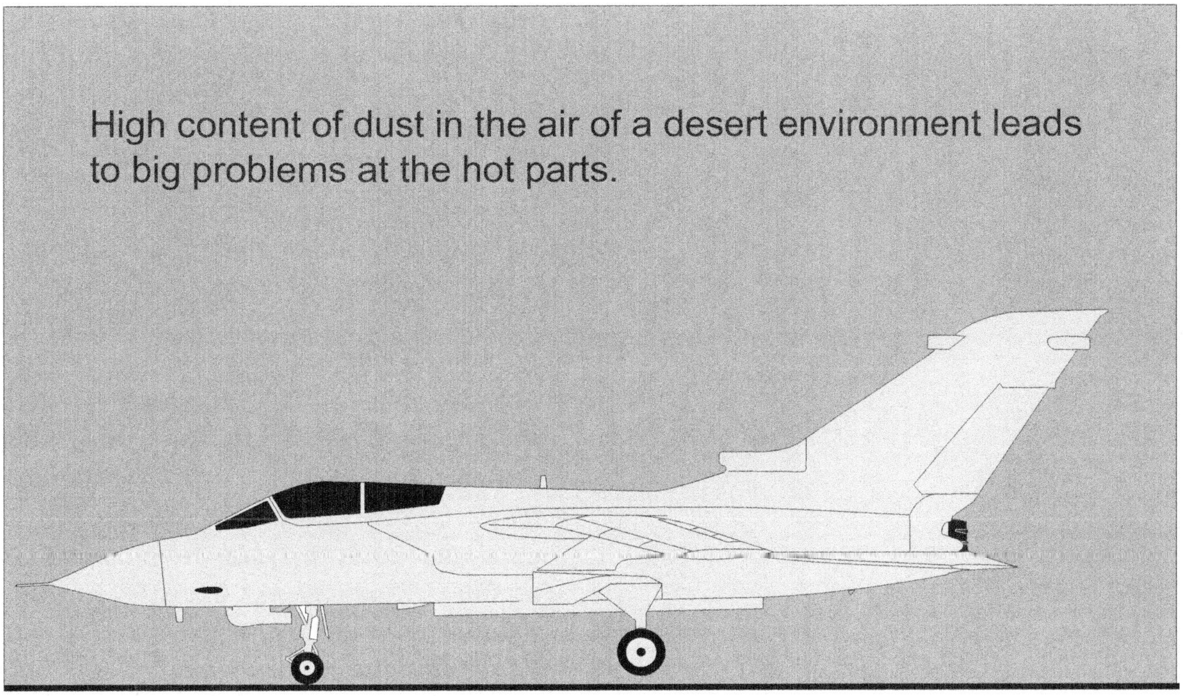

Example 5.3.2-5 (Ref. 5.3.2-8, Ill. 5.3.2-13):

Excerpt: "...(in) North American **desert environments**. Throughout these operations there was no indication of any adverse engine effects associated with **erosion or corrosion** of turbo machinery components or accessories.

However, problems were already being experienced with the **HP and IP Turbine blades**, both being cast equiaxed IN100 material.

The HPT blade was suffering **shroud loss and aerofoil failures** due to **thermal fatigue** and creep. Regular boroscope inspections were carried out to reduce the instances of blade failures.

The **IP turbine blade** was suffering failures due to **High Cycle Fatigue** (HCF), thermal fatigue and in some cases airfoil creep.

...(Use in near east desert environment) The first engine rejection occurred at unexpectedly low life due to **High Pressure Turbine** (HPT) blade Leading Edge (L/E) **burning** and (later)...a significant number of engines had been rejected due to similar problems.

Laboratory examination of failed...**HPT blades** showed that both the external leading edge and internal film cooling **holes were being restricted**. Measurements of L/E cooling hole outlet diameters revealed varying degrees of restriction from the normal size of 0.3 mm width, in some cases **total blockage**.... The inlet diameters of the L/E film cooling holes could be restricted to as little as 30% of normal diameter."

Comments: These blockages of the cooling air ducts in the turbine blades caused local temperature increases up to 200°C. If one considers that an increase of about 15°C can halve the life span, then the life span reduction here is a factor of about 1000! Therefore, life spans of only a few hours can be expected for these components.

***Illustration 5.3.2-13** (Ref. 5.3.2-8): The described three-shaft engine type (top diagram) of a twin-engine fighter aircraft was used in a **desert environment**. The ingestion of sand and dust into the engine is a serious problem. Interestingly, the greatest problems do not occur due to erosion in the compressor, but in the **hot parts**. Sand particles are carried onto the turbine blading by the main gas flow (bottom left diagram). Here the bores for the cooling air film are blocked up by the sand, which has been made doughy by the combustion chamber. Further damage to the cooling air configuration of the hot parts is caused by **blocking of the cooling air ducts** (bottom right diagram) inside the blading. The sand layer stuck to the outside was removed with a **blasting device**. A probe with compressed air was inserted through the boroscope opening in front of the blades and blasted the leading edges of the blades with abrasive particles (Al oxide).*

This kind of cleaning method is naturally only recommended in extreme cases, since there is a risk of the bores for the cooling air film becoming blocked by the blasting particles.

Newer developed blasting processes (volume 5, chapter 19.2.6) can avoid this disadvantage.

In desert environments, unusually heavy sand deposits can build up in the hot parts.

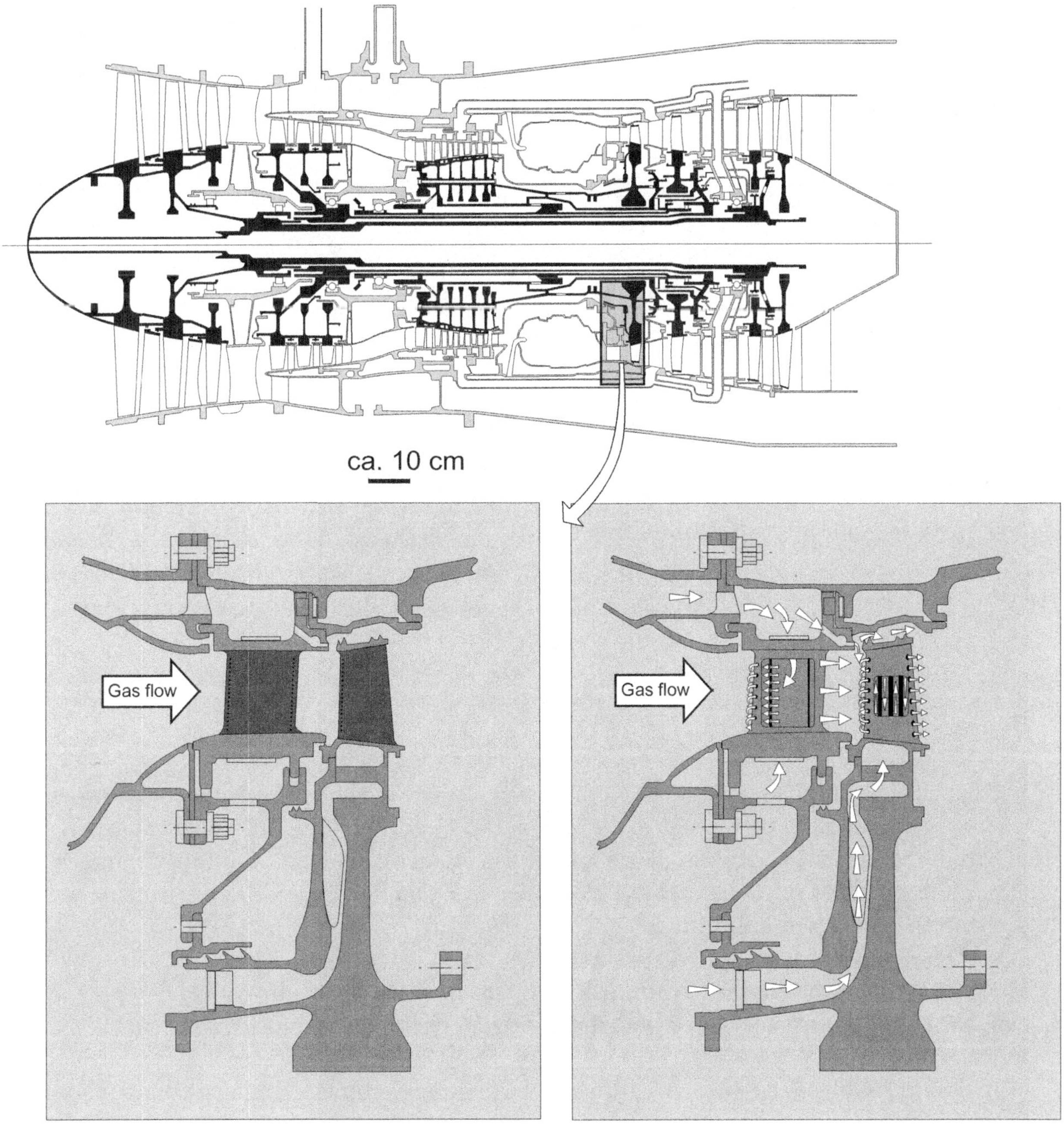

The darkened parts are those that are especially sensitive to sand deposits.

Cooling air flow in the high pressure turbine.

Ill. 5.3.2-13

Ill. 5.3.2-14

Illustration 5.3.2-14: *This diagram is based on a press photo (Ref. 5.3.2-9, Example 5.3.2-5). The light discolorations in the nose area can be seen clearly.*
In the photos it seems that it was not so much an erosion-induced roughening or delamination, but rather thick deposits of ***volcanic ash****. This is supported by the windows which, as far as can be seen, are partly covered, and not only blind due to the erosion roughening.*

Dangers of volcanic ash

The danger vulcanic ash poses to modern transport aircraft has been concretely known since the early 1980s. In a span of several weeks in the summer of 1982, two large four-engine aircraft over the eastern part of the Indian Ocean passed through the ash cloud of a volcano. All four engines failed. In these cases, the events were similar to those described in Example 5.3.2-5. It only became possible to finally restart the engines at low altitude.

The **damaging of the engines** by fine ash particles that float in the air (typical particle size, 0.005 mm, Ref. 5.3.2-10), but are very **hard and sharp-edged**, occurs in various ways:

- Erosion of blading and housings/casings in the compressor area, with worsening of compressor efficiency and operating behavior
- **Seizing of contacting surfaces** (Ref. 5.3.2-16) such as the clappers of fan blades
- **Blocking of air bores** and gills in the combustion chamber
- Blocking of air ducts in the turbine blades with resulting overheating damage (Ill. 5.3.2-15)

- **Blocking of the flow cross-sections in the turbine** due to sticking doughy and liquid ash. The **deposits** can be several centimeters (!) thick. Several hundred kilograms of hard-baked ash were removed from the four engines of one of the affected aircraft. The ash caused an unallowable pressure increase in the turbine, which the already erosion damaged compressor could no longer build up due to the thin air at high altitudes.
- **Damage to electrical systems** such as generators or ignitions.

The blocking of the hot part cross-sections with ash is especially dangerous, since some volcanic ashes have components that have **melting** points below 1000°C (the melting point of ash is usually between 1100°C and 1200°C, Ref. 5.3.2-16) and the typical gas temperature in and behind the combustion chamber is about 1300°C. The melt becomes tough or hardens into a **glass-like coating** on the cooled and therefore colder surfaces of the turbine blading. The smallest cross-section in the gas flow of the core engine is located in the turbine inlet stator. Even relatively small cross-section blockages (in the mentioned cases about 10%) cause the compressor to be unable to provide the necessary pressure increases in the relatively thin air at high altitudes. This results in a **compressor szrge** (Ref. 5.3.2-16) and **flame out** in the combustion chamber.

Damage due to ash is not the only danger that a cloud of vulcanic ash poses for aircraft engines. These clouds **contain gas-like concentration of sulfur and sulfuric acid,** the corrosiveness of which corresponds to PH values of 5.5 to 2.0 (Ref. 5.3.2-10). For example, volcanic eruptions such as those of El Chichon or Pinatubo (Ref. 5.3.2-11) released 10 to 20 million tons of sulfur dioxide, which resulted in corresponding amounts of acid in the atmosphere. Even if these gases are ingested by engines, it will not necessarily result in spontaneous damage or even engine failure. However, over long operating times serious corrosion damage can occur (such as **sulfidation**, Ref. 5.3.2-12, Example 5.3.2-6), especially in the hot parts. It is questionable how practical and technically possible cleaning the damaged parts is.

Preventing damage from volcanic ash:

The technical literature all agrees that damage can only be safely avoided by not flying through clouds of volcanic ash. The **onboard radar does not guarantee safe and sufficiently early detection** of the danger. For example, as described in Example 5.3.2-5, the cloud of volcanic ash was not recognized by the onboard radar. For this reason, early warning systems such as the Darwin Volcanic Ash Advisory Centre (VAAC, Ref. 5.3.2-13) were established especially in the high-risk geographic regions (for example, there are hundreds of active volcanoes in the Asia-Pacific region). These early warning systems also receive current reports of observations from flight crews in these areas.

Engines of **aircraft on the ground are also endangered by ash deposits**, which can be blown into the engine by wind and/or sucked in during startup. In these situations, engines must be covered in time. If ash has already entered the standing engines, they must be treated in accordance with manufacturer's specifications before startup due to the danger of a**sh burning into the hot parts**. Intensive cleaning is usually unavoidable.

In case the crew notices having entered an ash cloud during flight, engine power should be reduced as much as possible in order to lower the gas temperatures and prevent the ingested ash from turning into glass on the hot parts. However, this is not entirely risk-free, since it makes compressor stalls more likely, if significant deposits have already built up on the hot parts. The **ignition should be activated during the entire flight** through the ash cloud (Ref. 5.3.2-10). Activating all airbleeds enables the compressor to reach higher RPM and decreases the probability of a stall/surge.

Checking a suspicion, that vulcano ash got unnoticet into the aeroengines:

Because it is asbsolutely not certain, that there are external signs which point at vulcano ash, in case of doubt an inspection of the aeroengines after shut down is adviced. For this, depending from the aeroengine type the following possibilities offer, which analysis and assessment assumes expertise and experience.

Inspection of filters of the bleed air for the cabin ventilation. This should take place by photo-optical macroscopic (appearance and binoculars) and if the suspicion confirms, microscopic with a scanning electron microscope (=SEM). With a SEM, shape, structure and composition as characteristic features of particles can be identified. With this the expert gets important hints at risks for the aeroengines. Do the particles contain sulfur and phosphor, they can trigger in the hot parts hot gas corrosion as sulfidation. Glass (high SiO_2 content) can melt. With this the danger of blocking and overheating of cooled hot parts exists. This is especially true for the blades of the high pressure turbine.

Borescope inspections to identify changes of the compressor blading and of hot parts. To this belong unusual discolourations, roughness as well as deposits,

Healthmonitoring for the examination of the aeroengine behaviour. Especially interresting are effencies of compressor and turbine as well as if in this connection a possible increase of the gas temperatures exists.

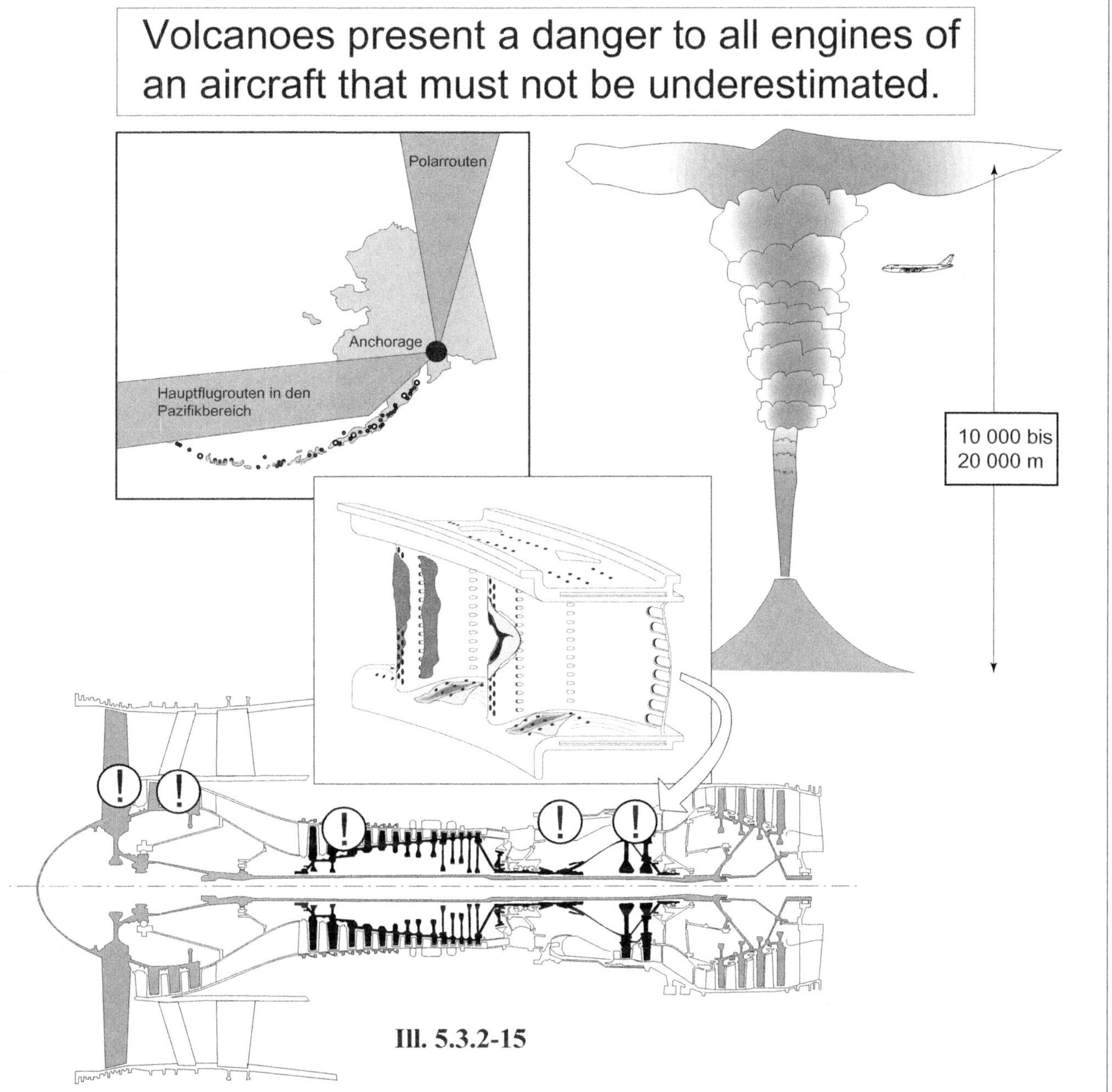

Ill. 5.3.2-15

Illustration 5.3.2-15 *(Ref. 5.3.2-15): In some regions in the world (in this example, Alaska) there are many* ***active volcanoes****. The dangerous* ***ash clouds*** *can threaten aircraft even at great distances (hundreds of kilometers) and* ***altitudes*** *up to 20,000 meters. In the lower sketch, the cross-section of the engine type from example 5.3.2-6 shows the areas especially threatened by the ingested ash. In this case* ***glassy depositions of molten dust*** *('glassing') accumulated at the leading edges of the 1st stage high pressure turbine stator (nozzle, sketch in the middle). This detail shows a nozzle vane with pronounced signs of overheating due to* ***blocking of the cooling air ducts****. However these did not occur in the case of the example 5.3.5-5.*

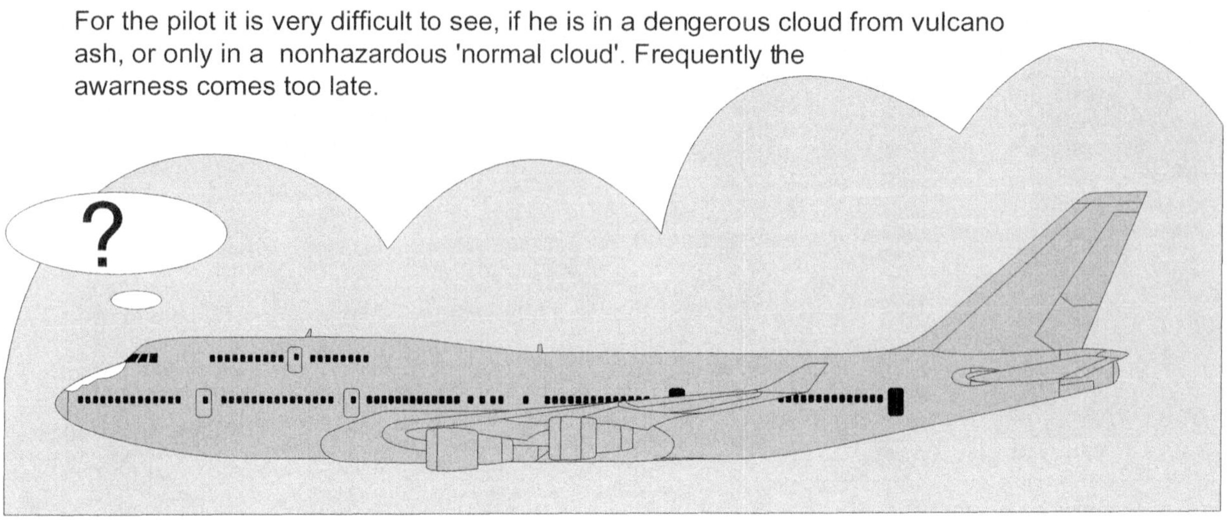

Example 5.3.2-5 (Ref. 5.3.2-14, Ill. 5.3.2-15):

Excerpt : "...At 11:50 local time, roughly 126 kilometers north of the airport (Anchorage, **Alaska**), when passing through an altitude of 7700 meters all four engines of the brand-new Boeing 747-400 suddenly failed. Several of the cockpit instruments went crazy or failed completely. The cockpit crew went into a gliding flight and sent out an emergency distress call.

The crew tried to restart the engines a total of seven times with no success. Only after eight minutes following the failure two of the jet engines (fan engines with large bypass ratios) started up again. By this time the aircraft had already lost 3700 meters of altitude. A short while later the other two engines also resumed operation. Even though many of the instruments were unusable, the crew was able to land the jumbo-jet in Anchorage. There it was discovered that the new **paint on the engine hull** was strangely roughed up (Ill. 5.3.2-14) as if it had been worked over with rough sand paper.

One day before the flight...took place, 3109 meter high (volcano) Mt. Redoubt had erupted after 21 years of dormancy. It spewed ash up to twelve kilometers up into the air for 45 minutes. The ash cloud passed over Anchorage towards the north...the cockpit crew later reported that the onboard radar did not register the ash cloud. From the cockpit windows it was also not possible to discern it from a normal water vapor cloud."

Comments: Closer looks at the press photos give the impression that it was not only a roughening of the hull area, but that there were also massive deposits of ash in this area.

It can be assumed that the engines failed temporarily not only due to the erosion damage at the compressor blading, but that the hot parts sustained overheating damage due to blockage of the cooling air ducts, and also that sticking ash seriously damaged/changed the blade profiles and flow cross-sections in the turbine.

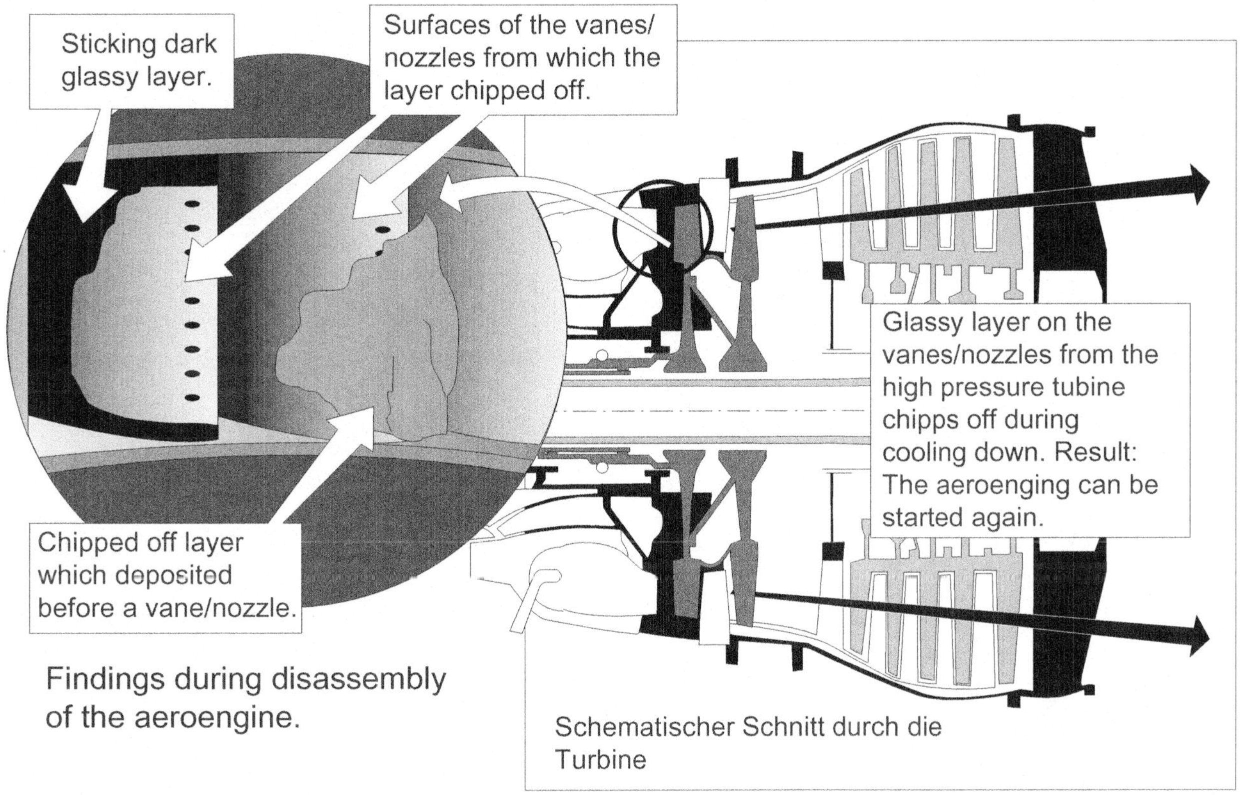

The fact that restart of the engines was only successful at lower altitudes can be largely explained by the higher air density, and therefore sufficient compression in the compressor with correspondingly larger air throughput at the startup RPM.

As later emerged, the erosion in the compressor was relatively light and did not explain the flame out of the aeroengine. Main reason for this was **sticking melt of the ash** at the **1st stage blades of the high pressure turbine** ('glassing'). An about 1,5 mm thick layer developed at the leading edge and pressure side. With this the **flow cross sections narrowed at the critical point**. So obviously the flame out of the aeroengines occurred. It is interresting, that nothing is reported about a sign for an aeroengine surch (bangs, shocks). This points at a **rotating stall** in the compressor (volume 3 Ill. 11.2.1.1-1) without a total flow disruption. The air flowrate however may be such reduced, that the combustion chamber in the high altitude quenched. A **restart of the aeroengines** was at least not successful, because of a lucky **property of the deposits**. These glassy layers **chip off large-scale** during cooling (Ill. 5.3.2-12.3) and have been transported with the gas stream, without damaging the following turbine blades (Ref. 5.3.2-23). This lead during the start trials to the **widening of the critical flow cross section**. Helpful may have been also for the restart the high air density during the start trials at low altitude. So a sufficient compression in the compressor with a correspondent higher airflow rate was possible.

Example 5.3.2-6 (Ref. 5.3.2-12):

Excerpt : "Corrosion caused by **sulphur-laden ash** from the eruption of Mt. Pinatubo may have cut deeply into the Philippine military's force of about 26 fixed-wing combat aircraft.

...As much as 10 in (25 cm) of ash, much of it the consistency of fine sand, 'got into everything' after (the) eruption.

...sulphur in the ash combines with moisture to create sulfuric acid, 'which gets into every crack and crevice,'..a consultant...said. He was charged with rehabilitating some of the equipment corroded by the eruption.

Comments: It was not possible to take the aircraft to a safe airport. The volcanic ash evidently also affected standing engines.

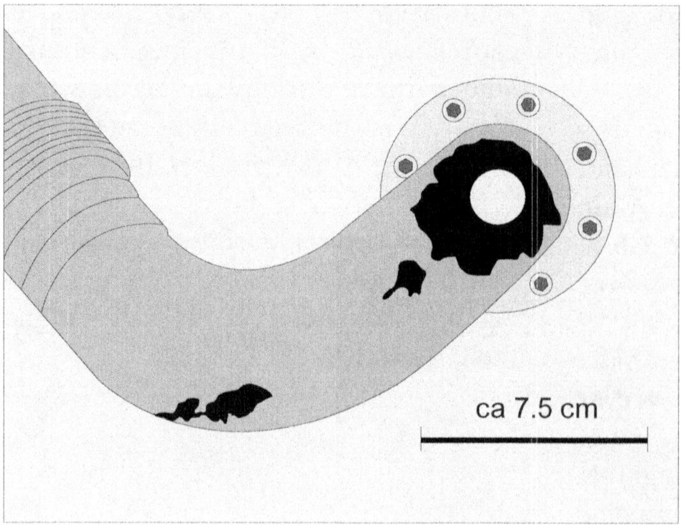

Not only the compressor blading is threatened by erosion.

Ill. 5.3.2-16

Illustration 5.3.2-16 *(Ref. 5.3.2-15): The walls of this air extraction duct in the compressor eroded at the bend behind the extraction hole during erosion tests on an engine of a middle performance class. The increased erosion attack at the outer radius of the pipe bend is characteristic, since centrifugal effects and the redirection of the flow concentrate the erosion particles in this area.*

What should be done if a potential danger through vulcano ash arises.

In larger amounts vulcano ash can cause aeroengines to fail/flame out within minutes. This can produce a dangerous flight situation (Ill. 5.3.2-15 and Ill. 5.3.2-16). Tereby are **ash concentrations** in the air which are **<u>not</u> clearly visible with the naked eye**. Contrary however seems to exist a connection between '**cloud hazes**' (e.g. cirrus clouds) and vulcano ashes/gases. Ash componentsm like sulfur (from SO_2) and particles form **aerosols**. Particles act as **condensation nuclei** for droplets and ice crystals. In this form seemingly little amounts of vulcano ash in the air which by far are **below the limit of visibility** have unacceptable **economical consequences** (Ref. 5.3.2-26). Already during a **flight of few minutes** are for the outsider seemingly harmless **deteriorations** possible. To these belongs especially the **constriction of cooling air holes**. However during further operation the locally increase of the component temperature **shortens the lifetime of the hot parts up to several orders of magnitude**. Thereby a for the flight safety critical spontaneous failure is rather relatively unlikely. So **at low ash concentrations no <u>acute</u> safety problem** is concerned. Precondition is, that the online monitoring and the specified aeroengine controls take place. Currently for a pilot it is not possible to identify such a situation sufficient certain. At first sight the statement amazing, that at the first glance obviously the **compressor blading is smoothened**/polished. So **even the compressor efficiency improves**. This becomes noticeable by slight **reductions of the EGT** (gas temprature behind the high pressure turbine). This effect is even increased if clogged cooling air holes of the hot parts **reduce the cooling air flow** and **additionally the efficiency of the engine**.

With a markedly drop of the efficiency and rise of the EGT is not to be reckoned before a distinct erosion (roughening and cange of the profile) of the bading.

However already at very low concentration of ash/gas a very serious **shortening of the hot parts lifetime**, especially the very **cost-intensive** blading of the high pressure turbine must be expected. The danger of a **long time deterioration by sulfidation** (Ill. 5.4.5.1-4) exists already by sulfur deposits which can hardly identified with the borescope. Get these failures visible usually concerned components must be exchanged. Such commercial risks may especially sudenly hit **repair shops** „Power by the Hour" contracts (flat compensation for every flight hour of the operator/contractual partner). In **contracts for repair/overhaul** and of **insurances** such a situation must be considered.

Illustration 5.3.2-17 *(Ref. 5.3.2-26): Shown are changes in civil fan engines of a 4-engined airliner. It flew about 7 minutes through a 'veil cloud' (**cirrus cloud**) contaminated with vulcano ash and gases. Subsequently concerned hot parts of the 4 aeroengines had to be repaired for a cost equivalent of the price from about 150 medium-class cars.*

*The assessment of the influence requires a **comparison standard**. This enable previous, documented routine investigations at the same aeroengine and/or sufficient experience.*

*As **measures** for the evaluation of the consequences offer itself:*

*- **Borescope inspections** of the blading from turbine („**S**") and compressors.*

*- Microscopic **investigations of air filters** like used in the air conditioning („**F1**" and „**F2**") or bleed air.*

*- **Health - monitoring** (volume 5 chapter 25.1) of the aeroengine, e.g. observation of a change of the EGT (see above).*

*- **Oil analysis** („**O**") spectrometric type (volume 5, chapter 22.3.4).*

Indications of vulcano ash in time before a failure of the aeroengine is to be expected, can be identified with targeted monitoring. Then however, already expensive repairs can become necessary.

free acc. Ref. 5.3.2-26

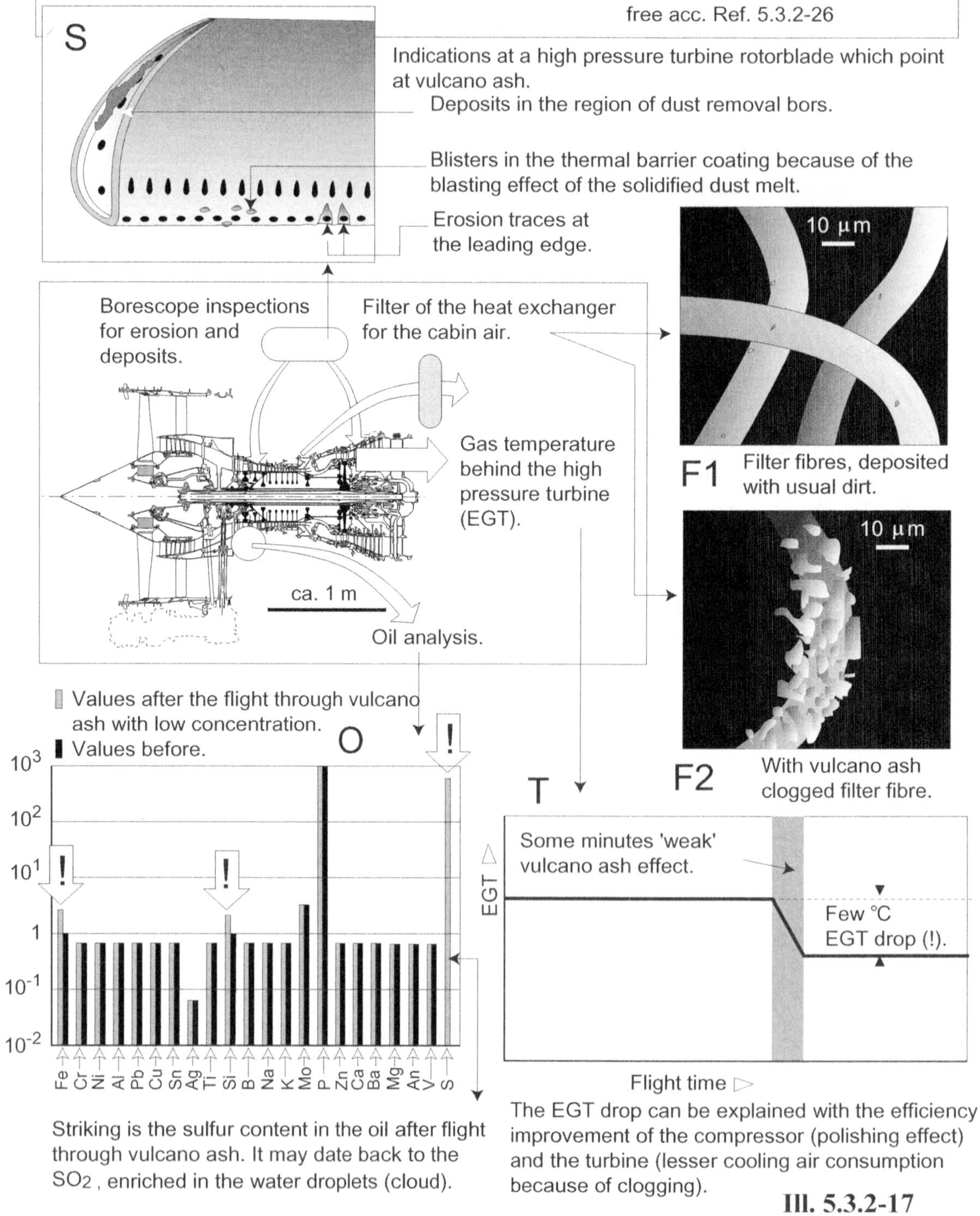

Ill. 5.3.2-17

Chapter 5.3.2 References

5.3.2-1 D.L.Mann, G.D.Warnes, "Future Directions in Helicopter Engine Protection System Configuration", AGARD-CP-558, proceedings of the conference "Erosion, Corrosion and Foreign Object Damage Effects in Gas Turbines", Propulsion and Energetics Panel Symposium, Rotterdam, The Netherlands, 25-28 April 1994, Chapter 4, page 8.

5.3.2-2 Rolls-Royce Aero-Engine Conference, contribution to the proceedings, pages 9 to 23.

5.3.2-3 B.C.Barry, T.C.Currie, "Erosion of T56 5th Stage Rotor Blades Due to Bleed Hole Overtip Flow", AGARD-CP-558, proceedings of the conference "Erosion, Corrosion and Foreign Object Damage Effects in Gas Turbines", Propulsion and Energetics Panel Symposium, Rotterdam, The Netherlands, 25-28 April 1994, Chapter 17.

5.3.2-4 V.R. Edwards, P.L. Rouse, "U.S. Army Rotorcraft Turboshaft Engines Sand & Dust Erosion Considerations", AGARD-CP-558, proceedings of the conference "Erosion, Corrosion and Foreign Object Damage Effects in Gas Turbines", Propulsion and Energetics Panel Symposium, Rotterdam, The Netherlands, 25-28 April 1994, Chapter 3, page 5.

5.3.2-5 Issue 10-14, periodical "Aviation Week & Space Technology, September 22, 1980, pages 143 and 144.

5.3.2-6 "Eagle Claw, The Raid into Iran", Villard Military Series, ISBN 0-394-74403-9, pages 85 and 86.

5.3.2-7 S.W.Kandebo, "USAF Weighs Alternatives In Wake of F100 Problems", Aviation Week & Sapce Technology, June 8, 1998, pages 22 to 24.

5.3.2-8 M.G. Down, M.J. Williams, "Out of Area Experiences with the RB199 Tornado", AGARD-CP-558, proceedings of the conference "Erosion, Corrosion and Foreign Object Damage Effects in Gas Turbines", Propulsion and Energetics Panel Symposium, Rotterdam, The Netherlands, 25-28 April 1994, Chapter 1.

5.3.2-9 H.Rademacher, "Luftverkehr: Riesige Aschewolken bringen Triebwerke zum Stillstand - Gefahr nach Vulkanausbrüchen", periodical "Flugrevue", 11/1991, pages 51 to 53.

5.3.2-10 D.Manningham, "Dust & Ashes", periodical "Business & Commercial Aviation", October 1997, pages 74 to 78.

5.3.2-11 T. Bührke, "Schwefelsäure zerstört Ozon", newspaper "Süddeutsche Zeitung" April 7, 1994.

5.3.2-12 "Volcanic Damage to Philippine Aircraft Imperils Anti-Insurgent Operations", periodical "Aviation Week & Space Technology", April 13, 1992, pages 60 and 61.

5.3.2-13 R.Cantor, "Complete avoidance of volcanic ash is only procedure that guarantees flight safety", periodical " ICAO Journal", September 1998, pages 18 and 19.

5.3.2-14 H. Rademacher, "Die Aschefalle", periodical "Bild der Wissenschaft", 10/1991,pages 80, 81.

5.3.2-15 M.G. Dunn, C. Padova, R.M. Adams, "Operation of Gas Turbine Engines in Dust-Laden Environments", AGARD-CP-421, proceedings of the conference "Advanced Technology for Aero Gas Turbine Components", Chapter 8, page 12.

5.3.2-16 D. Manningham, "Dust & Ashes", periodical "Business & Commercial Aviation", October 1997, Seite 74-78.

5.3.2-17 P.König, T. Miller, A. Roßmann, "Damage of High Temperature Components by Dust-Laden Air" , AGARD-CP-558, proceedings of the conference "Erosion, Corrosion and Foreign Object Damage Effects in Gas Turbines", April 1994, pages 25-1 to 25-12.

5.3.2-18 R.Renyer, "Inlet Protection Systems for U.S. Army Aircraft", proceeding (RTO MP-34, AC323/ AVT/TP14) des RTO AVD Symposiums, Ottawa, Canada, 18-21 October 1999, pages 6-1 to 6-4.

5.3.2-19 R.V. Narayana Murthy, V. Unnikrishnan, B.G. Amarnath, K. Srinivasa, "Erosion Problem of Axial Compressor Blades of a Turboshaft Engine", proceeding paper ISABE 87-7059, of the "Eighth International Symposium on Air Breathing Engines", June 14-19, 1987. Cinncinnati, Ohio, pages 619-626.

5.3.2-20 R.A.LeyesII,W.A.Fleming „The History of Norrth american Small Gas Turbine Aircraft Engines", AIAA, ISBN 1-56347-332-1, 199.page 551.

5.3.2-21 J.L.Labadie, „Volcanic Ash Effects and Mitigation", 1983, (www.crew.org, 23.11.2005) Seite 3.

5.3.2-22 ATSB Transport Safety Investigation Report, Aviation Safety Incident Report 200403110, „Engine failure - Melbourne Airport, Victoria 25-Aug-2004, Boeing Company 777-312, 9V-SYB", page 1-81.

5.3.2-23 Z.J.Przedpelski, T.J.Casadevall, „Impact of Volcanic Ash from 15 December 1989 Redoubt Volcano Eruption on GE CF6-80C2 Turbofan Engines", Proceeding of the 1st. International Symposium on Volcanic Ash and Aviation Safety, U.S.Geological Survey Bulletin 2047, 1990, page 129 - 135.

5.3.2-24 T.J.Casadevall, „Volcanic Hazards and Aviation Safety: Lessons of the Past Decade", Flight Safety Foundation, Zeitschrift „Flight Safety Digest", May 1993, page 1-9.

5.3.2-25 Prospekt der Fa. Straaltechniek International V.V., „Ice Blast Technology", www.straaltechniek.net, 3, page 1-9.

5.3.2-26 T.G.. Grindle, F.W. Burcham, Jr. „Engine Damage to a NASA DC-8-72 Airplane From High-Altitude Encounter With a Diffuse Volcaqnic Ash Cloud", Paper NASA/TM-2003-212030, August 2003, page 1-23.

5.3.3 Measures Against Erosion Damage

- Use of the most erosion resistant materials possible, the suitability of which must be verified with comparative **tests that sufficiently realistic for operating conditions** (angle of impact, speed, particle size, particle material, etc.).

- **Even during the development phase,** optimization and testing of sufficient erosion resistance of the engine with realistic erosion tests that simulate operating conditions.

- **Armoring/hard facing** blade areas that are especially prone to wear stress (for example, through sprayed **WC/Co**, see Ill. 5.3.3-4). Caution: for the rear compressor areas the coating must be tested for sufficient temperature resistance.
It must also be tested to determine whether these measures decrease the dynamic strength of the blades. This can be caused even by an undamaged coating (through residual stresses or notch effects at the coating edge, etc.). If possible the location and shape of armor should be determined only after a modal analysis (to determine vibration forms) in order to prevent the coating boundary from meeting with vibration-form dependent stresses. **With brittle coatings** there is a special **risk of crack initiation** through FOD or larger sand grains, which can result in an extreme decrease in dynamic strength

- **Galvanic coatings** with inert behavior can cause local cell action and pitting corrosion in compressor blades made from corrosion sensitive materials such as Al alloys or Cr steels. This can result in an unallowable decrease in dynamic strength. Even if no corrosion occurs, the tension residual stresses common in galvanic coatings (such as Cr or Ni coatings) are a threat to the dynamic strength.

- **Lacquers and coatings** with suitable ductile PU coatings give the best protection at high impact angles (Ill. 5.3.1-5), and their use is limited by their temperature sensitivity. Typical uses are on the spinner or the fan outlet stator vanes.

- **Constructive measures** (Ill. 5.3.3-5) **on the inside of cylindrical housings** with rotating internal flows and dust collections (Ill. 5.3.1-8)

- Use of **separators and filters** (Ills. 5.3.3.1 and 5.3.3.2). These have the disadvantage of creating resistance to the inlet air flow, worsening engine performance and fuel efficiency.

- Use of **vortex suppression devices** (**'vortex killers'** Ill. 5.2.1.2-9) to minimize the ingestion of foreign objects and dust from the runway.

- **Optimizing operation**, i.e. the taxiways (Example 5.2.1.2-4).

- **Suitable inspections** of the compressor (boroscopic inspections of the blading and housing walls).

- **Optimal constructive design of the inlet area** of the engine (Ill. 5.3.3-3). For example, the spinner shape and the fan geometry should be designed to work well together.

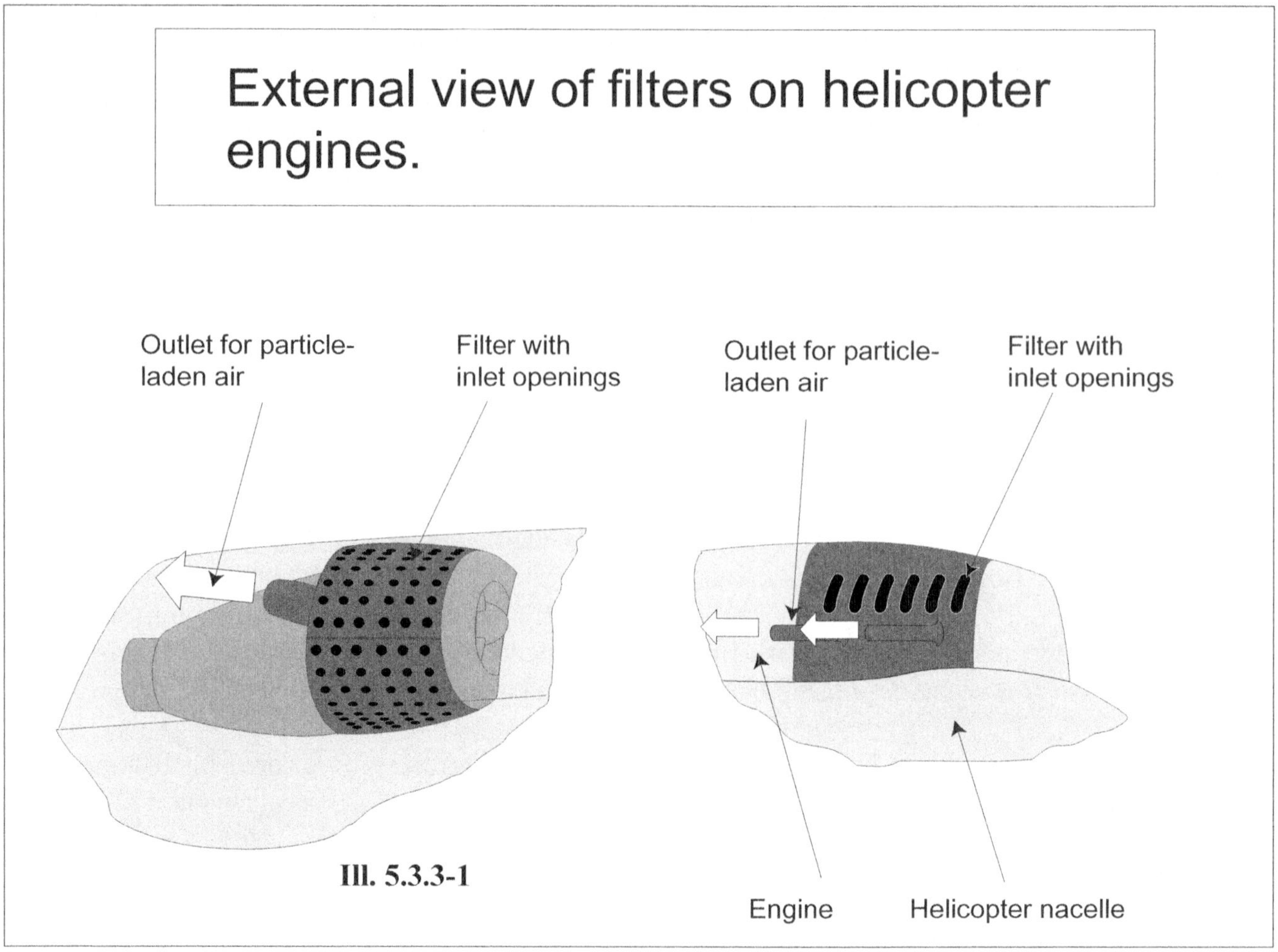

Ill. 5.3.3-1

Illustration 5.3.3-1: *The diagrams show **filter devices** in the inlet flow of helicopter engines. The intake openings and sockets for the particle-enriched exhaust air.*

Illustration 5.3.3-2.1: Filters *that rely on the filtering effect of **porous materials** are widely used in industrial gas turbines because there is usually a large amount of available space. These filters require sufficiently large surface areas in order to minimize the flow resistance.*
*In aircraft engines, especially in helicopters, **particle separators** that operate on basis of the principle of inertia have found widespread and successful use. While they do not reach the high separation rates achieved by filter materials, they have sufficiently low flow resistance at a compact size. However, it must be remembered that integrated particle separators in the inlet air flow will always increase the engine weight and worsen performance, i.e. increase the specific fuel consumption.*
*The top diagrams show separator types in which the air is diverted **tangentially** (e.g. through guide vanes, left diagram) and/or **radially** (right diagram). The heavy particles (diameter greater than 0.010 mm, see Ill. 5.3.1-9) are less likely to be diverted or are sufficiently centrifuged outward, resulting in a satisfactory separating effect.*
*It must be remembered, however, that the principle of these devices means that they **can not completely remove all particles** from the gas flow, and a certain amount of erosion stress in the compressor is to be expected.*
*However, in this case it is easier to realize erosion protection through **wear-resistant, hard coatings (TiN**, etc.), since the small, low-energy particles will not smash through the coating.*

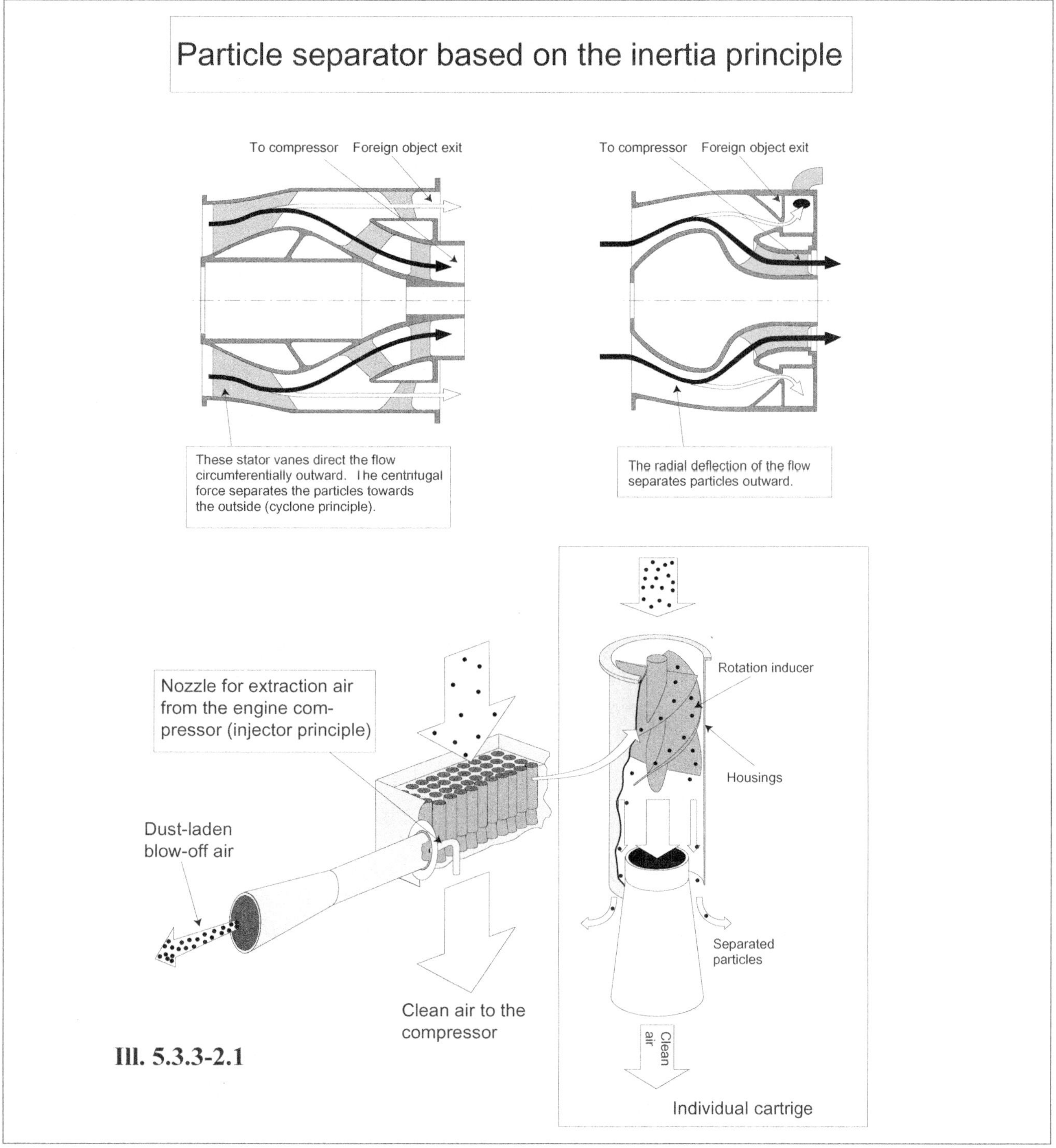

Ill. 5.3.3-2.1

The bottom diagram depicts a modularly constructed system with ***filter cartridges*** *in a parallel configuration. In these cartridges, guide vanes make the air stream rotate, and the particles which are centrifuged outward are carried overboard through a bypass by an exhaust air flow which is supported by air injected from the compressor. This configuration can be used as an optional retro-fitting, since it must not be integrated into the engine and the* ***modular construction*** *allows it to be adjusted to fit the individual air throughput of different engine types.*

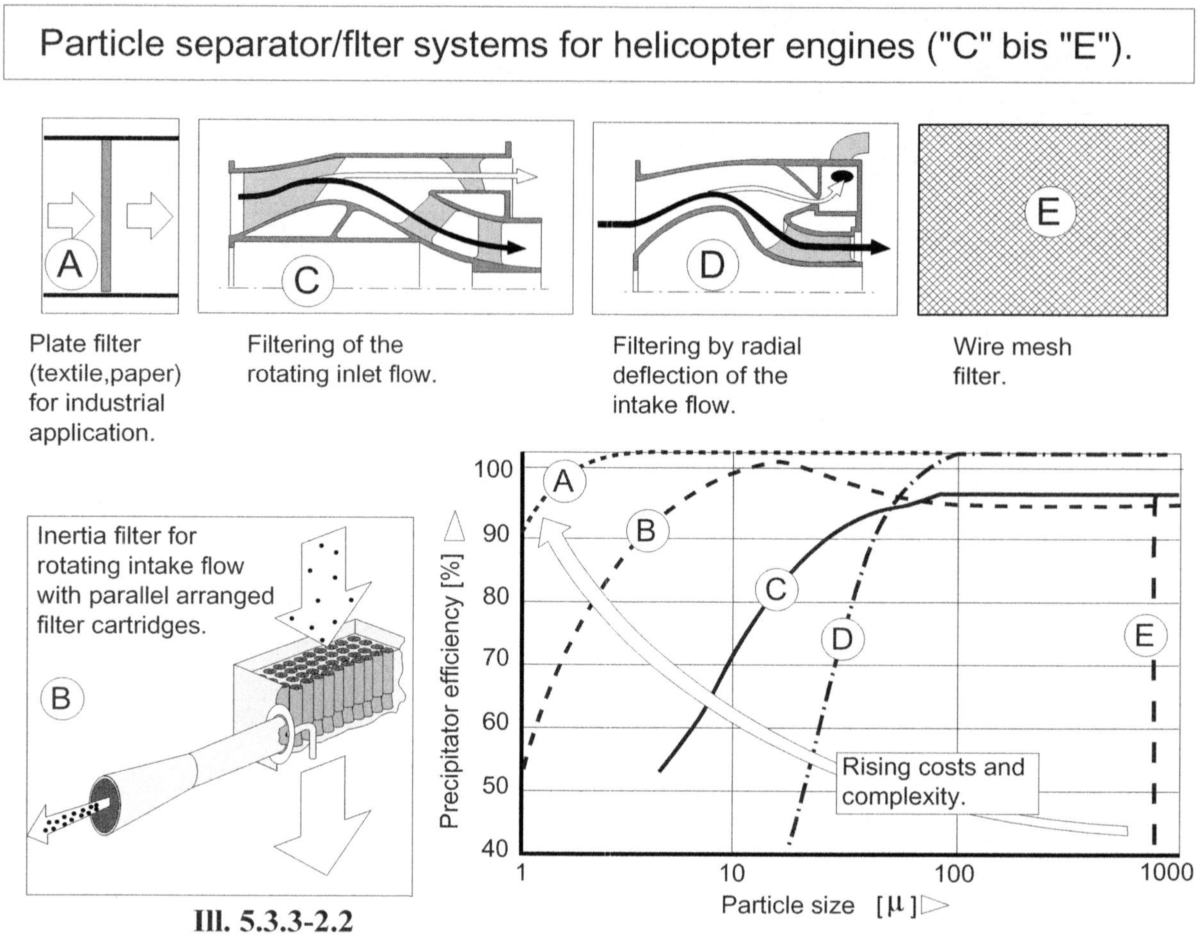

Ill. 5.3.3-2.2

Illustration 5.3.3-2.2 *(Ref. 5.3.3-1): The in Ill. 5.3.3-2.1 described and in use existing systems of filters/**separation systems** in the **intake air of helicopter aeroengines,** naturally **do not guarantee a totally particle/dust separation**. The so called **precipitator efficiency** is understandably depending from the particle size. It is markedly lower, that means the systems are more pentrable as plate filters („**A**"). We find them in industrial applications like for industrial gas turbines, compressors respectively power stations. These systems are too large and too heavy helicopter engines to get a sufficient low intake drag.*

*Sketch „**E**" shows an exeption. Here a low intake drag is achieved by a **wire mesh/sieve** with relatively **wide mesh size**. Naturally, with this the precipitator efficiency is corresponding bad.*

*For the application in helicopters most effective are **parallel applied filter cartridges** (Ill. 5.3.3-1), which are based on the inertia principle („**B**").*

*Then follow **integrated separator systems** („**C**", „**D**") at which the whole intake flow is brought into rotation and/or is radial deflected.*

The geometry of the spinner, the fan blade and axial clearances govern to a high degree the particle entry into the core.

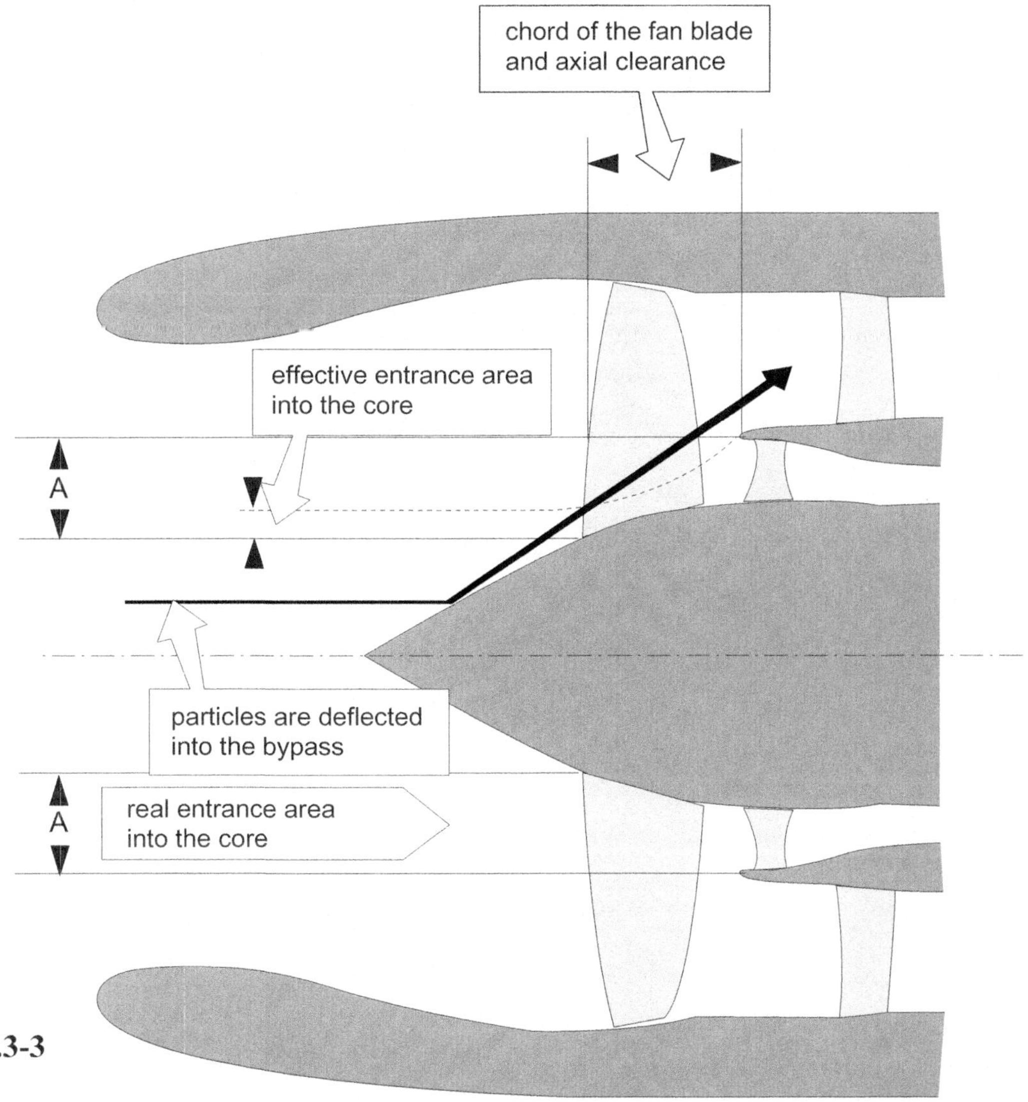

Ill. 5.3.3-3

***Illustration 5.3.3-3** (Ref. 5.3.1-12): In order to **protect the core engine from erosion and FOD**, and therefore also to reduce maintenance costs and service interruptions, the **fan area must be optimally designed**. Important factors include:*

*- Proper selection of the **cone angle of the spinner** can control the trajectories of ricocheting particles so that they pass through the bypass duct.*

*- The **distance of the fan blade leading edges from the splitter** must be coordinated with the cone angle of the spinner in order to direct the largest possible percentage of particles into the bypass duct.*

External Factors
Erosion: Measures

Practical operating experience is the deciding factor for the successful implementation of thermally sprayed coatings as erosion protection for compressor blades.

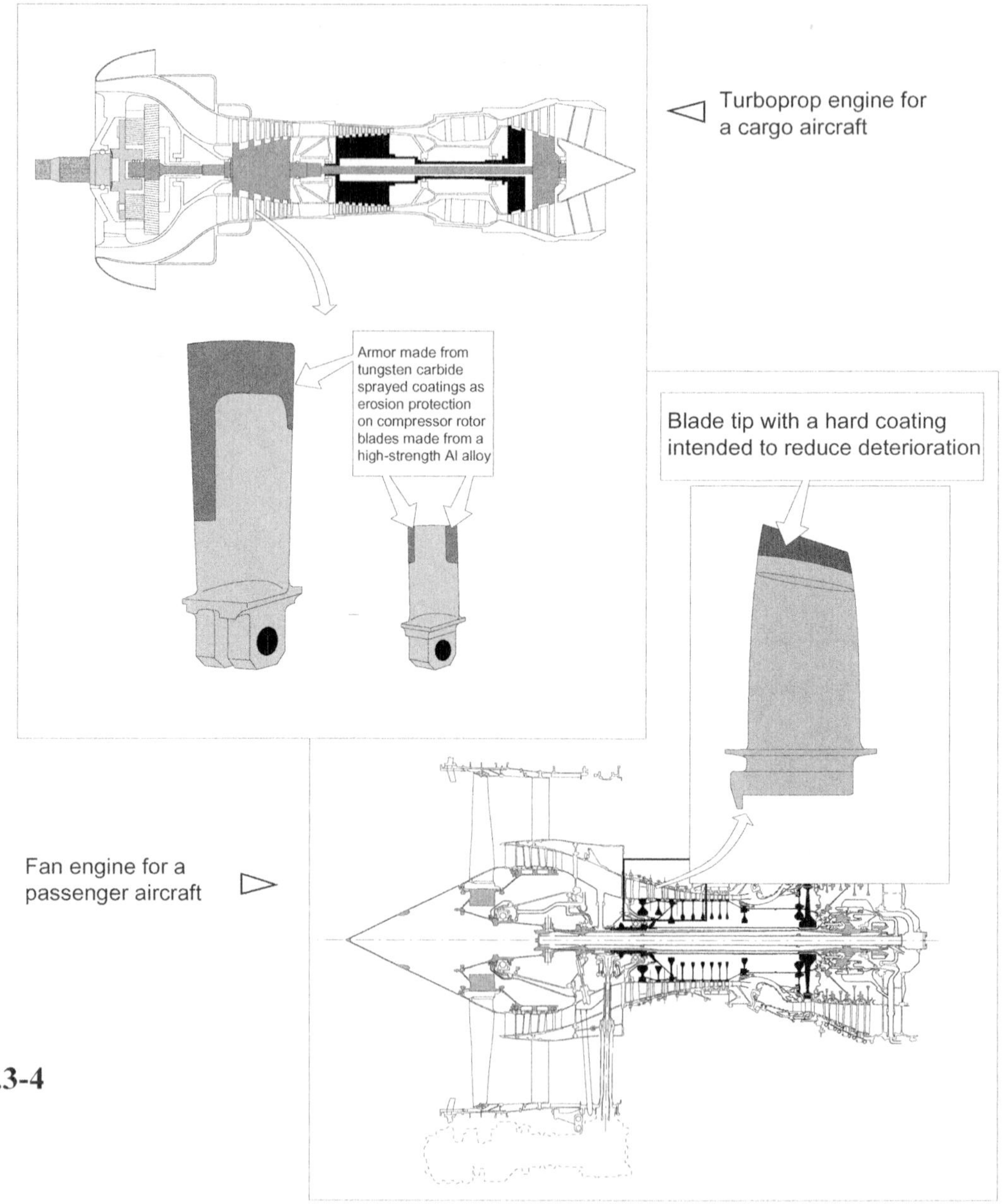

Ill. 5.3.3-4

Illustration 5.3.3-4: *In practice, the application of hard armor coatings/**hard facing** to **compressor blades** has proven effective not only in older engine types. However, with the narrow and sharp-edged blade profiles in modern compressors with high pressure ratios and high dynamic loads, this type of **erosion protection** can only find very limited use due to the lower tolerances for **roughness** increases, **dynamic strength losses**, and profile changes. Furthermore, no cell action must be allowed to occur between the armor and the blade material, which could cause dangerous **corrosion** in the transition zone (also see Ill. 5.4.1.1-8).*

Thin multilayer coatings have proved excellent erosion resistance during operation if not too much impact energy is expected.

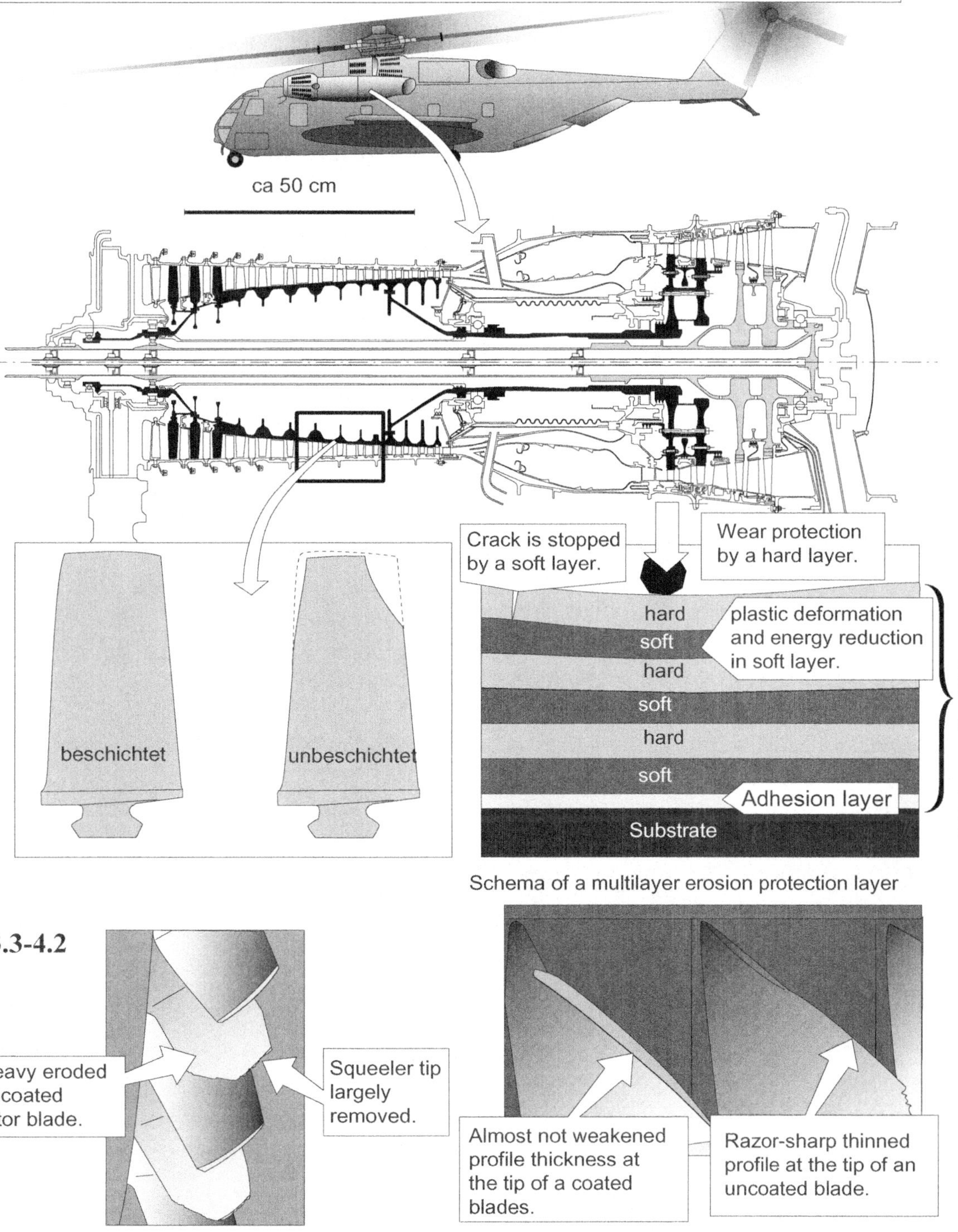

Ill. 5.3.3-4.2

Illustration 5.3.3-4.2 *(Ref. 5.3.3-8): Just in helicopters the compressors of the aeroengines are especially affected by* ***sand erosion*** *(Ill. 5.3.2-9). This leads to intense efforts to develop suitable*

protection coatings against sand erosion *(erosion resistant coating =* ***ER-coating)****. This is applied especially for the forward stages, usually made from high strength titanium alloy (Ti6Al4V = TiAl64). The coatings must be* ***hard****, to resist the abrasive wear of the dust at high and low inpact angle (Ill. 5.3.1-3). Its* ***thickness*** *is for modern compressors with thin sharp blading profiles markedly limited (range of 0,01 mm). A further requirement is, that the coating does* ***not trigger corrosion*** *(no developing of a corrosion cell, Ill. 5.4.1.1-8.1). All properties shall cover as far as possible the typical* ***operation temperatures*** *in modern compressors (-60 up to + 600 °C).*

It was aready early tried to deposit ***titanium nitride*** *(TiN, identifiable at the characteristic golden colour) because of its extremely high* ***hardness*** *(2800 - 3200 Vickers). However for single layer coatings it turned out during long time operation in desert enfironments, that these on rotor blades of the* ***front stages*** *are* ***penetrated*** *and* ***split out*** *by the relatively large and energy-rich sand particles (Ill. 5.3.1-3). This leads not only to abrasion/change of profiles, especially at the* ***leading edge*** *and an unwanted* ***roughness****. Critical is, that the fine cracks at the impact of the sand particle let* ***markedly drop the fatigue strength****. This also limits heavily the use. For the* ***rear stages*** *the single layer TiN coating has proved well. These are merely impinged from fine dust of the* ***splintered sand particles*** *(Ill. 5.3.1-3). Here the excellent wear behaviour at* ***flat impact angle*** *can be used.*

Not before a ***coating development in Russia****, obviously has proved excellent in a high number of shaft engines and fan engines in military and civil use.* ***The coating consists of several layers****. Thereby hard TiN alternates with tough, deformable metallic layers (of titan, sketch middle right). A special adhesion coating guarantees the binding at the substrate. The advantage of this* ***multi layer coating*** *is due to the high* ***tolerance against energy-rich sand particles****. A sintering of the coating does not occur. The roughening of the surface is obviously minimum. Even if it comes to* ***incipient cracks*** *in the heavily deformed zones of a hard coating layer, these will be* ***defused from the following tough/soft layer****. So dangerous fatigue cracks are prevented. This effect increases with the percentage of metal in the layer combination.*

The near-operation conditions testing in a typical shaft engine (sketch of the aeroengine) from a big transport helicopter (sketch above), confirmed the expectations. In contrast to an uncoated blading, the erosion caused portion of ***demounting aeroengines for overhaul up to 45 % could be lowered to 0****. Thereby the* ***separation rate*** *of rotor and stator blades* ***dropped from 70-80 % to 2-3%*** *. The most of these rest cases because of unacceptable FODs. The long time* ***efficiency drop*** *(deterioration)* ***decreased from up to 30 % to below 3 %****.*

Unfortunately from the literature on hand the behavior of the coating in the ***foreward compressor stages*** *at the testing during development and operation is not evident. These are exposed to especially* ***energy-rich particles*** *(size). Comparative pictures of coated and uncoated blades (sketches middle left and below) obviously derive from the middle and rear part of the compressor. Here also the single layer coating behaved during former testings promising.*

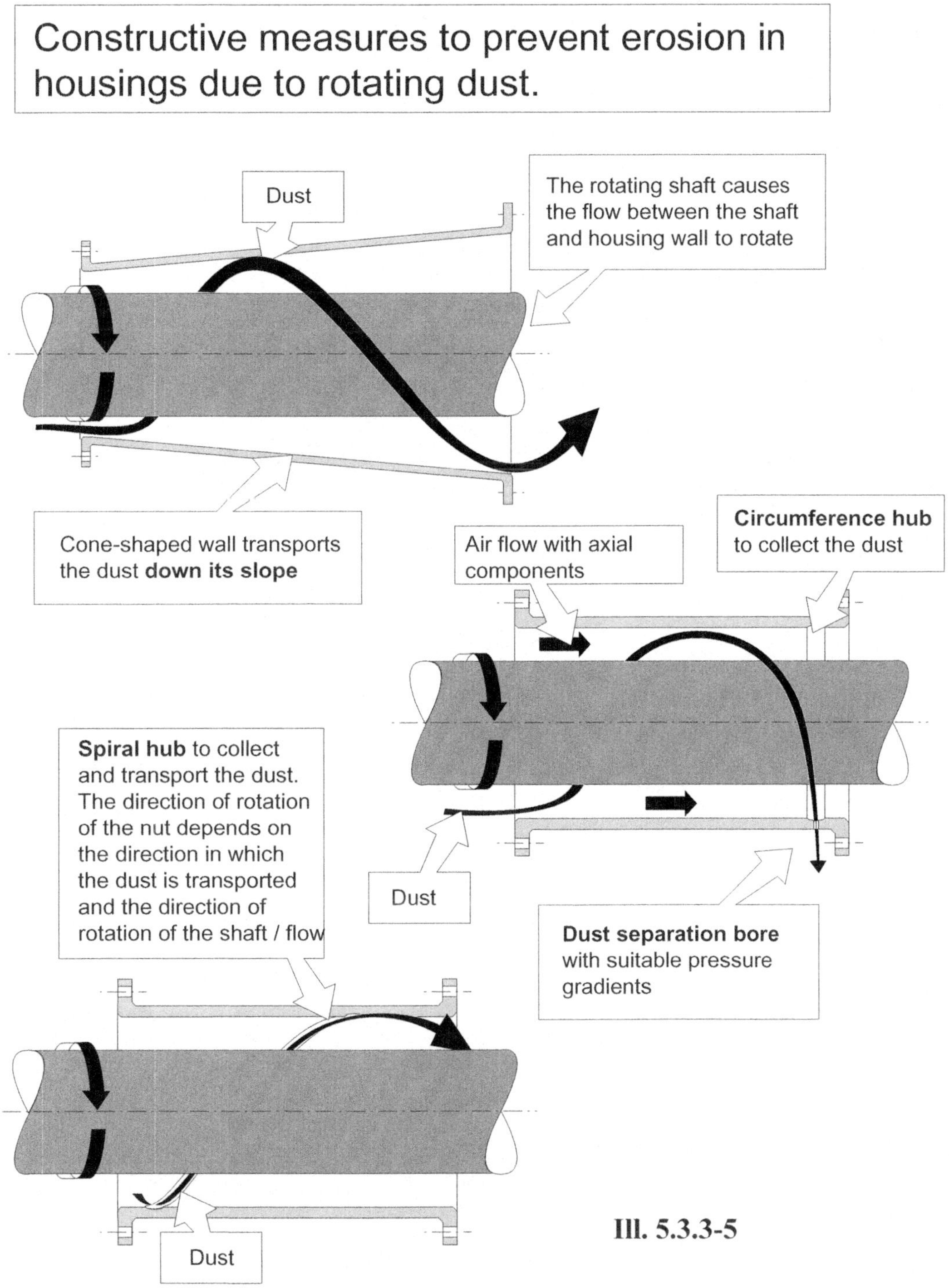

Ill. 5.3.3-5

Illustration 5.3.3-5: *Proper designing of the inner side of the housing/casing can cause* ***abrasive dust*** *(see Ill. 5.3.1-8), such as that from* ***crumbling labyrinth tip armor/hard facing****, hard rub-tolerant coatings, or* ***abrasive particles from brush seals****, to be carried safely out of the engine by a rotating flow. However, it must be ensured that this dust does not cause damage in other areas (wear, blocking of cooling air bores, etc.).*

Chapter 5.3.3 References.

5.3.3-1 D.L.Mann, D.V.Humpherson, „Helicopter Engine/Airframe Integration - The Way Ahead“, Paper des AGARD-Meetings on „Technology Requirements for Small Gas turbines“, Montreal, Canada, 4th-8th October 1993, page 2-3 up to 2-5.

5.3.3-2 „V2500 designed for maximum core protection“, Abbildung, Zeitschrift „Aircraft Engineering“, May 1993, page 3.

5.3.3-3 J. Jeffrey, M. Lenorovitz, „CFM56 Powerplant Fixes Based on DC-8 Operations“, Zeitschrift „Aviation Week & Space Technology“, February 14, 1983, page 32.

5.3.3-4 J.M. Lenorowitz, „CFM56 Powerplant Fixes Based on DC-8 Operations“, Zeitschrift „Aviation Week & Space Technology“, February 14, 1983, page 32.

5.3.3-5 „CFM56-2 Changed to Correct Problems“, Zeitschrift „Aviation Week & Space Technology“, November 1, 1982, page 35.

5.3.3-6 “NAVAIR Uses Russian Technology to Improve Marine Helo Reliability“, http://somd.com/news, Southren Maryland Online, 01 July 2006, page 1-5.

5.3.3-7 „Erosion Resistant Coatings for AGTR1500 Gas Turbine engine Compressoe System Powering Abrams M1A Tank Operating in Sand/Dust Environment“, http://ctmaideas.ncms,org, „CTMA (Commercial Technologies for Maintenance Activities)“, 7.7.2006, page 1-4.

5.3.3-8 P.Rodger, M.Duffles, „FCT (Foreign Comparative Testing Program) Erosion Resistant Coating Program“, NAVAIR.

5.3.3-9 United States Patent 5702829, „Mulitlayer material, anti-erosion and anti-abrasion coating incorporating said multilayer material“, http//:www.freepatentsonline.com, Anmeldung 04.36.1996, page 1-15.

5.3.3-10 E.Restrepo, V.Benavides, A.Devia, S.Olarte, M.Arroyave, Y.C.Arango, „Study of Multilayer Coatings of Ti/TiN/TiC Produced by Pulsed Arc Discarge“, Zeitschrift: „Brazilian Journal of Physics, vol. 34, no. 4B, December 2004, page 1748 - 1751..

5.4 Corrosion

5.4.1 Fundamentals

5.4.2 Stress Corrosion Cracking

5.4.3 Corrosion Fatigue

5.4.4 Hydrogen-Induced Cracks and Fractures

5.4.5 High Temperature Corrosion

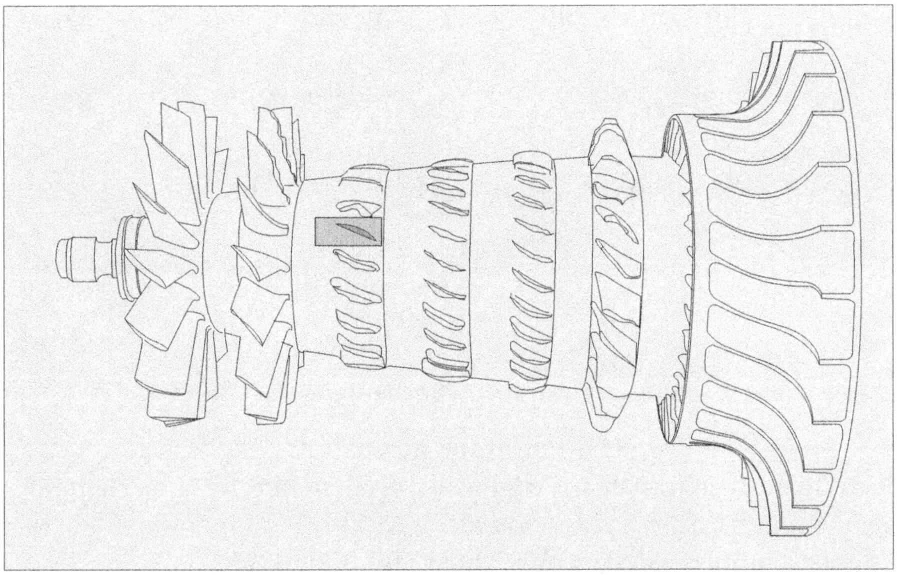

The increased use of titanium and nickel alloys reduces corrosion susceptibility in modern engines. Older engine types with compressor blades made from **Cr steels and aluminum alloys**, rotor disks made from steel, and housings/casings made from **magnesium alloys** or low-alloy steels, on the other hand, have a higher risk of corrosion.

Even in modern engines, **corrosion-sensitive steels** are used in casings and compressor blades in those areas where matched thermal expansions are necessary to maintain clearances. For example, in high-pressure compressors, **disks made from corrosion-sensitive Cr-steels** with "**balancing**" expansion behavior are used between titanium-alloy rotor disks and disks made from Ni-alloys. Corrosion damages additionally occur on special abradable coatings in the housing of the front compressor area. Extra attention is also given on hot gas corrosion in hot parts. An overview of the corrosion behavior of typical material families in engines is given in the table on page 5.4.1.1-5.

There are various types of corrosion with different **damage mechanisms** (Ill. 5.4.1.1-2).

External Factors
Corrosion Without Mechanical Loads: Fundamentals

Most corrosive media are brought into the engine along with the intake air. Upon entering the engine they can cause damage during operation. In the case of water corrosion, however, the **standstill times** of the engines are the deciding factor. Corrosion begins when condensation water is present and dissolves contaminants in the engine. Uncovered engines on aircraft that stand out in the open during the daytime and nighttime for longer periods experience air exchanges due to temperature variations and wind. This allows corrosive media (such as in ocean atmospheres) and/or humidity to enter the engine. When **condensation water** is created, the conditions for standstill corrosion are met. **For this reason, experience has shown that engines of aircraft with longer standstill times in areas with ocean atmosphere are subject to the greatest corrosion stress.**

Operating data that are important for corrosion influences	Military	Civilian
Operating times with constant loads	short	long
Takeoffs and landings per hour of operation	many	few
Flight altitude	frequently low	usually high
Operating times between ovehauls	10^3 hours	10^4 hours
Overhaul intervals (operating and standstill)	up to several years	
Standstill times	long, many	short

The above table shows that engines in **military aircraft** are exposed to much worse operating conditions with regard to corrosion than are engines in civilian aircraft. For example, an engine in a commercial passenger aircraft had less corrosion after 12,000 (!) flight hours than another engine of the same type in a fighter-bomber had after only about 500 flight hours.

Corrosion can also occur due to **improper storage** of engines (such as in containers with condensation water formation) or individual components that are insufficiently conserved in non-climatized storage areas (such as preassembled transmissions).

If one includes etching damage, especially, an additional possibility for corrosion damage is during the **production** or **repair** of engine parts if, for example, unsuitable cleaning or etching processes are used. Quality controls are not always sufficient to prevent the installation of parts damaged in this way.

During **mounting** and **maintenance** there is a danger that unsuitable **auxiliary materials** (washing media, lubricants, cleaning materials) may be used.

External Factors Corrosion Without Mechanical Loads: Fundamentals

Corrosive media:

There are many corrosive influences that affect engine compressors. These influences can be of very different types and have various increasing effects on one another. For example, one might consider an erosive influence that removes protective oxide layers or lacquers (abrasive cleaners, etc.), or perhaps a process in which **corrosive deposits** build up on certain coatings that are not particularly corrosive on their own (soot, oil remnants, etc.).

Typical corrosive media in the intake air:

• **Sea salt** in a watery solution in various concentrations or even as dry deposits. These contain primarily NaCl, $MgCl_2$, $MgSO_4$, and $CaSO_4$.

• **Industrial atmospheres**, i.e. deposits created by air contaminants from industrial processes. Especially strong corrosive attack can be expected from acidic watery solutions, such as sulfuric acid resulting from dissolved SO_2 and/or soot deposits.

• **Soot deposits due to recirculation** of the engine exhaust gases and oil remnants.

• **Washing material remnants** from compressor washing.

• **Remnants from mechanical compressor cleaning procedures.**

• **Dust deposits:** The corrosive effect may depend on the different compositions of the ground where the dust originated (e.g. high content of iron oxides or sulfur compounds such as gypsum). An especially corrosive atmosphere is that in hot dry regions such as deserts and steppes which are irrigated with ground water with a high saline content. The salt deposits remaining after the water evaporates are thrown up into the air with the dust, then settling on the engine components and corrosively attacking them.

• **Exhaust gases from munitions** (cannons, rockets)

Typical corrosive media from manufacture and overhaul:

• **Remnants and reactive coatings from cleaning baths** (overhaul) such as chlorides on titanium parts or acid remnants in the cooling air ducts of turbine blades.

• **Insufficiently reacted coating components** (such as in multi-component coatings)

• **Deposits and wear products** from manufacturing processes such as grinding dust and brush wear products. Tool wear products from cutting and non-cutting forming processes. Wear products from assembly tools.

• **Contaminants from abrasive and hardened peening materials.** These wear products can be caused by the peening particles themselves, such as micro-bonding or adhesion to the surface, as well as by peening material contaminants such as lead (from a protective lead cover).

• **Unsuitable lubricants** (e.g.on bolts and adjustable components) and sealing materials. These can have corrosive effects without thermally decomposing (for example, sea salt acting on lubricants containing MoS_2 on joints made from steel parts), or only after decomposing (for example, bolts in hot parts under the influence of high temperatures).

• **Unsuitable conservation oils** or use of wrong oil type (for example, cutting oil instead of conservation oil).

• **Writing** on parts with unsuitable felt-tip or oil **markers**.

Typical corrosive media from inside the engine itself:

• **Construction related:** Unsuitable material combinations (for example, cell action related to silver-plated parts in the hot part area), use of corrosion-sensitive materials (see table on page 4.5.1.1-5), or material properties (for examples, Cr steels with excessive hardness with regard to SCC).

• **Operation-specific:** deposits and decomposition products from leaking bearing oil.

• **Decomposition of oil with additives:** Transmission casings made from Mg alloys can be attacked by acids in the oil. Non-protective oxide layers can promote sulfidation in nickel alloys.

• **Corrosion products from other components** in the form of deposits in the downwind engine areas (for example, in the cooling air ducts of hot parts).

External Factors
Corrosion Without Mechanical Loads: Fundamentals

Corrosion behavior of important materials used in engine construction

Material	Type / example	Semi-finished product		Blades		Disks	Housings	Behavior during watery corrosion	Primary corrosion types in different corrosive media
		Knet Schm.	Cast	Rotor	Stator				
Low-alloy steels	Heat-treated stäeel	*	*			*	*	--	Surface, troughs, pitting
Cr steels	10-13% Cr	*	*	*	*	*	*	-	Pitting, SCC, corr. fatigue
Cr/Ni steels	18/9, A-286	*	*		*		*	+	Pitting, IG
Containing Fe with high Cr and Ni content	IN 718	*	*			*	*	+++	IG (in etching baths)
Ni alloys	Nimonic Waspalloy	*		*	*	*	*	+++	IG (in etching baths) pitting (Ag influenced)
Ti alloys	TiAl6V4	*	*	*	*	*	*	+++	SCC (in Cl above 300°C) sea salt above 500°C
Al alloys	AlCuMg	*	*	*	*	*	*	0	SCC, pitting, surface
Mg alloys	MgTh		*				*	---	Surface, pitting sea salt influenced
Be alloys	BeCu	*			*			0	SCC, pitting, surface (corresponds to Al alloy)
Carbon fibers with organic matrix				*			*	+++	Acts as a cathode during contact corrosion ("noble")

Very good +++ ++ + 0 - -- --- poor

Illustration 5.4.1.1-1:

Compressor rotor and stator blades "K1": *Pitting corrosion in blades made from Cr steels. Corrosion in soldered blades can be increased by cell action with the nobler solder (gold, silver, copper; Ill. 5.4.1.2-2). In blades made from Ti alloys there is a potential danger of hot salt corrosion at temperatures above 450 °C (Ill. 5.4.2.1-8).*

Compressor disks "K2": *Steel compressor disks such as are found in older engine types can exhibit pitting corrosion (Ill. 5.4.1.2-4). In some modern engine single disks made from steel (requiring corrosion protection) with a suitable expansion coefficient are used between disks made from Ti alloys and disks made from Ni alloys in the hot rear engine area.*

Compressor housing "K3": *Compressor housings of older engine types made from Mg alloys are susceptible to corrosion through ocean atmosphere and humidity (Ill. 5.4.1.2-9). In these engine types, rear compressor housings and combustion chamber casings are often made from low- and high-alloy steels that require corrosion protection.*

Rub coatings and abradables "K4" and "K5": *Components of these coatings, such as Al powder used as a filler in a polyester matrix, can be*

continued on page 7

External Factors

Corrosion Without Mechanical Loads: Fundamentals

Typical corrosion-stressed engine parts

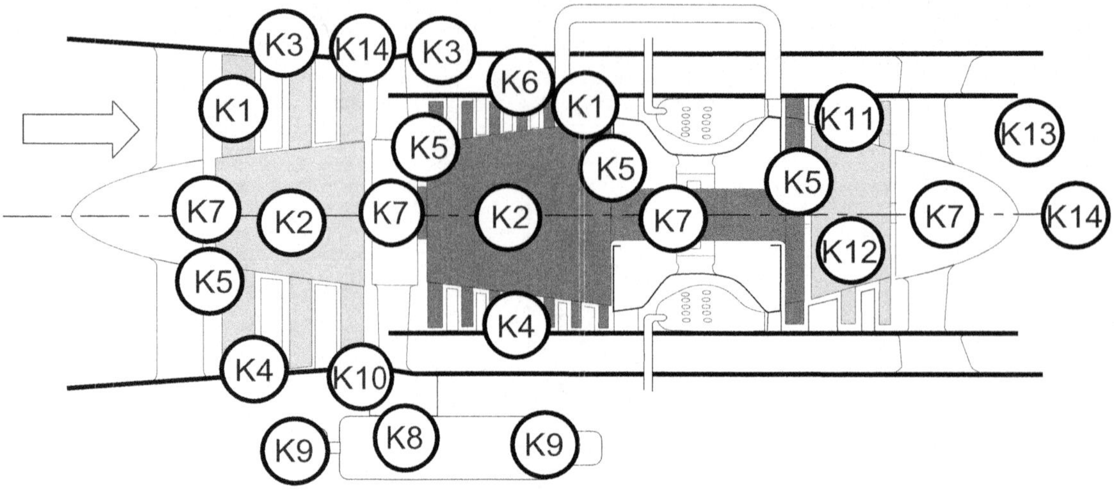

K1 Compressor runner and stator vanes of older engine types (steels)

K2 Compressor disks of older engine types (steels)

K3 Compressor housings of older engine types (Mg alloys and steels)

K4 Rub-coatings and abradable coatings in housings

K5 Abradable coatings in labyrinths

K6 Compressor variable vane adjusters

K7 Bearings

K8 Transmission housings (Mg- and Al-alloys)

Ill. 5.4.1.1-1

K9 Multiple-splining on shafts

K10 Threaded connections and clamping bolts (steel)

K11 Hot parts (sulfidation, hot gas corrosion)

K12 Turbine disks and blades (under the influence of silver)

K13 Turbine outlet housing, especially in civilian turbofan engines (steel)

K14 Joints (such as thrust reverserts, thrust nozzles, variable vanes) continued from page 5

seriously damaged by corrosion (Ill. 5.4.1.2-3), even causing them to break out or blister.
Compressor blade adjuster "K6": *Components of blade adjustment systems such as bearings, connectors, axles, and cables consist primarily of steels that can suffer corrosion damage (Ills. 5.-4.1.2-7, 5.4.1.2-8 , 5.4.2.2-1, and 5.4.2.2-3)*

Roller bearing "K7": *In roller bearings there is corrosion danger due to condensation water during standstill. Before installation, bearings, especially disassemblable ones, must be inspected for corrosion due to hand sweat.*

Gearbox housing "K8": *The housings of the auxiliary gearboxes are usually made from Al alloys or Mg alloys. Housings made from Mg alloys, especially, are prone to corrosive attack from the outside due to ocean atmosphere and from the inside due to aged oils (acidic).*

Multiple splining on shafts "K9": *These parts, which are usually made from low alloy steels, are susceptible to friction corrosion when affected by moisture in combination with micro-movements of the contact surfaces that are typical during operation.*

Threaded connections and clamping bolts "K10": *When these components are typically made from high-strength steels, improper heat-treating creates the possibility of corrosion-induced crack initiation and pitting corrosion followed by fatigue (Ills. 5.4.1.2-4 and 5.4.1.2-6).*
In threaded connections and bearing surfaces made from nickel alloys, there is a danger of damage due to decomposing lubricants that contain sulfur and/or sulfur from the fuel, which increase the corrosion process and often act in connection with silver coatings (Example 5.4.5.2-1, Ref. 5.4.5-2).

Hot parts"K11": *Sulfidation and hot gas corrosion under the influence of contaminants in the hot gas. Under certain conditions (temperature, composition of deposits) all materials with a Ni- or Ci-basis are at risk (Chapter 5.4.5).*

Hot parts in contact with silver "K12": *Silver on coated parts such as bolts and nuts can dissolve when attacked by acidic electrolytes (for example, due to condensation water that has been contaminated by an industrial atmosphere) and build up deposits on disks and blades. There it can cause dangerous pitting corrosion at operating temperatures.*
Sulfidation can be considerably increased by silver in the form of a wear product, coating material, or splashed melt.

Hinge bearing "K14": *These bearings are usually made from Martensite steels and are prone to rust in ocean environments in case of insufficient lubrication or corrosion protection. Experience has shown that in these cases* ***grease containing*** $\boldsymbol{MoS_2}$ *can evidently increase the corrosion between glide surfaces and lead to jamming. In this area, normal engine oil still proves to be effective.*

Illustration 5.4.1.1-1.2: *The potential corrosion media are already at the beginning of this chapter discussed in detai. Therefore in the following the presentation will be limited at the kind of* ***ingestion into the aeroengines****.*

Ingestion during flight*: Also in high altitudes the danger exists, that aeroengines will intake deteriorating media. Primarily concerned is vulcano ash (Ill. 5.3.2-15) and gases (e.g., rich in sulfur). Also sand storms can transport fine dust in high altitudes. The deterioration potential depends especially from the composition of the particles. So for example, a high content of gypsum (sulfates) intensifies the triggering of sulfidetion in hot parts (Ill. 5.4.5.2-2.1 and Ill. 5.4.5.2-2.2).*

In low altitudes during start and landing as well as flight missions *of military airplanes like fighters and for surveillance (e.g., for the control of submarines) the ingestion of the corrosion medium takes place by intake. Further the danger of corrosion comes from ingested residues of combustion (gases, particles) during firing of weapons like missiles and machine cannons (volume.3 Ill 11.2.1.2-10. During flights in low heights with high speed in the aeroengines of fighter aircraft occur especially high operation temperatures. So aging and decomposition processes at materials (e.g., plastics) and auxilaries like oil can be intensified. Aging products can act for example with condensate as corrosion mediumet and attack corrosion susceptible materials like magnesium alloys (gear boxes or aluminium filled rub in coatings, Ill. 5.4.1.2-1) and/or seals from elastomeres.*

On the ground during running aeroengines*: Here the special danger exists, that dusts, water and foreign objects are sucked in (Ill. 5.2.1.2-4). The intaken amount can increase during landing by activation of a thrust reverser (Ill. 5.2.1.2-3). The experience shows that durimng* ***test and certification runs*** *on the ground an increased danger of ingesting corrosive media exists. To these belong e.g., deteriorating* ***industrial gases****, especially from* ***chemical production processes, cleaning/washing, painting and electroplating****. So, in one case pitting corrosion at turbine disks of the rear stages made from a Ni-alloy was identified. The attack could be traced back to ingested exhaust gases from an electroplating plant, together with condensate during stand still over night.*
A special situation arises from outdoor so called ***field test rigs*** *in the military application. Here an increased potential danger exists, that deteriorating media from the environment are ingested (Ill. 5.5-1). To these belong media* ***from agriculture like fertilizer and pesticide****.*
A not so seldom situation occurs, if ***fire extinguisher*** *(Ill. 5.5-2) gets into aeroengines and there, especial in the hot part, discharges its typical aggressive components.*

Helicopter engines *are especially endangered by dispersed dust. This can be increased in two manners. By* ***erosion effect*** *arise fresh reactive surfaces which are especially prone after the shut down of the aeroengine. Also a recirculation of the exhaust gases with the ingestion of soot is especially likely at helicopters as well as vertical take-off and landing airplanes.*

At the ground parked airplanes*: It is this situation, at which usually the major part of the corrosion deterioration (idle corrosion) arises. Especially when immediate before, erosion produced reactive surfaces. For this reason special aeroengines from* ***military application with long stand still periods*** *and not necessarily long operation times are especially concerned. Also during stand still corrosive media can enter. Are the aeroengines not sufficient covered, the danger of a pollutant entry exists by wind and* ***air exchange****.* ***Temperature changes*** *(day/night) lead to corresponding change of the air volume*

External Factors
Corrosion Without Mechanical Loads: Fundamentals

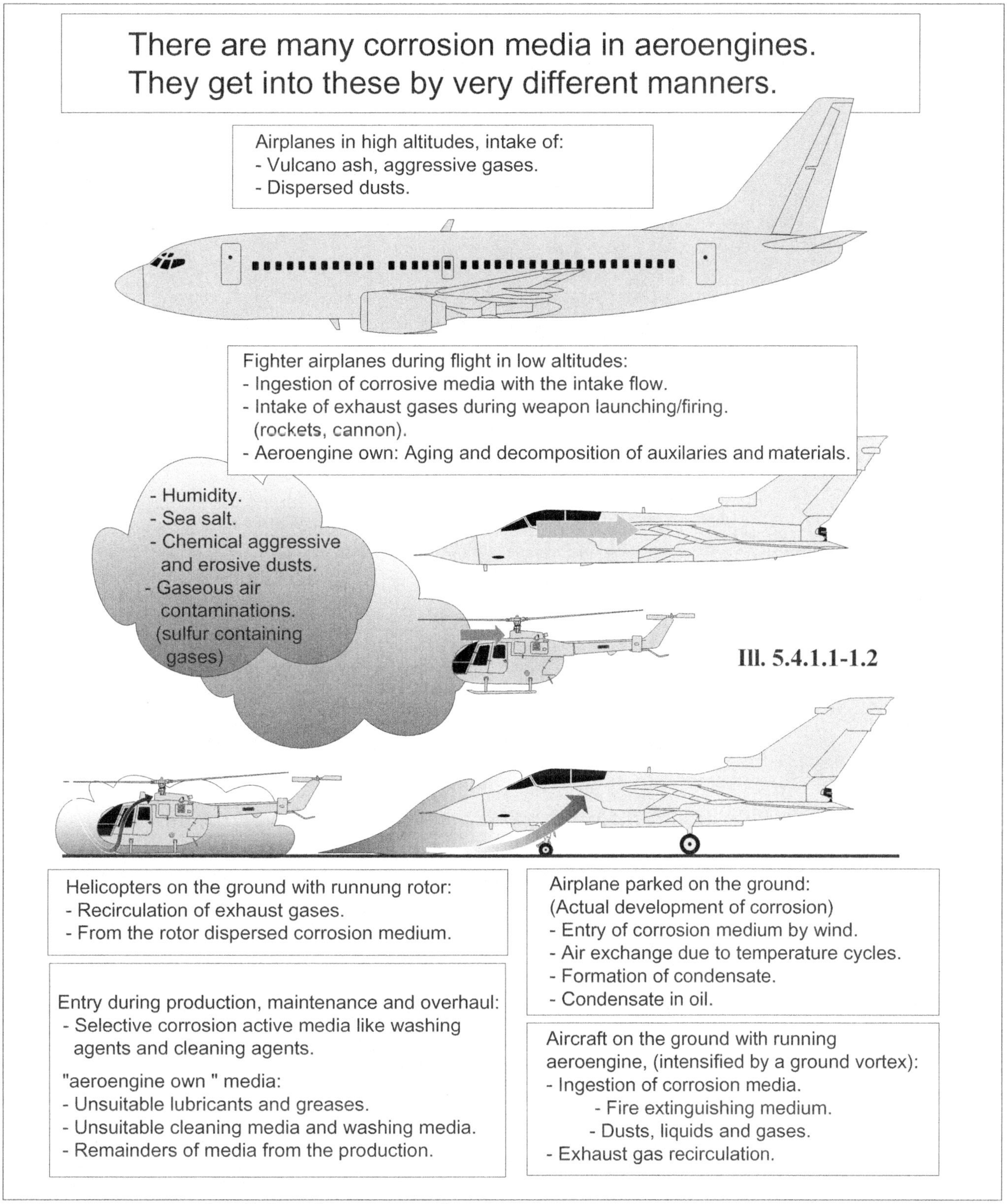

Ill. 5.4.1.1-1.2

in the aeroengine and with this alternating inflow and outflow.

External Factors
Corrosion Without Mechanical Loads: Fundamentals

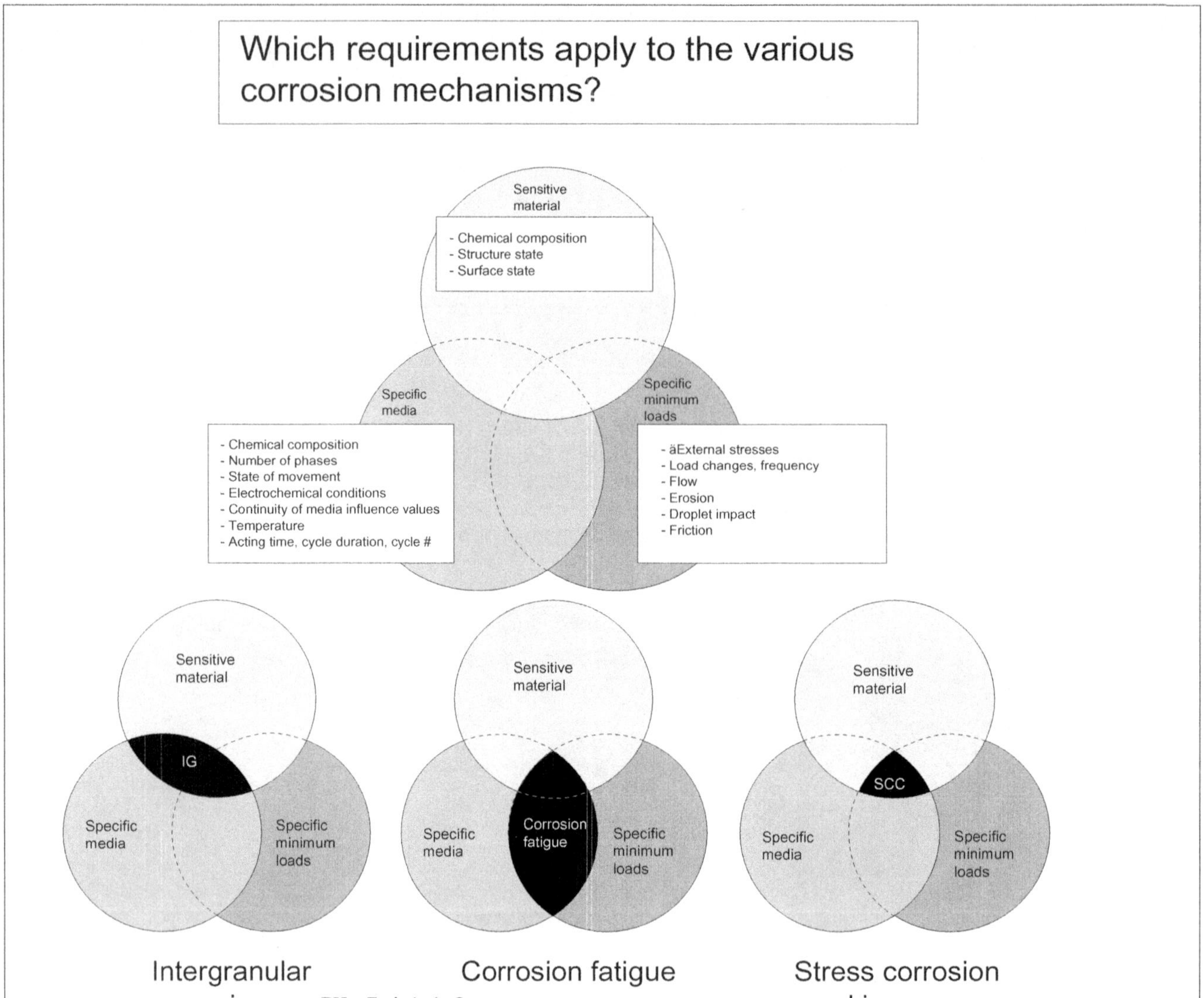

Ill. 5.4.1.1-2

Illustration 5.4.1.1-2: *Three main factors, i.e. corrosive media, materials, and mechanical loads, determine the behavior of a* ***corrosion system****. The depiction with the overlapping circles is based on Ill. 5.4.1-5. The three main influences are themselves affected by many different conditions (top diagram). This shows the complexity of corrosion processes.*

The bottom three diagrams show the conditions for three typical ***crack-initiating corrosion types*** *(black areas). As shown, the occurrence of* ***stress corrosion cracking*** *(SCC) requires the combination of all three main influences. If one of these influences is missing, such as sufficient tensile stress levels or stress concentration, then SCC can not occur. With* ***corrosion fatigue****, a specific medium and sufficient (dynamic) loads are necessary, but a special material state of the affected material (such as 13% Cr steels in the typical state during operation) is not a prerequisite.*

In order for ***intergranular corrosion*** *to occur, a sensitive material and a specific medium are necessary, but there is no specific minimum load level necessary for this type of corrosive attack.*

External Factors
Corrosion Without Mechanical Loads: Fundamentals

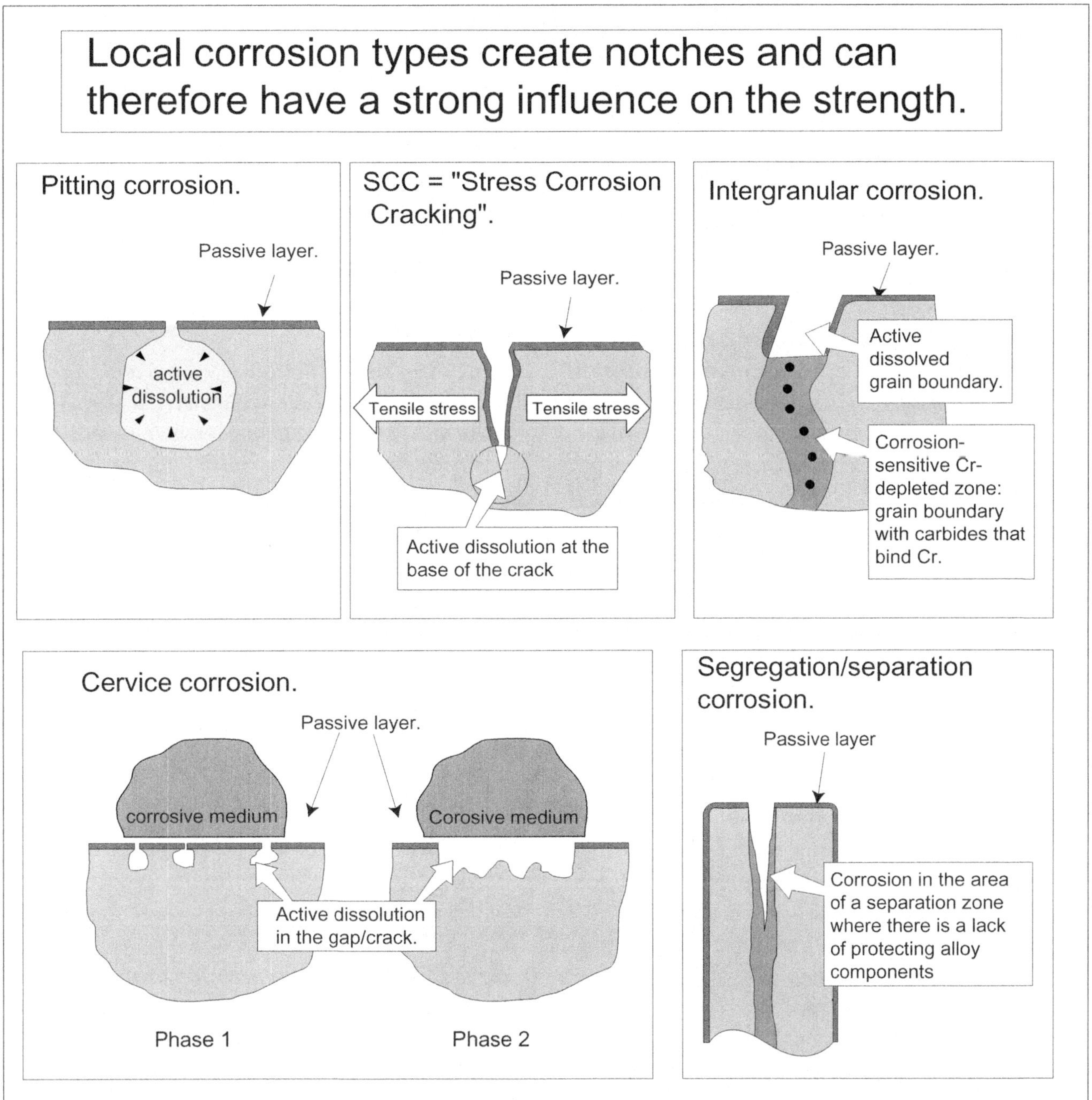

Ill. 5.4.1.1-3

Illustration 5.4.1.1-3 *(Ref. 5.4.1-1): The illustration is based on the mentioned literature. Locally acting corrosion types form notches, increasing the loads on the part. This leads especially to a noticeable decrease in the dynamic strength of the part.*

Pitting corrosion: *This type of corrosion occurs when Cl ions break through the inert surface layer and activate the base material. These areas of weakened inert surface layers are created by surface contaminants such as smeared foreign metal or inhomogeneities in the material. Pitting corrosion (top left) occurs especially in Cr steels*

External Factors
Corrosion Without Mechanical Loads: Fundamentals

(such as in the compressor blading of older engine types), if there is no sufficient (cathodic) corrosion protection.
A unique problem is posed by ***thin oxidation layers (tarnishing)*** *due to heat development during manufacturing processes or operation. In CrNi steels these oxide layers have a high chromium content which was taken from the surrounding base material, increasing its sensitivity to corrosion. The structure of these oxide layers has micro-cracks and allows the Cl-rich electrolytes to act on the corrosion-sensitive base material, resulting in pitting corrosion.*

Stress Corrosion Cracking: *This type of corrosion (top center diagram) can be either transgranular (austenitic Cr/NI steels of type 18 9) or intergranular (e.g. typical 13% Cr steels). The requirements for SCC are described in Ill. 5.4.1.1-2. Active material dissolving occurs in the base of the crack. In many cases typical signs of the influence of hydrogen can be seen at the micro-level (Ill. 5.4.4.1-2). The hydrogen is created during the chemical dissolving process and enters into the opened area at the tip of the crack, which is not protected by an inert coating, in a process in accordance with the illustrated model. If the crack growth is promoted by the influence of the hydrogen, it can lead to apparently spontaneous fractures. In general, the SCC sensitivity of a material increases along with its strength (hardness).*

Intergranular Corrosion: *This corrosive attack of grain boundaries (top right diagram) is due to the sensitivity of the grain boundary region. In this zone, the alloy component necessary for corrosion resistance is usually not sufficiently present in its corrosion resistant form. In the grain boundary area of sensitized CrNi steels (usually CrNi 189), it is typical for Cr to be bound in the form of carbides and no longer able to form an inert protective layer. This type of sensitization can occur at high operating temperatures (Ref. 5.4.2.2-1, Ills. 5.4.1.1-10 and 5.4.2.1-2), as well as during manufacture with improper heat treatment or in areas affected by heat from a weld. The prerequisite for sensitization is a composition-dependent sensitivity of the material. The addition of small amounts of strong carbide forming materials such as Ti, Nb, or Ta can ensure that enough Cr remains free to provide corrosion protection.*

Crack corrosion *(bottom left diagram): In order for a sufficient protective passive layer to form, there must be sufficient oxygen in the surrounding atmosphere. This is usually not given inside cracks. For this reason, the corrosive medium (such as Cl ions) can locally break through the weakened passive layer (phase 1). At first pitting-corrosion-like pittings are formed. These expand (phase 2) after the resulting corrosion products further activate the crack area.*
Crack corrosion is promoted by cell action, i.e. contact between surfaces that are located considerably far apart in the electrolytic series (see Ill. 5.4.1.1-7).

Separation corrosion: *When an alloy melt hardens, a partial segregation, i.e. concentration of alloy components occurs, at least on the micro level. With forged parts, the fine granularity and the deformation allow corrosion to occur in many micro-areas in a larger zone. Endangered parts include, for example, milled products in which a separation zone occurs at the surface (Ill. 5.4.2.2-1). In cast parts and also in welds, the zones of uneven alloy composition can be much larger.*

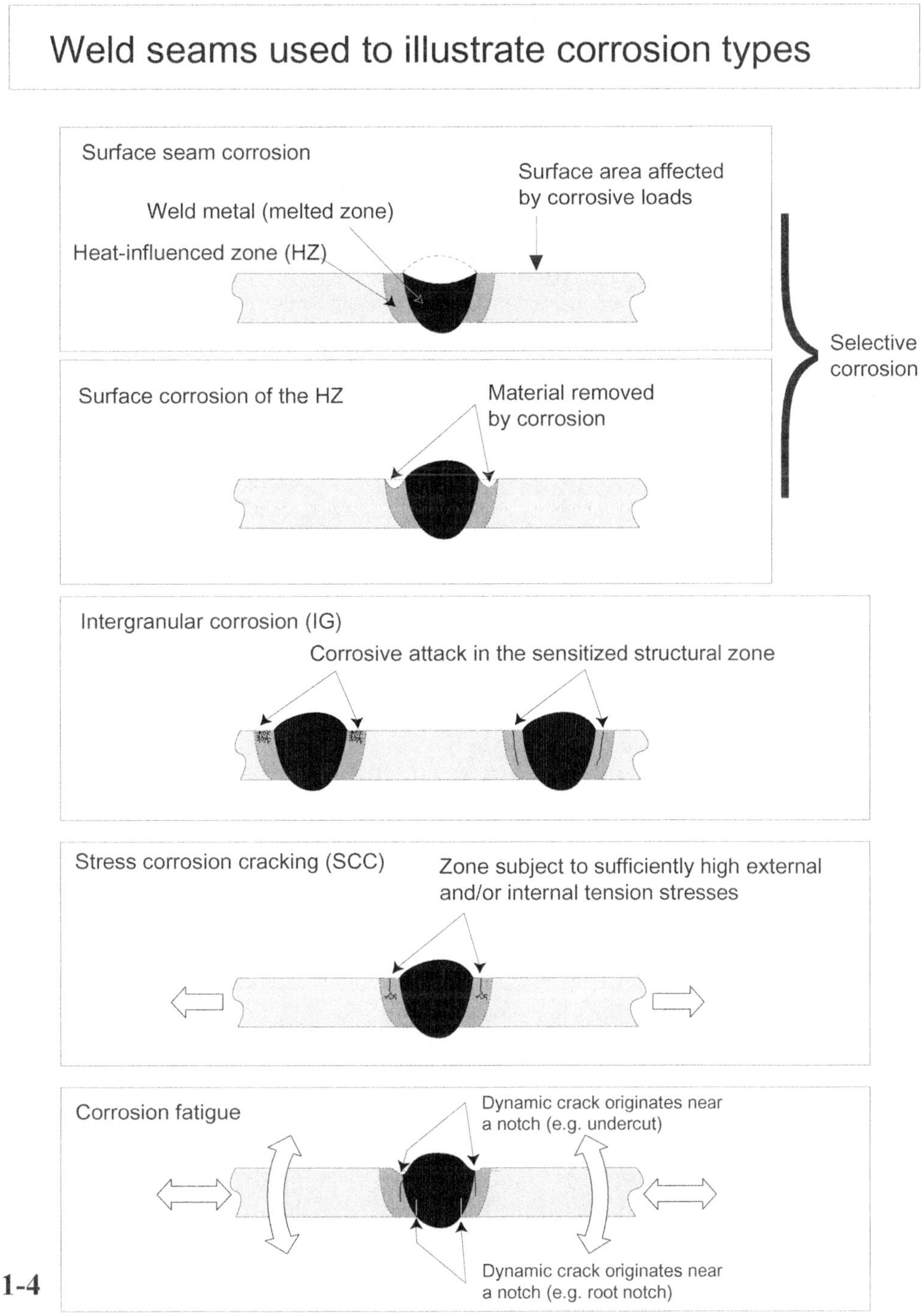

Ill. 5.4.1.1-4

Illustration 5.4.1.1-4: *Often, several different types of corrosion can be observed around weld seams. For this reason, this area is especially suitable for describing common corrosion mechanisms.*

The top three diagrams show corrosion forms that can occur even without noticeable mechanical loads. ***Surface corrosion*** *occurs in areas that are sensitive to the acting corrosive media. However, the affected surface does not show any* *pronounced sensitivity of individual components at the micro-level, resulting in largely even wear. This can occur either in the seam itself (top) or the heat-influenced zone.*

*If there is locally concentrated corrosive stress, such as due to cell action around contaminants or structural inhomogeneities, it can result in a corrosive attack that causes pitting (**pitting corrosion**).*

In crack-like corrosion forms, the crack characteristics largely depend on the mechanical tension in the corrosion area (Ill. 5.4.2.1-4).

***Intergranular corrosion** occurs along corrosion sensitive grain boundaries. This sensitivity can be caused by a "sensitization process", which requires a specific temperature range. Intergranular corrosion can be promoted by tension stresses.*

***Stress Corrosion Cracking** (SCC, Chapter 5.4.2) requires not only a material or material state that is sensitive to a specific corrosive media, but also sufficiently high tensile loading.*

***Corrosion Fatigue** (Chapter 5.4.3) occurs as a combination of a corrosive attack with vibrating loads. Fractures and cracks usually originate in notches and corrosion pittings.*

5.4.1 Corrosion Without Mechanical Loading

5.4.1.1 Fundamentals of Corrosion

5.4.1.2 Corrosion Damage

5.4.1.3 Measures against Corrosion Damage

Corrosion without mechanical loading is understood to be corrosion forms in which mechanical loading is not relevant to the damage caused by the corrosion process itself. This does not mean that no mechanical overloading can occur during or after the corrosion process; for example, overloading resulting from **sectional weakening** or the **notch-effect** of **corrosion pittings**. Corrosion forms without mechanical loading often occur in connection with **cell action** due to contact with another engine part and/or at contaminants (e.g. wear products) or a **material inhomogeneity** near the surface. The most common corrosive medium is moisture with **deposits** from an ocean atmosphere. However, maintenance and manufacturing processes can also result in attacks corrosive media. During manufacture or repair, there is an increased danger for unprotected parts which have not yet have had **coatings** applied or have had coatings removed.

Through the use of new materials such as carbon fiber-reinforced plastics or synthetic based, metal powder-filled **rub coatings**, corrosion can occur even in modern engines with blading and housings made from titanium and nickel alloys.

The use of chemically noble tip armor (to minimize rubbing wear; **WC/Co**, etc.) and erosion protection coatings (such as **TiN**) on steels and Al alloys can result in cell action and pitting corrosion.

The greatest corrosive stress in both the cold and hot areas of an engine can be expected in marine atmospheres.

Ill. 5.4.1.1-5

Illustration 5.4.1.1-5: Sea atmospheres *contain most of the corrosively acting substances that can cause corrosion in engines. Aside from chlorine, the most corrosive substance, sea air also includes compounds such as NaCl,* $MgCl_2$*, and* $MgSO_4$*, both in watery media as well as in dry deposits. After chlorine, sulfur and sodium are important causes of* ***high temperature corrosion*** *forms (such as sulfidation, Chapter 5.4.5).*

5.4.1.1 Fundamentals of Corrosion

Typical types of watery corrosion without mechanical loads

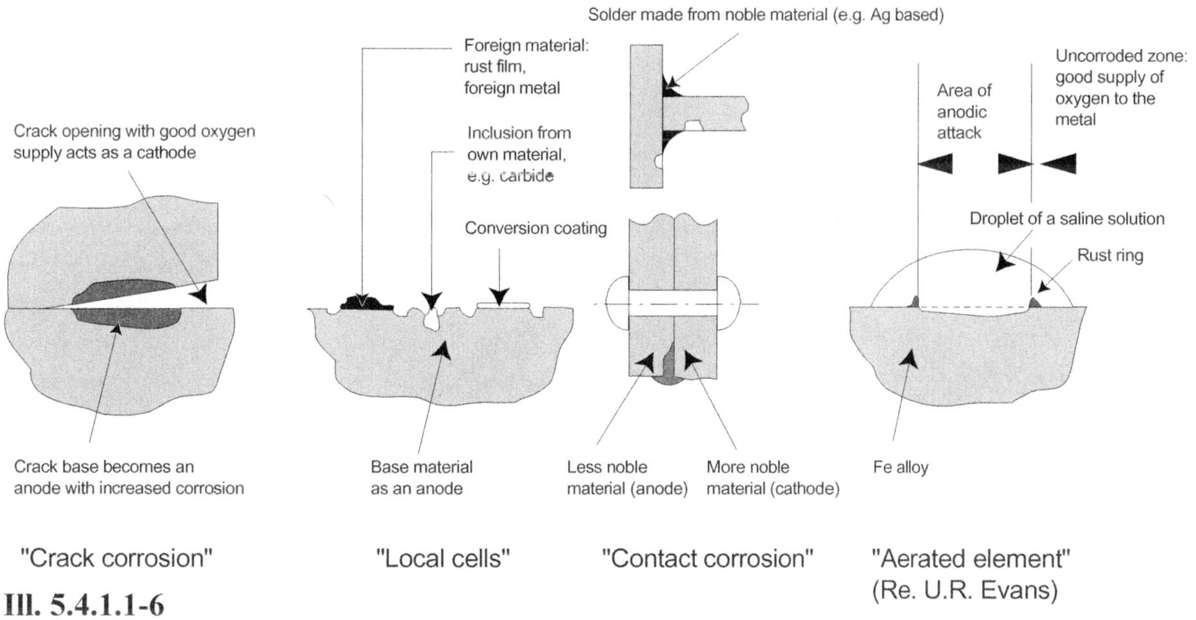

Ill. 5.4.1.1-6

Illustration 5.4.1.1-6: *If corrosion-stressed parts are made from materials with differing **corrosion sensitivity**, then there is a special danger of **crack corrosion** in the edge areas (left diagram). This can be additionally promoted if elastic deformations result in micro-movements between contact surfaces/fretting, damaging insulating conversion coatings. Loosely contacting surfaces and weld gaps (Ill. 5.4.1.1-11) can suck in the electrolytes through capillary attraction and trap them, leading to **corrosion of hollow structures** from the inside out. This results in the extent of the damage remaining unnoticed until very late, sometimes even after sectional weakening has occurred and led to spontaneous part failure. The strength of the crack corrosion effect is also dependent on the **surface area ratio** in the corroding area between the corroding (small) and corrosion-resistant (large) surface.*

*Contaminants can create local cells. A **local cell** (second diagram from left) is a small-surface corrosive cell that has an effective electrode surface of fractions of a square millimeter. Local cells can be local material contamination or structural differences, or even protective coatings, deposits of foreign material, locally hardened or altered areas in the surface structure.*

***Contact corrosion** can always be expected when parts made from different materials have*

electrical contact. The corrosion is generally stronger, the farther the positions of the materials (or their cells) are apart in the electrolytic series. This material combination then forms a galvanic cell with an anode (mainly anodic metal dissolution), cathode (little or no metal dissolution compared with the anode material), and electrolyte (third diagram from left, also see Ill. 5.4.1.1-7).
The metallic contact can be created by the contact between the surface, but it can also occur in welds and solders. However, it may also result from foreign metals smeared onto metal surfaces by repair or assembly tools.

*A **differential aeration cell** (right diagram) is caused by an electrolyte droplet on a metal surface. Increased corrosive attack occurs underneath the drop, while the greater presence of oxygen around the periphery creates an immune area here. **Hand sweat** and condensation water corrosion on roller bearings are typical for this corrosion form, and have **characteristic corrosion rings** around each droplet.*

***Illustration 5.4.1.1-7: Contact corrosion** plays an important role in parts that are connected by positive fitting, friction (threaded connections, etc.), and material bonding (soldering/brazing etc.). Contact corrosion is related to cell action. If two metals are in metallic contact with one another through an electrolyte (usually a conducting liquid), then the dissolution of the less noble metal increases. Dissolution increases with the potential difference of both metals.*
*The increased use of carbon **fiber-reinforced plastic** materials must be mentioned in connection with contact corrosion. Carbon fibers behave like noble metals in their electrochemical properties and can, for example, cause serious corrosion if they come into contact with Al alloys. For this reason **machined carbon fiber surfaces** (exposed, cut carbon fibers) must be separated from corrosion sensitive materials by an intermediate insulating layer, or must only be connected to corrosion-resistant metals (such as Ti alloys).*

Contact corrosion can be prevented through suitable material selection as shown in the diagram and/or through an electrically insulating separation of the metal surfaces. This is achieved through proper constructive design which avoids corrosion-promoting configurations (Ill. 5.4.1.1-11) which would lead to crack and/or contact corrosion. For example, especially corrosion-increasing configurations are those that include contact between parts made from metals with different potentials that result in direct metal contact. In contact corrosion, the potential difference between the pure contact metals is often not the deciding factor, but rather potential differences that, through so-called "practical series", can be named as potential corrosives in practical media. Influences such as conversion coating formation, electrolyte composition, and electrolyte temperature are especially important.

External Factors Corrosion Without Mechanical Loads: Fundamentals

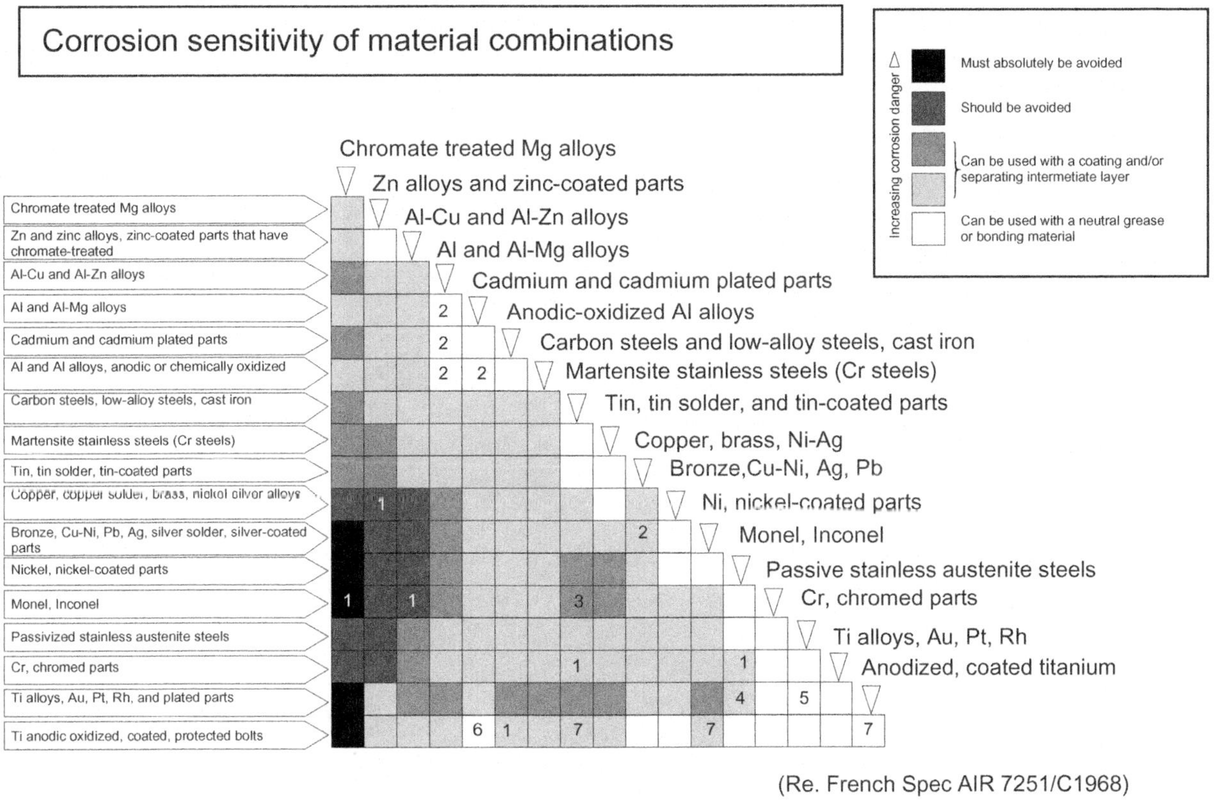

Ill. 5.4.1.1-7

Notes regarding the chart:

"1" Despite this, this metal combination is used in engine construction (even without protection).

"2" The parts should be lacquered.

"3" Is frequently used without protection in threaded connections in rotor and housings.

"4" Used in engine construction in threaded connections in rotors without any known problems.

"5" Dense coatings are a prerequisite for successful use.

"6" Danger of embrittlement through diffusion processes.

"7" Problematic with permeable coatings (damaged, cracks, porosity).

External Factors

Corrosion Without Mechanical Loads: Fundamentals

Illustration 5.4.1.1-8.2: *Because the compressor* ***blades made from an Al alloy*** *in this engine type exhibited heavy erosion despite an organic coating, it was tried to increase the erosion resistance with a nickel coating. This diagram shows a typical blade after a salt blasting test in the laboratory. Extreme corrosion occurred on the blade, especially at the edges and artificially created FOD notches. Despite the excellent corrosion strength of the coating and acceptable corrosion resistance of the base material, this type of erosion protection was unsuitable due to* ***cell action****. The following factors especially promoted corrosion:*

- "Nobler" coating material with high corrosion resistance on a less corrosion resistant base material.

- The coating had to be precipitated onto the metallic (i.e. unprotected) base material surface. This necessitated direct contact between the Ni coating and the Al alloy, which is very corrosion sensitive due to the lack of an oxide coating.

- An initially relatively large corrosion-resistant surface of the nickel coating compared with the smaller corroded base material surfaces under the coating damages. This poor surface ratio increases the corrosion effect.

- Foreign object impacts and erosion can locally damage the coating and make it permeable for electrolytes (sea water).

- Crack initiation in the nickel coating due to differences in heat strain relative to the base material.

- Sub-surface migration and delamination of the nickel coating.

Illustration 5.4.1.1-8.1 *(Ref. 5.4.1-2): Experience has shown that* ***corrosion damage due to cell action*** *is to be expected when one is "fixated" on another problem, such as erosion protection, and selects a coating based solely on this condition. For example, with an apparently corrosion-proof material such as 17/4 PH (17% Cr-steel type), ocean atmosphere can promote pitting corrosion due to cell action with erosion protection coatings. Merely the notch effect of the resulting corrosion pittings can considerably reduce the dynamic strength of the engine part. Cell action can also occur at the edge of armor/* ***hard facing*** *made from electrochemically noble materials (such as WC in a Co matrix, top diagram), or in areas where the coating has been broken out, such as are typical after foreign object strikes on thin hard coatings (e.g. TiN; bottom diagram).*

Failed erosion protection

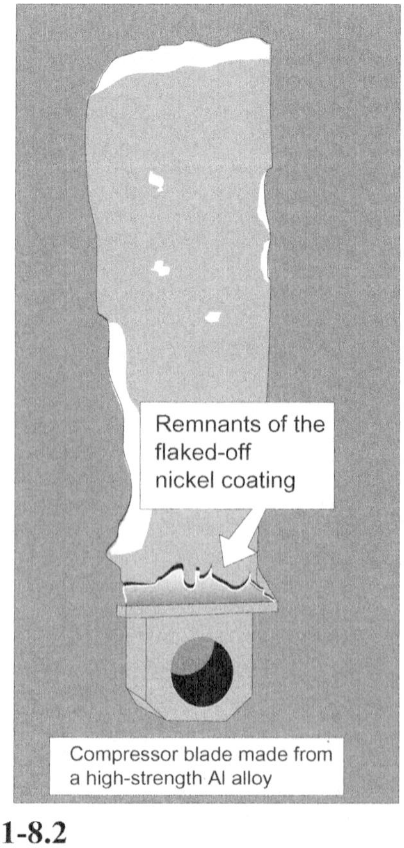

Ill. 5.4.1.1-8.2

Erosion protection coatings can promote corrosion through cell action.

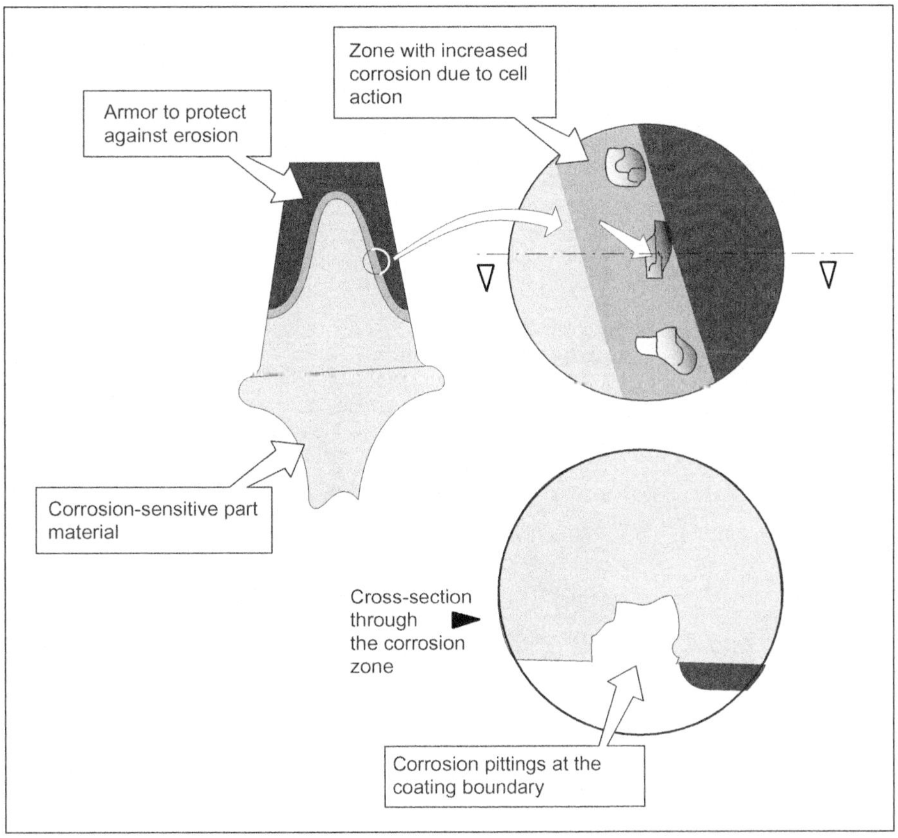

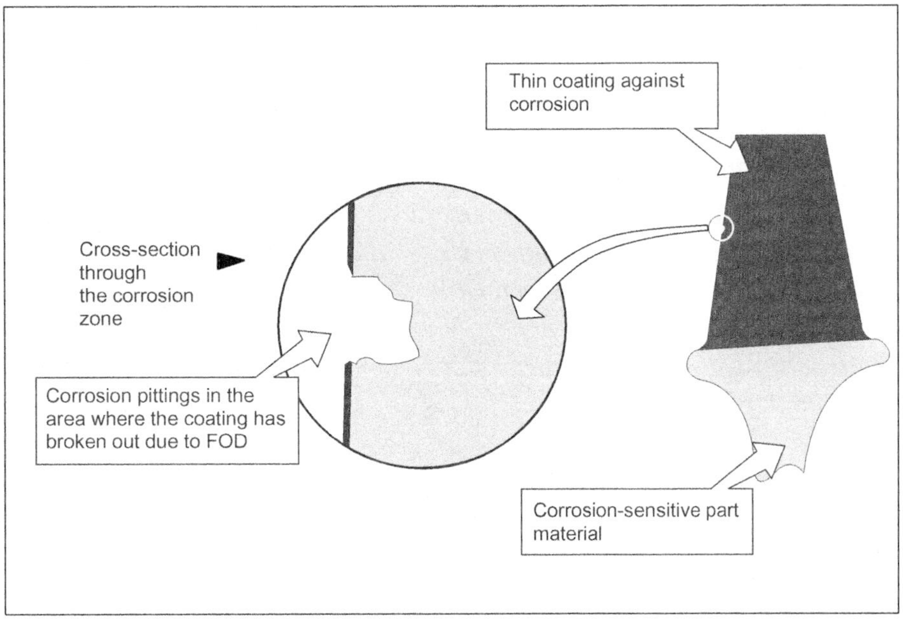

Ill. 5.4.1.1-8.1

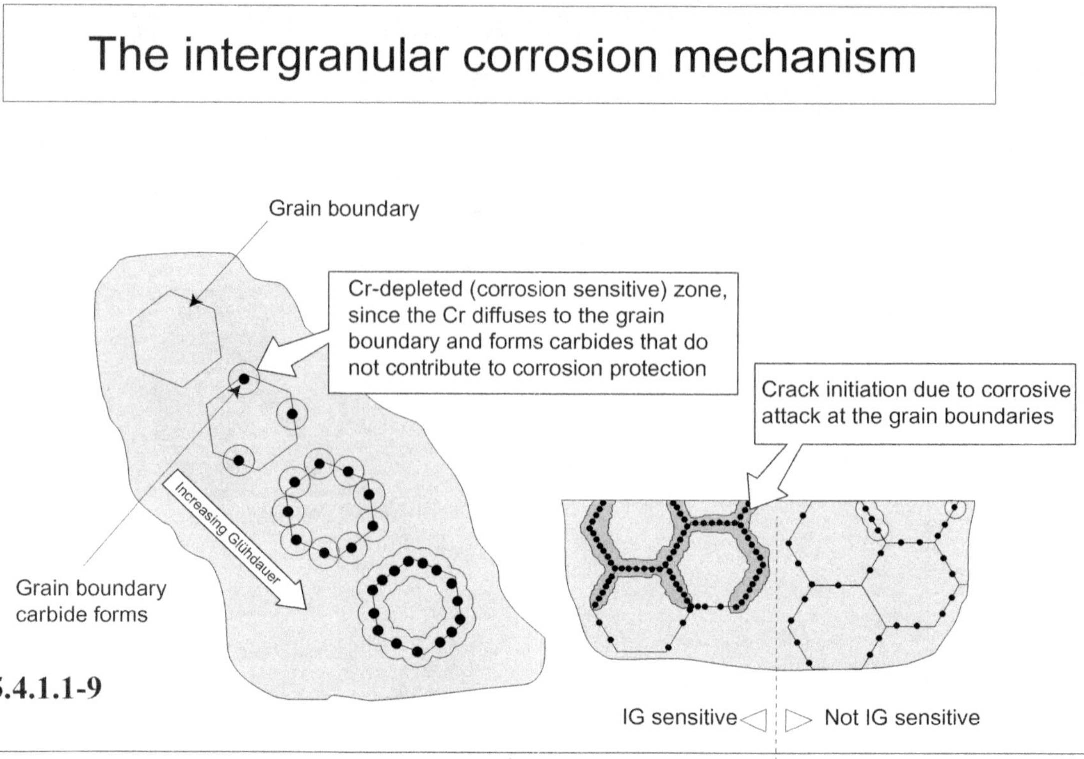

Ill. 5.4.1.1-9

***Illustrations 5.4.1.1-9** and **5.4.1.1-10** (Ref. 5.4.1-3): Endangered materials are insufficiently stabilized, sensitizable CrNi steels that should no longer be used in modern engine construction, but may be used due to material selection errors or improperly controlled vendor parts (Ill. 5.4.2.2-2). "**Sensitized**" non-rusting steels are especially sensitive to corrosive attack along the grain boundaries. The sensitization is due to Cr depletion of the grain boundary area due to Cr carbide formation (see Ills. 5.4.1.1-3 and 5.4.1.1-4). This is true both for **intergranular corrosion** and pitting corrosion at the **grain boundaries**. Sensitizing can occur in non-**stabilized materials** (without additional carbide formers such as Nb or Ta). The corrosion-sensitive structural state depends on the temperature and the time taken by the temperature introduction (top diagram). If the temperature is very high, such as is in a weld, then correspondingly short times (seconds) are sufficient for sensitization. If the temperatures are relatively low, such as operating temperatures (Example 5.4.2.2-2), then it may take several hundred hours for sensitizing to occur. These processes can cause long-term damage to engine parts that, as new parts, met all requirements with regard to intergranular corrosion attacks. It is also plausible that certain **etching procedures** that were successful with new parts could cause problems with run-in ones. For example, in an **overhaul**, parts with many operating hours can show unallowable etching due to a long-term structural change. An example of this is turbine blades made from an Ni-based cast alloy, which showed extreme intergranular corrosion after it was given an etching treatment in the course of an overhaul. The corrosion was so serious, that large grains fell out of the surface. Where this was not the case, crack initiation could not be safely detected by a penetration material test because the cracks were evidently filled corrosion products. The parts were noticed purely by chance and replaced. The inner damping of the blades had increased due to the cracks. When the blades were tapped, it resulted in an unusually dull sound. During heat treatment of sensitive*

Some materials can become sensitive to corrosion due to structural changes during production or operation.

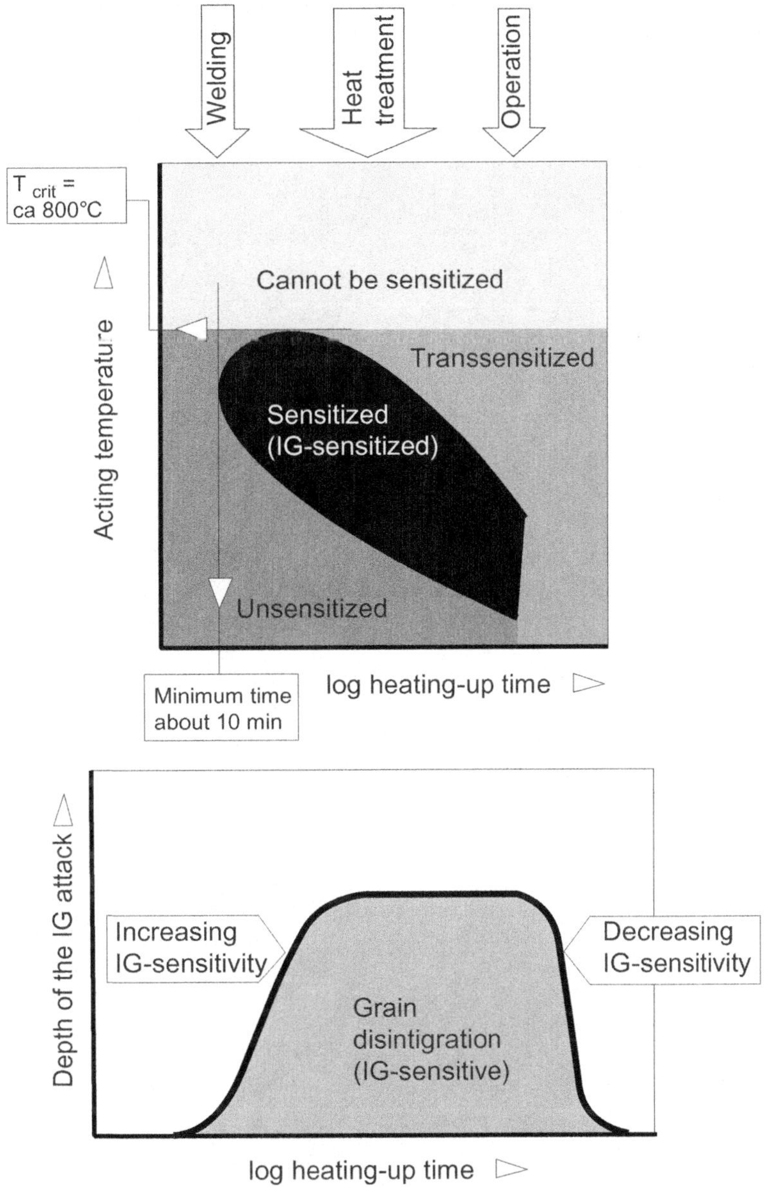

Ill. 5.4.1.1-10

Schematic depiction of the IG-sensitivity (sensitized by grain boundary areas with low amounts of Cr) of austenite unstabilized CrNi steels (type 18/8) caused by the influence of temperature.

materials at a middle temperature range, a span of minutes or hours is enough to cause sensitization. The damage risk can be minimized by observation of the sensitivity limits (bottom diagram).

External Factors
Corrosion Without Mechanical Loads: Fundamentals

The housings of older engine types are welded constructions made from steels with a large number of sheet, cast, and forged parts. This creates a special danger of corrosion.

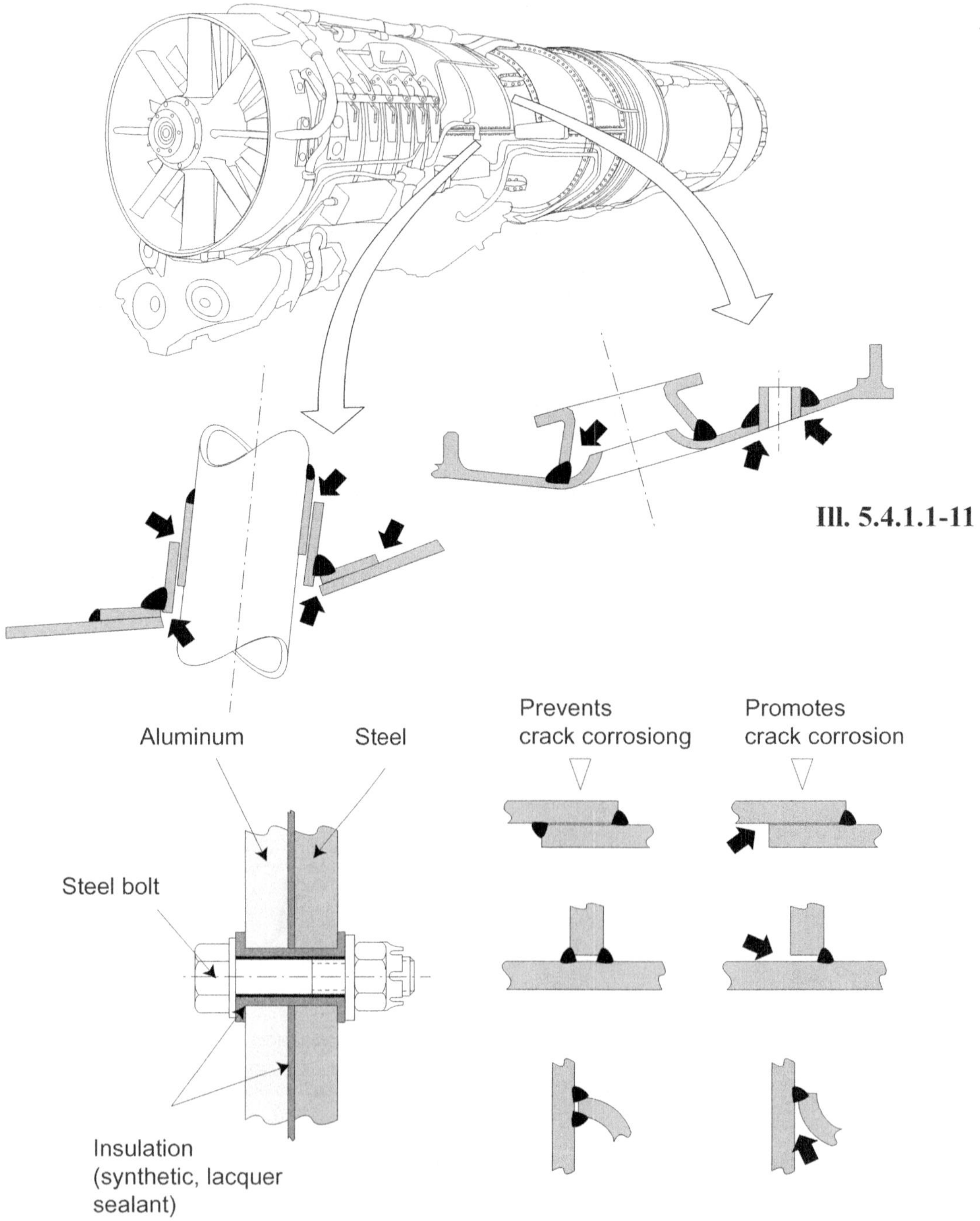

Ill. 5.4.1.1-11

Illustration 5.4.1.1-11: *The housings/casings of older engine generations, especially, are often* ***welded constructions made from steels*** *that are not corrosion-resistant. For these constructions, it is important that no outward-facing open gaps or hollow spaces area are created/remain; these are collecting areas for corrosive media and can not usually be sufficiently protected from corrosion. Sufficient corrosion protection can be achieved by, for example, welding on both sides. Threaded corrosion-sensitive contact surfaces (such as threads on Al and Mg cast parts) or plain bearings (such as the bearings of adjustable compressor stator vanes made from Cr steel in compressor housings made from Mg alloys; also see Ill. 5.4.1.2-7) should be insulated from one another through non-conducting inserts such as synthetic materials, sealing compounds, or lacquers. These can both seal gaps and* ***prevent metallic contacts****.*

Constructions should be designed in a way that unallowable damage to corrosion protection coatings can be avoided. This is true, for example, for contact surfaces between threaded parts with different stiffnesses which are subject to temporally changing loads (such as vibrations and heat strain) with relative movements.

Chapter 5.4.1.1 References

5.4.1-1 A.Bäumel, "Überblick über Korrosionsschäden an metallischen Werkstoffen unter besonderer Berücksichtigung der nichtrostenden Stähle", J.Grosch "Schadenskunde im Maschinenbau", TAE Kontakt & Studium, Volume 308, Edition 1995, ISBN 3-8169-1202-8, page 250.

5.4.1-2 H.J.Kolkmann, "Gas Turbine Compressor Corrosion and Erosion in Western Europe", AGARD-CP-558, proceedings of the conference "Erosion, Corrosion and Foreign Object Damage Effects in Gas Turbines", Propulsion and Energetics Panel Symposium, Rotterdam, The Netherlands, 25-28 April 1994, Chapter 30, page 6.

5.4.1-3 ASME, Metals Handbook Ninth Edition, Volume 11"Failure Analysis and Prevention", 1986, page 428.

5.4.1-4 P. Forchhammer, "Korrosionsschäden an metallischen Werkstoffen ohne mechanische Belastung", contribution in "Systematische Beurteilung technischer Schadensfälle", published by G. Lange, DGM Informationsgesellschaft Publishers, pages 263-289.

5.4.1-5 John Hickling, "Korrosionsschäden bei zusätzlicher mechanischer Beanspruchung", contribution in "Systematische Beurteilung technischer Schadensfälle", published by G. Lange, DGM Informationsgesellschaft Publishers, pages 291-311.

5.4.1.2 Damage due to Corrosion without Mechanical Loading

Extensive **corrosion damage not caused by mechanical loads** has been reported especially in older engine types that still used low- and high-alloy steels for housings, stators, and rotor parts. For this reason, large corrosion studies were conducted, especially in the 1950s. In all cases the corrosive medium was moisture in the air, and frequently involved ocean environments. The studies also showed that the **salt content** of the air above oceans/the sea was largely dependent on the **weather conditions**, the height of the waves, and the distance between the aircraft and the surface of the water. Moisture in the air is carried into the engines of standing aircraft both by wind (dynamic soak) and by an airflow caused by the engine cooling off after shut-down (static soak). In the worst cases, 200 ppb (10^{-9}) salt concentration was measured in the air inside an engine (Ref. 5.4.1.2-1).

The damage process during **moisture corrosion** usually involves the corrosion occurring primarily during standstill. When the corrosion has sufficiently weakened supporting cross-sections of engine parts, it can cause dynamic/fatigue cracks or spontaneous part failures.

A special corrosion danger with watery media is present when, on parts made from corrosion-sensitive materials, mating surfaces (such as at holes for clamping bolts/reamed bolts or mating zones of disks and spacer rings) cannot be coated with sufficient corrosion protection due to extremely close tolerances. This are for example **thight fit tension bolts** through the **hols of rotor disks**.

External Factors Corrosion without Mechanical Loading: Damage

"**Dry corrosion**" caused by **fused salt deposits** from the air flow can occur in the hotter engine areas during operation. This requires sufficiently high operating temperatures (for sulfidation on hot parts made from Ni-alloys, see Chapter 5.5; for pitting caused by salt on titanium alloys, also see Chapter 5.4.2).

Illustration 5.4.1.2-1: *The components of engine compressors typically affected by corrosive attack are the rotor blading, the stator vanes, and the inner housing walls, especially around abradable coatings.*

Rotor blading ("A"): *Blades made from Al- or steel alloys (such as typical 12% Cr steel) without corrosion protection are prone to pitting (Ill. 5.4.1.1-2). This causes considerable local decreases in dynamic strength and increases the danger of dynamic fatigue fractures in the blading.*

Abradable and rub-tolerant coatings ("B"): *These relatively soft and therefore porous coatings can absorb condensation water and corrode, and/or carry electrolytes to the base material and cause it to corrode. The polyester spray coatings filled with Al powder that are used in the front compressor area at operating temperatures up to about 100 °C have shown themselves to be especially vulnerable. The corrosion occurs in layers and can be recognized by blistering. The most sensitive area is the transition of the coating edge to the housing, since electrolytes can easily enter into the gaps.*

Soldered compressor stator vanes("C"): *If* ***cell action*** *occurs due to the metallic contact of different metals that are far apart on the electrochemical series (Ills. 5.4.1.1-5 and 5.4.1.1-6), corrosion will be especially pronounced. A typical example is* ***compressor stator blades/*** *vanes made from Cr-steel sheeting and joined with silver or copper solder. Chains of corrosion pittings occur in the root area on the* ***edge of the braze*** *parallel to the chord. This is also where the largest dynamic loads are to be expected in case of a flexural mode in the blade. Therefore, this type of corrosion is especially dangerous with regard to dynamic* ***fatigue fractures*** *in vanes.*

Due to the corrosion danger, today compressor blades made from Cr-steels are generally coated with inorganic ***lacquers filled with Al-powder****, which can be used at operating temperatures up to about 500 °C. Coatings that are sintered or* ***shot-peened with glass beads*** *become electrically conductive and provide the base material with* ***cathodic corrosion protection.*** *This means that, around small scratches, the base material is protected by the greater allowable dissolving of Al-coating components.*

External Factors
Corrosion without Mechanical Loading: Damage

Engines are most susceptible to corrosion during standstill.

Compressor

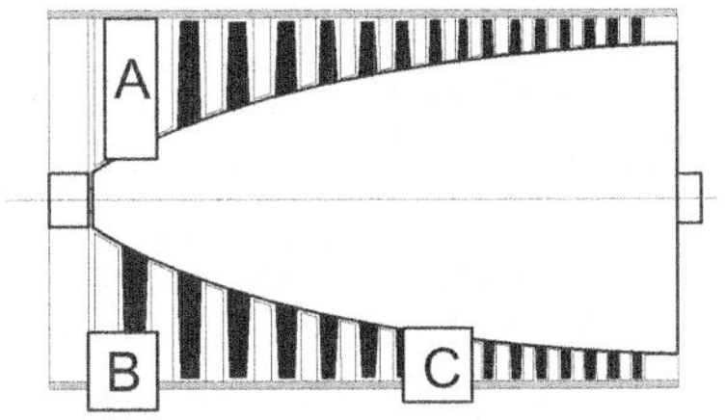

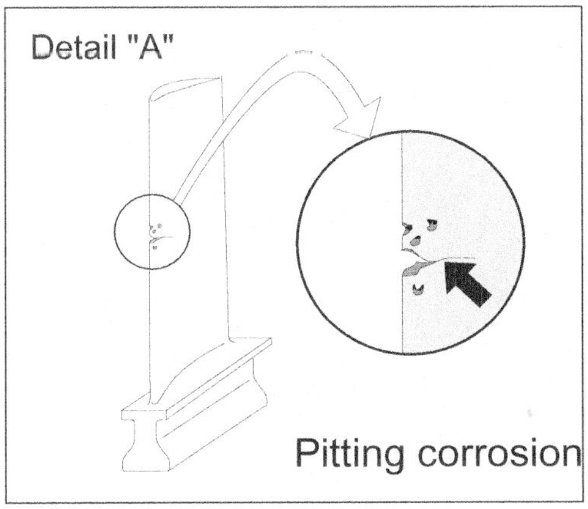

Corrosive attack on unprotected surfaces

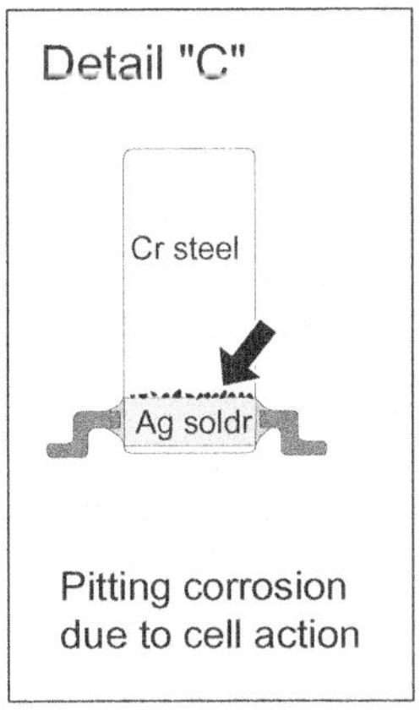

Increased corrosion on solder on compressor stator vanes

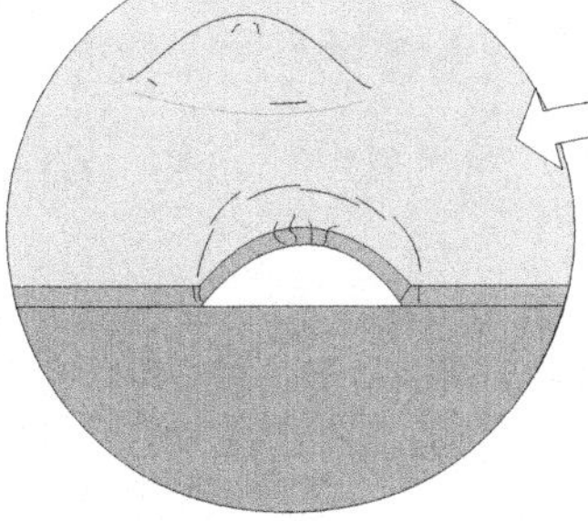

Cross-section of a corrosion blister

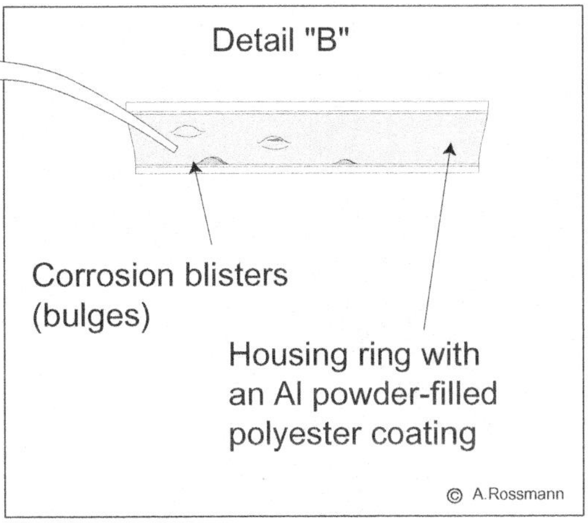

Ill. 5.4.1.2-1

Corrosion beneath abradable coatings

External Factors
Corrosion without Mechanical Loading: Damage

Corrosion in the compressor area is a special problem in older engine types.

Examples on compressor housings

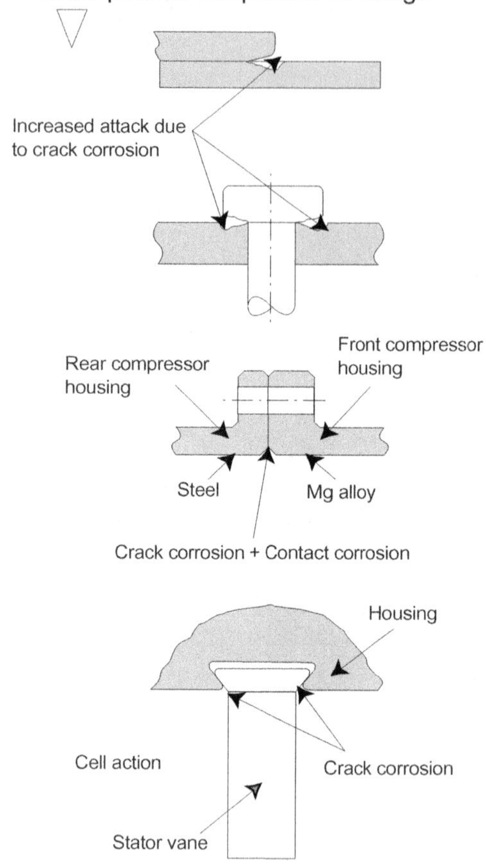

Examples on compressor blades

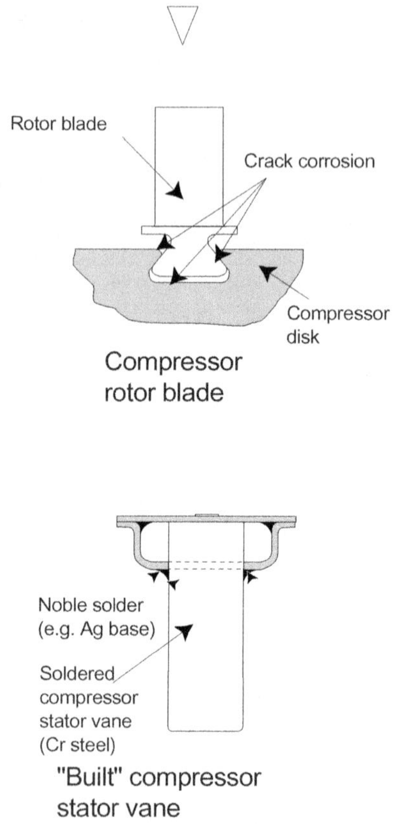

Variable vanes made of steel in a housing made from an Mg alloy

Ill. 5.4.1.2-2

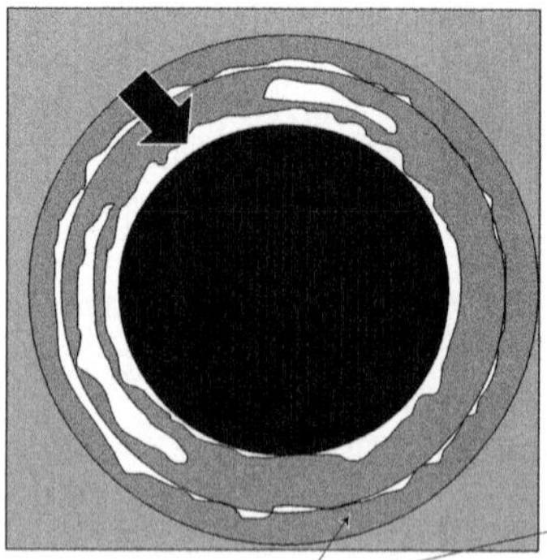

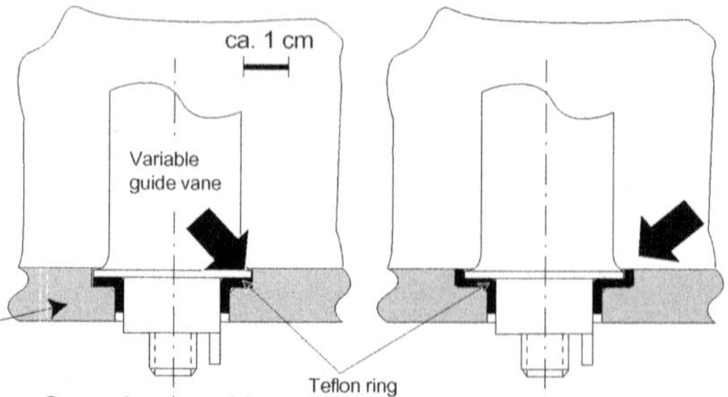

Illustration 5.4.1.2-2: *In the compressor areas of older engine types there are still various* ***corrosion-sensitive materials*** *in use. These are primarily:*

- low-alloy steels (boiler steels) in housings and disks

- high-alloy steels (Cr-steels) in bladings, disks, control systems (such as spherical bearings) and threaded connections. In the middle compressor area of modern engines, disks and housing rings made from steels with low thermal-expansion coefficients are frequently ***placed between parts made from Ti-alloys*** *and parts made from* ***Nickel alloys****. These parts must be equipped with corrosion protection that is suitable for the operating temperatures.*

- ***Al-alloys*** *in* ***rotor and stator blades****, as well as gearing and regulator housings*

- ***Mg-alloys*** *in compressor and gearing housings/ casings*

Today, these materials are all coated with corrosion protection layers, resulting in a considerable improvement. However, experience has shown that corrosion is to be expected under influence of ***condensation water*** *in ocean/sea environments, especially when various materials are combined (cell action, top left diagram) and at separable mating surfaces (crack corrosion, e.g. on flange surfaces). This corrosive attack, which usually takes the form of pitting, often originates in* ***damaged corners/edges*** *or worn areas of seating surfaces. A common area is the* ***seating surfaces of dovetail blade roots*** *(top right diagram).*

Compressor stator vanes made from steel sheeting that have been joined by ***brazing*** *are subject to increased corrosion* ***along the braze seam due to cell action*** *(silver or Cu solder). With adjustable stator vanes, the conditions around the adjustment mechanisms promote crack corrosion in the* ***magnesium housings/ casings*** *(top diagram).*

External Factors
Corrosion without Mechanical Loading: Damage

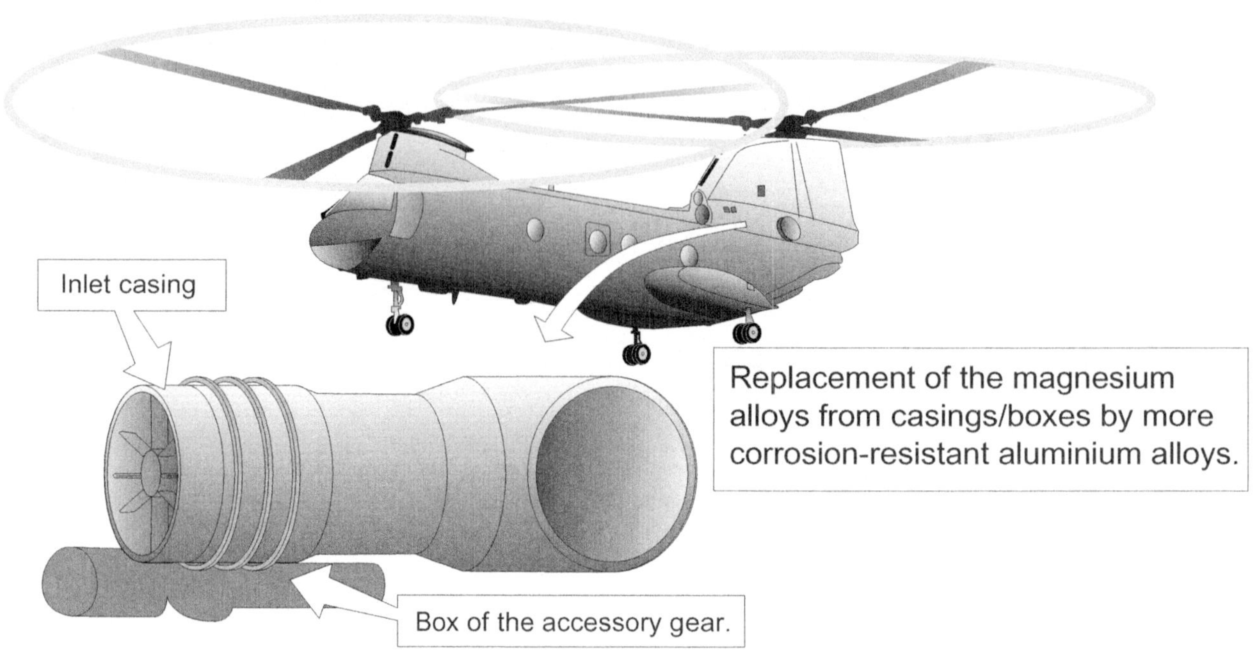

Example 5.4.1.2-1 (Ref. 433):

Excerpt: "Other changes include removing **magnesium** from the engine, which reduces the powerplants susceptibility to **saltwater corrosion** and erosion. ..the engine's **magnesium inlet and gearbox** housing were changed to aluminium, and the magnesium compressor housing was changed to less expensive stainless steel."

Comments: This case concerns an increased-performance version of an engine in a large, twin-engine military helicopter that has been in use for a considerable time. Evidently the material changes were necessary in the light of the bad experiences with the earlier versions.

External Factors Corrosion without Mechanical Loading: Damage

Corrosion caused the spalling of the rub in coatings in the casings of the high pressure compressor. It came to an about. 5 % drop of the aeroengine efficiency and a multitude of blade damages.

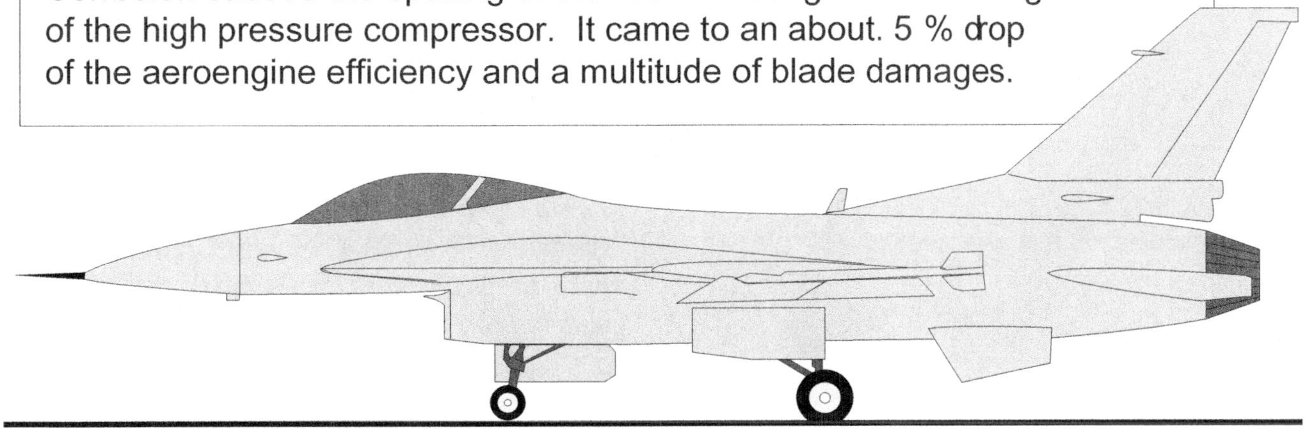

Example 5.4.1.2-2 (Ref. 5.4.1.2-1 and Ref. 5.4.1.2-3):

Excerpt: " ...Corrosion- prone hardware identified as a result of the ...test included the blade-tip rub seals employed in the first three stages of the high-pressure compressor. Corrosion within the bond coating of the seals resulted in widespread delamination or the seals partway through the program. The resulting loss of the seal material **decreased engine performance by more than five percent and produced impact damage to over one hundred compressor airfoils**.

Comments: These findings are from a very elaborate testing program in which the entire engiens were subjected to proper load cycles in a simulated ocean atmosphere (a **sea salt solution** was sprayed into the compressor). The test duration was up to 1200 hours with longer periods in which the engines were running. This became necessary after it was realized, that the testing conditions of the 1960s were insufficient. In the earlier tests, the salt was inserted into the engines as they ran at a mere few hundred **RPM**, which did not sufficiently reproduce the corrosion processes (e.g. sulfidation) in the hot parts.

***Illustration 5.4.1.2-3: Rub coatings** designed to minimize tip clearances in compressor housings are usually made from **porous spray coatings**, which can let through watery corrosive media or soak them up like a sponge. These coatings are attached to bond layers. A typical composition of an bond layer is 5% aluminum and 95% nickel. This **bond layer** can be damaged by corrosion, causing parts of the rub coating to delaminate (middle diagram). Consequential damages in the compressor cause the compressor efficiency to unallowably worsen.*

*This sensitivity was discovered in very extensive and realistic **corrosion tests** (Ref. 5.4.1.2-1), and countermeasures were implemented (Example 5.4.1.2-2).*

*The bottom diagram shows typical **corrosion symptoms** on sprayed coatings. Sprayed coatings in housings/casings can be made from an organic matrix (polyester, etc.) with metallic filling (Al-powder, etc.) and have a relatively dense structure, or they can be made largely from metal with a porous structure, such as is typical in Ni/graphite spray coatings. These coatings are usually attached to the base material by an bond layer.*

Spray coatings on rotor spacer rings can be made from relatively dense ceramic coatings on an bond layer. The most common failure mechanisms are:

*- **Corrosion of the abradable coating** itself:*
*The corrosive medium is carried into gaps and cracks by the capillary effect and leads to a layered attack along corrosion-sensitive filler materials. This is followed by **blistering** and **layered breakouts** within the coating. This type of corrosion is most common in abradable coatings with a synthetic matrix and filler materials.*

*- **Corrosion of the bond layer**:*
The corrosive media travels to the bond layer through porosity, cracks, and gaps in the abradable coating. It then damages the bond layer and causes the abradable coating to delaminate. This corrosion form is more common in porous metallic and non-metallic spray coatings.

External Factors
Corrosion without Mechanical Loading: Damage

Corrosion of rubbing coatings in the compressor occurs especially during marine operation.

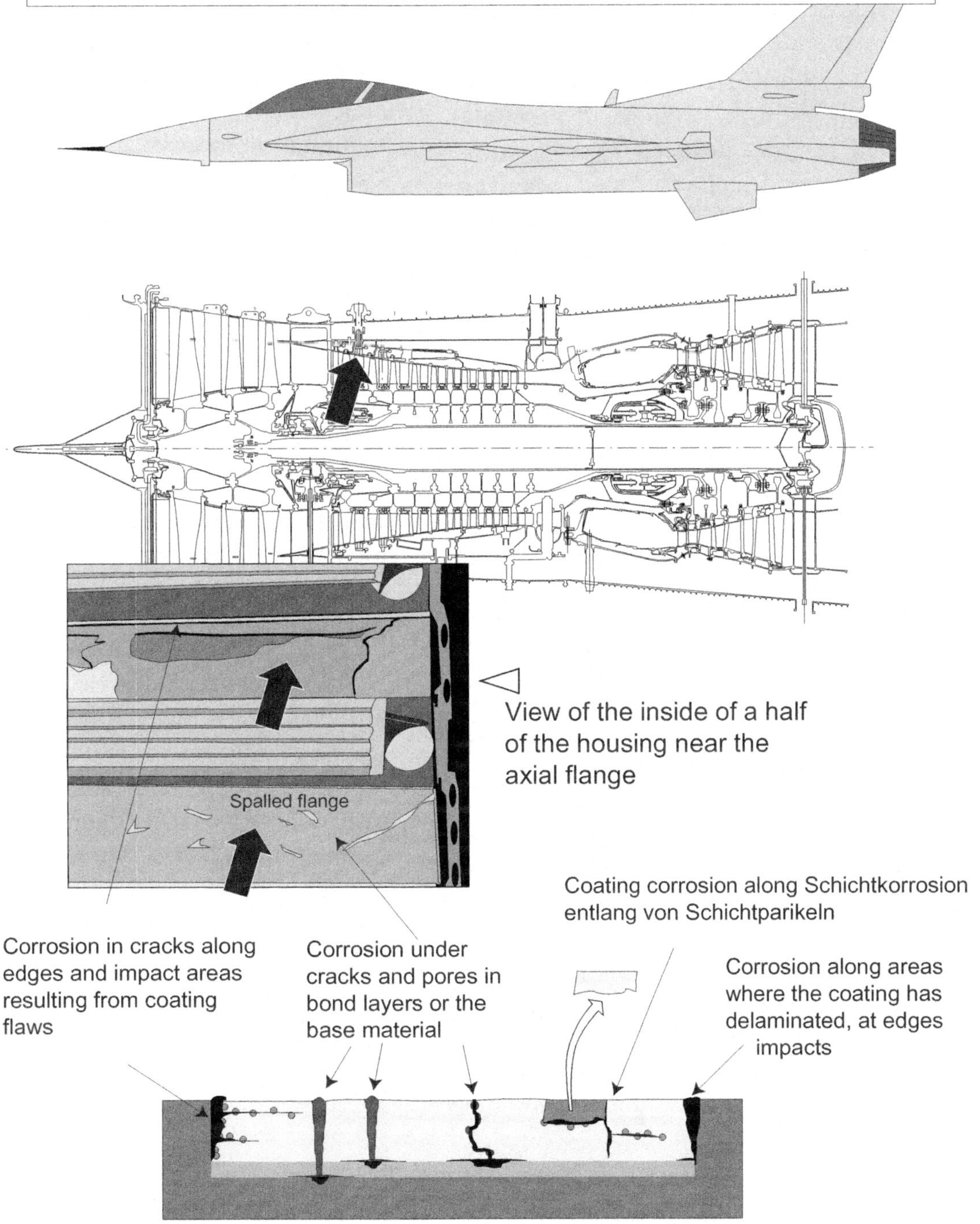

Ill. 5.4.1.2-3

Typical corrosion areas on sprayed coatings

External Factors Corrosion without Mechanical Loading: Damage

Example 5.4.1.2-3 (Ref. 5.4.1.2-3, compare with Ill. 5.4.1.2-4):

Excerpt: "The U.S. National transportation Safety Board is urging the FAA to mandate more frequent inspections of certain high-pressure compressor discs in .... engines to preclude **uncontained failures induced by corrosion pitting.** ...the action which centers on ...engines **with low-utilization HPC disks**, is prompted by the rupture and subsequent desintegration of the 9th stage HPC disk ...The aircraft had departed...when the No.1 engine's 9th-stage HPC disk rim separated and shrapnel exited the high pressure compressor case...As a result, fragments from the disk struck the left wing leading edge forward of the main spar, puncturing the "A" system hydraulic lines, which led to a complete loss of pressure in that system...

Metallurgical analysis of the fractured disk in... laboratories revealed a fatigue crack that had originated in the rim and migrated inward until reaching a critical length... In addition there was 'extensive corrosion pitting in multiple sites that were concentrated on the outer web and rim areas of the disk'... Further examination indicated that the failure originated in a dovetail slot. '**The large number of pits present suggests that corrosion pitting was the most probable cause of the failure**!....

(the engine manufacturer) defines an HPC disk as 'low utilization' if it accumulates fewer than 1,300 flight hours or 900 cycles per year, and it remains in that classification until it is recoated or replated, according to the company. A service bulletin currently requires that these disks be inspected every 10 years after being replated with nickel-cadmium. But NTSB... said that investigation of the...failure indicates that the 10-year inspection interval 'is too long' and should be reduced to significantly less than eight calendar years...

Both the BEA and the NTSB have 'found no evidence to suggest that the failed...9th-stage disk had been improperly inspected or replated eight years before the accident'...

The NTSB suspects that low-utilization HPC disks are more vulnerable to corrosive attack chiefly because they are not operated at a minimum number of hours or cycles per year. **Corrosion pits eventually penetrate into the disk material and develop fatigue cracks**....the safety board said corrosion pitting has been a factor in at least one other uncontained failure or (engine of the same type), low utilization HPC disk....(nearly 5 years before an aircraft of an other type) was destroyed by fire after the crew aborted takeoff ...

Analysis of the disk showed that it had ruptured from a fatigue crack on the disk rim that had originated at a corrosion pit measuring 0.010 in. (0.25 mm) deep. The rim also exhibited heavy surface corrosion and pitting..."

Comments: At least a few compressor disks of this older engine type are still typically made from a pitting-sensitive steel.

The different assessment of the safe inspection intervals is interesting. There is also a remaining question as to why no sufficient measures to prevent future damage were taken after the first damage, which occurred years before.

The ultimate moment of damage, when the blade rim separated, typically occurred during start-up or shortly afterwards.

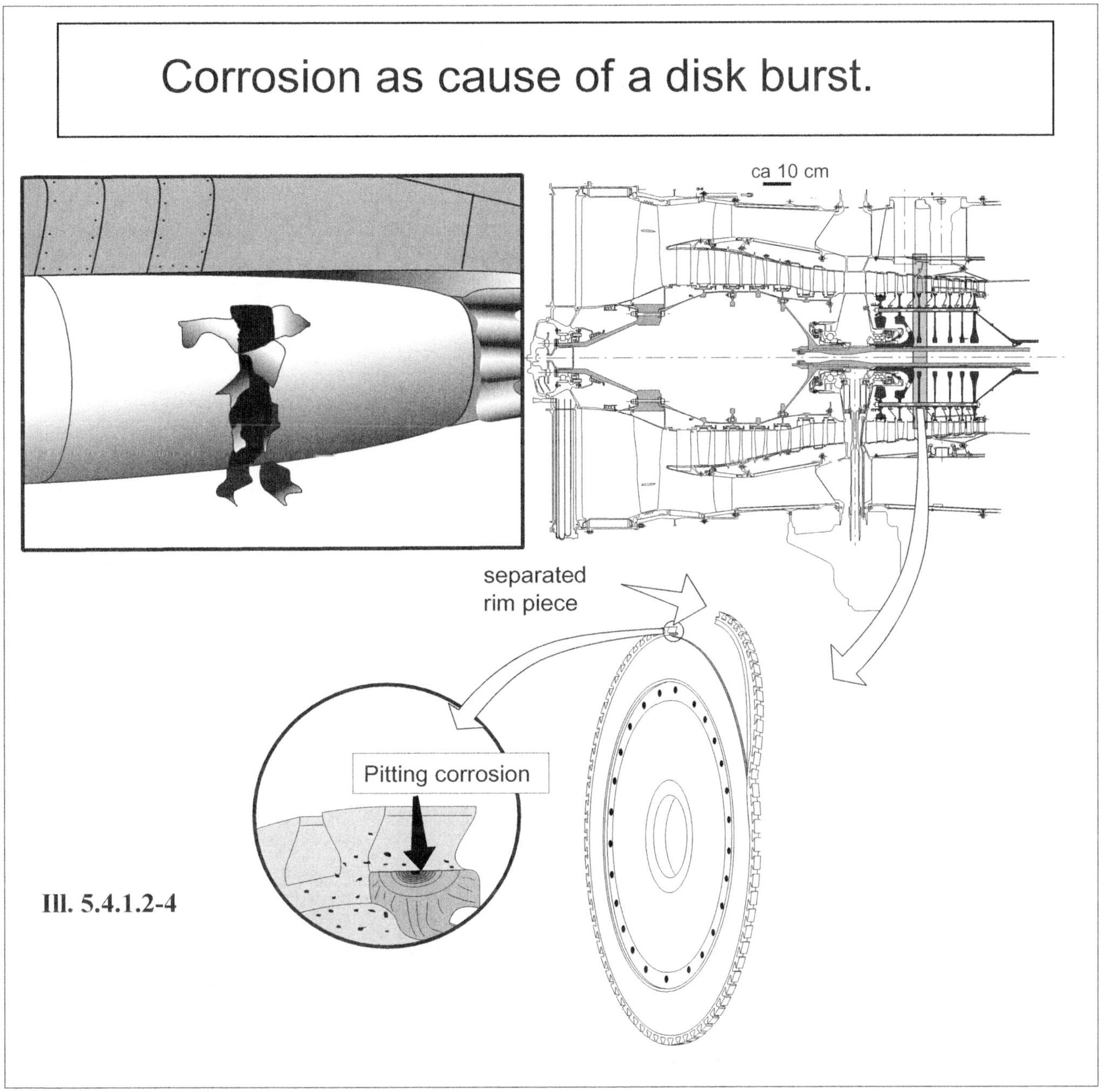

Illustration 5.4.1.2-4: *This damage to a* ***compressor disk*** *made from a corrosion-sensitive steel was evidently caused by pitting resulting from* ***condensation water*** *that attacked the engine during relatively long standstill times. The damage process is described in Example 5.4.1.2-3. A similar failure in the turbine region describes Ill. 12.6.3.3 in volume 3.*

Fracture of the spacers following corrosion damage.

Ill. 5.4.1.2-5

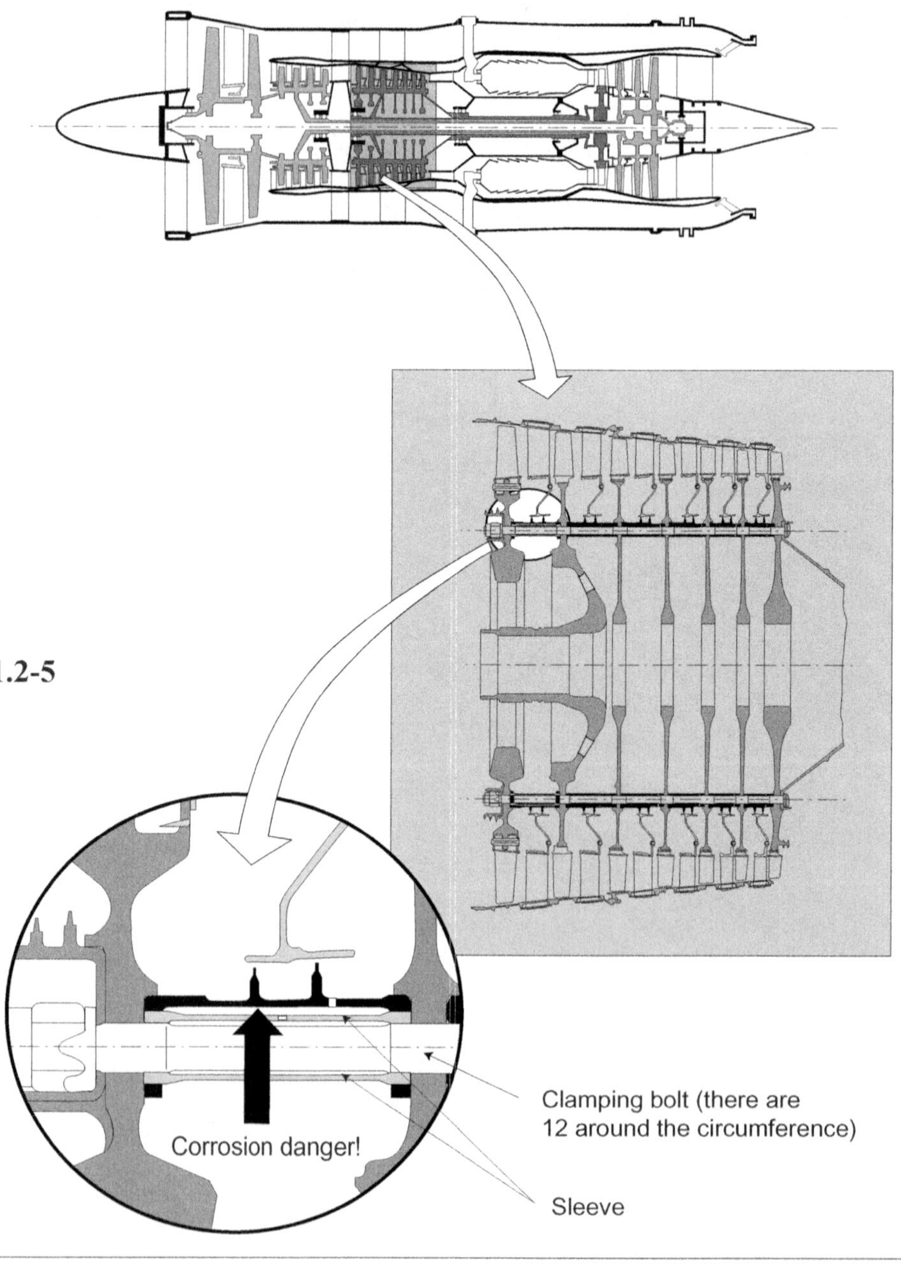

Illustration 5.4.1.2-5 *(Refs. 5.4.1.2-4 and 5.4.1.2-5): These corrosion damages (see Example 5.4.1.2-4) can be traced back to an unsuitable design combined with corrosion-sensitive materials and* ***maintenance problems*** *(Examples 5.4.1.2-5.1 and -5.2). It is not entirely clear, to what degree the corrosion was caused by condensation water during operation or through cleaning solutions that had entered during maintenance work.*

Example 5.4.1.2-4 (Ref. 5.4.1.2-4):

Excerpt: "...investigators discovered pieces of a spacer lying alongside the runway. Its location indicated it the left engine at approximate aircraft liftoff, penetrating engine case, fan duct and cowling.

...According to an FAA notice of proposed rulemaking...compliance with the inspection order will be required within two years, or 4,000 engine cycles...For certain high-time engines, inspection is sought within 1,000 cycles.

The notice also calls for the replacement of the suspect spacers, located between the seventh and eighth and also the ninth and tenth compressor discs in the engine's high-pressure section, as well as four other spacers in the same area, with an existing version of a more recent design. In most cases the substitution is to be made during the engine's next scheduled overhaul. The spacer problem affects 1,840 engines on domestic airliners and another 1,838 engines worldwide.

...(according to a representative of the FAA) the spacer problem is 'not viewed as an immediate, safety of flight' matter. He said (the engine manufacturer's) procedures require removal and inspection of older version spacer tie-rod tubes during overhaul, though the FAA uncovered some instances where this was not done. The engine manufacturer also published service bulletins in 1972 and 1981, both of which emphasized proper inspection and, where necessary, repair of...high-pressure section spacers.

As used in the (engine type), a spacer is a 3.in.-long (7.5 cm, see arrow 5.4.1.2-5) metal alloy sleeve that slips over the engine's rotating central shaft and separates turbine discs from each other. There are six spacers in the high-pressure section...A 'stack' of discs and spacers is bolted together using 12 tie-rods, which run through the discs and spacers parallel to the engine shaft.

Older (engine type's) high pressure section spacers incorporate 12 tubes, all mechanically pressed into place, to accommodate tierod passage. **Undetected corrosion between and around these tubes has resulted in eventual cracking and failure of the spacer.**

(the engine manufacturer) began exclusively shipping a new design...spacer with corrosion-defeating integral tie-rod passages over two years ago. Many of these are already operating in aircraft in the field. However, as replacement of the old-design spacer was not mandatory in the past, both new- and old-design spacers may be mixed on an individual engine....A total of 4,950 of the older- version spacers have yet to be 'attritioned out' domestically, while another 4,939 are still flying...

Since the (related engine type) was introduced into commercial airline service in the middle 1960s, 46 spacer-related engine failures have been reported. Thirty resulted in excessive vibration or in-flight engine shut-downs. The remaining 16 spacers separated from the engine shaft with eight having enough energy to penetrate the engine case, fan duct and cowling, in some instances causing airframe damage. Each replacement spacer costs approximately $3,500."

Comments: The large number of parallel damages (?) and the long time period over which the problem developed are astounding. Evidently even several uncontained failures were not considered sufficiently dangerous to mandate an earlier removal of the potentially dangerous parts.

Separation danger of aeroengines caused by corrosion.

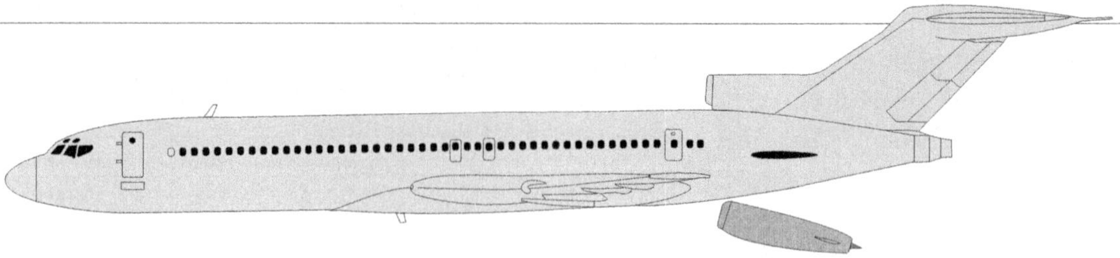

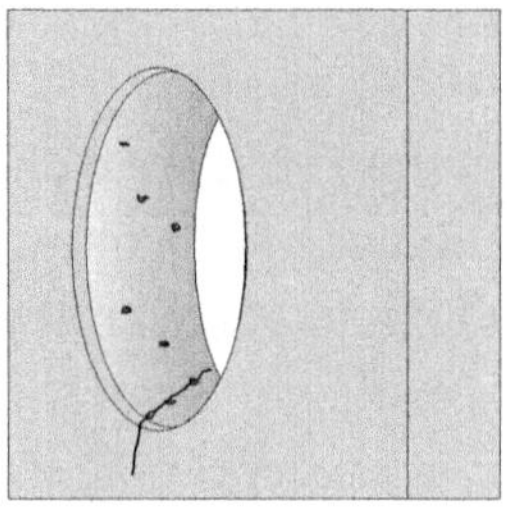

Corrosion in bores is a typical example for Murphy's law'. Everything gets together:
- Corrosion medium (salt water).
- Capillary action.
- Bad testability from the outside.
- Sensitive material.
- High dynamic load.

Example 5.4.1.2-5.1 (Ref. 5.4.1.2-9):

Excerpt: "The FAA is calling for detailed inspections to detect **cracks in engine support fittings**... that could result in separation of the Nos. 1 and 3 engines from the aircraft...Two of the reports noted that the cracks originated from the **large fastener holes** next to the fuselage, and a third report described a crack that had propagated almost completely through the fitting...**The cracks were caused by fatigue induced by corrosion pitting on the surfaces of the fastener holes** in the fittings."

Comments: Engine suspensions are often made from high-tensile steels that are sensitive to **pitting corrosion**. In the inside of bores, especially, capillary action can cause corrosive media (salt water) to collect and, over time, attack fresh metal surfaces that have been created by friction wear/**fretting**. A further problem is the fact that it is usually very difficult to visually inspect bolt connections from outside without disassembling the engine (voilume 3 Ill. 12.6.1-12).

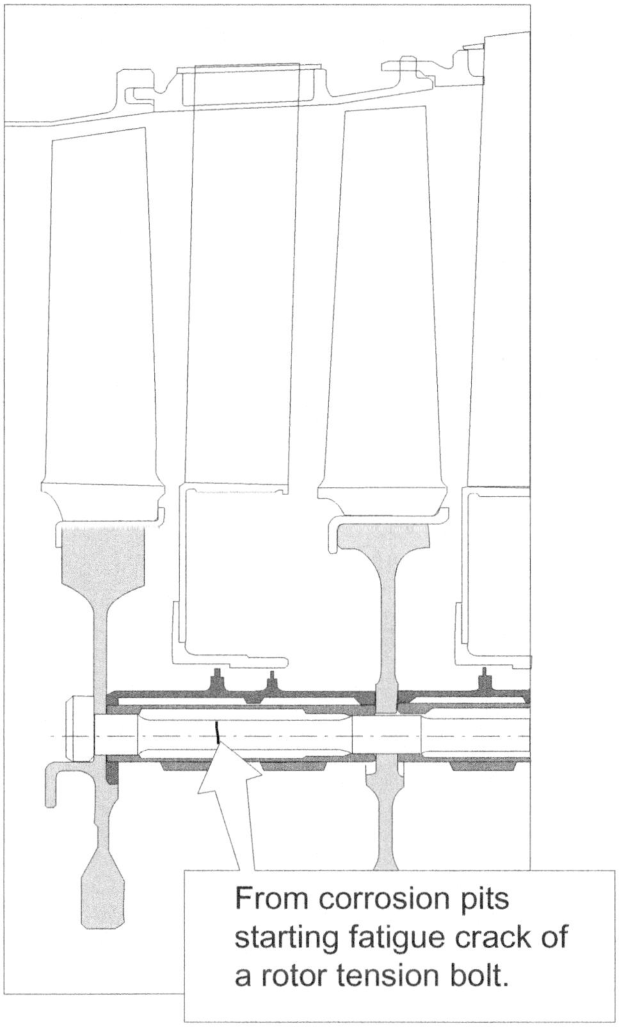

Example 5.4.1.2-5.2 (Ref. 5.4.1.2-6, see Ill. 5.4.1.2-6):

Excerpt: "The (NTSB) board's recommendation follows discovery of broken tie rods on an engine...The aircraft experienced an **uncontained failure** of the low-pressure turbine...during takeoff roll...

The board recommended...that all ...engines be disassembled at next shop visit and that the low-pressure compressor **rear tie rods** be **removed. cleaned and inspected**, recoated with antigallant and reinstalled or replaced. The rear tie rods are long bolts that hold together the four stages of the ...**low pressure compressor** (LPC). One of the tie rods...was fractured and had surface corrosion, and because of overstress, the **11 remaining tie rods also were fractured**.

Maintenance records showed the engine, including the LPC, had been repaired ...(about 12 and 8 Years before)...the board said. NTSB said it concluded that an antigalling coating was not applied to new LPC rear tie rods..(by the OEM) at the time of manufacture.

The board said that while... (the OEM) stated that ...(this) engine failure has been the only one that involved the fracture of an LPC tie rod that was supposed to have been **coated with the antigallant**, the lack of the antigalling coating on all 12 of the LPC rear tie rods... suggests the problem may be widespread.“

Comments: This does not necessarily mean that the failure of a single tie rod will cause the entire rotor connection to fail, before vibrations caused by imbalances indicate damage early enough to avoid a catastrophic failure.

Because this engine type is very widely used, the quoted reference literature estimates the potential costs for necessary preventive measures to be about \$$10^9$.

External Factors
Corrosion without Mechanical Loading: Damage

The fracture of a corroded clamping bolt in the compressor led to overspeed and disk failure in the turbine.

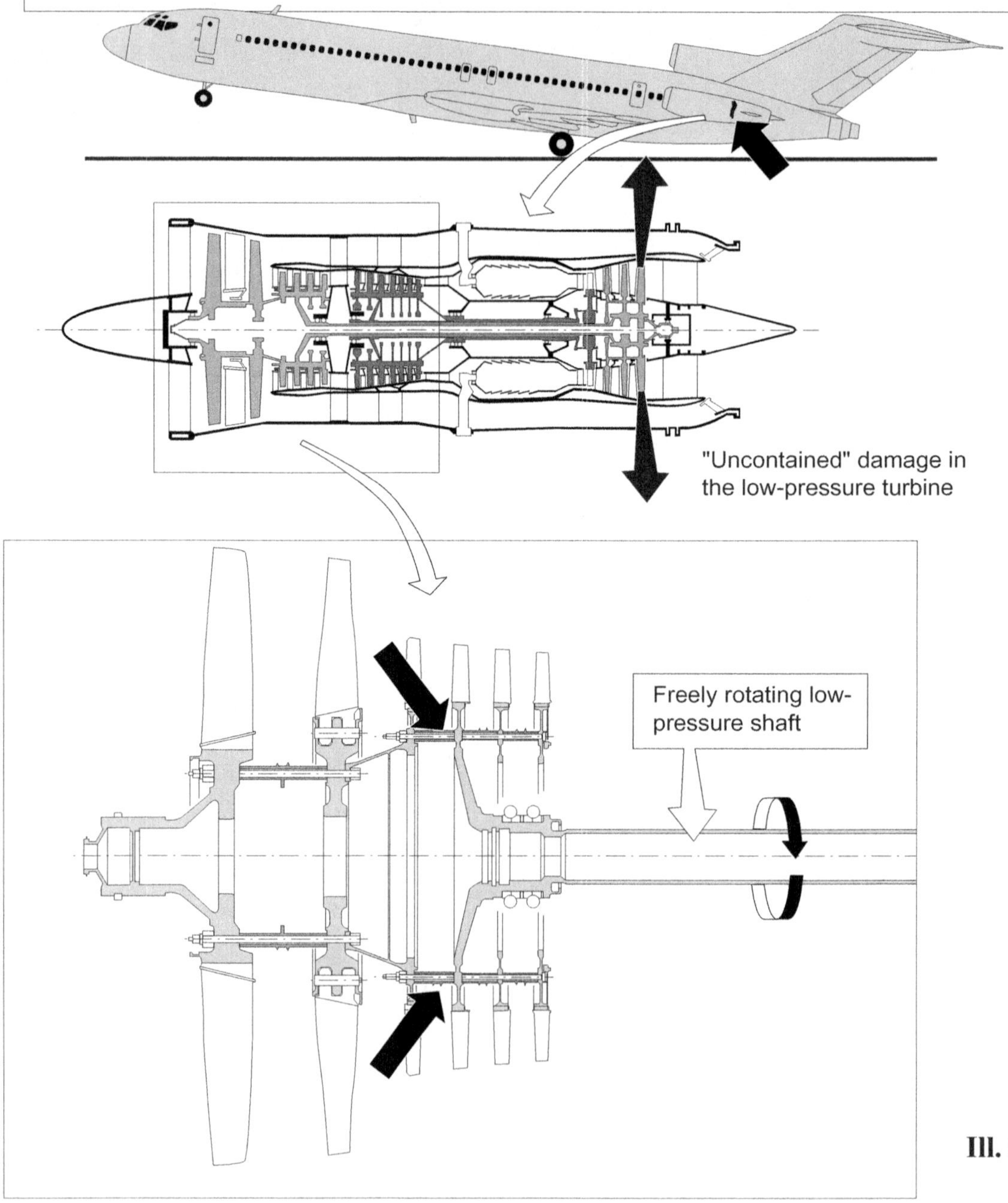

Ill. 5.4.1.2-6

Illustration 5.4.1.2-6 *(Ref. 5.4.1.2-6): Surface* ***corrosion*** *caused a* ***clamping/tensioning bolt*** *in the low-pressure compressor to fail (black arrows in bottom diagram, Example 5.4.1.2-5.2). After this, the remaining clamping bolts failed due to overloading and the shaft separated. This resulted in the now free-running low-pressure turbine reaching overspeed and failing and creating uncontained fragments.*

External Factors
Corrosion without Mechanical Loading: Damage

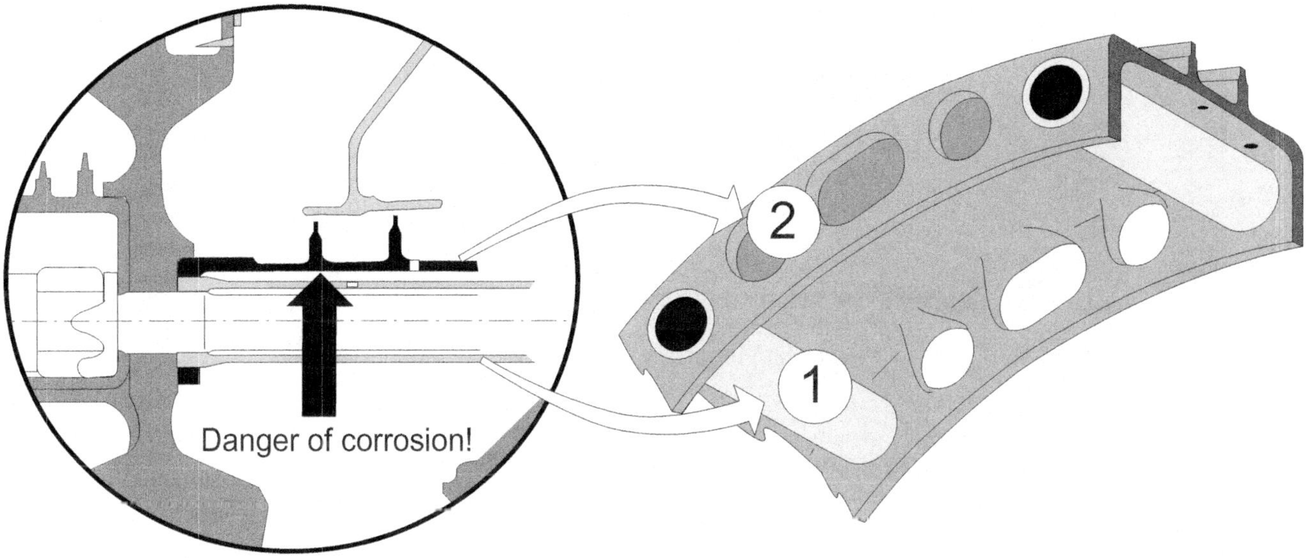

Example 5.4.1.2-6 (Ill. 5.4.1.2-5, Ref. 5.4.1.2-7):

Excerpt: "...However, the core problem seemed to be the compressor spacer rings...which have to be replaced in 1840 engines worldwide. The technical problem is very complicated in its details, but its fundamental principle is relatively simple: the seal system of an...engine (of the affected type) is held together by long bolts which pass through pipes („**1**") in the spacer rings („**2**"). **During inspections, these pipes must be removed in for cleaning and overhaul. However, this was often neglected in order to lower costs. As a result, cleaning solution collected between the pipe and spacer ring, causing rust to form.** ...The US company (engine manufacturer) maintains that it had insisted that all airlines inspect the spacer rings for cracks and replace them by early 1988 at the latest."

Comments: The above explanation, in which the corrosive media only entered into the damaged area in the form of cleaning solution during overhauls, cannot satisfactorily explain the life span-related inspection intervals in Example 5.4.1.2-4. Also, the involved media is a watery electrolyte which will evaporate after short run times in this rear area of the compressor. Therefore, in order for dangerous corrosion to take place, new electrolyte must always form during standstill. This can only be plausibly explained by **condensation water**, which dissolves dried cleaning solution remnants and makes them corrosive again. This probability seems rather remote when one considers that formation of condensation water requires an air exchange, making elements of the **marine atmosphere far** more likely as a corrosive media.

External Factors
Corrosion without Mechanical Loading: Damage

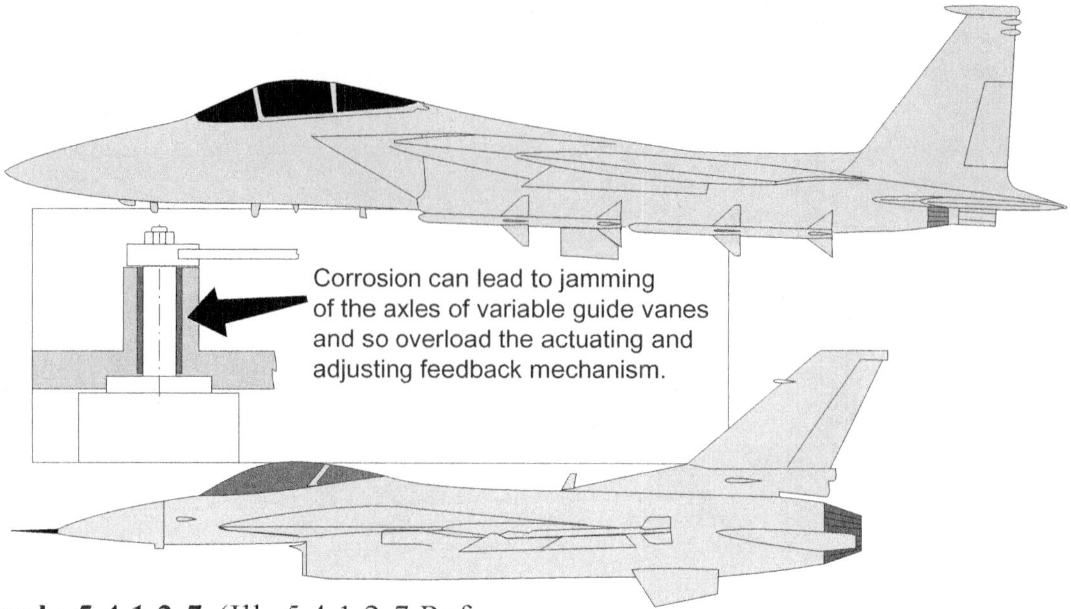

Example 5.4.1.2-7 (Ill. 5.4.1.2-7,Ref. 5.4.1.2-8):

Excerpt: "...Both training accidents...(of an military aircraft type with **one engine**) resulted in a successful crew ejection following engine stagnations at low altitude. The ...crash brought on worldwide inspections of the engine rear compressor **variable vane synchronizing arm bearing and control cable** for freedom of movement and the rear **compressor push-pull control cable**. Out of 271 powerplants checked, 12 were rejected. Six more have been taken out in inspections being conducted 10 hr. after the first check.

Sixteen days before the crash, it (the engine) had undergone a 50 hr. phase inspection at 84.9 total operating hours...

Inspection of the engine showed that:

The rear compressor variable vane (RCVV) feedback cable to the unified fuel control had a **broken link** at the RCVV bell crank. The cause was fatigue failure....

(the accident investigation on the crashed engine showed)...The RCVV feedback push-pull cable was found broken at the bell crank end because of **fatigue failure**.

The report said inflight failure of the cable would cause engine operation problems because of misscheduling of the vanes. Experiences with (an other military aircraft type with two engines of the same type) have shown six instances of **cable failure**, five of which resulted in **stagnation- two at part power and three in augmentor operation**. Three of those stagnations had successful airstarts."

Comments: The repeated fracture of the replaced feedback cable after a short run time indicates that overload conditions that were not eliminated by the replacement were already present. This may have been, for example, some stiffness. The literature does not indicate how much of a role corrosion played in this case.

Illustration 5.4.1.2-7 : Compressor stator vane adjustment systems *usually contain steels, which tend to corrode in marine environments, in bearings "1", **seals** "2", (also see Ill. 5.4.1.2-2) **hinge eyes** "3" and **cable connections** "4". It has been discovered that lubricants containing* MoS_2 *can even promote the sticking of plain bearings in marine environments. This can result*

Typical critical areas of a variable guide vane system.

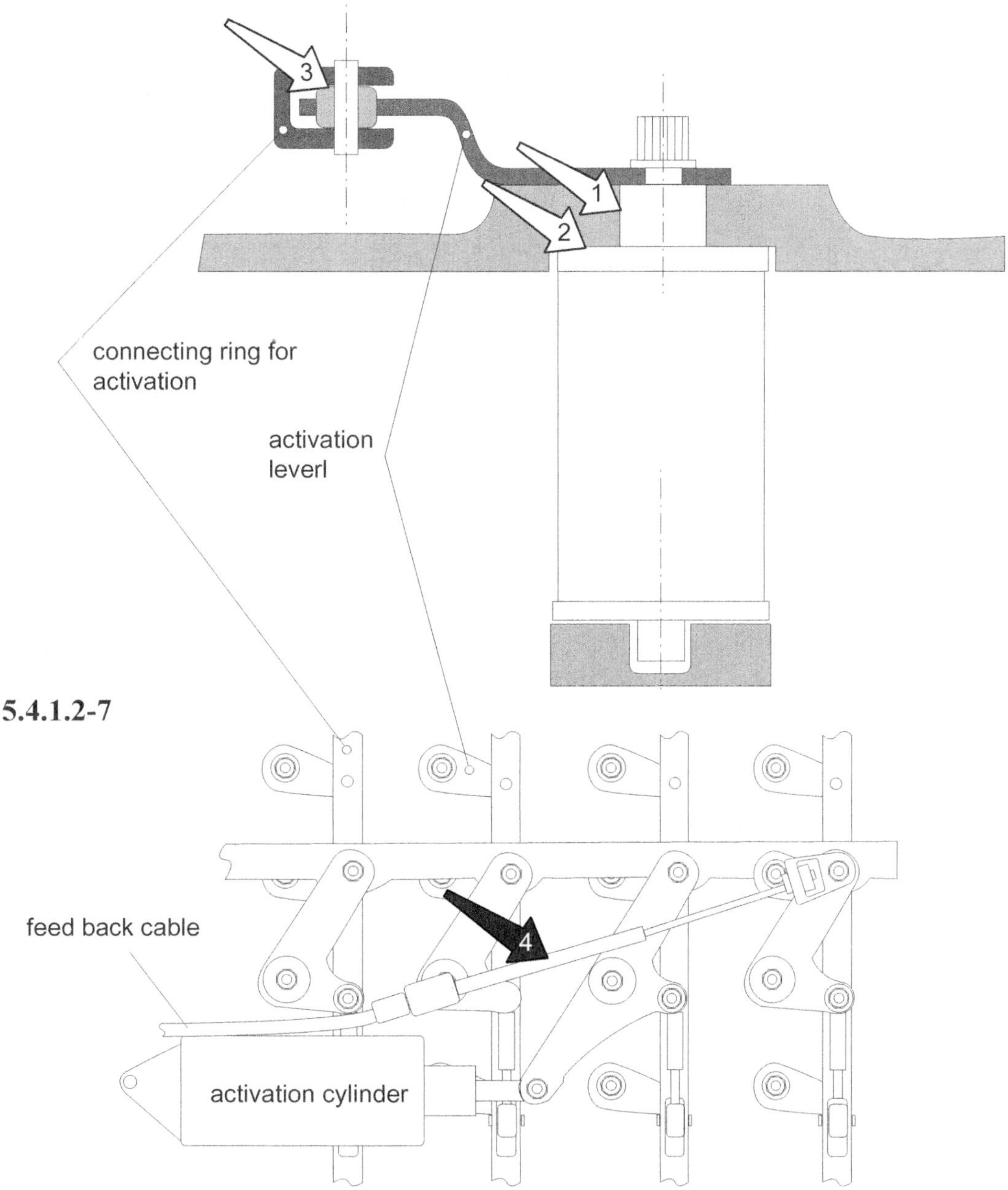

Ill. 5.4.1.2-7

*in overloading of the actuator components. Other possibilities include pitting corrosion, resulting in **dynamic fatigue fractures**. Additionally, the high strength of the steels used in these parts makes them susceptible to stress corrosion cracking (see Chapter 5.4.2). Failure of the adjustment system or its control and regulator systems can cause **compressor surges/stalls**, resulting in engine failure (Example 5.4.1.2-7).*

External Factors
Corrosion without Mechanical Loading: Damage

Problems with variable vanes can easily lead to dangerous situations.

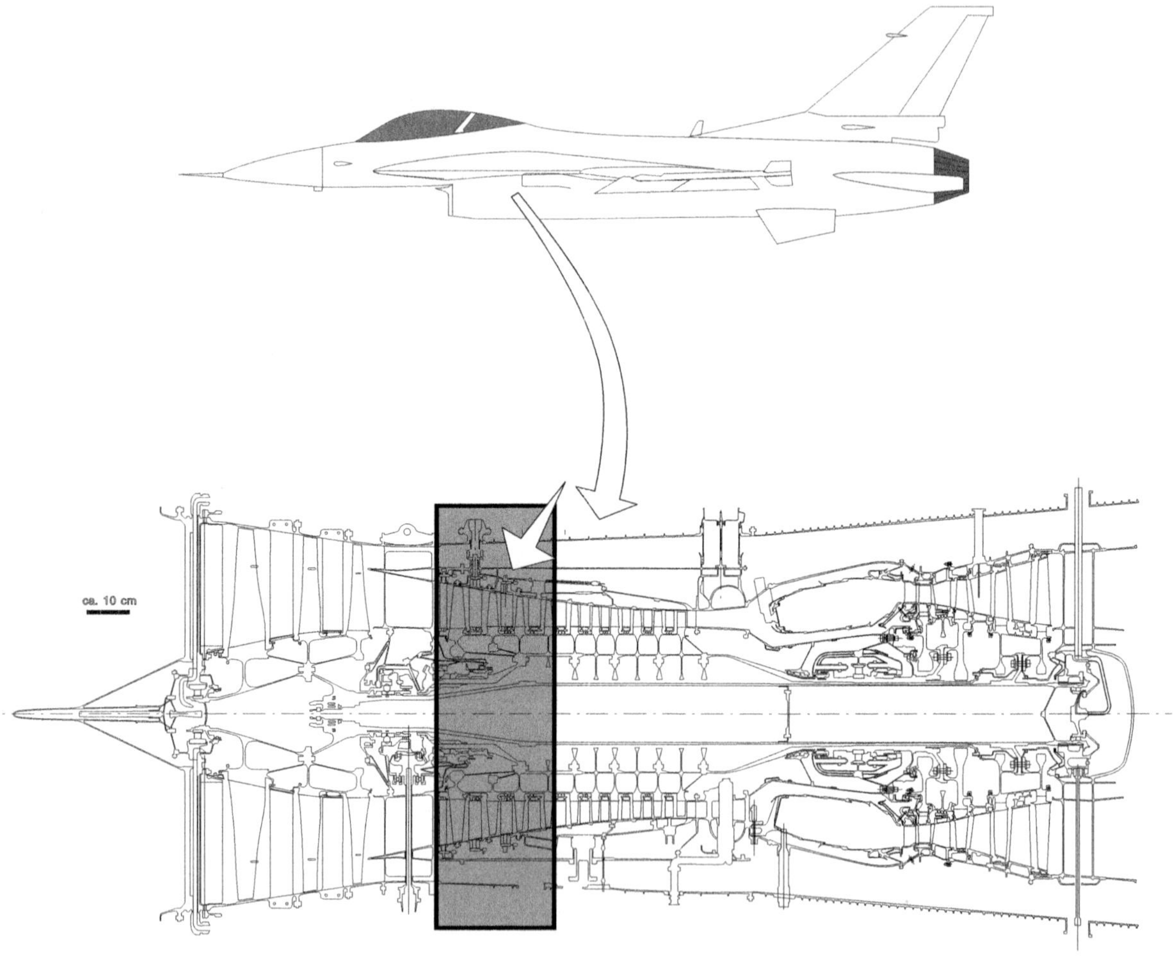

Ill. 5.4.1.2-8

Illustration 5.4.1.2-8 *(Ref. 5.4.1.2-8): The failure or* ***malfunction of the stator vane adjuster*** *in the compressor is an extremely dangerous situation. It can lead to immediate compressor stalls and engine failure (Example 5.4.1.2-6). If only one vane is affected, it can lead to dynamic fatigue fractures in the neighboring stages after a reasonably long run time. This dynamic overstress is due to the turbulent flow caused by the improperly positioned vane.*

Typical damages and damage causes in variable stator vanes are:

- Complete or partial shutting of the variable vanes of the affected stage

- Fractured transmission links

External Factors Corrosion without Mechanical Loading: Damage

- Sticking and/or fracture of bearings and joints

- Mechanical or electrical damage to the regulator/feedback cables

- Fracture of the vane bearings (shafts/axle lugs)

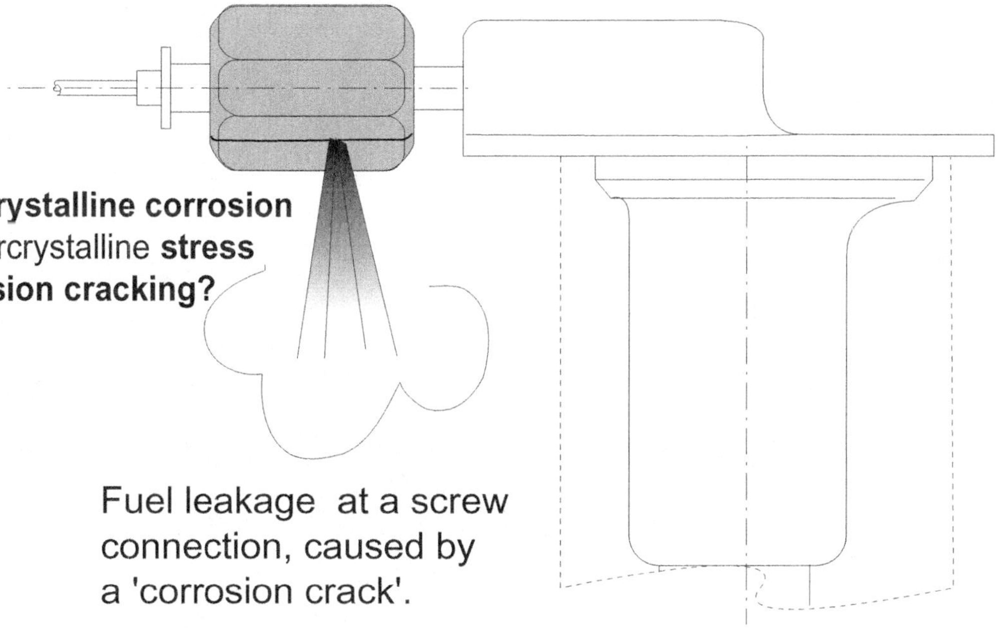

Example 5.4.1.2-8 (Ref. 5.4.1.2-10):

At the endcap of the fuel distributor a considerable **fuel leak** developed. An investigation showed, that the producer of this component did **not met the specified production process after the heat treatment**. This had as result a susceptibility for **intercrystalline corrosion**. An operation failure was the consequence, which can lead to a fire and the shut down of the aeroengine with a emergency landing.

Comment: Crack formation at the coupling nuts of **pipe line screw connections** can be traced back predominantly at a material structure, susceptible for a crack corrosion. Usually the cracks run axial, correspondent the structure orientation of the rod material from which the parts are produced. At steels an **unusual high hardness** (above 32 HRc) is a certain feature for an unsufficient heat treatment during production. It is not clear how far the intercrystalline path of crack is also influenced by operation caused tensile stresses and/or residual/internal stresses from the heat treatment process (Ill. 5.4.2.1-2). In such a case the corrosion can be rather assigned a stress corrosion cracking (Ill. 5.4.2.2-1 and Ill. 5.2.2-2). The sketch was reconstructed from the scarce informations of the literature on hand. Deviations are possible.

External Factors
Corrosion without Mechanical Loading: Damage

The edges of engine parts made from Mg alloys are especially prone to corrosion in marine environments.

Ill. 5.4.1.2-9

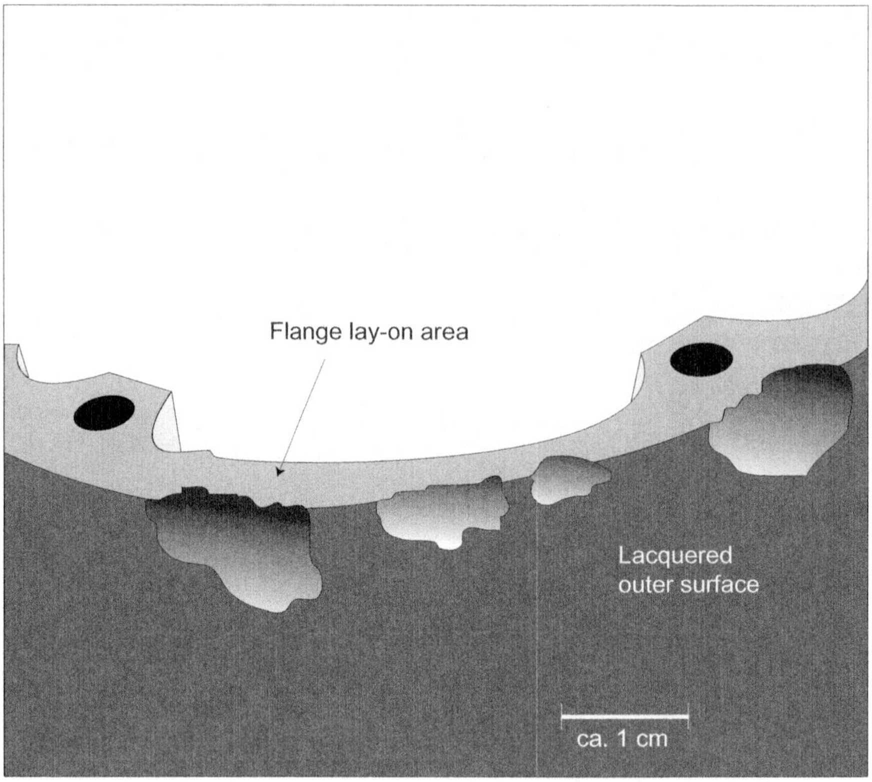

***Illustration 5.4.1.2-9: Corrosive attack on the edge of a lacquered/coated gear casing** made from a cast **Mg alloy**. The pitting corrosion under the influence of marine atmosphere is typical (Cl ions). Repair is possible in areas subject to low mechanical loads by filling them with synthetic material.*

Proper corrosion protection in accordance with regulations on magnesium parts has been proven to ensure long damage-free operating times. Therefore, if corrosion occurs, the cause is most likely defects or damage in the coating.

Corrosion of edges usually occurs at mechanical damage to the lacquer coating, insufficient coating thickness, or coating defects. Corrosion in the flange area is promoted by corrosion cracking conditions and cell action in contact with the mating surface.

Corrosion pittings on lacquered/coated surfaces are often related to improper preparation of the surface to be lacquered, such as corrosive remnants of cleaning baths and shot-peening materials.

This results in typical blistering due to corrosion between the coating and the base material.

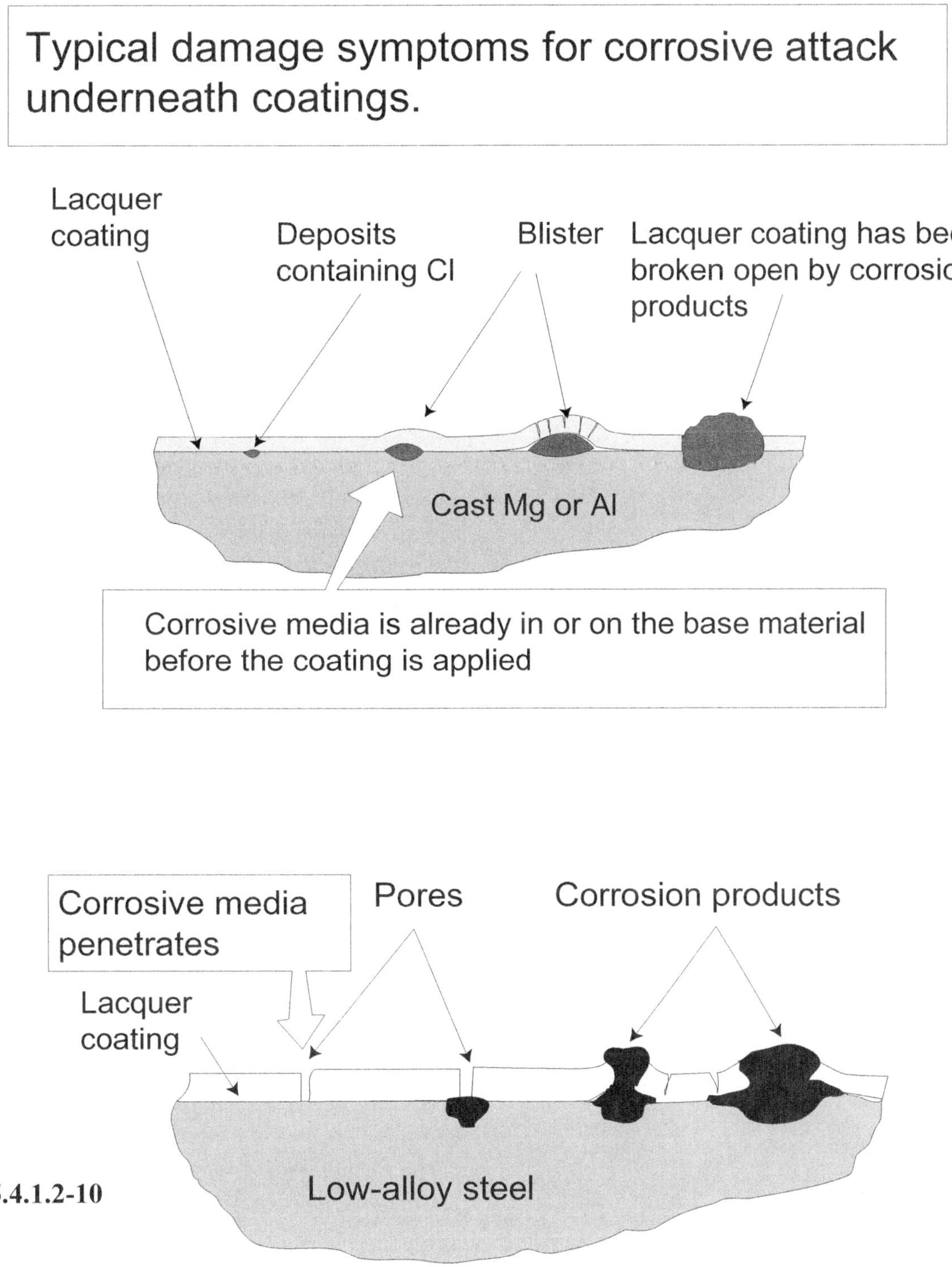

Ill. 5.4.1.2-10

Illustration 5.4.1.2-10: Corrosion of lacquered/ *coated parts has typical damage symptoms that can indicate the cause of the damage. The final stage of the corrosion is usually* ***blistering*** *due to the increased* ***volume of the corrosion products.***

***Illustration 5.4.1.2-11**: The experience shows, that especially corrosion must be expected if **splash water can get into the nacelles of the aeroengines**. Such situations develop during rain at the landing approach with extended landing gear or at at puddles on the runway. Prone are from experience airplane types with unfavorable aeroengine position in ralation to the nose landing gear (sketch below, volume 2 example 10-10.1). That is also true for turboprops when the main landing gear is swung out of the aeroengine nacelle (sketch above).*

Especially endangered by corrosion are components at and around the aeroengine as

*-mountings of the aeroengine and the pylon. Here the **shear bolts** (attachment bolts), mounting bolts from **high strength steel** tend to the **formation of corrosion pittings**. The result is a dangerous **drop of the fatigue strength**. Fatigue cracks cause the **separation of the aeroengine** (volume 2 Ill. 10-10.1) withm potential catastrpohic consequences.*

*- **Elektrical lines, contacts, connectors** (volume 5 Ill. 19.2.1-1.1 and Ill. 19.2.1-1.2) lo0se by corrosion through **breaks**, **slack joints** and **shorts** temporary or permanent its function. Consequences are malfunctions or the outage of components like control units as well as dropped out or **false alarm signals**. With this serious incorrect decisions of the pilot are pre-programmed. Typical are **erroneous fire alarms** with unnecessary in flight shut down (IFSD) of the aeroengine. Even worse may be, if fire is not signaled.*

*- The **variable guide vane actuation** is positioned at the compressor outside. So humidity from rain and splash water can enter. With this, corrosion (Ill. 5.4.1.2-7 and Ill. 5.4.1.2-8) can trigger **fracture** (Ill. 5.4.2.2-3.2), **jamming or unacceptable wear**. In an extreme case this causes **surging of the compressor** with thrust loss of the aeroengine during a delicate situation (start).*

External Factors
Corrosion without Mechanical Loading: Damage

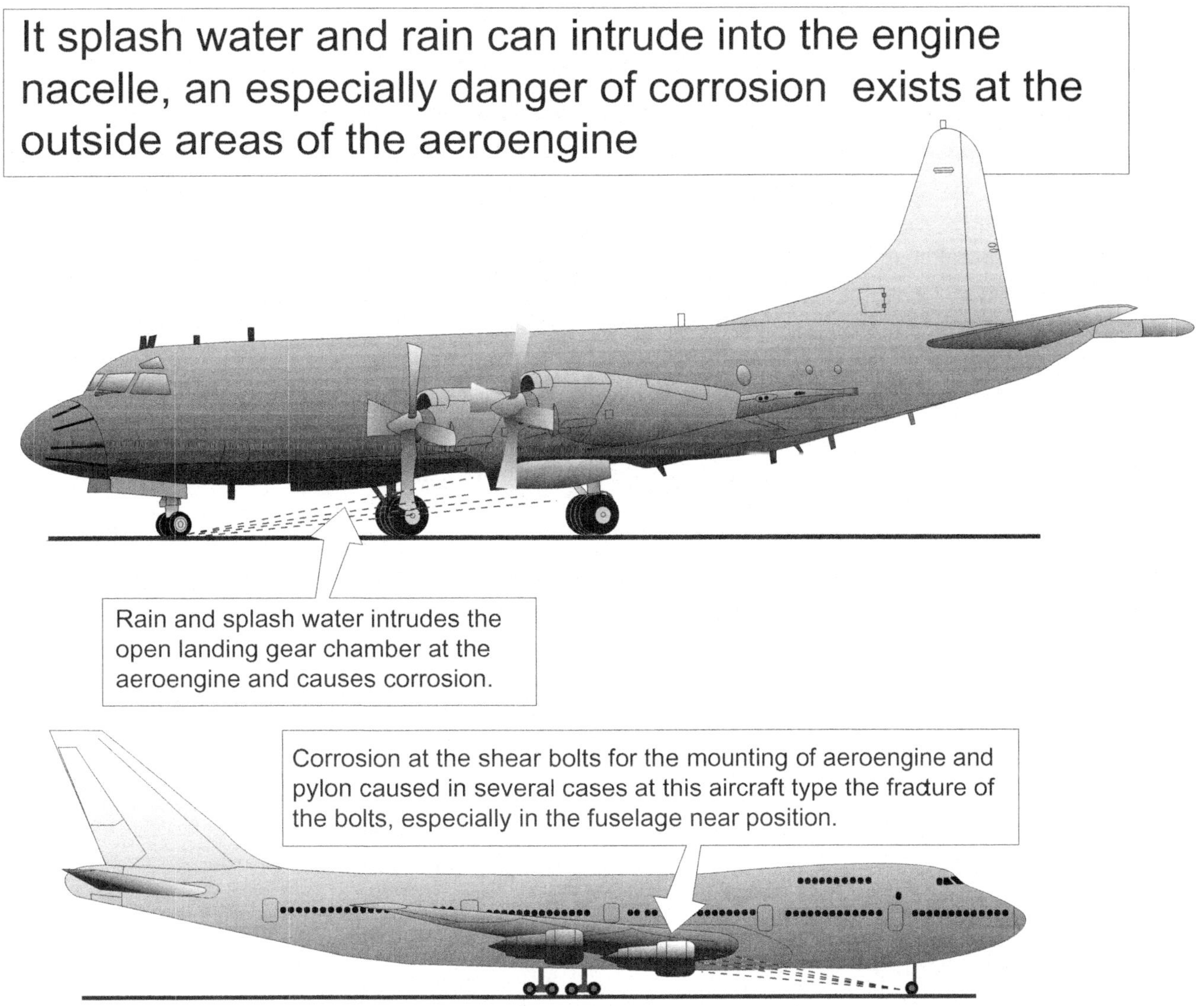

Ill. 5.4.1.2-11

Chapter 5.4.1.2 References

5.4.1.2-1 F.T. Carroll, D.R. Parish, "Testing Considerations for Military Aircraft Engines in Corrosive Environments (a Navy Perspective)", AGARD-CP-558, proceedings of the conference "Erosion, Corrosion and Foreign Object Damage Effects in Gas Turbines",Rotterdam, The Netherlands, 25.28 April 1994, Chapter 37, page 37-10.

5.4.1.2-2 J.Broichhausen, "Schadenskunde - Analyse und Vermeidung von Schäden in Konstruktion, Fertigung und Betrieb", pages 344-368.

5.4.1.2-3 E.H.Phillips, "NTSB Targets JT8D Cracks", periodical "Aviation Week & Space Technology, February 3, 1997, page 47.

5.4.1.2-4 "FAA Proposes Inspection of JT8D Spacers After Crash Investigation", periodical "Aviation Week & Space Technology", January 1986.

5.4.1.2-5 "FAA acts on JT8D", periodical "Flight International", 11. January 1986, page 5.

5.4.1.2-6 "NTSB Seeks JT8D Inspections", periodical "Overhaul & Maintenance", April 1999, page 73.

5.4.1.2-7 "Rost an den Flügeln", periodical "Wirtschaftswoche Nr. 9, 21.2.1986, pages 92 and 94.

5.4.1.2-8 D.R. Griffiths, "Crash Investigations Spur Engine Recommendations", periodical "Aviation Week & Space Technology", September 15, 1980, pages 86 and 87.

5.4.1.2-9 "FAA Orders Inspections of Boeing 727 Fittings", periodical "Aviation Week & Space Technology", March 17, 1997, page 101.

5.4.1.2-10 "Blow-dry your jet", periodical "Flight International", 18 May 1985, page 49.

5.4.1.3 Measures Against Corrosion Damage

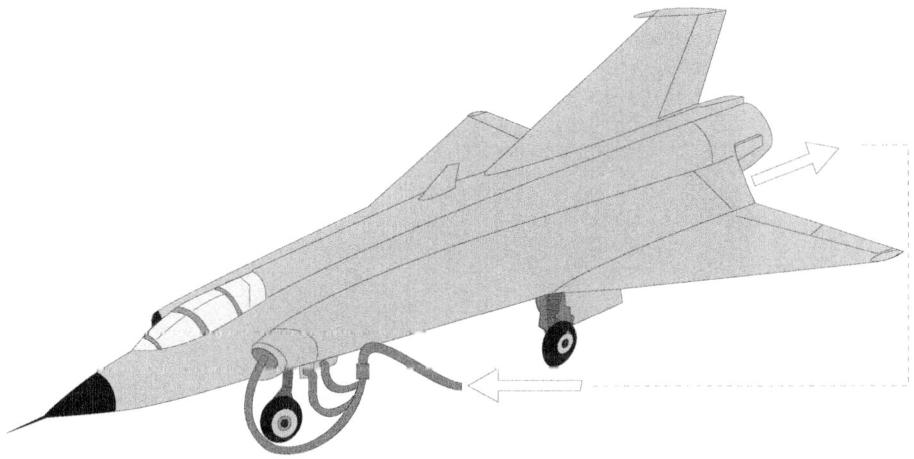

In order to **prevent corrosion** damage without a noticeable effect of mechanical loads, there are many different preventive **measures** which can be selected to provide optimal corrosion protection in each individual case.Repairing or minimizing damages to corroded parts difficult due to the high safety demands on aircraft engine parts and necessitate sufficiently **realistic testing** to ensure satisfactory operating behavior.

Storage and Parking:

• Proper storage of parts, units, and engines in containers and packaging with and without conservation. Parts that are stored for a long period (years) should be at least **spot-checked at suitable intervals** for possible corrosion damage. Special attention must be given to the roller surfaces and bearing tracks of roller bearings. Tarnishing and rough areas in the rolling element are typical signs of **standstill** corrosion under influence of condensation water or humidity. Use of the prescribed auxiliary materials such as conservation oil and **dehumidifiers** is necessary. It must be ensured that the packaging provides the required seal.

• **Conserving compressors** with corrosion-sensitive areas (abradable coatings, Cr-steel or Al-alloy blading in older engine compressor types) before longer standstill periods.

• Connecting the engine to **an air-conditioning/dehumidifying unit** (Ref. 5.4.1.2-10). This measure is especially suitable for **long standstill** times in marine environments.

External Factors Corrosion without mechanical Loading damage: Remedies.

Illustration 5.4.1.3-11 *(Ref. 5.4.1.3-1 and Ref. 5.4.1.3-2): Corrosion primarily occurs by* ***condensate formation*** *in standing aeroengines. Temperature differences and with this the volume change of the air inside the aeroengine* ***during stand still*** *provide in the mounted condition a sufficient air exchange. The experience shows. that air with a* ***relative humidity below 50%*** *reduces the danger of corrosion essential. Therefore the* ***dehumidification*** *of the air in a standing aeroengine is an effective method to minimize corrosion deteriorations in aeroengines. Such conditions will be achieved by continual* ***blown dry air into the aeroengine*** *(open system, sketch below right) or into the storage room (closed system, sketch below left). As air dehumidification in the practical application a continual operating regenerative system according the ljungstroem principle prevailed (sketch in the middle). Centerpiece ia a* ***rotating ceramic disk with honeycomb structure channels which by absorpiton extract humidity from the process air****. After a certain rotation the hunidity of the gathered air is detracted by by a separaten flow of heated air („Reactivation Air“) from the disk.*

External Factors
Corrosion without mechanical Loading damage: Remedies.

With a relatively simple device dangerous and expensive corrosion deteriorations at aeroengines can be prevented.

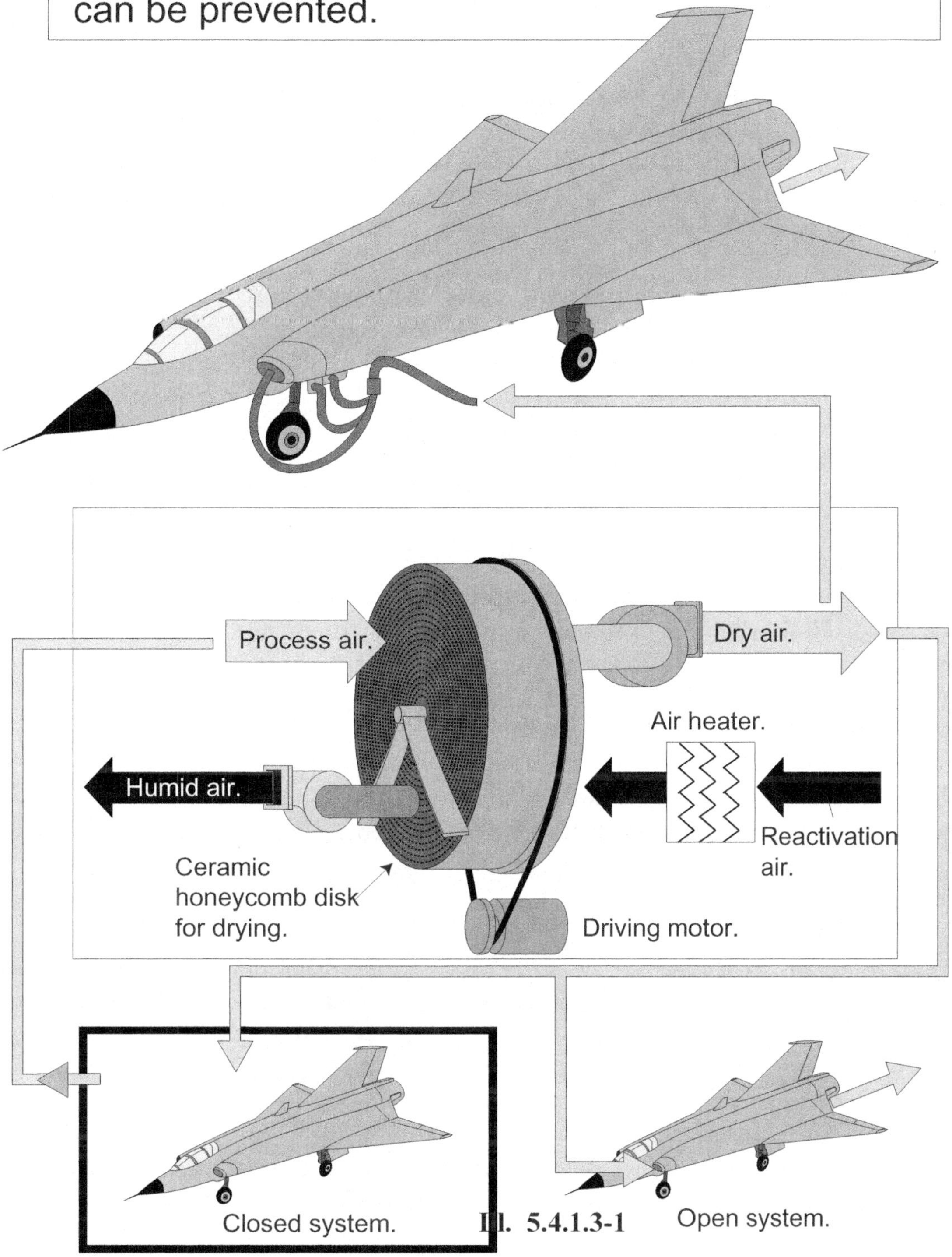

Ill. 5.4.1.3-1

External Factors Corrosion without mechanical Loading damage: Remedies.

Construction/design:

• The constructive **design** (i.e. the designer) largely decides the corrosion behavior of the engine parts during operation. For example, exposed sharp edges, gaps, and hollow spaces are typical corrosion-sensitive weak points that should be avoided. Otherwise, experience has shown that in later serial operation, repairs or retrofitted improvements will be very costly.

• Use of suitable **protective surface coatings**. These must be selected not only with regard to corrosion, but also different operating loads such as erosion and resistance to auxiliary materials and cleaning solutions (oil, methanol, fuel, etc.). Ideal coatings have a protective cathodic effect, which ensures that in case of local coating damage (such as through FOD), there is still corrosion protection from the surrounding coating areas.

• As far as possible, **corrosion-resistant materials** should be used. Coatings that protect against erosion and wear must not have cell action with the base material.

• Preventing **metallic contact** between materials with different corrosion sensitivity with possible electrolyte contact through non-conducting intermediate layers. Contact between cut **carbon fiber** layers (if there is not insulating resin layer, the fiber-reinforced carbon material behaves like a noble metal) and corrosion-sensitive metals, such as light metal alloys and steels, presents a special danger of cell action.

• In joined parts made from corrosion-sensitive materials such as steels, gaps and hollow areas should be avoided if possible (for example, in welded constructions). If **hollow areas** are unavoidable, they should be outfitted with corrosion protection and openings to ensure sufficient air exchange (drying).

Development, Testing, and Verification:

• Realistic testing of corrosion protection systems. This includes parameters such as time periods, load size and progress, temperatures, surrounding media, representative media, and material condition.

Manufacture and Repair:

• **Manufacturing processes** should be selected and optimized to ensure the least possible risk of corrosion. This includes factors such as **galvanic baths**, **cleaning baths**, and **etching fluids**.

• After "risky" manufacturing processes, proper **control measures** must be implemented. Depending on the type of expected damage, these may include, for example, control of the bath, inspection of the part surface (visually or microscopically with use of synthetic imprints), or non-destructive crack inspection.

• In case of **storage during manufacture**, corrosive influences must be avoided by, for example, suitable containers or coverings. Parts in the process of being manufactured should be stored in a position that prevents fluid from collecting and remaining.

External Factors Corrosion without mechanical Loading damage: Remedies.

• Use of proper **tools** (chipping, handling, fixing, mechanical cleaning, etc.) that do not deposit any corrosion-promoting foreign material. If corrosive deposits are unavoidable (such as brushing or a peening process with steel shot), the part must be properly cleaned afterward.

Maintenance, Mounting, and Repair:

• **Washing compressors** in suitable (prescribed) intervals with suitable (prescribed) media.

• Use of **mounting tools** that do not leave unallowable deposits or damage protective coatings on corrosion-sensitive parts (such as gearbox casings made from Mg alloys).

• If corrosion is discovered, it must be **repaired** or corrected sufficiently early (in accordance with guidelines).

• Development of proper (practical and sufficiently safe) **testing methods** and specifications (e.g. within the guidelines of overhaul manuals) to evaluate and separate reparable from irreparably damaged parts (e.g. sample sheets).

• **Repairing corroded cross-sections** on parts with low mechanical loading (such as deposit welding on housings made of steel or light alloys), or patching housings made from light metal alloys (especially Mg alloys) with filled synthetic resin pastes.

• **Sealing porous** (e.g. with increased porosity in light metal housings/casings through lacquer-stripping baths) cast housings with suitable **impregnation systems** such as water glass or synthetic resins.

• **Fading out** damaged areas and applying suitable coatings.

Chapter 5.4.1.3 References.

5.4.1.3-1 „Blow-dry your jet“, Zeitschrift „Flight International“, 18 May 1985, page 49.

5.4.1.3-2 Prospektangaben der Fa. AB Carl Munters in Sollentuna (Schweden) zu Vorrichtungen für „Dehumidification“. 2004.

5.4.2 Stress Corrosion Cracking (SCC)

5.4.2.1 Funtamentals of SCC

5.4.2.2 Damage due to SCC

5.4.2.3 Measures for Preventing SCC Damage

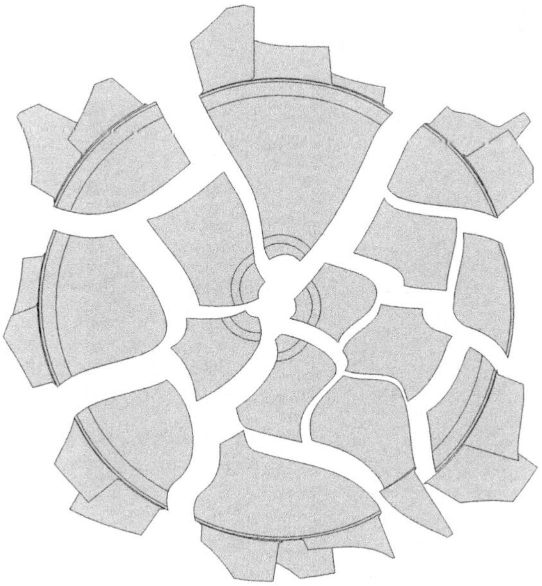

Stress corrosion cracking (SCC) is characterized by the following damage symptoms

• Unusually **rapid crack growth** (Ill. 5.4.2.1-1).

• **Embrittlement**, i.e. decreased contraction at fracture

• Decrease in **operating life** under static loads

• Decrease in **fracture energy/fracture toughness** ($K_{IC} > K_{ISCC}$)

Therefore, crack initiation and fracturing of engine parts, along with their associated consequences, are likely damage types.

A **stress corrosion cracking system** consists of a material sensitive to the specific acting media under sufficiently high tension. The most susceptible are hard, i.e. **high-strength, material** variants that are subjected to high loads due to their high strength. This can lead to spontaneous catastrophic

External Factors
Stress Cracking Corrosion: Fundamentals

Media that cause SCC in engines

Material type	Media that can cause SCC in engines	Characteristics of the SCC mechanism
Low-alloy steels	Nitrate and carbone solutions, chloride	Intergranular
Cr steels	Chloride	Intergranular
CrNi steels	Chloride ions, sulfite ions	Depending on the chloride content, pH value, stress levels, and temperature: transgranular SCC until the entire surface is attacked by pitting corosion. May be intergranular at Cr-depleted grain boundaries
Ni alloys	Sulfur	Mostly SCC-resistant with regard to watery electrolytes, but SCC may occur under the influence of sulfur in an airtight zone
Al alloys	Chlorine ions	Intergranular
Mg alloys	Chlorine ions	Intergranular
Ti alloys	Chloride	• With smooth surfaces required temperature > 300°C • With notched or cracked parts, SCC can occur at room temperature (Ill. 5.4.2.1-8)
Be alloys	Halogen ions (e.g. Chloride) and sulfate ions	Pitting corrosion

failure of the affected parts. Stress corrosion cracking can be **prevented**, if one of these criteria for the development of this corrosion type can be eliminated through suitable measures.

The typical **SCC systems in compressors** are shown in the above table. It can be seen that almost all material families can suffer SCC damage under certain conditions, although the probability of it occurring varies. The most common and aggressive corrosive media are **watery media containing chlorine** ions. SCC can be caused without watery electrolytes by **hot salt on titanium alloys** (Ill. 5.4.2.1-7), and by **sulfur as a decomposition product on Ni-alloys**.

5.4.2.1 Fundamentals of Stress Corrosion Cracking

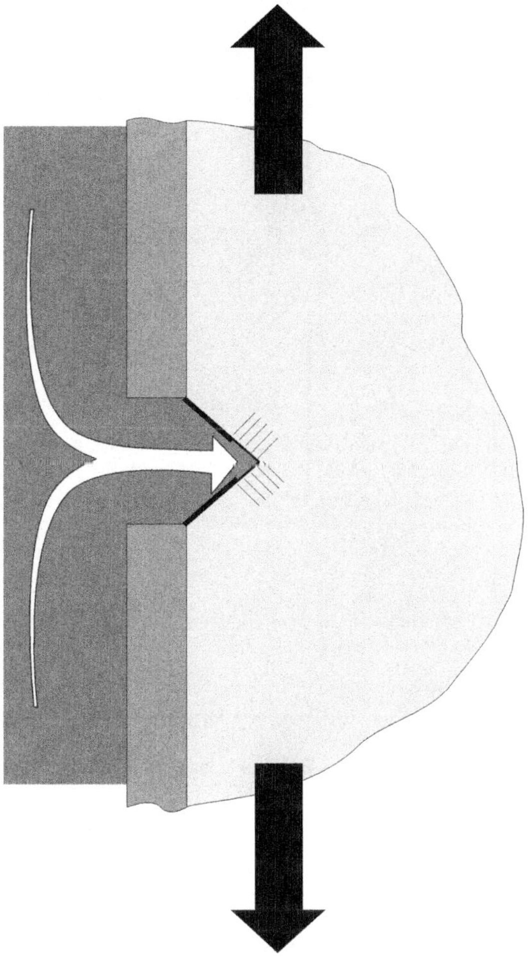

Stress corrosion cracking encompasses a variety of **crack-inducing mechanisms** (see page 5.4.2.1-4 and Ill. 5.4.4.1-4) that only occur in case the following three conditions of the corrosion system are met:

- Specific **corrosion-sensitive material** (Table 5.4.2.1-2)

- Specific **corrosive media** (Table at page 5.4.2.1-2)

- Tensile **stress above a threshold** value, i.e. above a critical expansion rate (see Ill. 5.4.2.1-1)

Depending on the corrosive, stressing, and material influences, SCC can be either **intercrystalline** or **transcrystalline**. SCC can be categorized into two main groups, according to the damage mechanism: **tension-induced** and **expansion/strain-induced** SCC. The damage symptoms can vary, depending on the surface behavior and the size of the loads (Ill. 5.4.2.1-4).

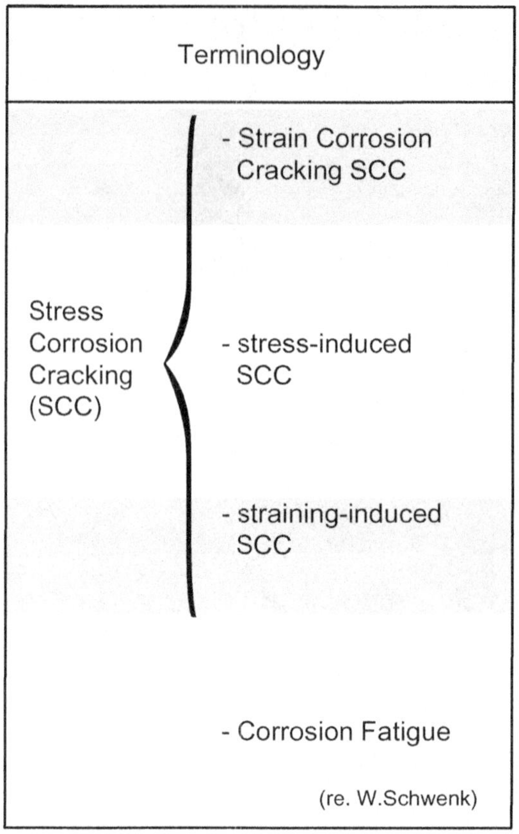

In **stress-induced SCC**, the crack forms **under the influence of a static load** (expansion speed = 0). The protective coatings are torn open (Ill. 5.4.2.1-2), allowing the corrosive media to act on locally exposed metal surfaces. Here, as well, expansion is ultimately the deciding cause of the local destruction of the protective coating (Ill. 5.4.2.1-2). The destruction occurs **before** the crack-initiating corrosive attack. Therefore, the creation of a protective coating at the phase boundary between the metal and the corrosive media is an important requirement for this type of SCC. This type of corrosion has characteristic damage symptoms (Ill. 5.4.2.1-5). **Threshold stresses** (Ill. 5.4.2.1-1) and **threshold potential** are typical parameters for SCC and must be determined with suitable testing procedures. No SCC occurs below these values. A distinction is made between **anodic** SCC (as described above) and **cathodic** SCC ("hydrogen-induced crack initiation", also see Chapter 5.4.4), in which atomic hydrogen enters into the expanded metal grid during the corrosion reaction. This type of area is typically the bottom of a nick or a crack tip with high stress concentration. In these areas in austenite CrNi steels (for example), under the influence of Cl ions above 80°C, transcrystalline or intercrystalline (e.g. in heat-treated steels) crack growth occurs. With anodic SCC it is plausible, that the discharge of Hydrogen ions in the crack may create H-atoms, the diffusion of which has an embrittling effect and influences periodic crack growth (Ill. 5.4.4.1-4).

Strain-induced corrosion cracking (Ref. 5.4.3-1) can be expected when, as a result of the plastic deformations, the brittle reaction layer (Ill. 5.4.2.1-2) cracks, and there is not sufficient time for the protective coating to regenerate. This is the case when the **coating growth rate is lower than the expansion rate** (Ill. 5.4.2.1-6).This type of corrosion is common in high-strength titanium alloys. In stress-induced SCC there is no lower critical expansion rate.

External Factors
Stress Cracking Corrosion: Fundamentals

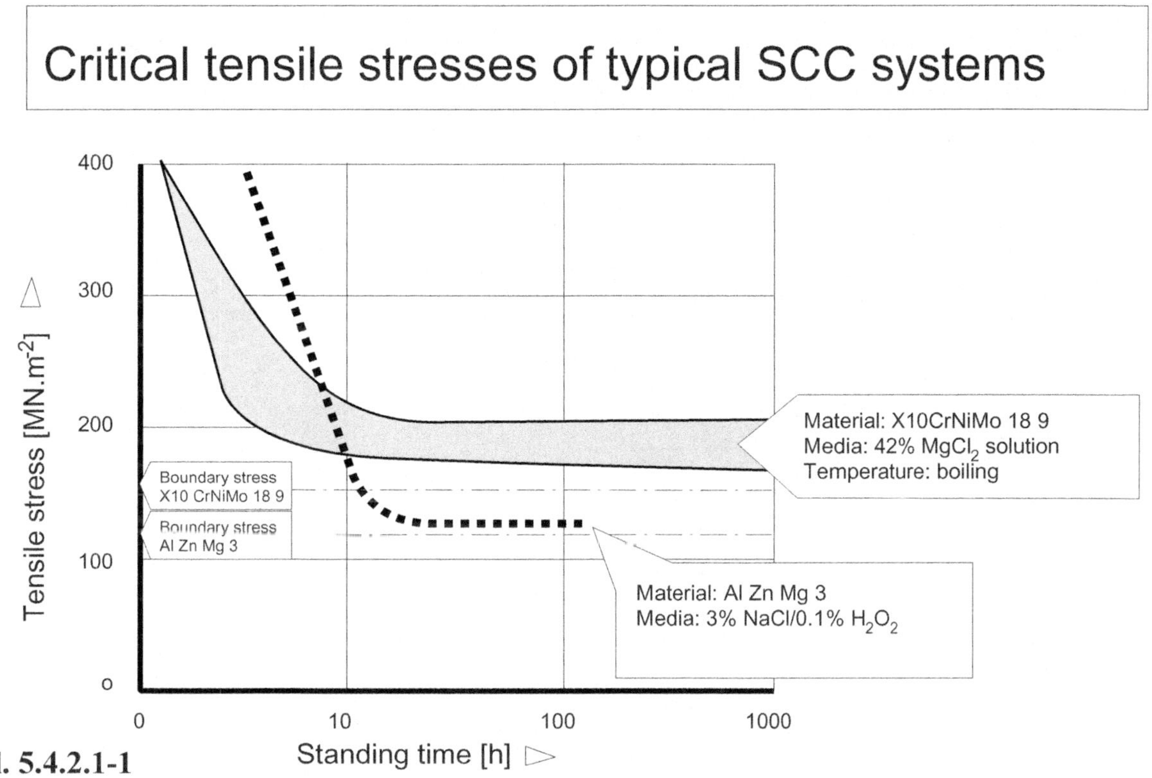

Ill. 5.4.2.1-1

Illustration 5.4.2.1-1 *(Refs. 5.4.2.1-1.1 and 5.4.2.1-1.2): The diagram shows the typical* ***behavior during stress corrosion cracking*** *of two material families. If the static* ***tensile stress*** *in the CrNi steel is less than 150 $MN.m^{-2}$ and in the Al alloy less than about 120 $MN.m^{-2}$, then no crack will form with the given corrosive media. This shows that suitable tension-decreasing measures (such as heat treatment) and/or induction of* ***compressive stress*** *(e.g. through* ***shot peening)*** *can be effective in preventing SCC damage.*

Conversion coating damage due to deformation of a base material (schematic).

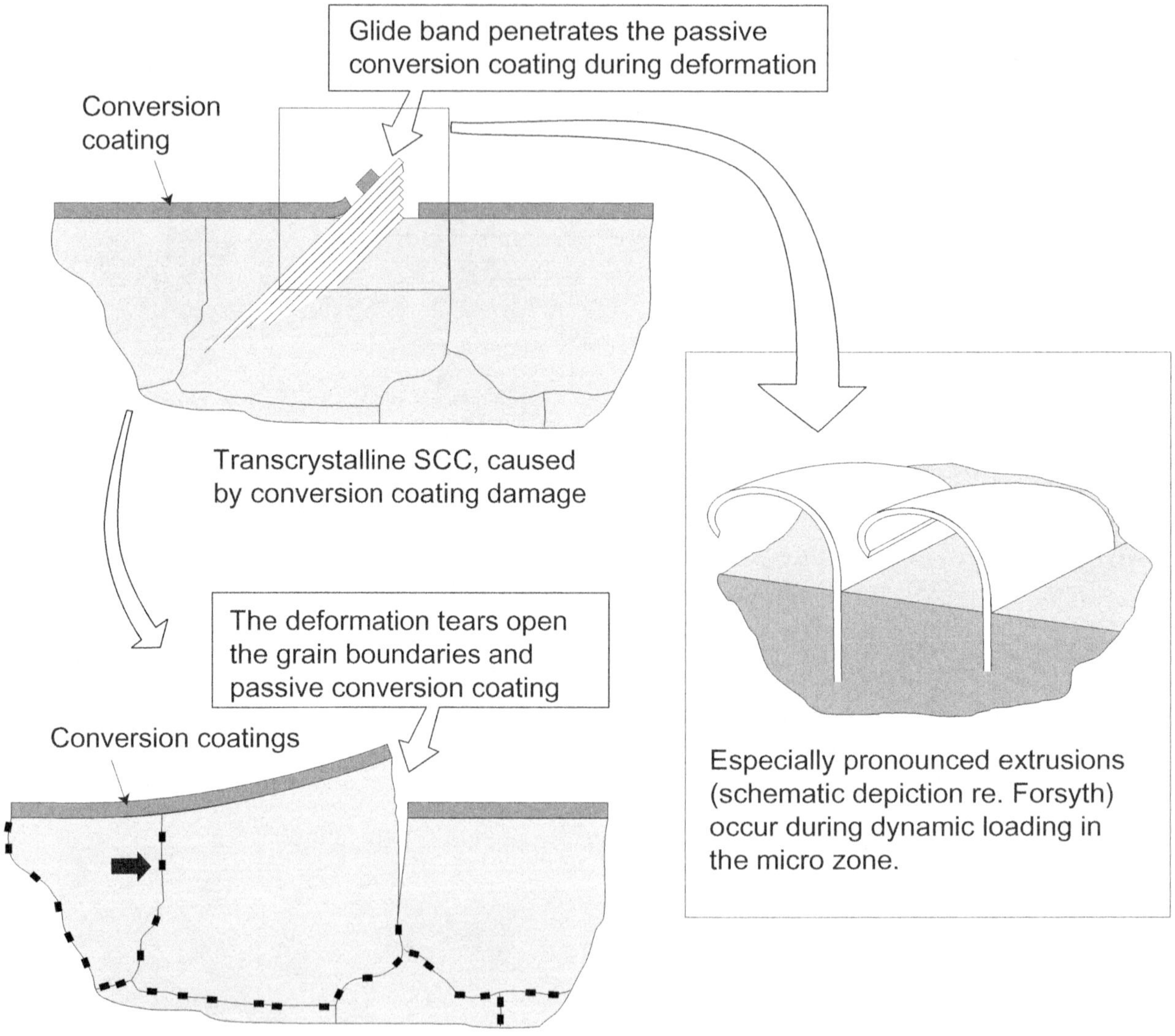

Intercrystalline SCC caused by conversion coating damage.
Grain boundary carbides (arrow) indicate the sensitized material (diagram based on H. Klingele)

Ill. 5.4.2.1-2

Damage to a passive conversion coating due to glide bands and coating cracks (re. H.Klingle)

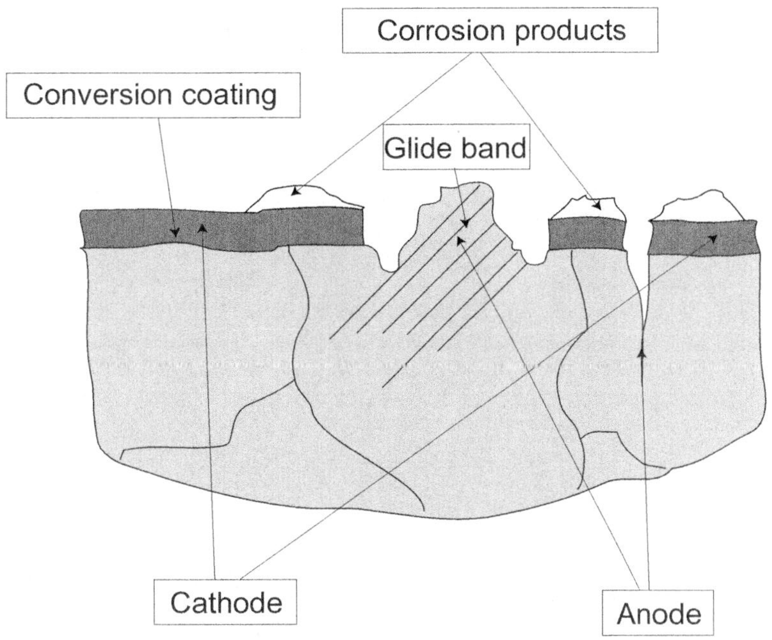

Ill. 5.4.2.1-3:

Illustrations 5.4.2.1.2 *and* ***5.4.2.1-3*** *(Ref. 5.4.1.2-2): For every corrosion type, especially for SCC, damage to the* ***passive layer*** *(cathodic zone) is important for the creation of a corrosion-sensitive (anodic) area (also see Ill. 5.4.4.1-4). Depending on how the damage in the* ***grain boundary*** *area is created through crack initiation (bottom left), such as along the grain boundary surfaces or from the inside of the grains to the surface in the shape of extrusions (top and middle diagrams), either intercrystalline SCC or transcrystalline SCC are promoted.*
Extrusions *are created in the micro-zone through dynamic changing loads. The surface deformation this causes is massive passive layer damage (singularity).*

SCC crack growth mechanism

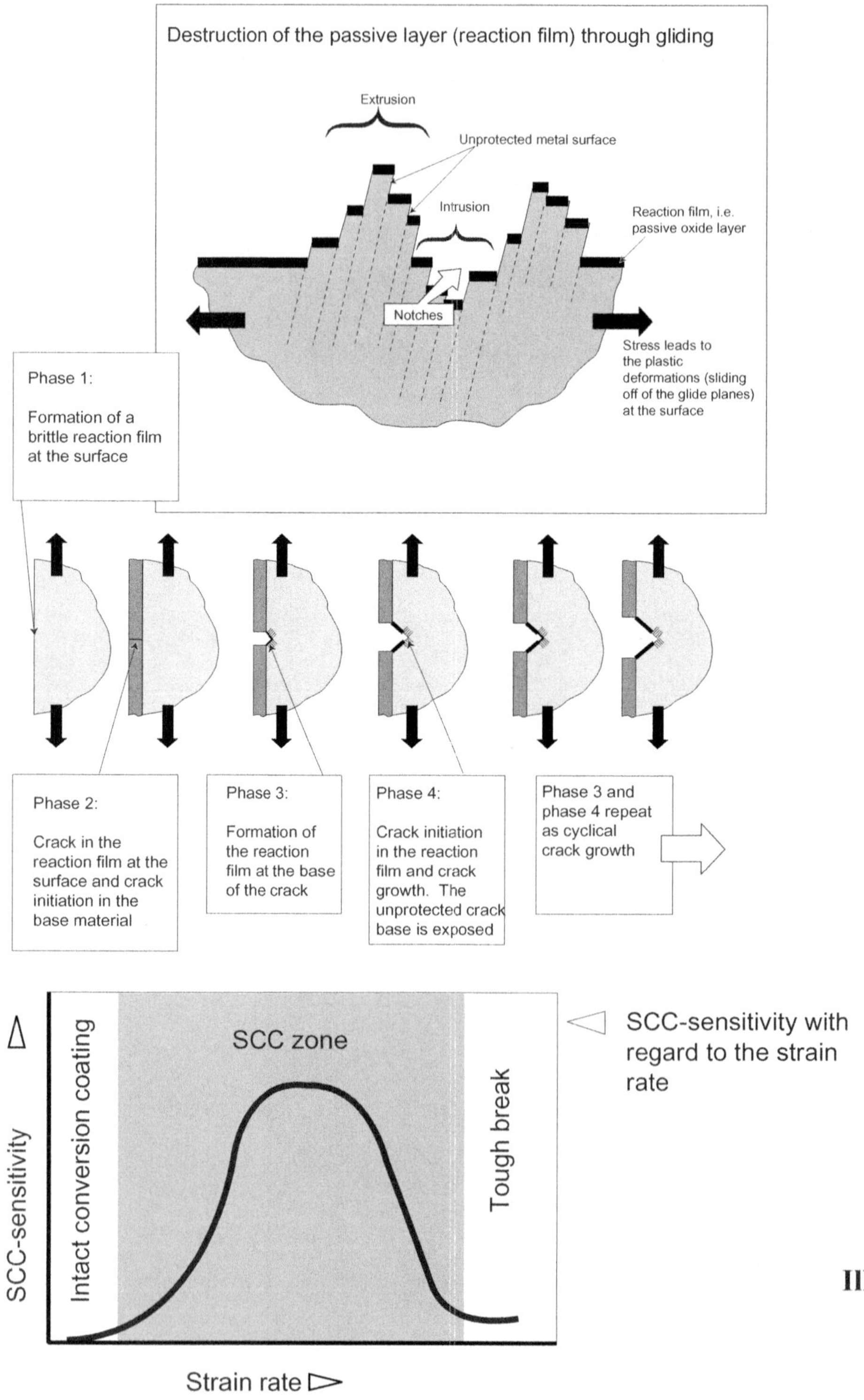

Ill. 5.4.2.1-4

Illustration 5.4.2.1-4: *In* ***stress-induced SCC*** *the following damage mechanism takes place: at the surface of a normally corrosion-resistant material a brittle* **reaction layer** *forms (phase 1 in middle diagram), such as an oxide layer that prevents corrosion. Static expansion/strain of the material (expansion rate = 0) can cause this brittle coating to crack (phase 2). This can occur even due to relatively minor elastic macro-expansion, causing* ***micro-slipping*** *in the grains due to the different orientation of the grains (top diagram), which break through the protective coating as* **extrusions** *or expose the base material as* **intrusions***. This process destroys the protective layer so that local corrosion can occur. Once the protective surface coating is damaged once, the brittle reaction layer again forms in the base of the crack, which cracks again and causes crack growth. This crack mechanism leads to* ***step-by-step crack growth*** *that leaves characteristic progress lines in the micro-zone which, along with the reaction layers, indicate the damage mechanism of the SCC (see Ill. 5.4.1.2-5).*

Strain-induced SCC *(bottom diagram) on the other hand, occurs under increasing loads. The damage model is the formation of ductile, regenerative reaction layers, which plausibly explains why this SCC occurs increasingly in a limited expansion rate range. If the expansion rate is greater than the speed at which the reaction layer regenerates, it will not heal and corrosion will increase. If the expansion/strain rate is further accelerated, the SCC effect near the tough force fracture decreases.* ***Titanium*** *alloys exhibit typical behavior for strain-induced SCC (Ill. 5.4.2.1-8).*
Different expansion rates occur during engine operation during startup (due to, for example, increased centrifugal force in disks or pressure increases in the housings/casings, etc). The damage case described in Example 5.4.2.1-1 and Ill. 5.4.2.1-7 should be considered in connection with this phenomenon.

Illustration 5.4.2.1-5: *For SCC, the crack expansion/**strain rate** can be depicted in a similar way as with cyclical **fatigue processes** (top diagram). However, the dependency on the tension intensity at the crack tip is depicted, and not the nominal stress. This is understandable when considering the crack growth mechanism in Ill. 5.4.2.1-3, which **depends largely on the deformations** at the crack tip.*

*Above a threshold value below which no crack growth occurs, the crack growth speed increases rapidly with the **stress intensity** (area "I").*

In zone "II" the crack growth speed remains constant over a large range of the stress concentration deflection. When the critical crack length is reached, ultimate fracture occurs.

*The bottom diagram is a schematic depiction of a **fractured surface with typical characteristics of SCC**, such as steps and cracks in the brittle reaction layer (without signs of the influence of hydrogen; also see Ill. 5.4.2.1-4). The markings of step-by-step crack growth are interesting, and must not be confused with the striations of a dynamic fatigue fracture. The cracked brittle reaction layer on the fracture surface clearly differentiates this damage from other similar crack types.*

External Factors
Stress Cracking Corrosion: Fundamentals

Material behavior under influence of SCC

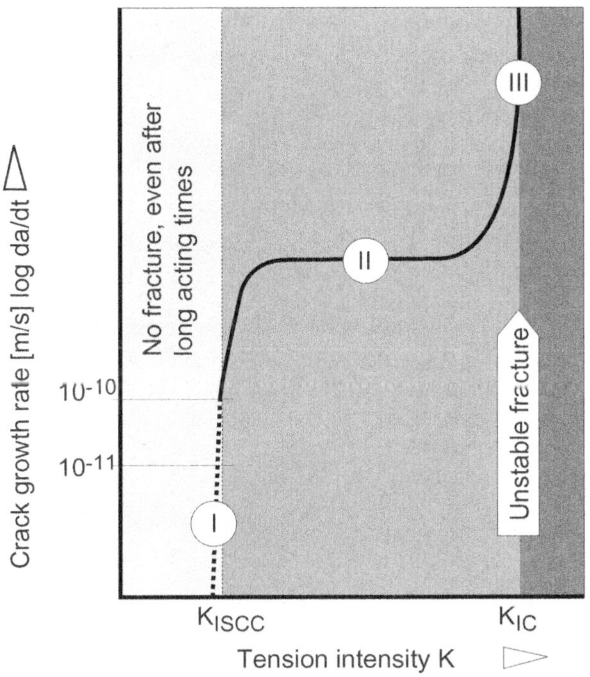

Crack growth rate da/dt under the influence of a corrosive media with regard to the tension intensity K

Typical stressed zones:

- (I) Strong influence of the tension intensity K on the crack growth rate da/dt
- (II) Crack growth rate largely independent of K
- (III) Large increase in the crack growth rate with K

Note: for a sufficiently large plate with crack going all the way through,

$$K = R_0 \sqrt{\pi . a}$$

R_0 = specific tension, a = half of crack length

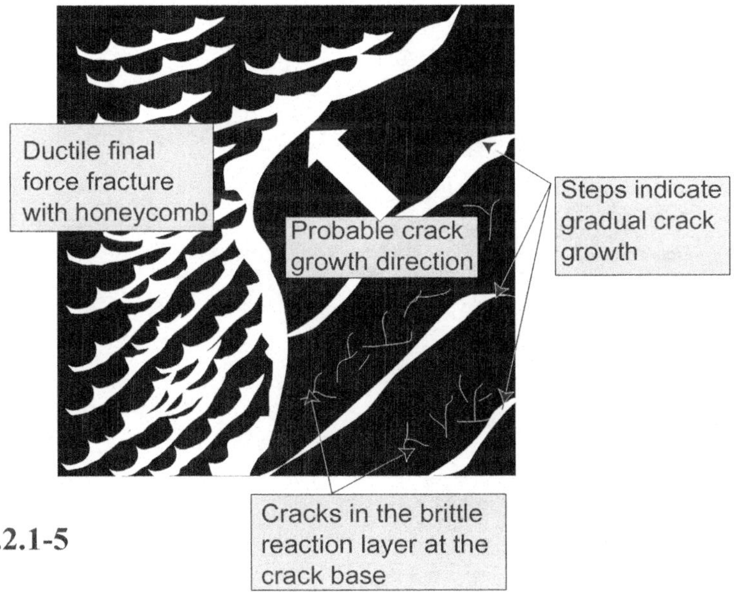

Ill. 5.4.2.1-5

External Factors
Stress Cracking Corrosion: Fundamentals

Crack types depend not only on the material and the acting media, but also largely on the strength of the tensile stress.

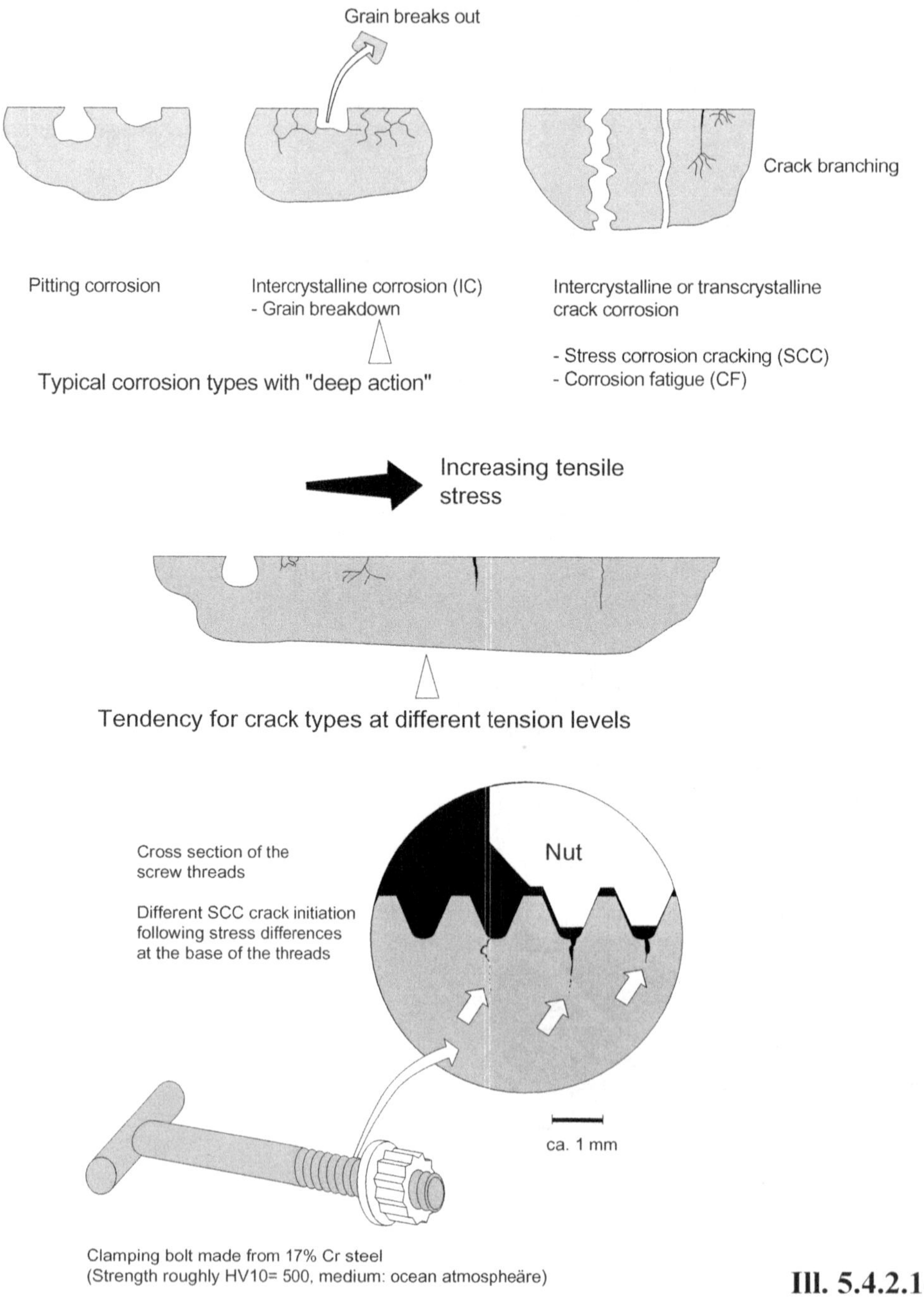

Ill. 5.4.2.1-6

Illustration 5.4.2.1-6: *The **form**, in which the corrosive mechanism attacks (top diagram), depends on parameters such as the sensitivity of certain structural components, the corrosive media, and the effective tensile stresses. For example, finely branching cracks indicate*

External Factors
Stress Cracking Corrosion: Fundamentals

Fracture of an integral compressor wheel (Blisk), caused by stress corrosion cracking.

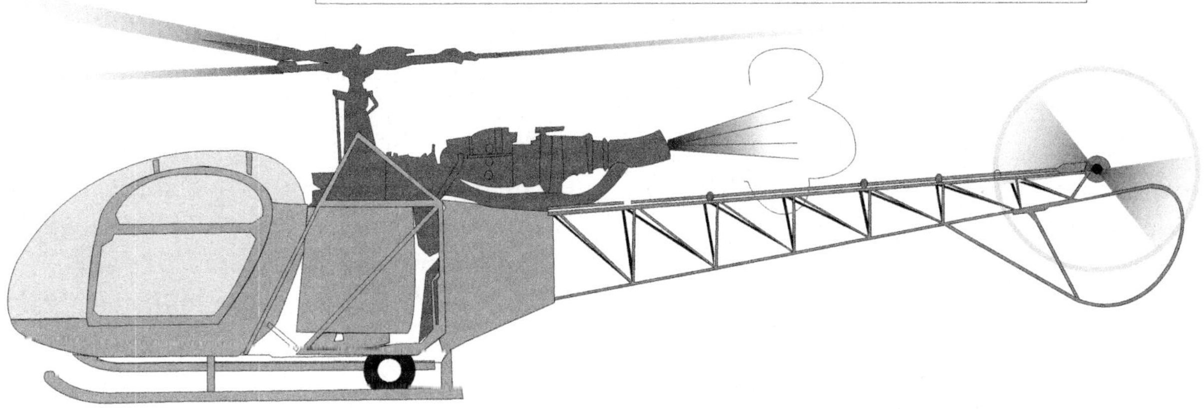

Example 5.4.2.1-1 (Ref. 5.4.2.1-3, see also Ill. 3-5 and Ill. 5.4.2.1-7):

Excerpt: "...In a helicopter turbine, after about 1000 flight hours, the axial compressor disk broke at 44,000 RPM during idling on the ground and was completely destroyed (see Ill. 5.4.2.1-5).

After hardening, (The) axial compressor disks made from turbine blade steel X15Cr13 were not tempered at the usual 725°C, but only at 540°C, in order to achieve the greatest possible strength. This treatment made the steel especially sensitive to corrosion, since a web of chrome carbides precipitates on the former austenite grain boundaries. The disks still achieve the required life span, since the high centrifugal force throws off the moisture from the incoming air that acts as a corrosive media. However, an apparently harmless constructive change -tightening of the balance ring- caused the moisture to build up, resulting in the disks failing due to SCC and corrosion fatigue."

Comments: This damage case, which was investigated and described by Prof. G. Lange of the TU Braunschweig, shows the influences on the development of SCC in an exemplary manner. The **increased hardness** achieved by the lower tempering temperature did not increase safety against disk fractures, but led to a strength decrease and catastrophic part failure under corrosion conditions that are typical during operation. During engine startup, especially, conditions for **strain-induced SCC** (see Ill. 5.4.1.2-2) and corrosion fatigue were most evidently present.

relatively low tensile stress levels, while deep single cracks indicate high tensile stress levels (middle diagram).
*In **threaded connections**, the tensile loads on the thread base change at the first threads. This has a pronounced effect on the development of the corrosion cracks (bottom diagram).*

Illustration 5.4.2.1-7 *(Ref. 5.4.2.1-3, Example 5.4.2.1-1):*

This example is an impressive depiction of the influences on SCC (compare with 5.4.4.2-2). Special ***tests*** *are necessary to determine the SCC-sensitivity of a material. A sample test is shown in the bottom diagram.*

Problems with corrosion tests during damage investigations:

If this type of test is intended to determine or verify the ***corrosion sensitivity*** *of a system, then even minor deviations from the damage case should unallowably falsify the results.*
If, for example, the material sample does not sufficiently correspond to the part, or if there is a difference in the ***structure or orientation*** *relative to the direction of the tensile stress, the SCC behavior of the sample may be unallowably altered relative to the actual damaged part. For example, in cast parts, it must be assumed that, due to the casting and freezing conditions, different part areas may have very different* ***grain sizes*** *and that there are eliquation-dependent deviations among the alloy components. This behavior overlays with the operating similarity of the electrolyte and the tensile stress levels. If the part is coated, the coatings must be properly reproduced.*
If, in a ***sample****, a necessary threshold value of the stress concentration is not exceeded, as would be expected with smooth samples, then the sample is unsuitable for determining the SCC sensitivity of certain materials, such as titanium alloys (see Ill. 5.4.2.1-8).*
If the damage mechanism concerned is a strain-induced process, in which the induced ***expansion/ strain rate*** *is important, then a test such as this one, with static loading, is of little or no value (see Ill. 5.4.2.1-8).*
Therefore, the corrosion sample of an affected part should, if possible, be taken from the ***same production batch****.*
Surface hardening and induced compressive stresses can prevent SCC, while cold deformations with induced tensile stresses and/ or structural changes (such as martensite formation) can promote SCC.
Therefore, in tests that are designed to permit conclusions as to the operating behavior of engine parts, it must be ensured that the ***surface condition*** *(e.g. treatment) of the sample or test part is sufficiently similar to the original part.*

External Factors
Stress Cracking Corrosion: Fundamentals

Stress Corrosion Cracking (SCC) only occurs when three conditions are met:

- Sensitive material
- Specific media
- Sufficiently high tension stress, i.e. sufficient stress concentration

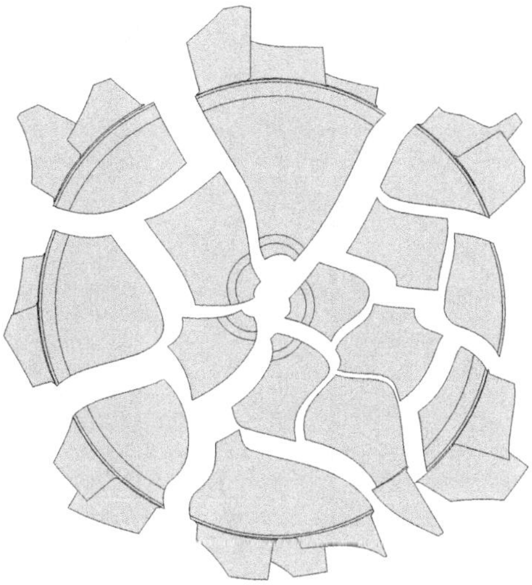

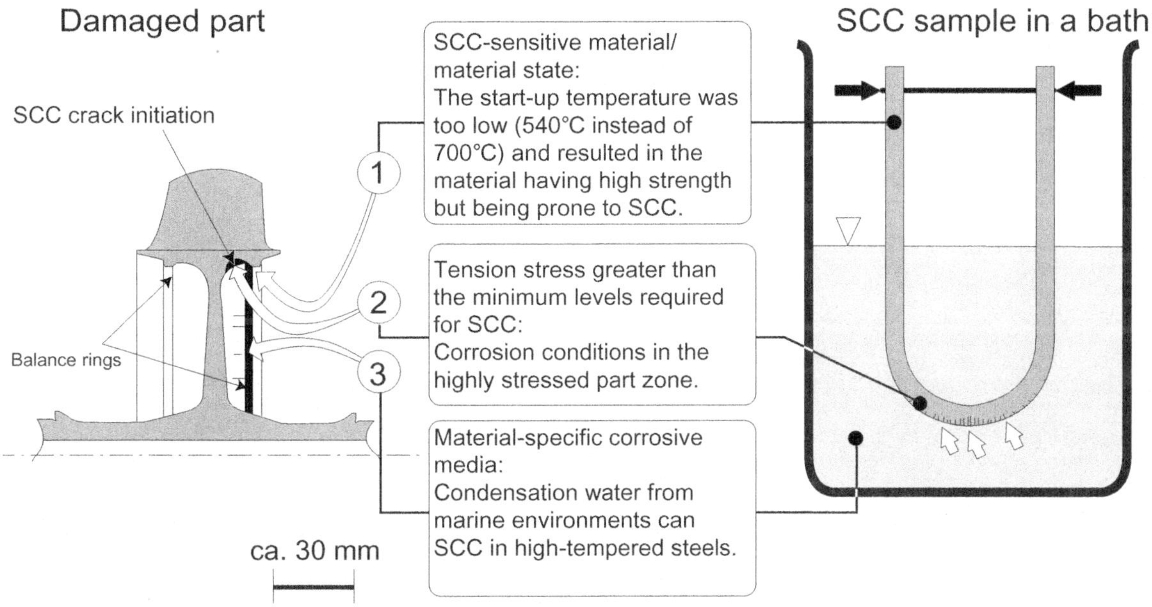

SCC damage to an axial compressor disk in a small gas turbine (re. G.Lange)

Ill. 5.4.2.1-7

External Factors
Stress Cracking Corrosion: Fundamentals

Titanium alloys are extremely sensitive to the influence of hot salt at temperatures above 450 °C.
However, there have been no reported damages of this type during operation. Why not?

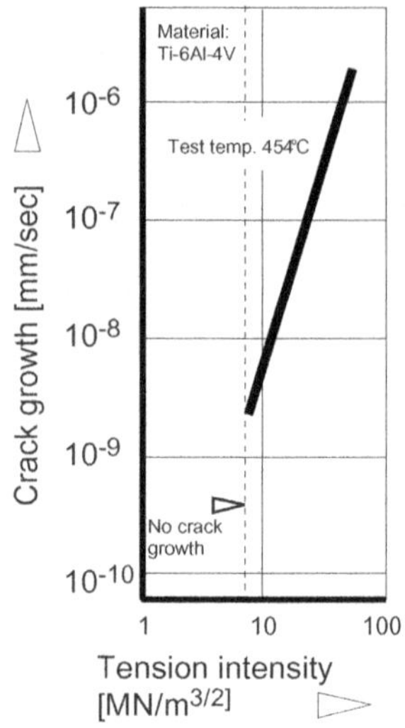

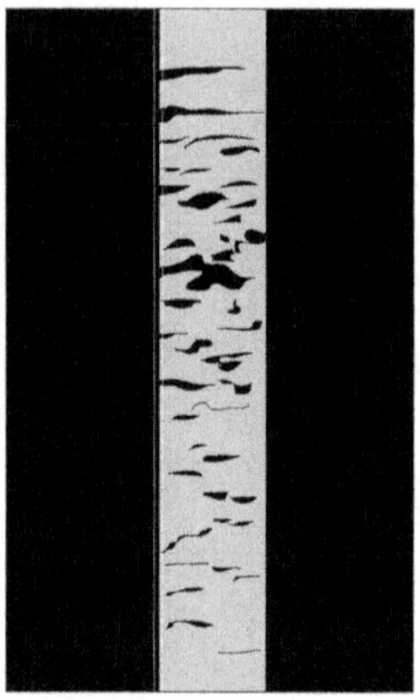

Test sample:
Ti 6-6-2 coated with salt
Standing time 50 h at 50 kp/mm^2
Test temperature 450 °C

Crack initiation in a high-strength Ti alloy after 30 minutes in a perchloroethylene degreasing bath at 90°C. The specific tension was 80% of the tension strength

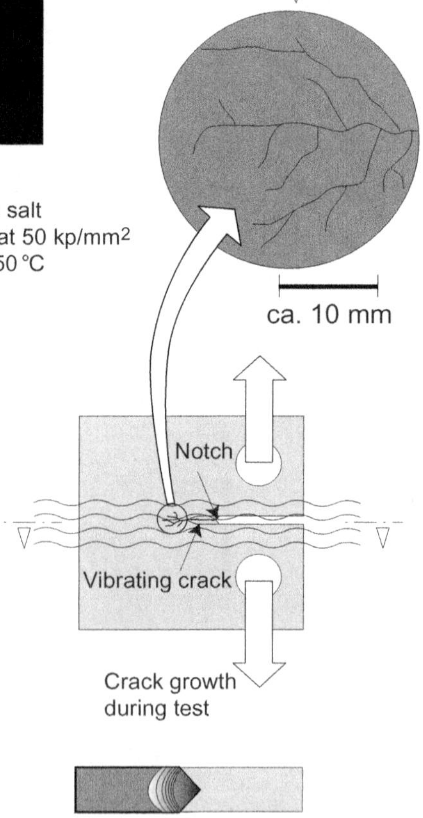

Verification of SCC-sensitivity in Ti alloys through measurement of crack growth in notched or cracked samples.

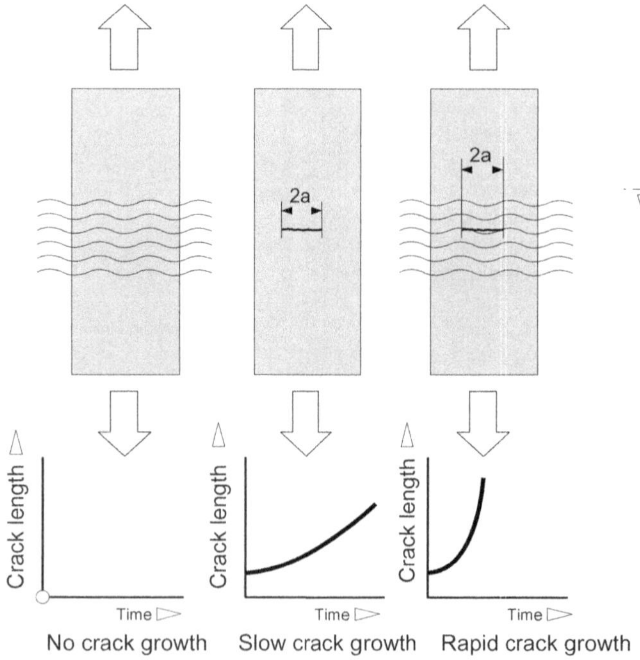

SCC investigation on titanium alloys

Ill. 5.4.2.1-8

Illustration 5.4.2.1-8: *The problems associated with* ***reproducing SCC*** *damages in laboratory tests on samples is especially pronounced with* ***titanium alloys.***

For example, Ti-alloys show serious damage in ***laboratory tests with hot salt*** *above about 300 °C, and especially above 450°C to 500°C (top diagrams). However, there has been no known engine damage that was determined to have been caused by hot salt corrosion, even though highly-stressed titanium parts in modern engines, such as bladings, are struck by marine atmosphere and often exhibit salt deposits. A satisfactory explanation for this behavior might be the deviation of laboratory tests from actual operating conditions.*
Cases of SCC due to hot salt attack have been reported ***during the manufacturing process*** *(Ill. 5.4.2.2-6, Ref. 5.4.2.2-5). In one case,* ***hand sweat*** *reacted with a titanium alloy fan blade during heat-treatment.*
Cases of SCC have also been reported in connection with reaction ***coatings containing Cl*** *(such as in etching, cleaning, and degreasing baths), contaminants (synthetic wear products containing Cl), and labeling.*

In laboratory tests on titanium alloys it is important to simulate a realistic tension intensity. During manufacture, after damage was observed on the untreated seam head of a electron beam weld seam (Ill. 5.4.2.2-2), even though smooth samples in the same media (hot perchloroethylene degreasing bath at about 90°C) showed no comparable damage, the SCC-sensitivity of a high-strength Ti alloy was finally determined (bottom diagram) with the aid of ***CT samples*** *(bottom right diagram).*
According to Ref. 5.4.2.1-4, chloride compounds, which are created, for example, through surface reactions in ***cleaning baths containing Cl*** *(HCl splits off and reaction creates chlorides), can later (e.g. during following manufacturing processes or operation) lead to stress corrosion cracking in titanium materials at temperatures above 150°C.*

Chapter 5.4.2.1 References

5.4.2.1-1.1 W.Gruhl, "Zeitschrift für Metallkunde" 53, (1962) page 670.

5.4.2.1-1.2 H.Spähn, H.Steinhoff, periodical "Werkstoffe und Korrosion", 20 (1969) page 733.

5.4.2.1-2 L. Engel, H. Klingle, "Rasterelektronenmikroskopische Untersuchungen von Metallschäden", Gerling Institut für Schadensforschung und Schadensverhütung GmbH Köln, 1974, ISBN 3-9800043-0-9, page 182.

5.4.2.1-3 G.Lange, "Zerstörung von Hubschrauberturbinen durch Einsatz eines Stahls in korrosionsanfälligem Zustand bei gleichzeitig nicht werkstoffgerechter Konstruktion", "Zeitschrift für Werkstofftechnik", 5. Jahrg. 1974, Nr 1, pages 9 to 13.

5.4.2.1-4 H.Simon, "Oberflächenreaktionen an Titanwerkstoffen", periodical "Metall Oberfläche" 5-1982, pages 211-217.

5.4.2.2 Damage due to Stress Corrosion Cracking (SCC)

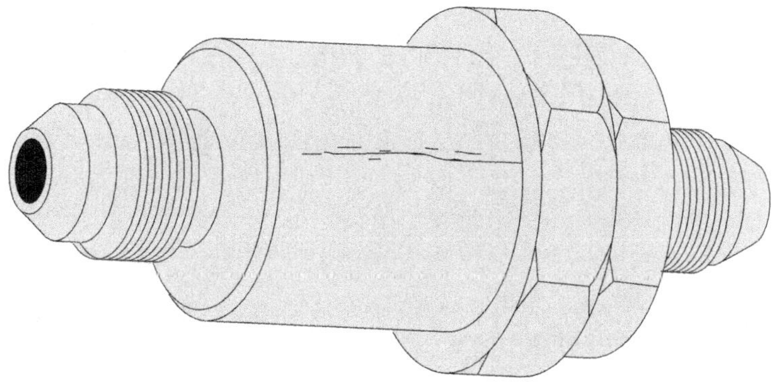

SCC can cause engine parts to fail prematurely through **crack initiation and fracture**. It must be remembered, that SCC not only increases the **subcritical crack growth rate**, minimizing the remaining life span in the crack growth phase, but also decreases the critical fracture toughness considerably, shortening the critical crack length and further decreasing the crack growth phase.

Assuming that the requirements for stress corrosion cracking are sufficiently understood, and that construction/design, material specifications, and the allowable manufacturing processes reflect this, then SCC damage should only be expected in case of **unallowable deviations**. Experience seems to confirm this. In most cases, this type of damage occurs when the material of specific batches or lots is in a SCC-sensitive state due to certain deviating **manufacturing steps** or **parameters** (Ill. 5.4.2.2-1). These deviations can be due to, for example, steel **heat-treatment** processes. Especially sensitive materials are those with deviations towards higher s**trengths/ hardness** and simultaneously have sensitive structural areas (such as grain boundaries, Ill. 5.4.2.1-5). In hard Al alloys, such as are used in threads of fuel lines, deviations in the hardening process can cause unallowable sensitivity to marine environments. If the damage can be limited to specific lots or batches, the chance of achieving targeted and limited measures for replacement or risk assessment.

During manufacture and overhaul, changes to **auxiliary materials** such as cutting, etching, and cleaning fluids and/or soiling (wear products from tools, shot peening media, etc.), or changes to the surface (such as hardening, internal stress) can promote unexpected SCC damage.

However, it is also plausible for operating temperatures, acting over a long period of time, to cause damaging changes in the material (Ill. 5.4.2.2-2). This has been observed not only in superalloys (such as turbine stator vanes made from Co alloy) and steels, but also in Al alloys. SCC damage may then occur either during operation or during overhaul procedures, even though there was no unallowable deviation from regulations.

The SCC behavior of materials used in engine manufacture:

Cr steels:

Cr steels have very different sensitivity to SCC, depending on heat treatment processes (structure, hardness), carbon content, and alloy additions. For example, it has been observed that with Cr steels (13% Cr steel), of the type which are frequently used for threading on tension bands (Ill. 5.4.2.2-1), a **material-dependent minimum hardness** is required in order for SCC to occur under the influence of chlorides in a watery solution. In this way, in compressor stator vanes (Ill. 5.4.2.2-1) made from this type of material SCC occurred only above a hardness of about 40 HRC. The SCC-crack sensitivity is promoted through directional material impurities such as are typical for free-machine materials that contain sulfur. The use of this type of material in aviation should be avoided today.

Cr Ni steels:

These materials are sensitive to SCC if the austenite is in an unstable condition due to an improper heat treatment. Depending on the nickel content, sensitivity increases with higher carbon content. For example, the material type CrNi 18 9 is sensitive above a carbon content of about 0.045 %. Cold deformation increases the SCC sensitivity, as long as no deformation martensite has formed. Even very minor plastic cold deformation can make CrNi steels with over 10% Ni prone to cracking, even if they are SCC-resistant in the elastic zone. Small amounts of martensite provide cathodic protection against SCC (base potential of the martensite). If, however, much more deformation martensite is created (on the gliding plane), the SCC-sensitivity is increased. If ferrite is present, it provides cathodic protection for the austenite, lowering the danger of SCC. Carbide precipitations (on the grain boundaries) decrease the danger of transcrystalline SCC. However, along with “normal intercrystalline corrosion“ on passive surfaces, intercrystalline SCC can occur on largely active surfaces. If the structure is so unstable, that (unusual?) operating temperatures cause undesired changes, then parts may only become sensitive to SCC during operation and suffer corresponding damages (Ill. 5.4.2.2-2).

If, in these steels, the adjacent tensile stresses are not sufficiently great to cause crack initiation, they often cause pitting corrosion (Ill. 5.4.2.1-4). The table on page 5.4.2.1-2 is an overview of the SCC-sensitivity of different steels, as well as Ni-based alloys.

SCC has been found on austenitic steels in engines in the following **watery media**, the corrosive components of which are primarily Cl and sulfite ions:

- damp air
- sea water
- marine air
- calcium chloride
- sulfuric acid
- zinc chloride
- magnesium chloride
- sodium chloride
- damp hydrogen sulfide

Cl-ion contents of greater than 60 mg/l at temperatures above 70°C have been shown to be generally damaging.

Al alloys:

Intercrystalline SCC can affect Al materials in watery electrolytes containing halogen, in organic solutions, and in damp air. The **high-strength** and temperable, but **corrosion-sensitive** Al alloys of type AlZnMg and AlZnMgCu have been shown to be susceptible to SCC. Even normally corrosion-resistant AlMg alloys have been affected by SCC in cases where the structure was in a critical state. AlCuMg and AlMgSi alloys have proven to be largely immune to SCC, but they tend to intercrystalline brittle (force) fractures in their fully hardened states. However, alloying additional "stabilizers" (such as Mn, Cr, or V) and suitable **heat treatment** (delayed quenching after solution annealing, gradual quenching) permits sufficiently safe use of high-strength Al alloys. The primary cause of SCC-sensitivity in Al materials is believed to be grain boundary precipitations, which cause the typical intercrystalline crack progress that occurs when a potential difference is created between the precipitations and the alloy components of depleted grain boundaries. The crack progress runs along the **grain boundaries** that are perpendicular to the direction of the loads. Therefore, the crack growth rate is largely dependent on the specific grain shape and the manufacture of the semi-finished product (as far as this affects the structural orientation), and also on the intensity of the stress. However, the damage symptoms indicate that, aside from the precipitations, a further SCC-promoting mechanism is involved. This mechanism may be cold-hardening zones, which are caused by single-phase segregation at low temperatures.

"Pre-corrosion" in an unloaded state can drastically reduce the time to fracture due to SCC under loading. The critical tension, above which SCC occurs, may only be 10% (!) of the tensile strength of the material, depending on the "direction of fiber" of the material. The critical tension can lead to crack initiation, but it will not usually be sufficient to cause crack growth.

Ti alloys:

Titanium alloys without cracks or sufficiently sharp nicks are not SCC-sensitive in watery media that contain chloride. This is evidently also true with regard to crack-initiating **hot salt corrosion** (top left diagram). Crack growth only occurs above a **stress concentration threshold**. Sensitivity to SCC is especially pronounced in tests with constant strain rates (see page 5.4.2.1-15), which indicates strain-induced SCC (Ill. 5.4.2.1-2). Critical crack growth rates (at which the influence of SCC becomes noticeable) have been observed between 0.02 mm/h und 1.5 mm/h. No SCC damage is to be expected outside of this range, unless the loads reach the local tensile stress, such as in a nick. However, this is not true for all SCC-inducing media. SCC fractures were observed on flat samples in halogenide-containing alcohols with low water content (10^{-1} to 10^{-2} %).

Although no operating damages to high-strength Ti alloys due to hot salt stress corrosion (above 300°C, see Ref. 5.4.2.2-4) have been reported, there is a potential crack problem in case of hot salt contact (top middle diagram). This reduces the bearable loads, increases the creep strain, and the contraction at fracture at room temperature is reduced after the attack. However, the increased creep strain is less due to any ductile material behavior than to an accordion effect following the large crack initiation. In tests with the alloy Ti6A14V, the hot salt SCC-sensitivity considerably increased with increasing grain size (from 0.025 to 0.1mm).

It can be assumed that, under hot salt SCC action, the hardnesses perpendicular to the deformation direction are considerably worse than those that are parallel. In case of perpendicular loading, fractures are usually crystalline. As the temperature increases, the time to an SCC fracture decreases considerably more than could be explained by creep effects alone. Cold hardening worsens the performance against hot salt SCC, contrary to normal SCC in watery solutions.

This brings up the question as to what extent hot salt SCC on titanium alloys can be seen as SCC in the strictest sense of the word.

***Illustration 5.4.2.2-1**: Experience has shown that parts/elements of **auxiliary units are especially prone to SCC** damage. So-called "**catalog parts**" evidently exhibit the following characteristics/weaknesses most often:*

- The parts are relatively inexpensive.

- The parts do not always meet the high quality demands of aviation technology.

- The designer accepts the technical data somewhat uncritically.

- These vendor parts are evidently not as closely monitored as "own parts" by the aviation companies.

Further reasons for SCC damage might lie in the use of materials such as strong Al alloys and heat treatable steels, which have a certain potential susceptibility to SCC, and are more likely to be used in operation in temperature ranges that cause structural changes that sensitize them to SCC. These materials cannot simply be replaced by less sensitive ones, since they have certain necessary properties (such as high strength in screws and ball joints).

***Pipe and hose screws** (middle left diagrams):*

*These elements are often made from bar stock with a pronounced **lengthwise structure** (see page 5.4.2.2-2). This promotes crack growth in case of SCC, because the threading and internal pressure makes operating loads act perpendicular to the structural direction (see page 5.4.2.2-3)*

Clamping and fastening screws (right top and middle diagrams):

*Clamping screws of the type frequently used on tension bands to fasten auxiliary components, are traditionally made from 13% Cr steels. Heat treatable steels are also used for ball joint sockets, which play an important role in mechanical force transmission. These materials become SCC-prone when they **exceed the intended strength**, which is usually due to improper heat treatment during manufacture. Experience has shown that the risk zones are usually in the first carrying threads (plane which lays on the nut, Ill. 5.4.2.1-4) or in formed notches, such as the forged-on head of the T-screws of tension bands.*

Compressor blading (older engine types):

Deviations during manufacture resulted in root boxes that were soldered to 13% Cr steel compressor stator vanes (bottom diagram) having a hardness above 40 HRC, which is dangerous for SCC. During operation, large lengthwise cracks formed in the bending radius. It is not entirely certain how great the influence of the high strength and resulting high internal stresses was on the crack initiation.

External Factors

Stress Corrosion Cracking: Damage

Stress corrosion cracking can occur during operation on very different engine parts.

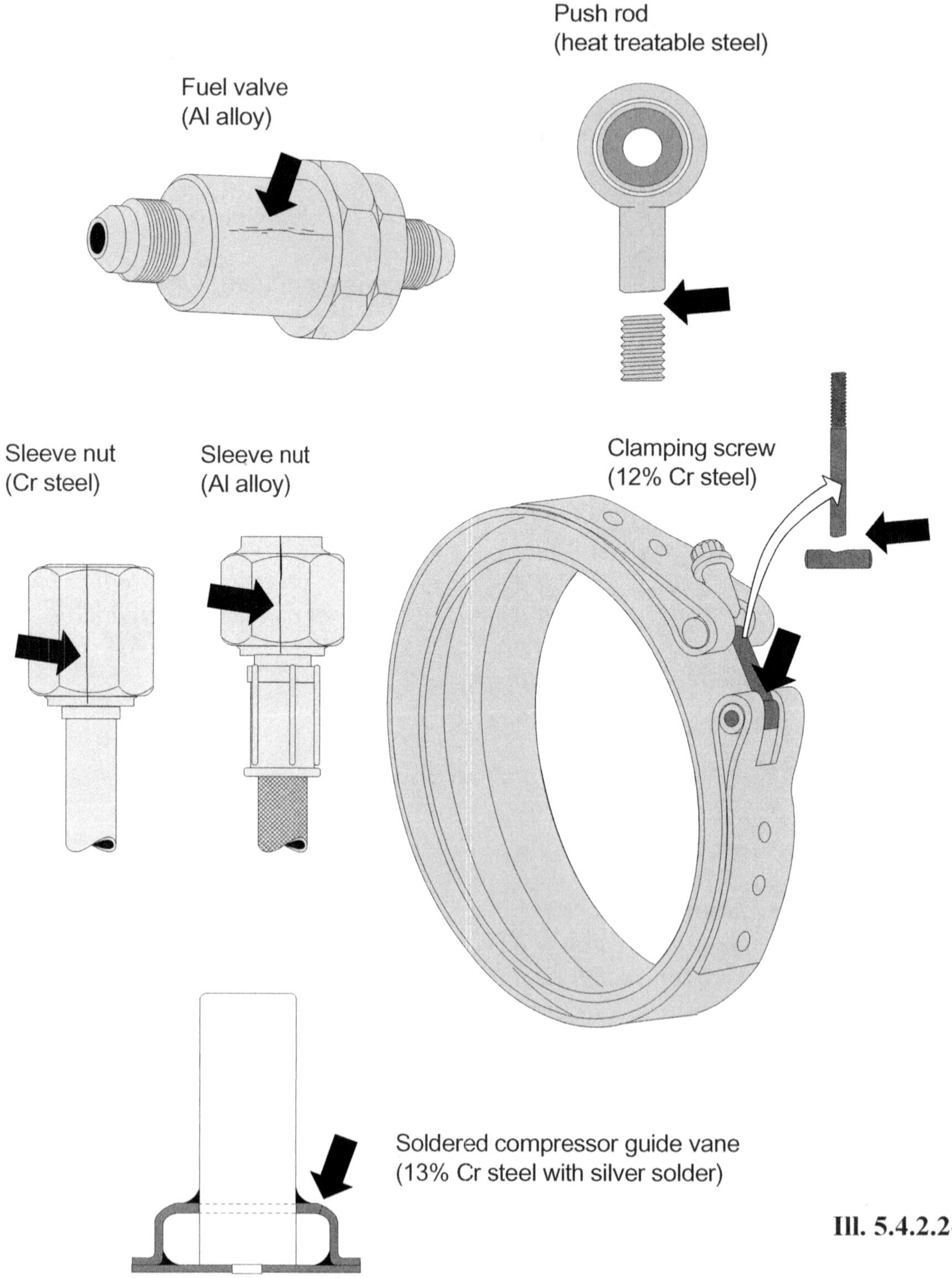

Ill. 5.4.2.2-1

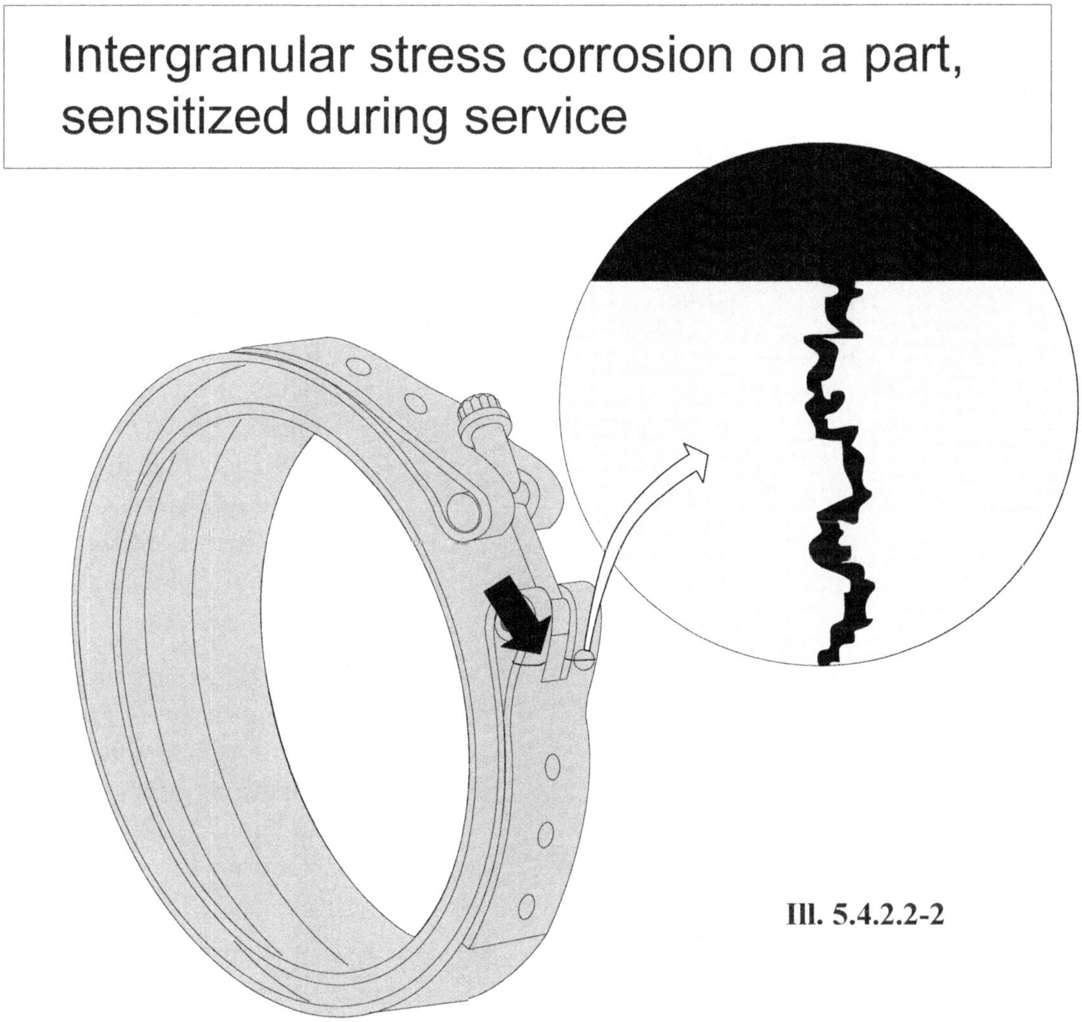

Ill. 5.4.2.2-2

Illustration 5.4.2.2-2 *(Ref. 5.4.2.2-1): This **tension belt** comes from a fighter aircraft engine and is used to fasten the hot air outtake pipe to the compressor. The sheet metal that was used for manufacture was made from a material of type CrNi 19 9 with 0.3%* carbon content. The operating temperature was between 425 and 540°C. After an **operating time of between 2 and 3 years**, the fastening flange for the tension bolt failed due to an intercrystalline fracture, even though there was no prescribed life span for the tension clamp (area indicated by arrow). The results of a metallographic inspection revealed that the damaged area contained coarse grain, which was probably due to a straightening process during manufacture. There were many carbides on the grain boundaries in this area. Under operating conditions, the **sensitization** caused by the cold-deformed structural instability resulted in characteristic crack progress under influence of an intercrystalline SCC attack in marine atmosphere (see detail, page 5.4.2.2-2).*

** the 0.3% carbon content of the affected material given in the literature seems much too high, and suggests that the wrong material was used: This would make the damage unavoidable.*

External Factors
Stress Corrosion Cracking: Damage

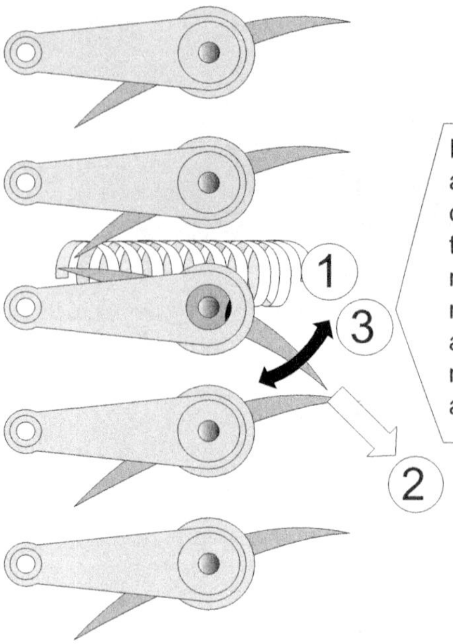

Example 5.4.2.2-1 (Ref. 5.4.2.2-2, Ill. 5.4.2.3-4):

Excerpt: " An aircraft crash in the Netherlands was caused by disintegration of a jet engine. Fractography showed that the chain of events started with **stress corrosion cracking** (SCC) of a pin attached to a lever arm of the compressor variable vane system. Such a lever arm-pin assembly costs only a few dollars. Investigation of **hundreds of pins** from the accident and a number of identical engines revealed that this was **not an isolated case**. Many pins exhibited various amounts of SCC. The failed pin in the accident engine happened to be the first fractured one.

..In this case the stress was a residual stress arising from the production method. There was a clear correlation between the presence of salt deposits on the levers and SCC of the pins.

...The corrosive **environment in western Europe** explains why the problem manifested itself in the Netherlands at a relatively early stage engine life. The main point is, however, that the manufacturer **selected an SCC-prone material** in the design stage. The solution has been to change the pin material."

Comments: It seems incomprehensible, that if the corrosion-sensitivity of the material was unknown, or if a corrosion prone material is selected in production (perhaps they hoped that the sealing would hold?), that it must be assumed that there will be no corrosive media in the surrounding atmosphere that might affect the part. The mentioned material change shows, there were evidently other material options available.

However, this example also shows that, when an engine intended for operation in a certain climate is operated in a different atmosphere, the suitability of the engine materials must be comprehensively determined and verified.

A wrong angle of attack causes:
„1" Stall and und vortex formation.
„2" Wrong flow off angle.
„3" Blade vibrations.

***Illustration 5.4.2.2-3** (Ref. 5.4.2.2-2): The compressors of modern fighter aircraft have several **variable vane** stages. Failure of the adjustment mechanism, even without fragments passing through the compressor, causes uncontrolled movement of one or more vanes and leads to flow instabilities and vibrations with corresponding consequential damages. In single-engine aircraft, this presents an immediate danger of a crash (Example 5.4.2.2-1).*

In the case at hand, the adjustment pins of the individual vanes (bottom left diagram) were made from a "corrosion-resistant" material type: an nitrogen-hardened stainless 18 8 steel, re. ASTM A276 (with alloy components of 16-18%Cr, 8-9% Ni, 7-9% Mn and 0.08% N; interestingly

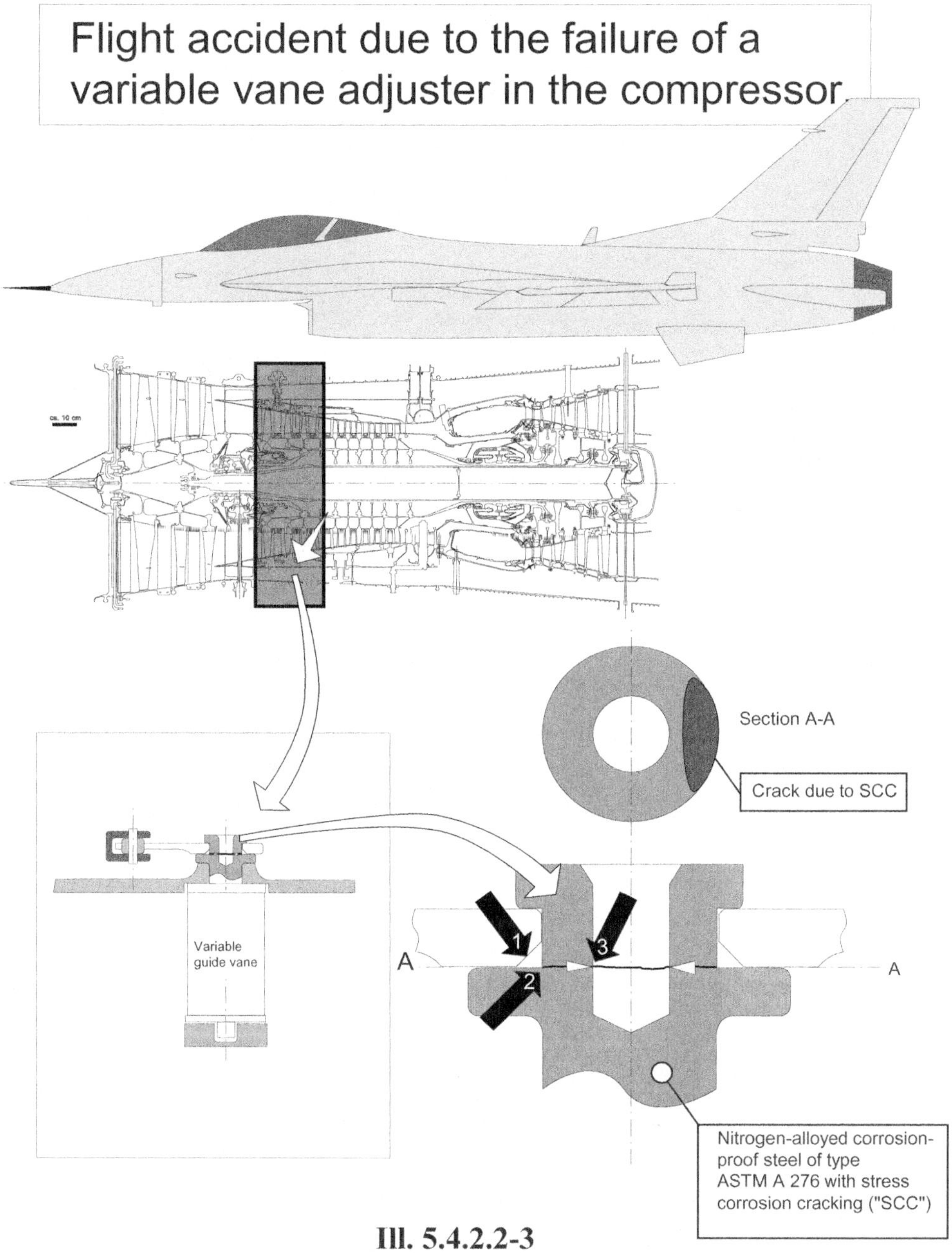

Ill. 5.4.2.2-3

enough, although this involved use in aviation, the material is not an aviation standard).

The ring duct (arrow 1), which is necessitated by the design, was evidently not sufficiently sealed off by the bearing, seating, and mating surfaces. ***Saline condensation water*** *(marine atmosphere) was able to collect in this space. During engine startup, it is to be expected that the concentration of this saline solution would increase as it* ***boils****, and that it would act on the pronounced notch in the pin (arrow 2) and have a very aggressive corrosive effect. The transcrystalline SCC crack (arrow 3) then traveled through the pin's cross-section from this notch.*

Aside from the unsuitable material, the damage was also promoted by an unfortunate constructive design.

External Factors

Stress Corrosion Cracking: Damage

Stress corrosion cracking due to the influence of baths during production and overhaul.

Stress corrosion cracking around a weld seam on a shaft made from a high-strength Ti alloy: After degreasing in a bath containing Cl with no inhibitor followed by a heat treatment.

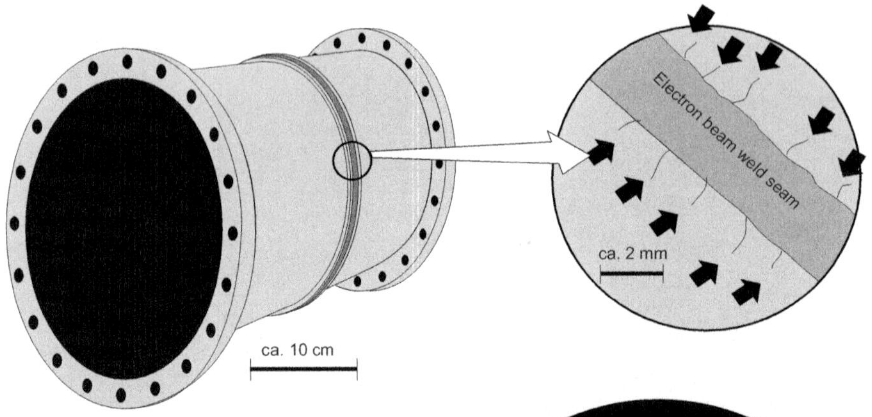

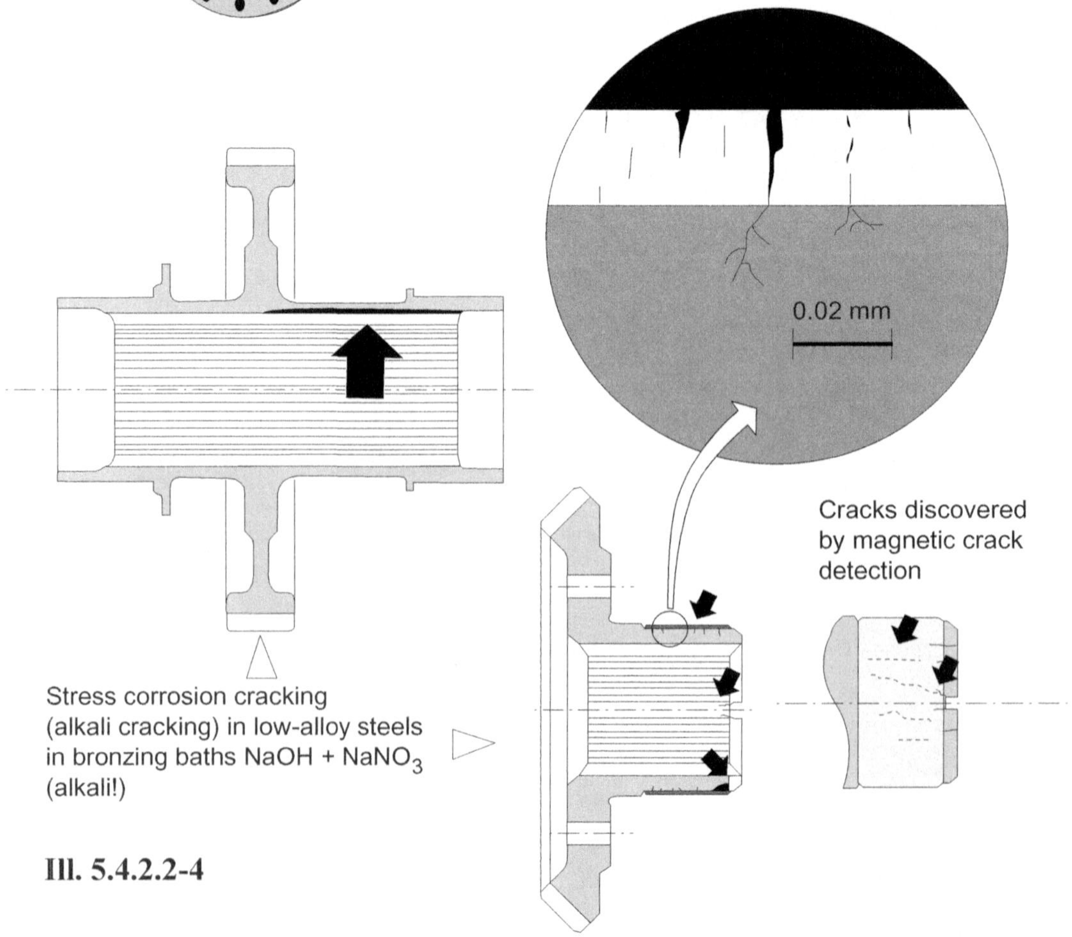

Stress corrosion cracking (alkali cracking) in low-alloy steels in bronzing baths NaOH + $NaNO_3$ (alkali!)

Ill. 5.4.2.2-4

Illustration 5.4.2.2-4** (Ref. 5.4.2.2-3)**: Stress corrosion cracking damage to titanium alloy** parts is extremely rare, and has only been known to happen in connection with **manufacturing processes**. For example, cleaning in hot degreasing liquids containing Cl, such as perchlorethylene or **trichlorethylene**, can cause SCC in areas with notches or high tensile stress, such as around a weld seam head (top diagram). Even if no spontaneous crack initiation occurs in the bath, it must be assumed that **reaction layers** on the surface containing Cl were created in the baths, and that they can lead to crack initiation during later heat treatment or even at normal operating conditions. It is also possible, that parts that have been degreased in the listed baths, and that have reaction coatings containing Cl, will tend to crack initiation during **welding**. With low-alloy steels of the type that are used as a standard material for transmission **gears** and **multiple-splined shafts** for powering auxiliary units, pronounced crack initiation due to SCC can occur (bottom diagrams). These damages have been observed not only during **new part production,** but also occur increasingly during **repair** and recoating of toothed gears. In these parts, worn bearing seats and seal ring tracks were ground down and chromed. The **Cr coating** is then ground and polished to the right size. If intercrystalline cracks are found in this coating during magnetic crack detection after bronzing, then it is possible that the cracks which are normally a part of the Cr coating are acting upon the base material, which is subject to grinding tensile stresses. Evidently, grinding stresses that were created during the grinding of the Cr coating also play a role. The **burnishing/ bronzing bath** is a hot alkali ($NaOH + NaNO_3 + NaNo_2$). For this reason, this state is referred to as **alkali cracking** or **alkali brittleness.

*If magnetic crack detection shows cracks in the Cr coatings, then these cracks are not the allowable and typical cracks associated with Cr coatings, as is often mistakenly assumed (the typical allowable cracks are not magnetic and will not be detected). In this case, it must be assumed that the **detected cracks are in the magnetic base material below the Cr coating. Shot peening of the entire part surface has proven to be effective**, especially the ground*

External Factors

Stress Corrosion Cracking: Damage

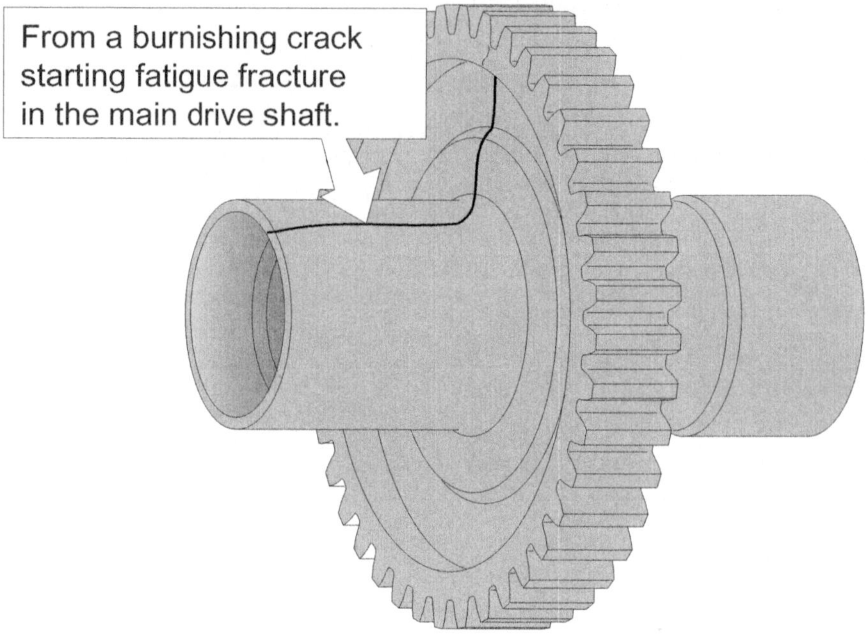

Example 5.4.2.2-2 (Ill. 5.4.2.2-4):

The gear shaft at the power takeoff of an engine from a twin-engine helicopter fractured due to a dynamic fatigue that originated in SCC cracks (**alkali brittleness**, see Ill. 5.4.2.2-4 bottom left). The cracks developed in a **bronzing bath**/burnishing**,** which is used during **manufacture** respectively overhaul of the parts. In this case, the parts were not sufficiently shot peened (introducing protecting compressive internal stresses). This allowed tension residual stresses created during hardening to initiate cracks.

Due to the poor control configuration in this engine type, the fracture of the power takeoff shaft led to the engine "**running away**", resulting in a **turbine disk failure** that damaged the parallel engine with catastrophic consequences. (volume 2, Ill. 8.1-16, volume 5 Ill. 21-10)

External Factors
Stress Corrosion Cracking: Damage

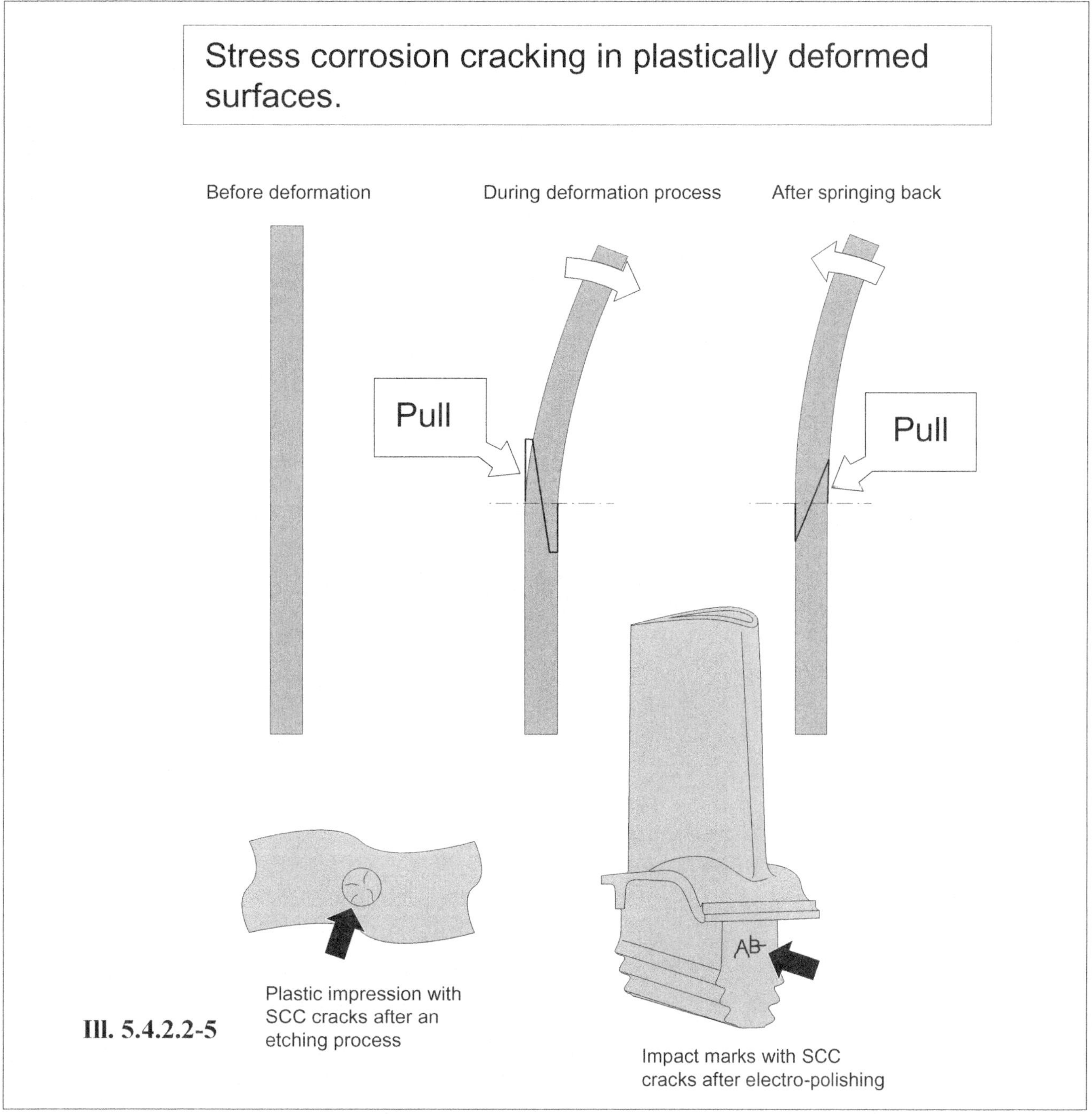

Ill. 5.4.2.2-5

Illustration 5.4.2.2-5: SCC cracks *can occur in unexpected part zones, such as in areas where compressive stresses would seem more likely. During the* ***deformation process****, localized* ***exceeding of the yield point*** *plastically compresses the material under compressive stress, i.e. it is shortened and then permanently stretched in the tensile stress zone. When the external loads stop acting and the material springs back, the stresses are reversed. Compressive stresses develop in the plastically stretched areas, while tensile stresses develop in the compressed zones. These tensile stresses can be a prerequisite for SCC. This type of damage has been discovered during repair work involving galvanic or chemical treatment of engine parts.*

External Factors
Stress Corrosion Cracking: Damage

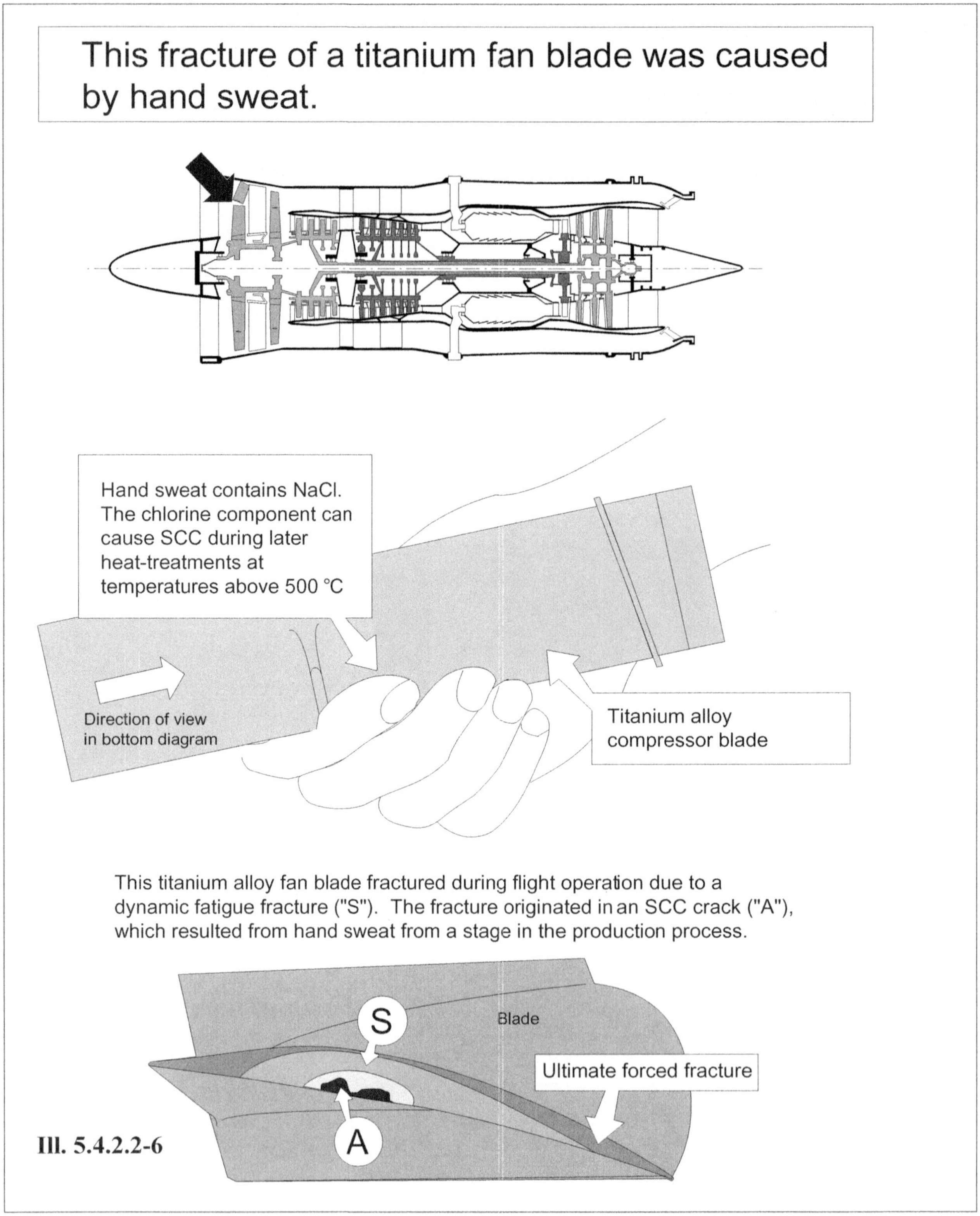

Ill. 5.4.2.2-6

Illustration 5.4.2.2-6 *(Ref. 5.4.2.2-5): The sensitivity of high-strength* ***titanium alloys*** *to* ***hot salt corrosion*** *has already been discussed (see Ill. 5.4.2.1-8). In the depicted case, fan blades made from the alloy Ti-8Al-1Mo-1V in several engines fractured during operation. Investigation*

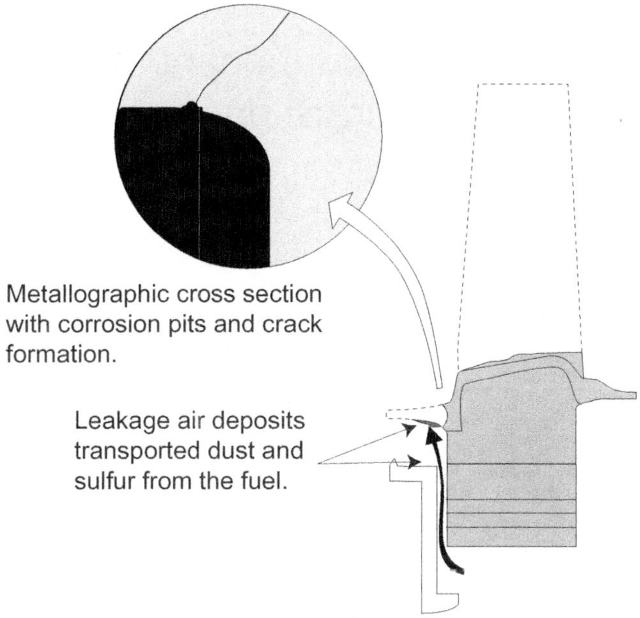

Example 5.4.2.2-3 (Ref. 5.4.2.2-9)**:**

During climb of a large twin-jet airliner at one aeroengine a loud bang occurred. The aeroengine was shut down. This was the second case at the same airplane (volume 5 Ill. 21-9). Because of a gap between blade root and cover plate, **sulfur containing dust** (calciuma sulfate = gypsum) **and sea salt could accumulate below the root platform** of rotorblades from the 2nd high pressure turbine stage. Obviously this dust composition was operator specific. Additionally hotgas with sulfur from the fuel had access. The endangered zone was protected with a **coating** against hotgas corrosion. After many thousand operation hours it came at two aeroengines tp corrosion pittings with an average depth up to 0.3 mm.

The investigations showed, that obviously the **certain Ni-cast alloy of these blades** is sensitive for such failures.

The cracks started from the pittings (notch effect) and lead at least to a blade fracture. In the literature at hand, once is spoken about „**stress corrosion failures**“ (usually stress corrosion cracking = SCC) and at an other citation about „**fatigue cracks**“. This term usually indicates vibration fatigue cracks. With this the failure mechanism and failure sequence is not clear. Thinkable is, that the cracks, starting from the sulfidation attack, have been accelerated by the sulfur (Ill. 5.4.2.2-7).

*revealed that the blades failed due to dynamic fatigue fractures that originated in typical SCC cracks. The operating temperatures, which never exceed 115 °C in this part, are not sufficient to cause SCC cracks due to hot salt action. Further investigations and tests showed that the blades were evidently **handled with bare hands** during manufacture. This transferred **hand sweat** onto the blade surface. During later heat treatment, the Cl component from the salt content was sufficient to initiate cracking in the shape of a crack zone. The crack lengths were up 5 mm (see bottom diagram), which means that these damaged blades did not undergo sufficient crack testing before being installed in the engine. The cracks were sufficient to allow the normal tolerable blade vibrations to cause crack growth until the blade ultimately suffered a force fracture. In this case, fragments even escaped from the fan housing. Of course, there must have also been high **tension residual stresses** in the affected blade area, if only temporarily. It is likely that the residual stresses were created during a **manufacturing process** that involved chipping.*

External Factors

Stress Corrosion Cracking: Damage

Sulfur from the environment, like air, dust or lubricant can trigger symptoms of stress corrosion cracking (SCC) at hot parts .

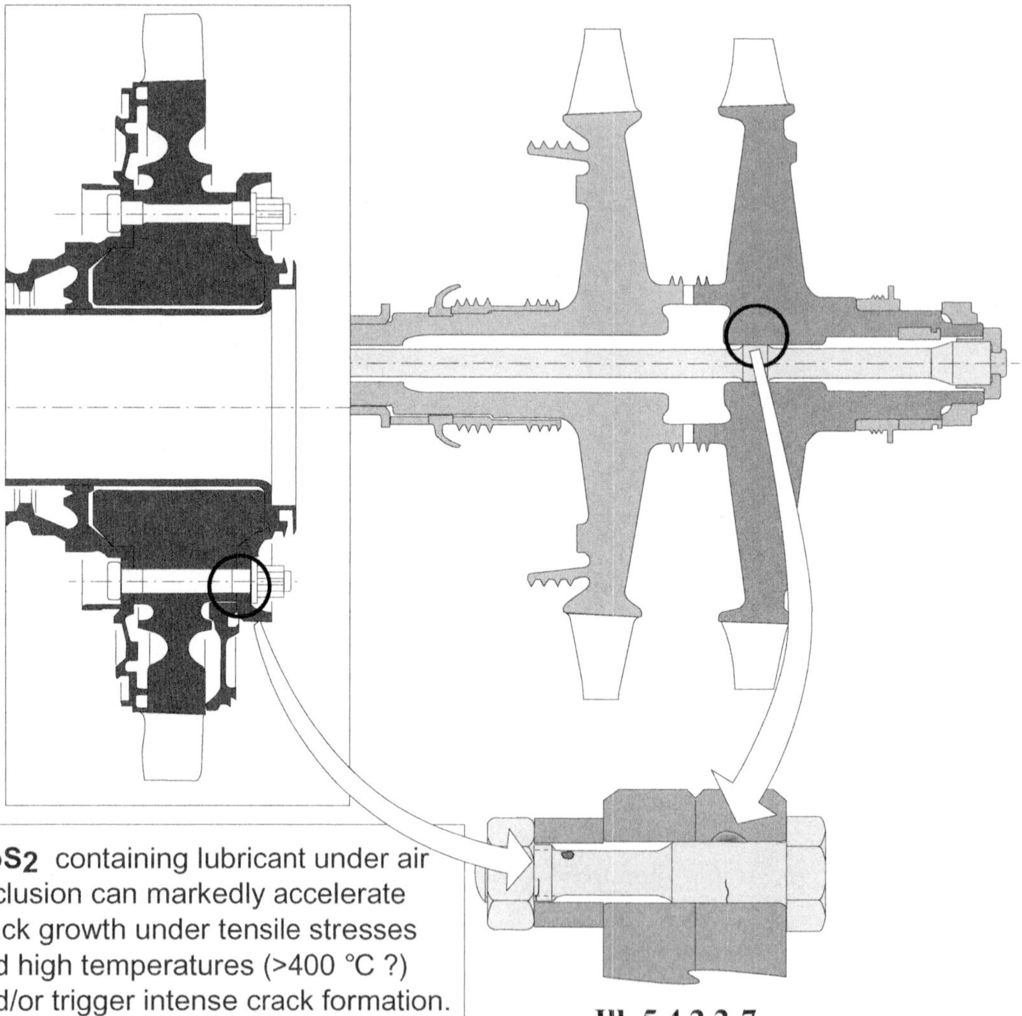

MoS2 containing lubricant under air exclusion can markedly accelerate crack growth under tensile stresses and high temperatures (>400 °C ?) and/or trigger intense crack formation.

Ill. 5.4.2.2-7

Illustration 5.4.2.2-7 *(Ref. 5.4.2.2-6, Ref. 5.4.2.2-7 und Lit. 5.4.2.2-8): Also at* ***hot parts*** *crack formation is known, which can be assigned a stress corrosion cracking. However here without watery electrolyte. From experience the effect of* ***sulfur*** *(volume 5, Ill. 23.3.1.2-14) is expecially dangerous. With* ***MoS_2 containing lubrication grease crack formation under air exclusion respectively lack of oxygen is triggered****. With this the lubrication grease obviously does not burn/oxidise at time. These conditions prevail in* ***threads*** *and in* ***fit bores****. Promotive are* ***notches*** *like micropores in castparts from Ni alloys (turbine wheel right sketch above),* ***thread turns*** *of bolts (sketch below, volume 5 Ill. 23.3.1.2-9) or* ***corrosion pits*** *(e.g., through* ***sulfidation****, example 5.4.2.2-3). Here fast crack development can occur. It can depend from the necessity of a minimum tensile stress. Often also a connection with silver is observed (volume 3 Ill. 12.4-14).*

Chapter 5.4.2.2 References

5.4.2.2-1 "Metals Handbook Ninth Edition" Vol. 11 "Failure Analysis and Prevention", American Society for Metals, (ASM), Metals Park, Ohio 44073, page 309.

5.4.2.2-2 H.J.Kolkman, G.A. Kool, R.J.H. Wanhill, "Aircraft Crash Caused by Stress Corrosion Cracking", ASME-Paper 94-GT-298, proceedings of the "International Gas Turbine and Aeroengine Congress and Exposition", The Hague, Netherlands- June 13-16, 1994.

5.4.2.2-3 H.Simon, "Oberflächenreaktionen an Titanwerkstoffen", periodical "Metall Oberfläche" 5-1982, pages 211-217.

5.4.2.2-4 F.T. Carroll, D.R. Parish, "Testing Considerations for Military Aircraft Engines in Corrosive Environments (a Navy Perspective)", AGARD-CP-558, proceedings of the conference "Erosion, Corrosion and Foreign Object Damage Effects in Gas Turbines" Rotterdam, The Netherlands 25-28 April 1994, Chapter 37, page 37-1.

5.4.2.2-5 E.U.Lee, R.G. Mahorter, J.D. Wacaser, "Fracture of Ti-8Al-1Mo-1V Alloy Fan Blade by Stress Corrosion Cracking and Fatigue". proceedings of the symposium on the 1-6 May 1977, Toronto, Canada, "Fractography in Failure Analysis", ASTM Special Technical Publication 645, pages 128-143.

5.4.2.2-6 R.S.BhattAchaRya, S.KrishnaMurthy, A.K.Rai, J.A.Kramer, „Threaded Fastener Coatings for Aerospace Applications", Zeitschrift „ Lubrication Engineering", Volume 52, No.3, March 1996, page 237-241.

5.4.2.2-7 National Transport Safety Bureau (Australia), A.Romeyn, Technical Analysis Report 17/03, „Bolt Fracture High Pressure Turbine Disk Assembly", 2003, page 1-21.

5.4.2.2-8 R.S.BhattAchaRya, S.KrishnaMurthy, A.K.Rai, J.A.Kramer, „Threaded Fastener Coatings for Aerospace Applications", Zeitschrift „ Lubrication Engineering", Volume 52, No.3, March 1996, page 237-241.

5.4.2.2-9 TSB of Canada, Aviation Investigation Report Number A01F0020, „Power Loss - No. 2 Engine, Skyservice Airlines Inc. Airbus A330-300 C-FBUS, Columbo, Sri Lanka, 15 February 2001", page 1-6.

5.4.2.3 Measures to Prevent Damage due to Stress Corrosion Cracking

SCC can not develop if one succeeds in removing one of the following three prerequisites from the corrosion system:

- sensitive material
- critically high tensile stresses
- material-specific corrosive media

All remedies must be oriented to this requirements.

• Selection of materials that are SCC resistant, i.e. determining semi-finished product production, manufacturing processes, and heat treatments that prevent the development of structures that promote SCC. Heavily soiled materials such as free-machined materials (with impurities as chip breakers). This is especially true for corrosion types that involve hydrogen creation.

• **The highest possible material strength should not be used**, neither through loading or production (e.g. shaping, heat treatment). For example, in Cr steels, hardness above 30 HRC is an indicator for possible SCC-sensitivity.

• Depositing **hand sweat** before the material is subject to temperatures above 500°C must absolutely be avoided. This may occur during manufacture before heat treatment, and should be preventable by wearing suitable gloves.

• Care must be taken during corrective **cold-forming** of sensitive materials (such as when repairing damages through straightening and non-chipping shaping). Dangerous local residual stresses and/or structural changes may occur.

• The material should not be loaded **across the direction of fiber** (rolling, forming direction when forging). This is especially true for SCC-sensitive Al alloys and Cr steels.

• **Reducing tension residual stresses** from manufacture as much as possible through suitable heat treatment (low-tension baking).

• **Inciting compressive internal stresses** at the surface, such as through shot peening or rolling. Using chipping procedures with set parameters (e.g. chipping rate, chip depth, tool geometry, cool cutting media), for which compressive stresses have been verified after chipping. No changes to procedures or parameters without verification of comparable compressive stresses in the surface being machined.

Experience has shown that **grinding** will create tensile stresses in surfaces. For this reason, grinding should, if possible, be avoided in systems that are prone to SCC.

• **Certifying and monitoring auxiliary materials** during manufacture (e.g. cutting fluid, cleaning and etching baths, etc.). Special caution with auxiliary **materials with halogen compounds** such as chlorine. If in doubt, these products should be certified in suitable SCC tests.

Experience has shown that the following combinations of processing bath and material require special caution:

- **bronzing/burnishing baths** with hardened and case-hardened steel parts

- **hot perchloroethylene degreasing baths** with high-strength Ti alloys

- **Rinsing** baths contaminated with chloride combined with parts made from high-strength Al alloys
and (martensite) Cr steels.

In every case, **sensitive parts** (such as parts made from case-hardened steels such as gears and shafts) should be sufficiently **shot peened** on all surfaces that are to be treated with a bath.

• **Realistic and suitable testing** of materials (e.g. goods receipt) and parts (patterns, part testing) for SCC-sensitivity to manufacturing and operating influences.

• **Weld seams** should be located in non-critical areas. These include areas that are not under high tensile stresses, or those in which no electrolyte fluid can collect.

• Suitable **realistic corrosion systems** (material, loads/sample shape, media) must be selected in order to verify SCC risks. Defining the characteristic (e.g. crack growth rate, time to crack initiation, time to fracture) is important for selecting a test for SCC-sensitivity. For example, **SCC testing of Ti alloys** frequently requires nicked or cracked samples, with the crack growth speed and crack image as criteria.

• Use of **electrochemical testing metho**ds (e.g. measure potential) in order to determine SCC-sensitivity.

• **Hollow spaces** should be avoided as early as the design phase. In rotor parts such as disks and spacers, these are conical lugs in which corrosive media can collect during standstill or under the influence of centrifugal force.

In **blind holes**, the hole and threads must be sufficiently sealed against operating fluids. It must be ensured that no corrosive remnants (e.g. from galvanic baths or etching and cleaning baths) remain in the threaded blind holes, which might later react with other fluids during operation (such as condensation water) and form dangerous electrolytes.

• Selection of **suitable coatings**, such as organic lacquers and inorganic coatings (Al powder-filled lacquer with a ceramic binder, enamel, metallic coatings). **Cathodic protective effects** are desirable, since they provide corrosion protection of the base material even if the coating is damaged all the way through.

• **Avoiding sharp corners** and edges, as well as crack-like flaws, around which stress peaks can be expected.

• The **surfaces** should be as **smooth** an **clean** as possible. Contaminants that can change the corrosion behavior (e.g. cell action), such as may be deposited during manufacturing or repair procedures (e.g. smeared tool material, remnants of peening material), must be avoided or safely removed.

• With “**catalog parts**“, special care is necessary to ensure that:

- the material is in the most SCC-resistant state possible with regard to the expected media

- no additional sensitivity is created due to a **combination with other parts** (e.g. cell action)

- **quality control** must meet aviation standards

- the unique **operating conditions** on the engine are taken into account during design and selection.

5.4.3 Corrosion Fatigue

5.4.3.1 Fundamentals of Corrosion Fatigue

5.4.3.2 Damages due to Corrosion Fatigue

5.4.3.3 Measures to Prevent Corrosion Fatigue Damage

Corrosion fatigue is a corrosive influence that promotes fatigue crack initiation and growth under dynamic loads. This corrosion type is primarily dependent on loads, unlike stress corrosion cracking, which is primarily time-dependent.

5.4.3.1 Fundamentals of Corrosion Fatigue

Corrosion fatigue is a type of low-deformation crack initiation under **alternating corrosion and mechanical loads** that are independent of the three influences that make up the corrosion system.

Unlike stress corrosion cracking, corrosion fatigue is not tied to a specific system, metal/media. Therefore it has a far **greater number of possible damages** (Ill. 5.4.3-1). Both actively corroding materials and materials that behave passively can be damaged. A possible reason for the "universal" appearance of this corrosion type is that, under dynamic loads, pronounced extrusions can form in the micro-zone (Ill. 5.4.2.1-2), which represent intense damage for protective coatings (passive coatings). If pitting corrosion or intercrystalline corrosion (Ill. 5.4.3-2) occur during the crack initiation phase, the relationships become very complicated. For this reason, practical experience and close investigation of operating conditions are indispensable for successful countermeasures (Ref. 5.4.3-2). Unlike with pure dynamic fatigue, there is a clear **influence of frequency** on noticeable corrosion fatigue with an SCC influence (Ref. 5.4.3-2). The material, corrosive media, and loads (e.g. frequency, mean stress, stress deflection) determine whether corrosion fatigue cracks are transcrystalline or intercrystalline (Ref. 5.4.3-1). The influence of corrosion fatigue can typically be seen in heat-treated steels as an intercrystalline component in the crack progress. In exceptional cases, even with "real" corrosion fatigue, intercrystalline crack progress has been determined in martensite and austenite steels in zones with middle stress intensity **amplitudes**. The intercrystalline component increases along with increasing **average loads.**

The term "corrosion fatigue" is differentiated from dynamic fatigue fractures in some literature, if the fractures originate in corrosion marks without another corrosive media acting. This seems impossible to the author when considering that virtually all metals have better dynamic strength in a vacuum than in an atmosphere, and that the crack growth speed can be up to four times as great as that in a vacuum (Ref. 5.4.3-2). This shows that a corrosion fatigue effect can always be expected, and especially when the surrounding media has already lead to pitting or intercrystalline corrosion. However, this effect is often not noticed, since the characteristic values for the material's fatigue strength are determined in air and/or the failure mechanism in sample tests is considered to be the first crack. For this reason, the crack initiation and growth influence does not usually stand out. However, the more dynamically loaded parts, in which the crack growth phase forms a conscious or unconscious factor of the life span concept, are used (such as in highly loaded rotors in modern engines), the more important it becomes to consider the influence of corrosion fatigue.

For example, it has become normal practice to determine the LCF life span of rotors in vacuum testing rigs due to the lower energy consumption. However, before conducting this type of test, one must be certain that the vacuum is not a falsifying factor, since this would otherwise mean that the determined usable life spans would be longer than those that could actually be realized in practice.

Behavior of different material groups:

Cr steels: in these heat-treated steels, corrosion fatigue usually occurs in connection with pitting corrosion. At first, crack progress is primarily intercrystalline, i.e. it runs along the former austenite grain boundaries that are still present in the heat-treated structure. This intercrystalline progress can already be seen under the influence of pure water. Even with high-frequency vibrations, such as are typical in compressor bladings, intercrystalline fracture components occur (Ill. 5.4.3.2-2). With "real" corrosion fatigue (no SCC component), intercrystalline fracture components are only to be expected at middle stress concentration amplitudes (re: Ref. 5.4.3-3). The sensitivity to corrosion fatigue (and SCC) increases with hardness. It is plausible that the damage is also promoted by the increased notch sensitivity at higher hardnesses. Above 40 HRC, these materials should not be used under corrosive conditions.

Cr-Ni steels: with these austenite alloys, which normally have pronounced passive corrosion behavior (Ill. 5.4.3.1-2), pitting corrosion and/or intercrystalline corrosion are often present in the crack origin. Even in the sensitized material, dynamic fatigue cracks are usually transcrystalline. However, high mean stress and low frequencies promote intercrystalline crack progress. Tests showed fatigue cracks that were transcrystalline in a vacuum, but had pronounced intercrystalline growth at low frequencies (10^{-3} Hz) under the influence of water (!)(Ref. 5.4.3-3). This was traced back to an increased SCC-influence (intercrystalline growth) during the fatigue process.

Mg- and Al-alloys: with these alloys, dynamic fatigue fractures often originate in corrosion pittings. Marine atmosphere is noticeable even at very low dynamic loads (Ill. 5.4.3.1-1).

Ti-alloys: these alloys show an influence of hot salt on the LCF and HCF behavior above about 250 °C. In both cases the life span/dynamic strength decreases considerably. The influence is especially noticeable at low-frequency loads in the LCF zone. The sensitivity to hot salt action depends on a multitude of other influences such as the structure of the material.

Ni-alloys: Corrosion fatigue effects were observed in certain media such as greases containing MoS_2 (at high temperatures) or special air contaminants. For example, MoS_2 in closed air (such as between mating surfaces with screws) can extremely accelerate LCF crack progress, whereby grain size seems to have an influence. Pronounced stage I fractures develop (crack surfaces), the size of which naturally corresponds to the grain size. With the large grains typical in cast parts, correspondingly large cracks develop. It can be assumed, that single crystal material may be especially to crack surface development, depending on the orientation of their crystals relative to the loads. Experience has shown that the same is true for fatigue processes in protective gases such as argon. Evidently, even small impurities in the gas (e.g. iron compounds) can cause accelerated crack growth.

Illustration 5.4.3.1-1: *Similar to marine atmosphere, salt water steam has a weakening effect on the dynamic fatigue strength of many materials (bottom diagram, Ref. 5.4.3-5), when compared with dry air. For example, a decrease of over 50% can be expected in Cr steels and high-strength Al alloys.*

In the top diagram (Ref. 5.4.3-4), curves "1" and "3" show this effect schematically. This also shows that, if corrosion fatigue is suspected, testing must be done in the operating medium. Pre-corrosion (curve "2") without dynamic loads in a corrosion chamber or in a submerged test is not sufficient and results in overly high fatigue strength relative to realistic operating conditions. Even the crack initiation phase is not satisfactorily simulated in this type of test. Fracture mechanics tests are necessary for considering the crack growth phase.

With complex operating damages, sufficiently safe conclusions are no longer possible from laboratory tests under standard conditions (Ref. 5.4.3-2).

External Factors
Corrosion Fatigue

In most cases, corrosion fatigue influences fatigue crack formation in the atmosphere, even if it is not usually noticed.

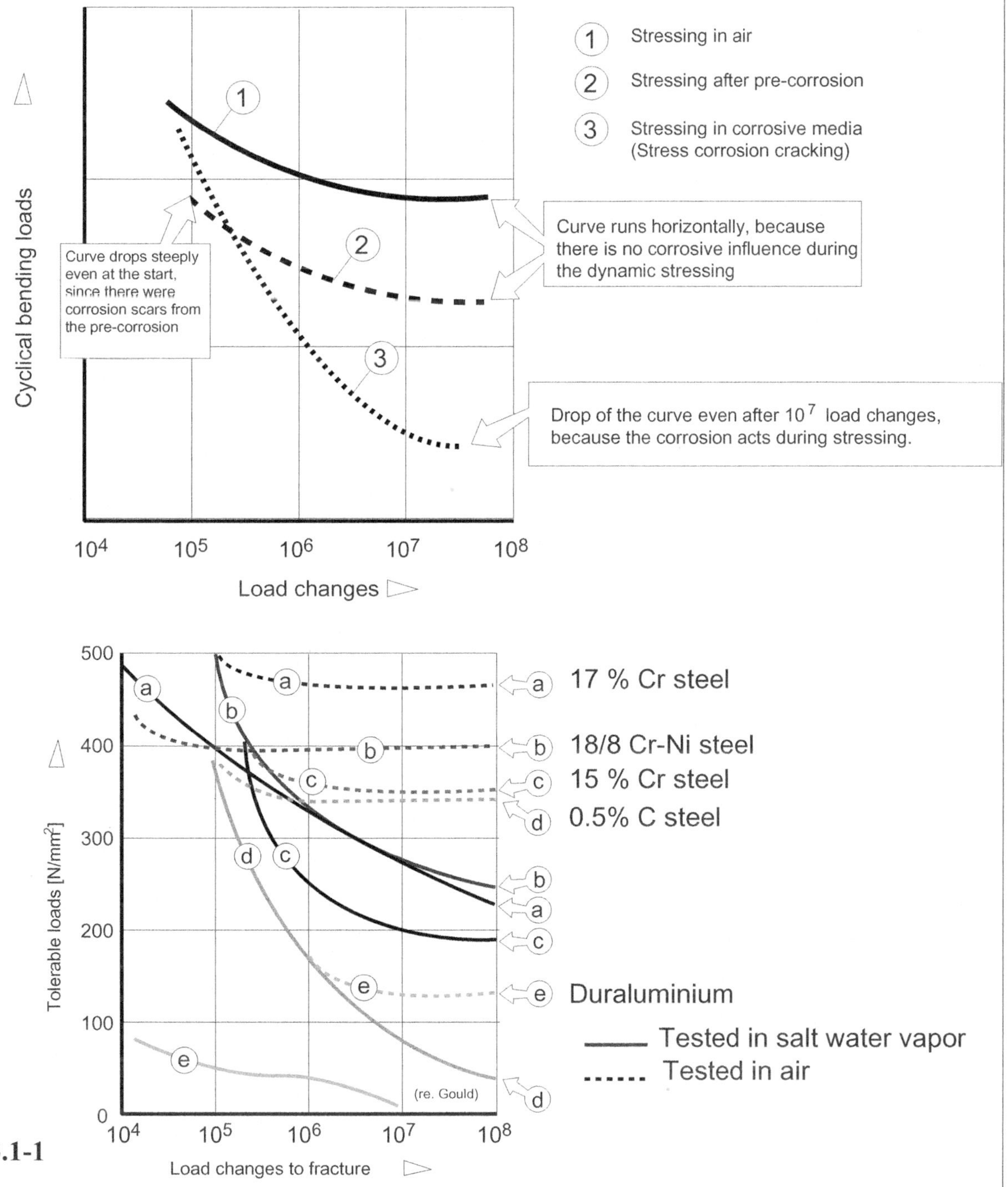

Ill. 5.4.3.1-1

Stress corrosion cracking can have many very different forms, even in the same material.

Ill. 5.4.3.1-2

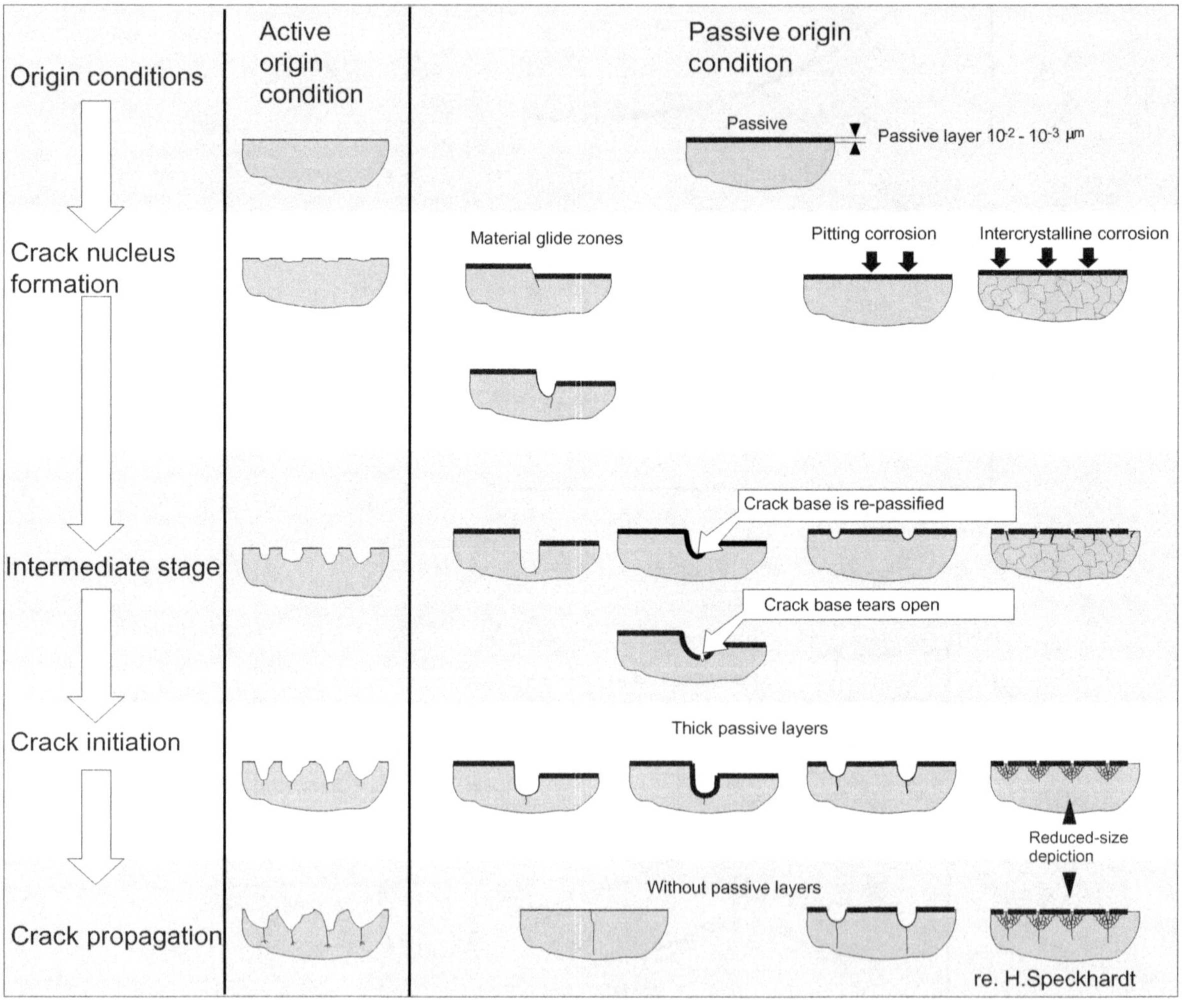

Illustration 5.4.3.1-2 *(Refs. 5.4.3-2 and 5.3.3-6): The diagram shows the typical phases of corrosion fatigue damage and gives an impression of the variety of damage symptoms.*

Origin conditions: *materials that have surfaces with active corrosion behavior (such as unalloyed and low-alloyed steels), as well as those with passive behavior (such as high-alloy steels; form protective coatings), can show corrosion fatigue.*

Crack nucleus formation: *glide planes, corrosion-sensitive phases, inhomogeneities at grain boundaries (e.g. selective attack of Cl ions on sensitized Cr-Ni steels), weaknesses in the protective coating (such as flaws at structural inhomogeneities and impurities), and crack initiation in non-metallic inclusions such as carbides are attack zones for corrosion fatigue. The protective passive layer is broken through and damaged due to slippings in the surface zone (compare with Ill. 5.4.2.1-2). This both exposes unprotected metal surfaces, and also creates notches in this zone with increased mechanical stresses that promote the corrosion process. In the passive state, the cracks often remain unfilled by corrosion products.*
In the active state, first local pittings develop.

Intermediate stage: *In the active state of a material group, the pittings expand, whereas in the active state the pittings become deeper, and their bases create new protective coatings and become passive. These, in turn, then crack open again during further load cycles (compare with Ill. 5.4.2.1-2). The process of the passive layer formation in the base of the pittings is a reason for the frequency-dependency of corrosion fatigue. If the grain boundaries are especially sensitive, i.e. suitable structural zones are present, then an initial attack begins here. The main growth direction of this attack is through the protective passive layer into the material.*
In the active state, no passive coating is created, and the corrosion spreads out to the sides across a larger area.

Crack initiation: *from the cracks and nicks that formed in the intermediate stage, relatively sharp cracks form with no notable expansion due to eroding corrosion.*

Crack propagation: *depending on the size of the loads, high stresses result primarily in straight single cracks that run into the material and low stresses result in branching at the end of cracks.*

Illustration 5.4.3.1-3: *Almost all compressor parts are subject to more or less high dynamic loads. This is especially true for the blading. Because the simultaneous presence of a corrosive media is necessary for corrosion fatigue damage to occur, the temperature, pressure, and air speed in the working compressor play an important role.*

The increasing pressure allows liquid water to be present even at about 140°C (top diagram), so stages with this compression temperature can be affected by it.

The typical salts of ocean water can release chemically combined water (bottom diagram) at higher temperatures (up to over 250°C), creating aggressive moisture in existing cracks at least for a short time.

As shown, damp marine air is sufficient for fatigue corrosion in Cr steels under dynamic loads (compressor blades in older engine types). The probability of corrosion fatigue crack initiation is lower in the hotter rear compressor zones than in the forward areas, where moisture can act even during operation. During startup and acceleration of the engine, however, there may temporarily be sufficient moisture in the entire compressor. This moisture can be in the form of condensation water that collects in structural hollow spaces (Ill. 5.4.2.2-3), gaps (such as the blade root area, constructed V-stator vanes with hollow boxes- Ill. 5.4.1.2-2, variable vane adjusters), corrosion notches, roughness, and porous deposits or coatings, and only dries out during operation, when its concentration is increased. Experience has shown that dynamic fatigue cracks and fractures of compressor blades made from Cr steels and Al alloys often show signs of corrosion fatigue.

It cannot be absolutely guaranteed that there will be no liquid water in the compressor during operation, if only temporarily.

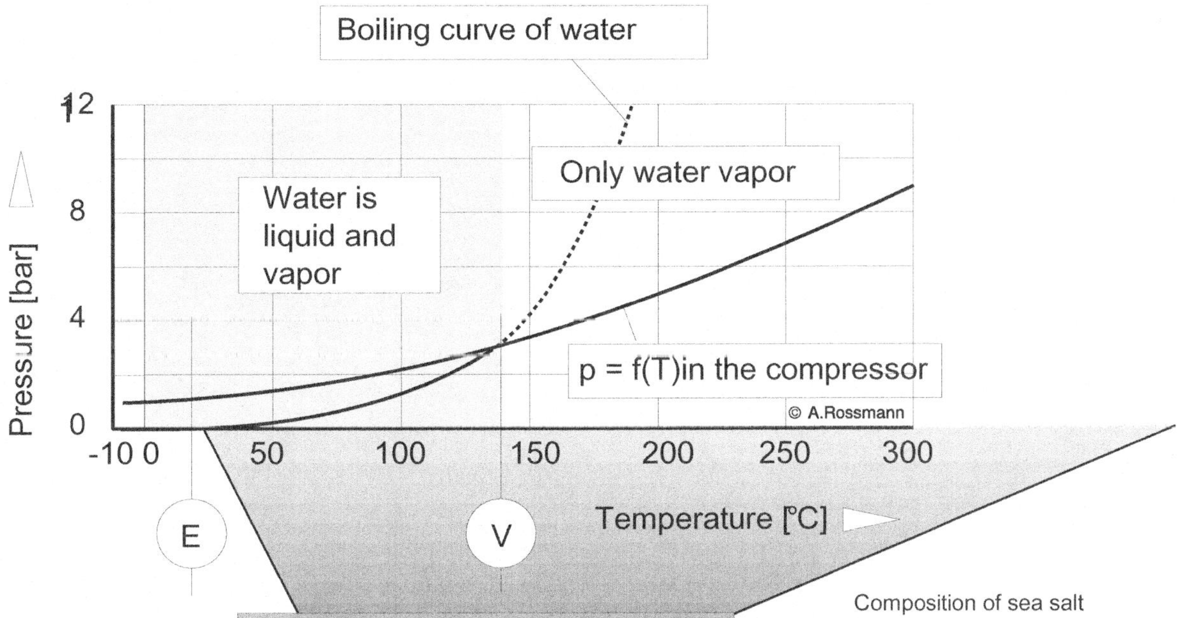

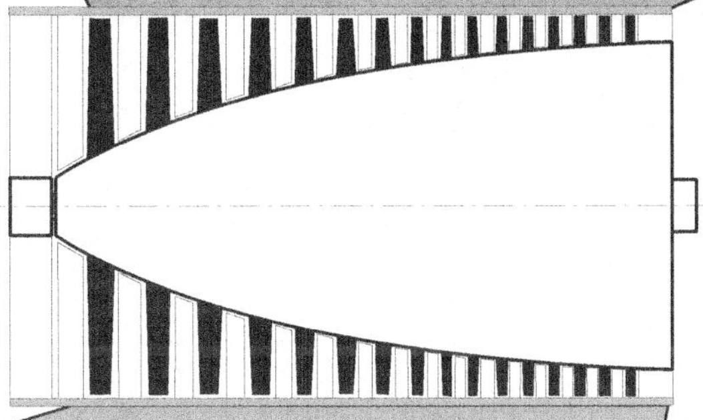

Ill. 5.4.3.1-3

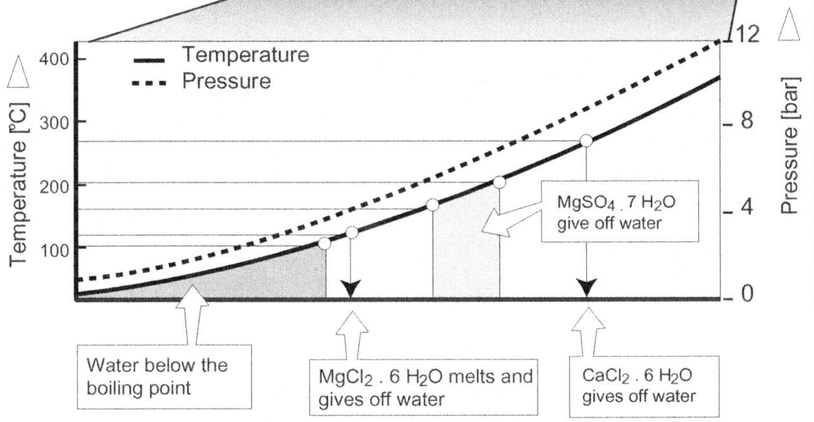

Possible causes for moisture in the compressor during operation

Composition of sea salt

Component	Cont-ent [%]	Temp. [°C] Above which chemically combined water is given off
NaCl	77	-
$MgSO_4 \cdot 7\ H_2O$	9	150 200
$MgCl_2 \cdot 6\ H_2O$	7	120
$CaCl_2 \cdot 6\ H_2O$	3	260
KCl	2	
NaH CO_3	<1	
NaB	<1	

The chemically combined water makes up about 50% by weight of $MgSO_4$, $MgCl_2$ and $CaCl_2$

5.4.3.2 Damage due to Corrosion Fatigue

Even though, judging by the damage mechanism, dynamic fatigue damage would seem to indicate an atmospheric influence and corrosive environments would seem to decrease dynamic strength and increase crack growth rates, there are very few cases in which damage is explicitly traced back to corrosion fatigue.

This is true both for the damages shown in Ills. 5.4.1.2-4 and 5.4.1.2-5, which have a causal relationship with pitting corrosion, and also for Ills. 5.4.3.2-1 and 5.4.3.2-2.

Illustration 5.4.3.2-1 *(Ref. 5.4.3-7): This compressor of a small helicopter engine contains cast bladed disks (blisks) made from a precipitation hardened high-strength steel (17-4 PH). In an unprotected state, this material tends to pitting corrosion under the influence of chlorine ions, which are a typical component of the salts in marine air (Ill. 5.4.3.1-3). Salt deposits on the blades are hygroscopic and absorb water from the surrounding air. With* $(NaH_4)_2SO_4$ *and NaCl deposits this occurs above 80% relative humidity, with* $MgCl_2$*, above a mere 32%.*

Corrosion protection for the compressor blading:

Regular ***washing*** *of the compressor with a suitable detergent has proven to be a suitable measure against the corrosive attack.*
Coating *the compressor with NiCd provides sufficient corrosion protection only when the coating is not damaged all the way to the base material, which experience has shown cannot be ensured under realistic operating conditions. Al powder-filled* ***inorganic lacquer systems****, however, provide active corrosion behavior with* ***cathodic corrosion protection*** *of the base material that is effective even in case of local coating damage that exposes the base material (e.g. scratches or FOD).*

Hard erosion coatings *with noble corrosion behavior such as TiN, which build up a potential relative to the base material, dangerously promote corrosion pittings through cell action. Because FOD damage or erosion notches can always be expected to break through or chip off hard protective coatings and expose the base material, these protection systems are unsuitable in cases where corrosion-sensitive base materials would be subject to erosive and corrosive loads.*

Serious compressor damage resulting from a local corrosive attack.

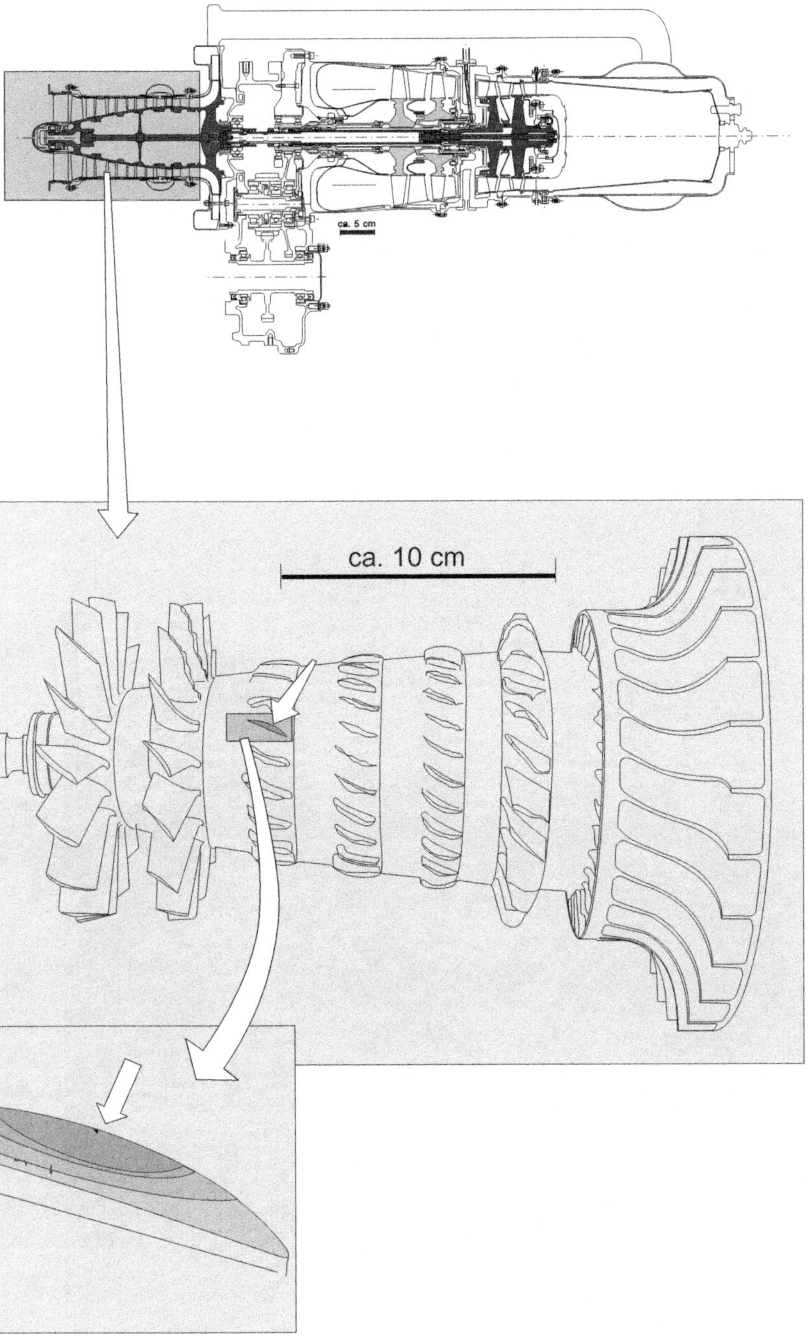

Ill. 5.4.3.2-1

Corrosion fatigue fracture of a compressor blade made from tempered 13% Cr steel.

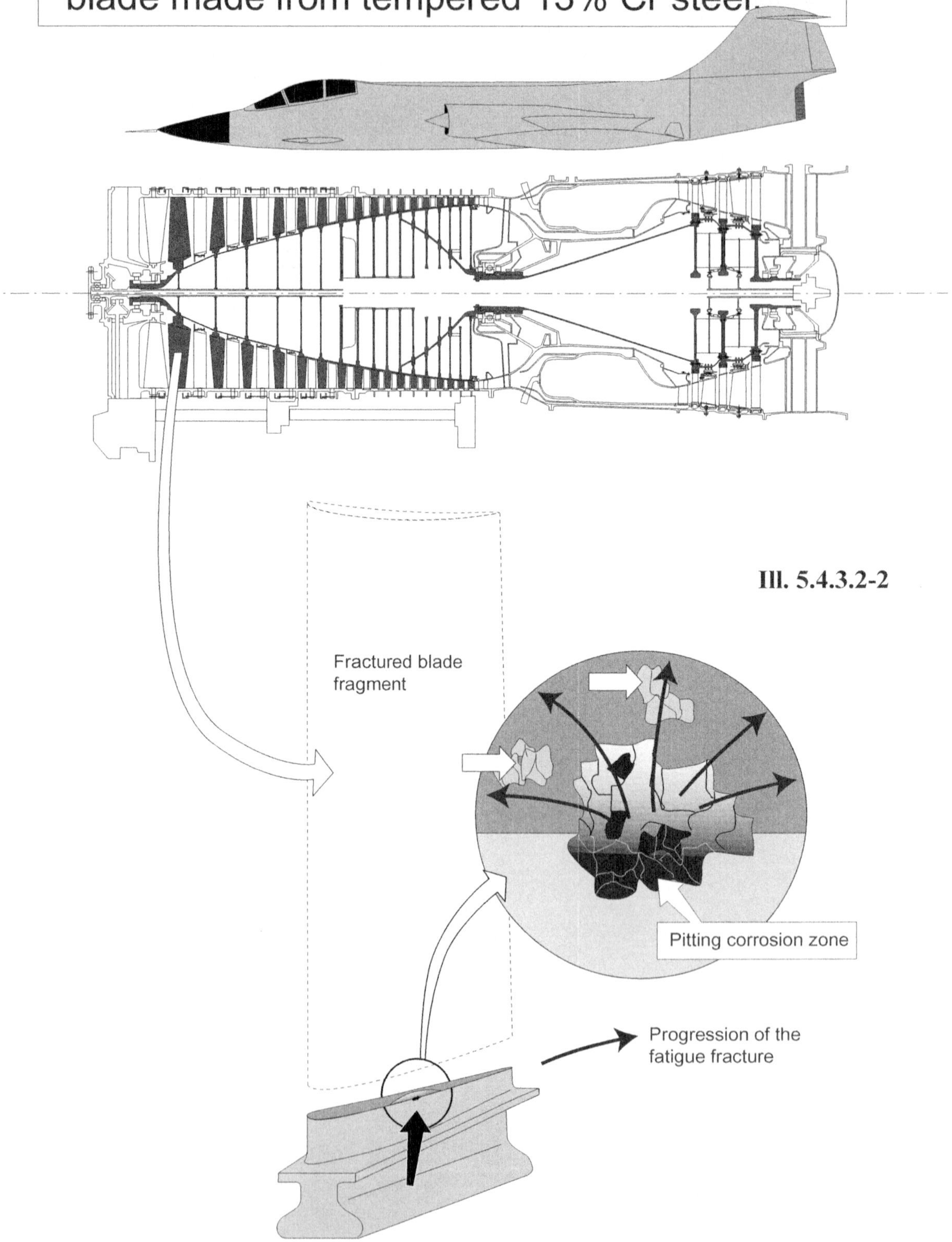

Ill. 5.4.3.2-2

Illustration 5.4.3.2-2: *The compressor blading of older engine types is often made from Cr steels. Only in a polished state do these steels (so-called knife steels) have sufficient corrosion resistance to Cl ions in watery media. In practical operation, however, experience has shown that after a relatively short operating time, erosion roughs up the blades and causes pitting corrosion in the base material. The sensitivity to cracking corrosion types (SCC and corrosion fatigue) usually increases with the hardness of these heat-treated materials.*
In the case at hand, a fighter aircraft crashed after the fracture of a forward compressor rotor blade due to a fatigue fracture that originated in a pitting corrosion notch. The hardness of the blade was clearly to high (45 HRC, rather than the maximum allowable hardness of 40 HRC). Since inspection of a large number of other blades did not reveal any more similar deviations, this case was classified as an isolated flaw, which has been borne out by following years of experience.
The isolated case showed the intercrystalline structure typical of pitting formation in Cr steels. The fatigue fracture has several intercrystalline regions (white arrows), which indicate the corrosive influence during crack growth typical for corrosion fatigue.

5.4.3.3 Measures for Preventing Corrosion Fatigue Damage

In general, the same rules for preventing pure dynamic fatigue fractures hold true for measures for preventing corrosion fatigue:

• lowering the dynamic loads and, if possible, also the mean stress

• avoiding stress peaks

• measures to prevent corrosive attack (i.e. the presence of corrosive media) during operation (i.e. during dynamic loading). This includes suitable protective coatings, but also means that constructive design must avoid gaps and hollow spaces that open outwards, and in which corrosive media can collect. Sufficiently large openings can also prevent the accumulation of the media during operation and contribute to corrosion protection.

Compressive internal stresses from rolling or peening do not prevent corrosion fatigue damage in an active state (material decomposition, pitting corrosion). With passive surfaces (without pitting corrosion), however, improvement is possible.

• Material selection must be supported by suitable, realistic testing methods. For example, testing parameters such as the load cycles and frequency must be considered. These must realistically correspond to the expected operating conditions.

• Sufficiently early and practical overhauling of at-risk parts must be planned and undertaken. This can, for example, prevent dangerous pitting corrosion that would result in a corrosion fatigue-promoting stress concentration.

• Material parameters must be determined under conditions that simulate those during operation as closely as possible (media, loads, time spans).

• The maximum attainable hardness values should not be taken advantage of, if this would dangerously increase the risk of corrosion fatigue.

Chapter 5.4.3 References

5.4.3-1 M. Karim-Khani, "Ermüdungsverhalten randschichtwärmebehandelter Stähle ohne bzw. mit Korrosionseinfluß", Mensch & Buch Verlag, Berlin 1999, ISBN 3-9333-46-55, pages 78 to 83.

5.4.3-2 H.Speckhardt, "Werkstoffverhalten bei Überlagerung von Schwingungsrisskorrosion und anderen spezifischen Korrosionserscheinungen", Bruchuntersuchung und Schadensklärung, Allianz 1976, pages 75-82.

5.4.3-3 M.O.Speidel, "Interkristalline Korrosionsermüdung in Stahl", Bruchuntersuchung und Schadensklärung, Allianz 1976, pages 83-87.

5.4.3-4 Kh.G.Schmitt-Thomas, Th.Kirner, "Form und Mechanismen von Rissen und Brüchen unter gleichzeitiger korrosiver und dynamischer Beanspruchung mit unterschiedlichen Frequenzen", periodical Werkstofftechnik 9, Verlag Chemie GmbH, D-6940 Weinheim 1978, page 407.

5.4.3-5 A.Tross, "Der Mechanismus des Dauerbruchs und seine Beeinflussung durch Korrosion". periodical "Werkstoffe und Korrosion", Volume 20, Issue 11/1969, pages 954 to 962.

5.4.3-6 H.Spähn , G.H.Wagner,"Corrosion Fatigue of Steels", Bruchuntersuchung und Schadensklärung, Allianz 1976, pages 59 to 74.

5.4.3-7 H.J.Kolkmann, "Gas Turbine Compressor Corrosion and Erosion in Western Europe", AGARD-CP-558, proceedings of the conference "Erosion, Corrosion and Foreign Object Damage Effects in Gas Turbines", Rotterdam, The Netherlands 25-28 April 1994, chapter 30 page 30-3.

5.4.4 Hydrogen Embrittlement

5.4.4.1 Fundamentals of Hydrogen Embrittlement

5.4.4.2 Damage due to Hydrogen Embrittlement

5.4.4.3 Measures Against Hydrogen Embrittlement Damage

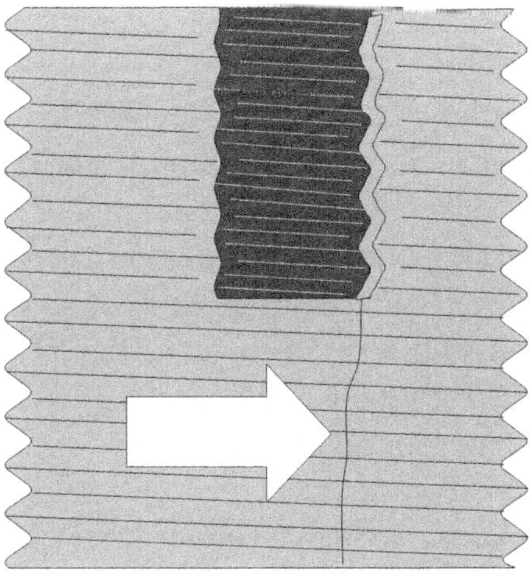

Hydrogen embrittlement (also referred to as hydrogen embrittlement cracking HEC, hydrogen stress cracking HSC, and hydrogen-induced cracking HIC) is a type of fracture that occurs in certain metals due to decreased toughness resulting from the absorption of hydrogen. The brittle failing of the material is due to diffusion processes (Ill. 5.4.4.1-1), and requires sufficiently low strain speeds. Contrary to other types of embrittlement, notching loads are unsuitable for verification (Ill. 5.4.4.1-6). If brittle hydrides are created, as with titanium alloys, then the fracture behavior will be brittle in tests with high strain rates.

The influence of hydrogen can increase the notch sensitivity and crack growth rate considerably (Ill. 5.4.4.1-1). In order to enter into the atomic lattice of a metal, the hydrogen must be atomic, since the molecule is already too large. Because hydrogen atoms immediately combine into molecules, the hydrogen atoms must form directly on the surface of the metal. This can occur in very different ways 5.4.4.1-3). The collection and recombining of hydrogen at flawed areas (such as non-metallic inclusions and inhomogeneities) in the structure creates a tensile stress concentration and causes the lattice to expand. The molecular hydrogen in the metal pores is under high overpressure and "bursts" the structure (Ill. 5.4.4.1-2). The tendency to hydrogen embrittlement usually increases with the strength of the affected materials (Ill. 5.4.4.1-5).

5.4.4.1 Fundamentals of Hydrogen Embrittlement

Influence of materials:

Many technical materials are sensitive to hydrogen embrittlement to some degree, whereby the important influences for the action of the hydrogen can be very different. **Al and Cu alloys** are relatively resistant to hydrogen embrittlement, but in high-strength Al alloys, especially, embrittlement caused by hydrogen action on crack development has been observed. Austenite steels are affected to a certain degree. **Ti alloys, ferrite steels, and Ni-based super alloys** can embrittle strongly (Refs. 5.4.4-1 and 5.4.4-2). In **high-strength steels**, hydrogen embrittlement can cause failures at only 1 ppm hydrogen content above the solubility limit (for iron at room temperature, roughly 10^{-3} ppm), in Ti alloys, at about 35 ppm (solubility limit for titanium is 20-20 ppm). Aside from roughness, development of oxide films, absorption of hydrogen due to impurities, flaws in the micro-structure, and the hydrogen transport related to dislocation movements are all important influences on the hydrogen absorption (amount and distribution).

The greater the strength of a material, the greater its sensitivity to hydrogen embrittlement usually is (Ill. 5.4.4.1-5). At the same time, it must be remembered that the alloy composition, as far as it influences strength, also affects the tendency to crack formation. For example, hydrogen embrittlement damage is especially pronounced in **spring steels and maraging steels**, whereby elastic properties and hardnesses are not influenced. Therefore, hydrogen embrittlement cannot be verified based on these characteristic values. The **dynamic strength** of steels at sufficiently high frequencies also seems to exclude any influence of hydrogen embrittlement (Ref. 5.4.4-13). With Ti alloys, however, there could be an influence if brittle hydrides are present on the surface and/or in the structure.

Fine-grained material is less sensitive to hydrogen embrittlement. This can be seen as due to higher ductility, lower density of "hydrogen traps" (larger sum of grain boundaries), and the more homogenous stress distribution in the micro-zone.

The **heat treatment state** of the material is of special significance. For example, the tendency to hydrogen embrittlement decreases along with the fineness of spherical carbides in steels.

The considerably lower solubility of hydrogen in titanium materials at room temperature than at high temperatures or even in the molten state (welding, casting) means that at higher concentrations with slow cooling, the solution in the lattice and the enrichment at micro-defects leads to the precipitation of hydrides. Hydride development can be suppressed through rapid cooling. In some cases, the hydrogen embrittlement is related to the breaking-open of the hydride platelets (Ref. 5.4.4-3). Therefore, it is understandable that the size and orientation of the hydride platelets is important for the embrittlement effect. The limit values for hydrogen in **titanium materials** are: in α alloys such as TiCu2, 100 ppm H; and in α/β alloys, 125 ppm H (Ref. 5.4.4-5). If hydride precipitations are detectable in metallographic cross-sections, then the hydrogen concentration is considerably greater than 100 ppm. Non-oxidizing acids on titanium surfaces create externally undetectable hydrides, which decompose at temperatures between 200°C and 300°C. The hydrogen that is freed during this process can enter into the metal. The absorption of hydrogen is blocked by oxide coatings. In titanium materials, suitable heat treatment in a vacuum can expel hydrogen (even recombined hydrogen). Relative to degasification (also called de-embrittlement) of steels, the required temperature is high (e.g. 700°C 6-24 h in a vacuum of 10^{-3} - 10^{-4} torr).

Unlike the behavior of steels (Ill. 5.4.4.1-6) **titanium alloys embrittle due to the development of brittle hydrides,** and not due to deformation of the lattice. The hydrogen embrittlement of **α-titanium alloys** (such as TiCu2) is determined by a pronounced tendency to form hydride platelets in the structure. This also causes pronounced brittle behavior during fracture **even at high strain rates.** Both a decrease in notch impact strength and decreased plastic strain in tension tests have been observed. Therefore, verification of hydrogen embrittlement in α-titanium alloys is done with notch impact tests. Time-dependent failure under steady stress is not pronounced.

Even **β- and α/β- titanium alloys** (such as Ti6Al4V) can form hydrides and embrittle due to hydrogen. However, with these alloys the hydride development is not as pronounced as in α alloys. In this case, as with steels, dissolved diffusible hydrogen in the lattice seems to be more embrittling than the hydride development. The hydrogen that influences the crack development forms primarily on the boundary surfaces of the α/β– phase. **Hydrogen diffuses at the crack tip, where brittle carbides form and promote crack growth.** This makes understandable the effect whereby decreasing expansion rates promote crack behavior. Depending on the strength of the affected material, the hydrogen content, and the sharpness of any present notches, there is a static fatigue limit below which no fracture will occur, regardless of the temporal duration of the loads. Verification of hydrogen embrittlement in α/β-titanium alloys is done with notching rig tests.

Influences of surrounding and operating factors:

Hydrogen embrittlement is especially pronounced at "medium" temperatures (Ref. 5.4.4-3). For example, hydrogen embrittlement is only observed in ferritic steels in the temperature region between -70°C and +140°C (Ref. 5.4.4-4). This can be explained by the temperature-dependant mobility of hydrogen. The following influences promote the effect of hydrogen on crack development in metals:

- slow loading speeds, static loads (Ill. 5.4.4.1-6)

- increasing hydrogen content

- type I internal stresses such as contraction strain or conversion stress

- type II and III internal stresses such as with cold forming and hardenings

- low temperatures (but above 70°C)

- multiple-axis stress conditions, such as in notches and material flows

- structural transitions, such as at welds

Surface influences:

The greater the surface roughness, the greater the surface area available for the inward diffusion of hydrogen. Therefore, rough surfaces also increase the tendency to hydrogen absorption and crack initiation. Hardened surfaces (e.g. through shot peening), on the other hand, generally have compressive residual stresses with metal lattice tightening that slows the inward diffusion of hydrogen. If, during hydrogen absorption, the surface relatively quickly precipitates largely hydrogen-proof coatings, it can considerably reduce the amount of inward-diffused hydrogen, relative to cases with slow or porous coating development. This can occur in identical processes merely by changing the precipitation parameters, for example.

Special attention should be paid to **surface contaminants**, if these are to be removed in baths. In one case, **iron oxide deposits on compressor blades made from a high-strength titanium alloy** embrittled due to hydride formation in an HNO_3 bath (Ill. 5.4.4.2-5). This is especially surprising since no major hydrogen formation was expected to occur in this bath.

Illustration 5.4.4.1-1 *(Ref. 5.4.4-3): Hydrogen that has been absorbed by metals can considerably increase the growth rate of cracks (e.g. also corrosion cracks, see Ill. 5.4.4.1-4). Experts worldwide do not agree in their estimations of the role of hydrogen on the embrittlement process. To the author, one, albeit old, process seems to describe the damage process in steels so plausibly, that it will be explained in the following. It must not be denied that at least the explanation of the pore development phase is not considered to be proven by much technical literature (Ref. .5.4.4-4).*

At the crack tip, a typically shaped plastic deformation zone forms (top left diagram), in which a lattice expansion occurs under the influence of the 3-axis tensile stress. Atomic hydrogen diffuses into this area. The hydrogen diffusion occurs (top right diagram, Refs. 5.4.4-3 and 5.4.4-6) in the direction of the greatest tensile stress conditions (largest lattice expansion) present in a "germinal point". These are flaws in the transition of the plastic zone to the elastic zone. The hydrogen collects at the relatively small "germination point". In less hard steels with hardnesses below about 30 HRC, the hydrogen can recombine to molecules that can no longer diffuse into the metal. In this way, hydrogen pressure p_n (10^4 - 10^8 bar) rapidly builds up. When the gas pressure in the expanded (toward the inner crack) pores around the "germination point" in front of the crack has risen sufficiently, the "explosive effect" causes it to combine with the crack tip and increase crack growth.

However, this damage model is unsatisfactory for explaining the behavior of high-strength and case-hardened steels. For example, in these materials, damage has been observed to decrease again after longer times of conditioning (Ref. 5.4.4-14). The hydrogen is trapped at lattice deformations in reversible or irreversible ***"Traps".*** *This model also explains why long times of conditioning diminish the embrittlement effect, since the hydrogen is able to distribute better into the lattice from the reversible traps. However, the reversible traps can also act as hydrogen sources and provide hydrogen to zones with lattice expansions (e.g. in front of a crack tip). This explains the grain boundary cracks typical for hydrogen embrittlement.*

The tensile stresses which cause hydrogen embrittlement must not necessarily be caused by external loads. Internal stresses from heat treatments, welds, coatings, and machining processes are frequently sufficient to cause hydrogen to collect in resting parts, such as during storage. Once hydrogen has collected at the "germination points" and created micro-cracks/pores, the damage will be irreversible.

As long as the hydrogen in reversible traps is atomically dissolved in the lattice, the damage is reversible (bottom diagram); suitable heat treatment (expulsion of hydrogen - degasification) can prevent the embrittlement process (Ill. 5.4.4.1-7).

Hydrogen embrittlement depends on diffusion processes. Therefore, time plays an important role.

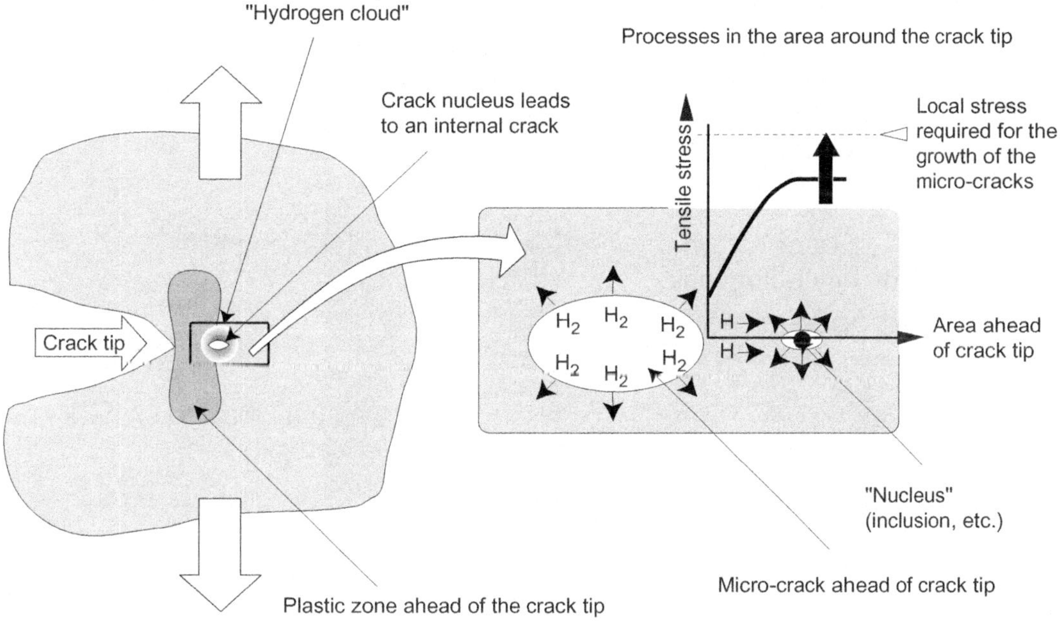

Crack growth and hydrogen influence (according to data from Tetelman and Mc. Evily)

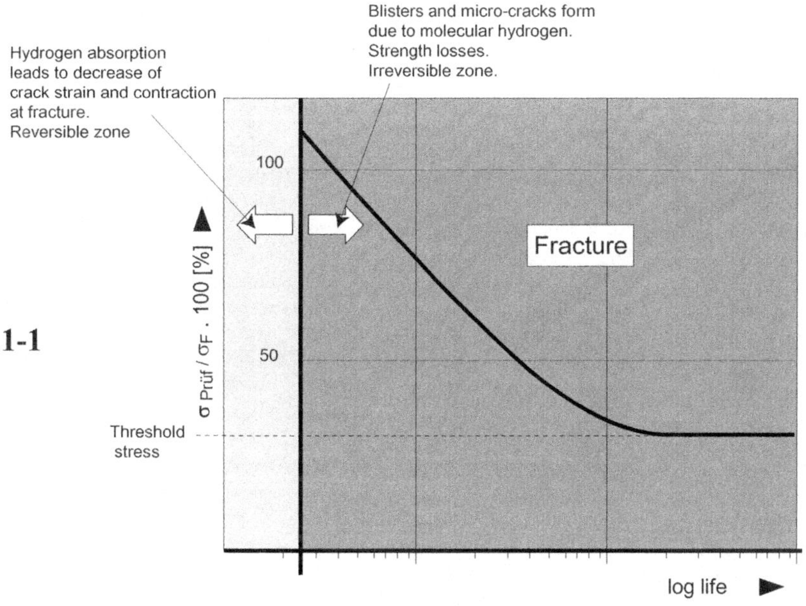

Ill. 5.4.4.1-1

Hydrogen embrittlement of steels in a solution containing H_2S (re. W.O. Bonner)

Effects of hydrogen in the structure of metal

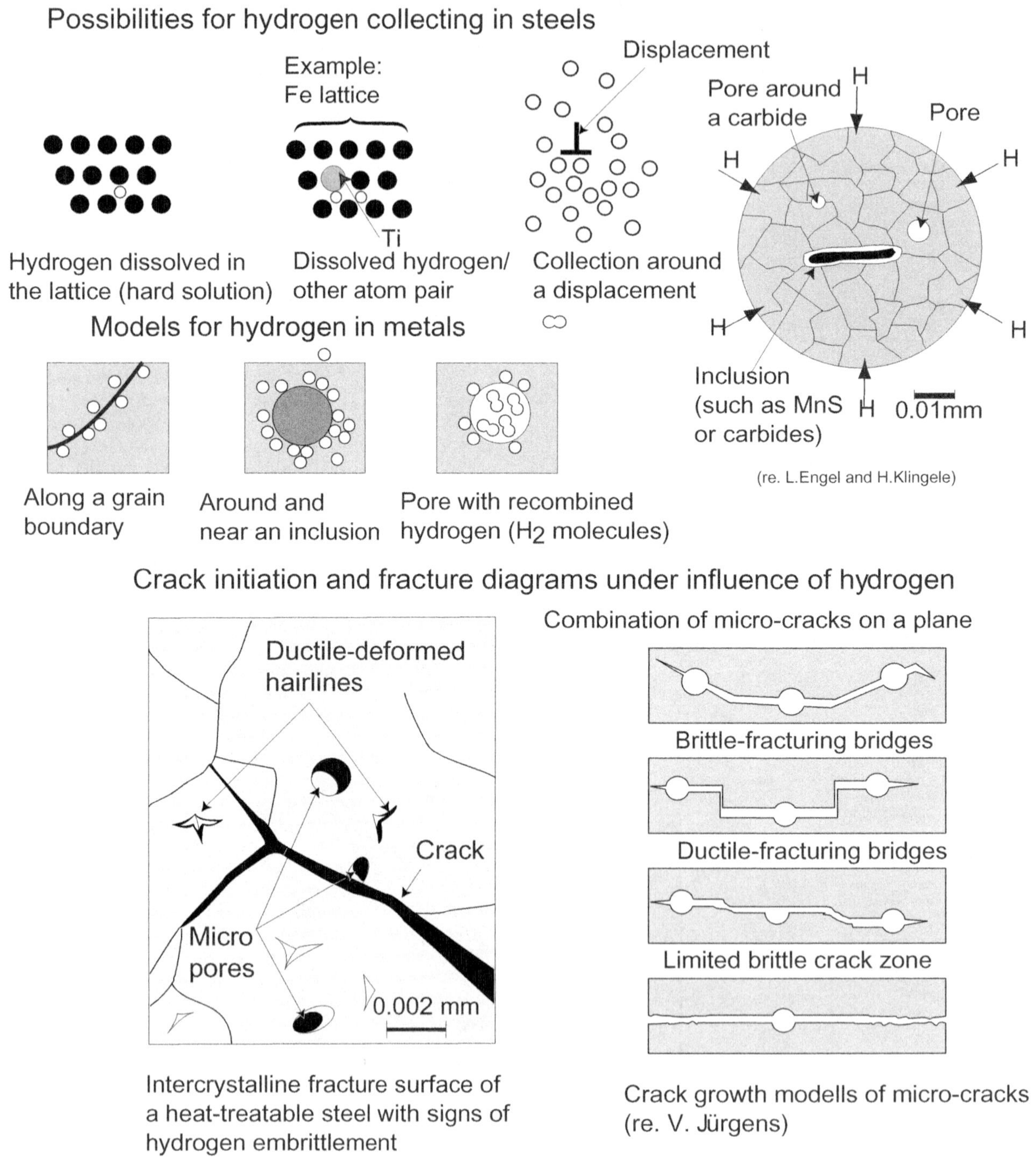

Intercrystalline fracture surface of a heat-treatable steel with signs of hydrogen embrittlement

Crack growth modells of micro-cracks (re. V. Jürgens)

Ill. 5.4.4.1-2

Illustration 5.4.4.1-2: *The crack-promoting effect of hydrogen on metals can be caused by different effects (top diagram) that can occur individually or in combination with one another:*

Deformation of the metal lattice: *This embrittlement mechanism is dominant in high-strength and case-hardened steels (Ref. 5.4.4-14). The metal lattice is deformed by included hydrogen atoms. This reduces the mobility of dislocations, resulting in a loss of ductility. The hydrogen is caught in reversible or irreversible traps, depending on whether or not diffusible oxygen can be released. These traps are caused by lattice inhomogeneities and are located primarily in the grain boundary zone. This explains the intercrystalline crack development typical for hydrogen embrittlement. The hydrogen diffuses primarily in already-deformed atomic lattices. These include, for example, areas that have been cold-deformed under mechanical tensile loading, as is the case at crack tips (Ill. 5.4.4.1-1). But even hardened martensite steels have suitable deformed lattices. These influences are especially effective in high-strength steels.*

Recombination and gas pressure buildup: *The hydrogen can diffuse out of the lattice into flawed zones (e.g. cast porosity, micro-cracks, deformation pores, inclusions, Ref. 5.4.4-6), around which the lattice is expanded due to the stress concentration, e.g. under internal stresses. In less high-strength steels (with hardnesses below about 30 HRC), pore formation is explained by a recombination to molecular hydrogen. This hydrogen gas can no longer escape through diffusion due to its relatively large molecular diameter. In the present or forming hollow spaces, extremely high pressures (top right diagram, Ref. 5.4.4-4) build up with an "explosive effect" which can cause brittle fractures without any major external loads.*
A typical phenomenon is the microscopically small pores on the grain boundaries (with no sign of ductility of the crack honeycombs) of the fractured surface (bottom left diagram), which is REM-detectable evidence that hydrogen influenced the damage mechanism (Ref. 5.4.4-4). Hydrogen embrittlement is usually intercrystalline (bottom right diagram, Ref. 5.4.4-7). However, there are also cases of transcrystalline fractures. Hairlines (so-called crow's feet) are flat crack honeycombs due to the remaining ductility of the material.

Hydride formation: *Hydrogen can be chemically bonded by metals as a brittle hydride. Hydrides in the form of surface coatings or platelets in the structure promote brittle fractures. This embrittlement form is especially well known in Ti alloys (see page 5.4.4-3).*

Illustration 5.4.4.1-3: *Atomic hydrogen can develop on metal surfaces and be absorbed by diffusion under various different scenarios:*

Metal melts ("1"): *At the high temperatures, the steam is split and atomic hydrogen develops at the surface of the melt. Especially large amounts of hydrogen can dissolve in melts, which leads to pronounced crack development in the frozen material.*

Welding ("1a"): *Contact between water and the melt is caused by*
- damp electrodes/electrode coverings resulting from improper storage and failure to bake out a sufficiently short time before welding.
- damp protective gas (e.g. argon). Hydrogen absorption resulting from unsuitable permeable steam feed hoses has been observed.
- part surfaces dampened by a film of condensation water (see Ill. 5.4.4.2-4).

Casting ("1b"): *In this case, also, hydrogen absorption occurs due to water from a damp environment. The causes are similar to those in welding. For example, as with damp electrode coverings, the use of damp shell moulds can provide a hydrogen supply.*

Processing baths ("2"): *In galvanic baths (e.g. coating baths for silver, chrome, nickel, cadmium, or zinc) or chemical baths (e.g. etching baths, electroless coating baths for nickel, etc.) with cathodic hydrogen creation at the part surface (cathodic reduction, $H_2 \rightarrow 2\ H^+ + 2\ e$). These reactions also occur during certain corrosion processes (Ill. 5.4.4.1-4). These include corrosive processes, especially at crack tips, where there is increased possibility for diffusion of the atomic hydrogen due to lattice expansion (Ill. 5.4.4.1-1) (elastic deformations, ductile cold strain).*

"Contact diffusion" ("3"): *Hydrogen diffusion can occur in case of intense metallic contact with a hydrogen-charged surface, such as screwed or shrink-fit part surfaces.*

Thermal decomposition of hydrogen compounds ("4"): *If a compound containing hydrogen, such as water, is heated to high temperatures, it may thermally decompose and result in the creation of atomic hydrogen. This hydrogen can diffuse into the metal walls of, for example, pipes or boilers.*
During case-hardening through carbo-nitriding (Ill. 5.4.4.2-3), the gas decomposes, creating atomic hydrogen (e.g. $3\ Fe + CH_4 \rightarrow Fe_3C + 4\ H$) that diffuses into the part being carburized.

Diffusion of catalytically dissociated hydrogen ("5"): *From the molecular state (e.g. natural gas or hydrogen as engine fuel), dissociation due to a catalytic effect can occur at the surface of a part (e.g. a pipe). The dissociation can be explained by plastic deformations in the micro-zone, in which active fresh metal surfaces are created. These plastic deformations can be expected, for example, under LCF loads.*

External Factors
Hydrogen Embrittlement

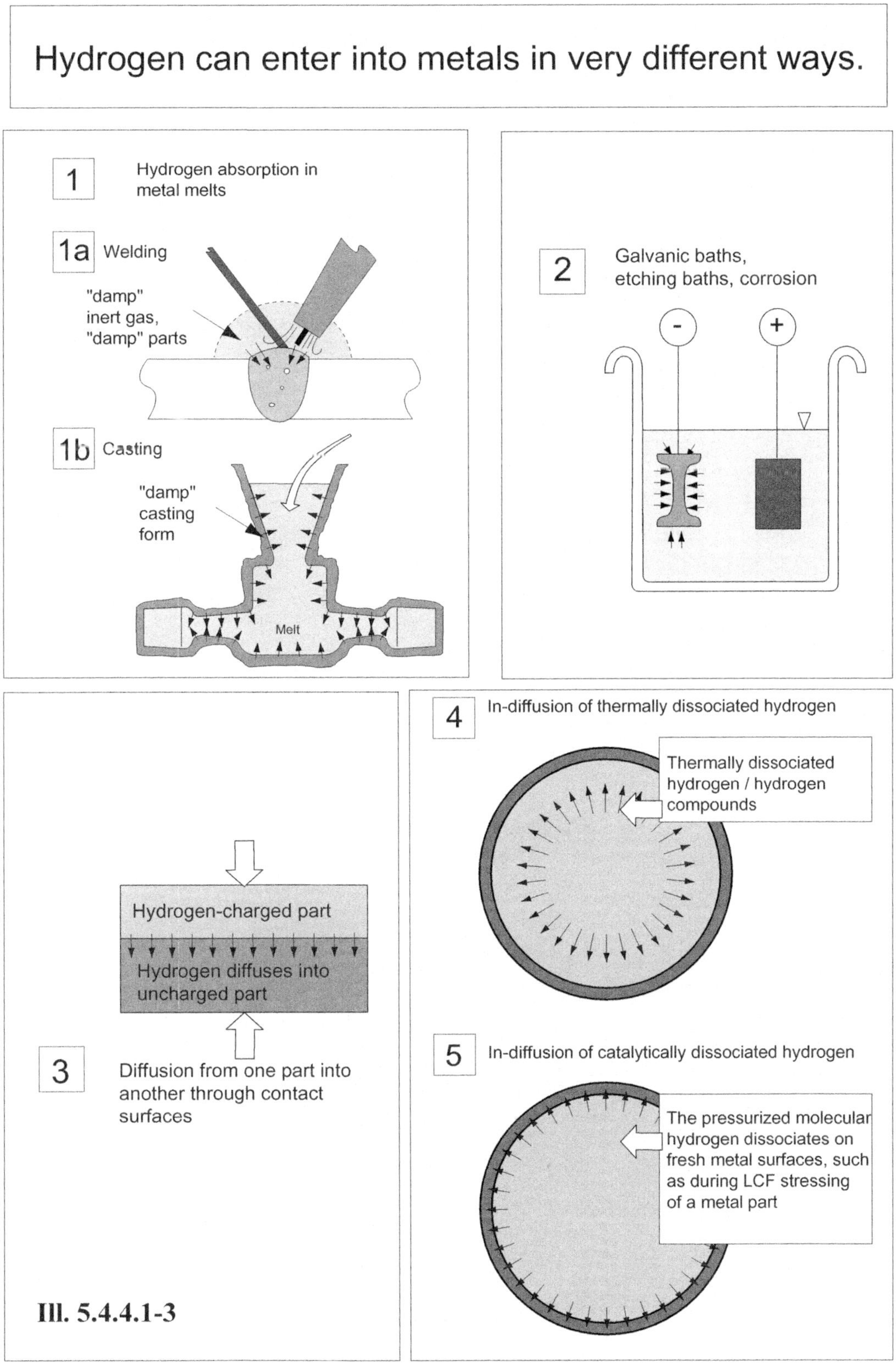

Ill. 5.4.4.1-3

Hydrogen embrittlement plays an important role in corrosion processes.

Anodic SCC

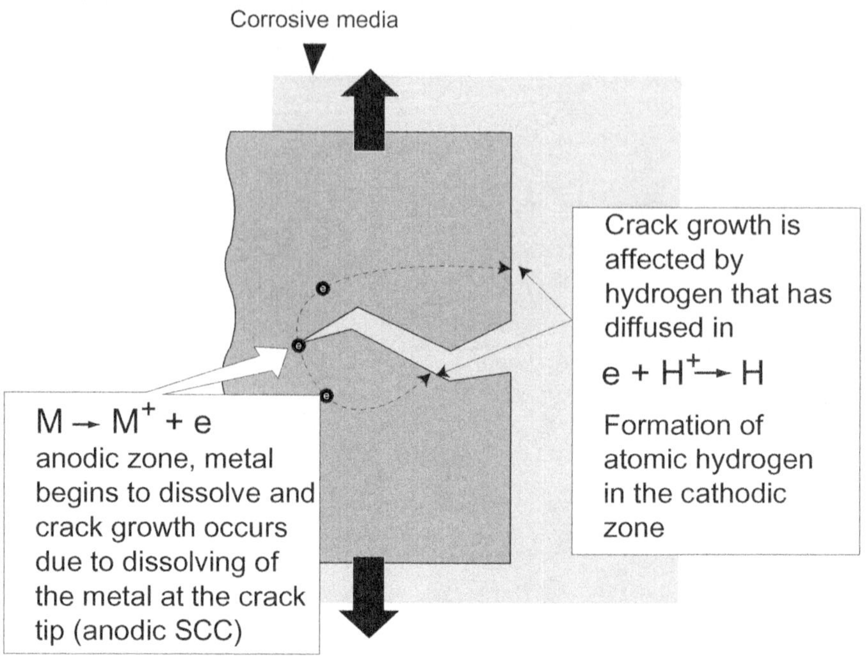

Crack growth due to dissolving of the metal at the crack tip (anodic SCC)

Cathodic SCC

Ill. 5.4.4.1-4

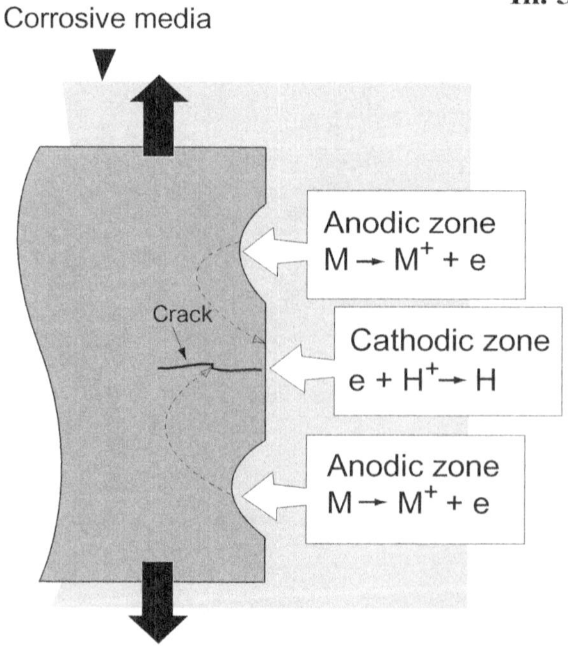

Crack initiation due to hydrogen-induced influences combined with corrosion (cathodic SCC)

(diagram based on B.F.Brown)

***Illustration 5.4.4.1-4** (Ref. 5.4.4-8): Hydrogen embrittlement plays a deciding role in the damage mechanism SCC (Chapter 5.4.2.1). During the anodic partial reaction, the actual erosive/dissolving corrosion process of the metallic material is accompanied by reaction coating development (Ill. 5.4.2.1-2). This creates positively charged metal ions when electrons separate (oxidation process).*

*In the case of **anodic SCC** the material is in a passive state, in which dissolving takes place only at the crack tip. At the surface of the part and crack, the negative charge of the material leads to an anodic partial reaction and atomic hydrogen creation due to the charge equilibrium with the positively charged hydrogen ions of the surrounding electrolyte (reduction process). This hydrogen can diffuse into the part and accelerate crack growth at the crack tip (Ill. 5.4.4.1-1). The anodic and cathodic partial reactions area always coupled. In corrosion, the cathodic partial reaction usually determines the speed.*

*In **cathodic SCC** the material dissolution at the part surface occurs in zones with active corrosion behavior (anodic zones). The dissolving of metal and creation of positively charged metal ions frees electrons, which move to cathodic (passive re. corrosion) surface zones where charge balancing with the hydrogen ions of the electrolyte creates atomic hydrogen, which diffuses into the metal lattice and promotes SCC.*

The creation of passive reaction coatings during the corrosion process can result in the development of protective conversion coatings (passive coatings). This allows the same metal to occur in both the passive and active state, depending on the corrosion conditions.

The influence of hydrogen embrittlement components in SCC can be seen on the fracture surfaces through characteristics such as micro-pore formation, quasi crack surfaces, and "rooster feet" (Ill. 5.4.4.1-2, Ref. 5.4.4-4). The passive state retards corrosion and is therefore usually desirable, but it is also more prone to deep localized corrosive attack, the dreaded pitting corrosion (Ill. 5.4.2.1-6).

*In **Ti alloys**, etching and signs of hydrogen embrittlement were discovered in connection with **the influence of typical aviation-use hydraulic oil above 130°C** (Ref. 5.4.4-13). At these temperatures, the hydraulic fluid evidently forms an acid that corrodes titanium and causes hydrogen embrittlement. According to the literature, only a certain beta-titanium alloy is immune to this damage. This alloy is used on the tail cone and thrust jet of the newest cargo aircraft.*

The absorption of hydrogen can lead to cracking, strength losses, and shortening of life spans.

Crack strain as a measure for the influence of hydrogen on an alloyed heat-treateable steel

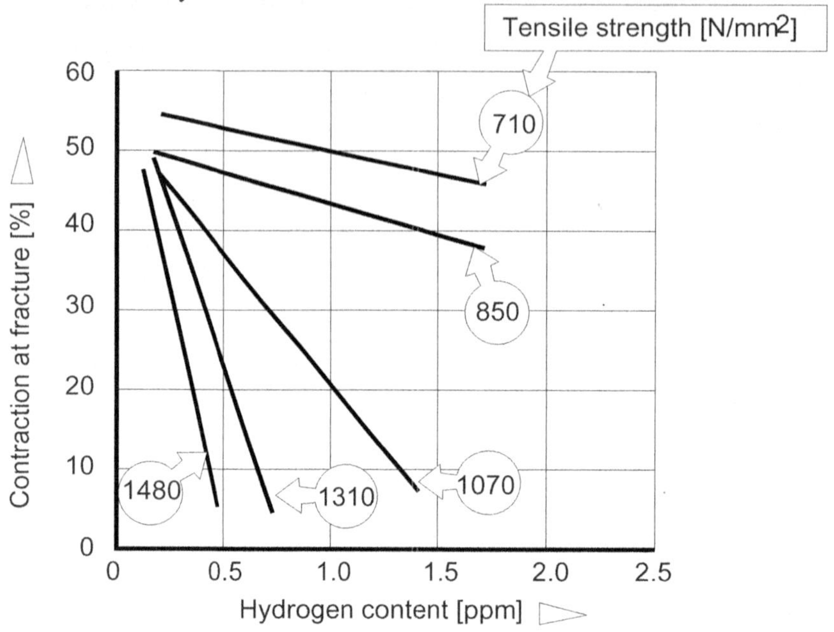

Material: low-alloy heat-treatable steel with σ_B = 1600 N/mm2

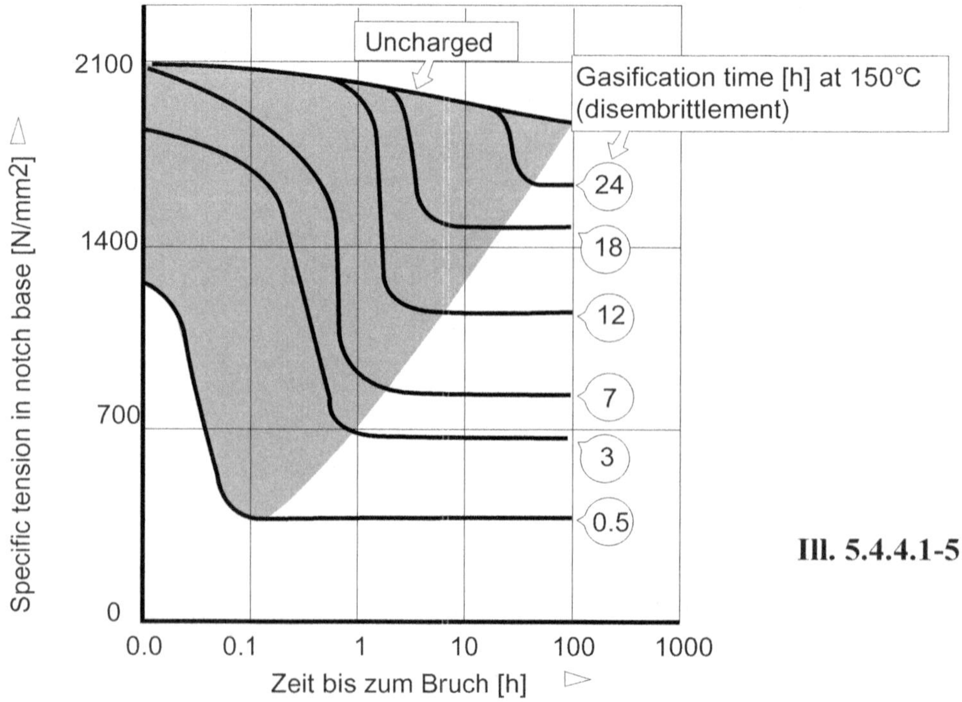

Ill. 5.4.4.1-5

Even if hydrogen embrittlement can cause serious damages, verifying it in technical tests is not an easy task.

Verification of hydrogen embrittlement in materials with no hydride creation

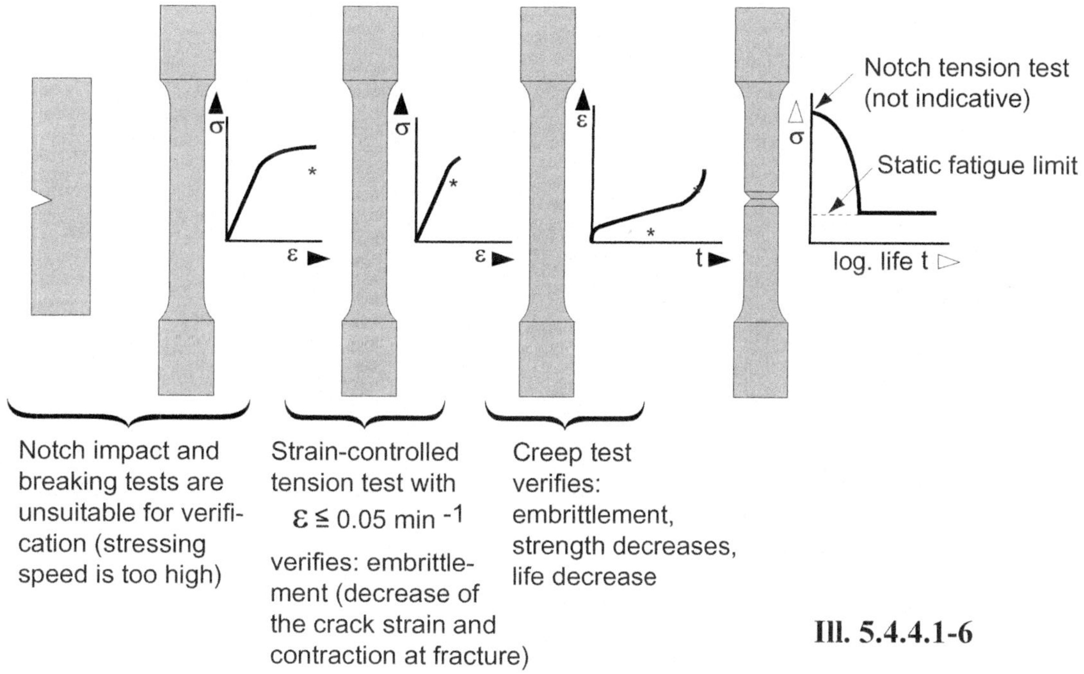

Ill. 5.4.4.1-6

***Illustration 5.4.4.1-5:** Generally, the greater the strength of a material, the greater its sensitivity to hydrogen embrittlement (top diagram, Ref. 5.4.4-9). With increasing hardness, only a fraction of the amount of hydrogen that is diffused into the material is necessary to have a pronounced effect in reducing the contraction at fracture.*

The bottom diagram shows the decrease in strength and life span in notched samples of a high-tempered steel under static loads. The samples were charged with hydrogen and subjected to varying degasification times at a heating temperature of 150°C. The influence of the heating times on the usable strength can be clearly seen.

***Illustration 5.4.4.1-6:** The effect of hydrogen on crack development in **titanium alloys depends on the creation of brittle hydrides** (see page 5.4.4-3). If relatively slow diffusion processes occur in metal (e.g. steels), embrittlement (expansion of the lattice) is delayed (Ill. 5.4.4.1-1). Therefore, hydrogen embrittlement is not usually (except for embrittlement of a-titanium through hydride formation) verifiable through **notch impact tests** or **tensile tests** (usually at high speeds). The high loading speed does not allow sufficient time for embrittling diffusion processes. **A decrease in crack strain and contraction at fracture can be expected from strain-controlled tensile tests with <0.05 mm/mm.min**.*

*This damage in a **creep test** can be recognized by the **decrease in strength and operating life.***

*The **tensile notch test** does not give any clear results.*

On the other hand, influence on the crack development (critical loads for crack initiation and crack growth rates) can be expected from

Disembrittlment of hydrogen-charged steels occurs over several hours at a temperature above 150°C.

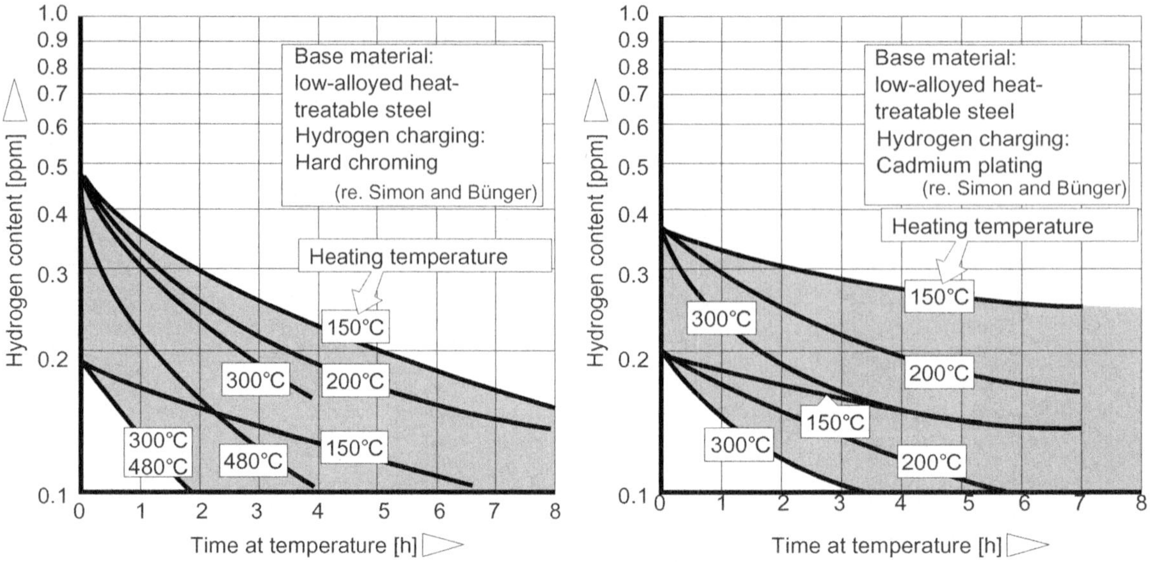

Ill. 5.4.4.1-7

*fracture mechanical **CT samples** with typical high stress concentrations (Ill. 5.4.2.1-6).*

***Illustration 5.4.4.1-7** (Ref. 5.4.4-5): As long as no recombination of atomic hydrogen to molecules has occurred (Ill. 5.4.4.1-1), several hours of heating in air at a few hundred °C can expel the hydrogen or distribute it in the lattice in such a way that it sufficiently prevents the crack development effect. It is also possible to "disembrittle" Ti alloys that have become embrittled through hydride formation (page 5.4.4-3) by pyrolizing the hydrides at high temperatures (about 200°- 300°C). Because of the danger of recombination and/or bonding in irreversible "traps", disembrittlement (see Ill. 5.4.4-2) must be done immediately (within a few hours) after hydrogen absorption in order to be successful.*

Experience has shown that this is where the most mistakes are made, when the parts are allowed to sit for too long before disembrittlement or degasification is forgotten entirely.

The diagrams show that with increasing degasification temperatures and longer hold times, considerable reduction of the original hydrogen content occurs. The remaining amount of dissolved hydrogen depends largely on the amount of dissolved hydrogen before the disembrittlement process. There is no great difference between the degasification behaviors of chromed and cadmium-plated parts. However, it is obvious that cadmium-plated steels are less able to expel hydrogen than are hard-chromed parts. The difference may lie in the gas permeability of the coatings with a different sealing effect.

5.4.4.2 Damage due to Hydrogen Embrittlement

Hydrogen embrittlement can become noticeable in several different ways:

- Decrease in ductility and strength, especially under slow tensile strain and stress concentrations.

- Accelerated crack growth rates in engine parts with cracks. This leads to brittle fractures which can occur even under very low external loads, depending on the amount of hydrogen. These fractures can occur several minutes after hydrogen absorption, or spontaneously after years of operation. Typical damages in aircraft engines are:

- Cracks and fractures in cadmium-coated, hardened/heat-treated steel parts such as thread inserts or disk springs (Ill. 5.4.4.2-1).

- Brittle fractures in cast parts made from high-strength steels (maraging steels, precipitation-hardened steels) in unfavorable material states (Ill. 5.4.4.2-2)

- Probable crack initiation inside titanium forging blanks around pores that were not forged out. This crack development has been confirmed both immediately after forging (when heat-treatment occurred the following day), and on finished heat-treated parts under operating loads.

- Inner crack formation in case-hardened gears (Ill. 5.4.4.2-3) due to hydrogen absorption during carbonizing in a gaseous atmosphere.

- Crack initiation in the welding seam area of low-alloy steels after water was absorbed during welding (Ill. 5.4.4.2-4).

Crack growth due to the influence of hydrogen. The stronger the material, the more sensitive it is.

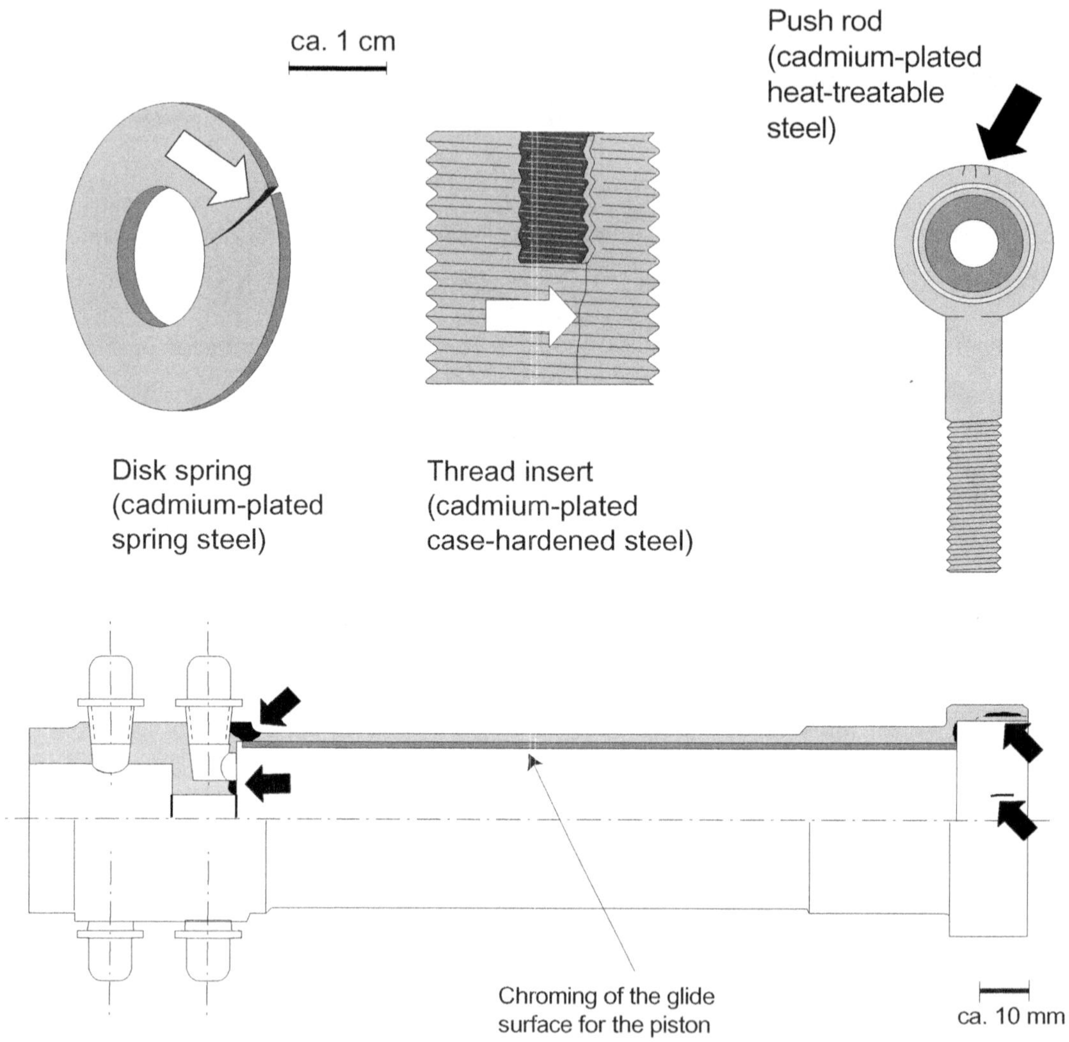

Ill. 5.4.4.2-1 Hydrogen embrittlement in a hydraulic piston made from a high-strength heat-treatable steel

Illustration 5.4.4.2-1: *Typical parts in which hydrogen embrittlement damage is frequently reported are hardened or high-tempered parts made from low- and high-alloy steels with galvanic coatings, especially cadmium coatings (Ref. 5.4.4-15). The supply of atomic hydrogen seems to be especially intense during this process. Typical parts include disk springs, thread inserts, spring washers, plain washers, and threaded rods (top diagrams).*
The lower diagram shows a hydraulic piston from a thrust jet adjuster made from a high-strength heat-treated steel. After a restorative chrome-plating procedure, certain areas

Fracture of a compressor disk due to hydrogen embrittlement.

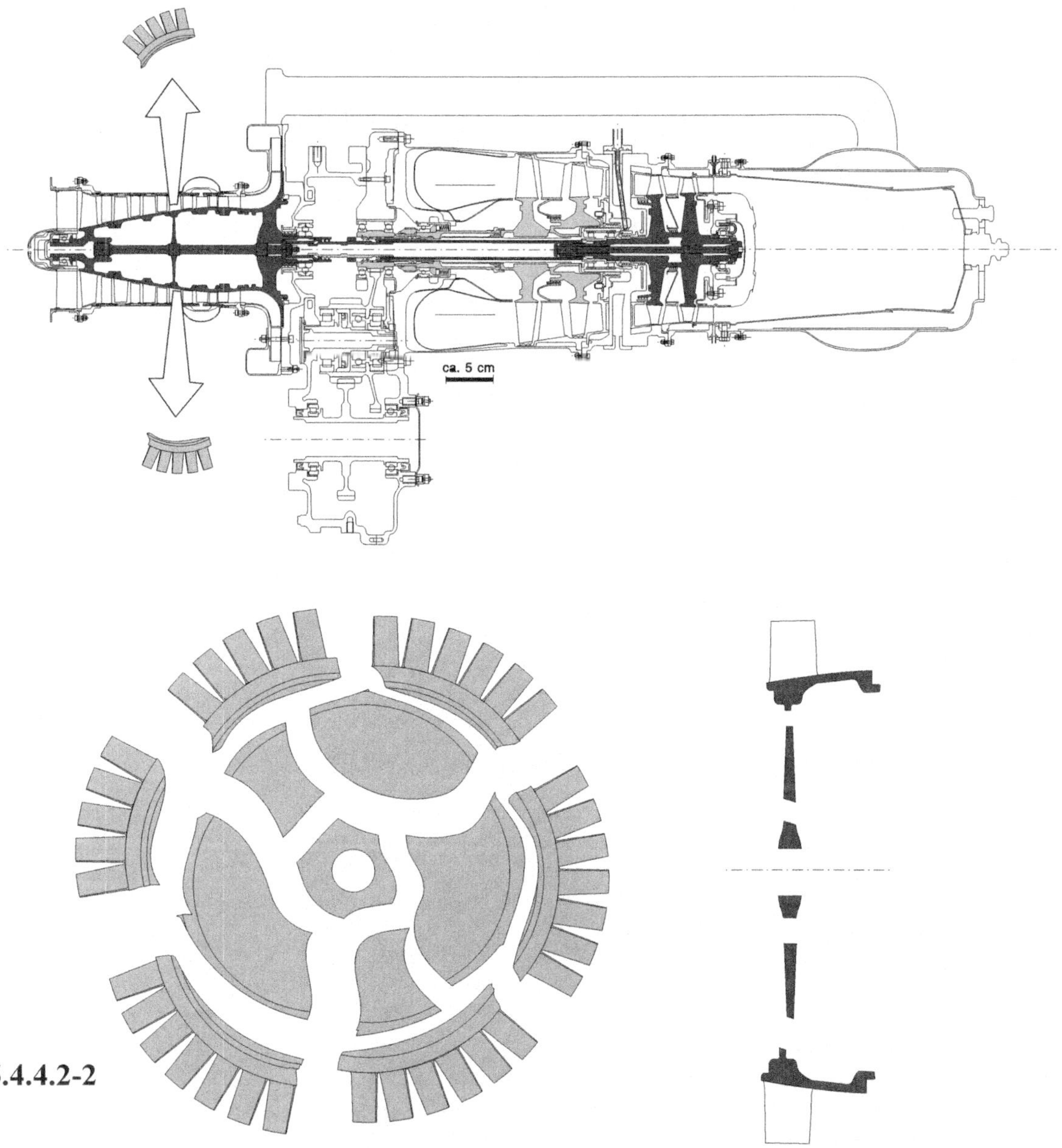

Ill. 5.4.4.2-2

Illustration 5.4.4.2-2: *This integrally cast compressor disk from a helicopter turbine made precipitation-hardened high-strength steel (type 17 4 PH) was in an unfavorable heat-treatment state with too great a hardness. Hydrogen absorption probably occurred in an etching bath during new part manufacture. The disk failure occurred spontaneously during operation, and the disk broke into many pieces (compare with Ill. 5.4.2.1-5). The crack development was transcrystalline with typical signs of hydrogen embrittlement such as micro-pores on material inhomogeneities.*

Tooth failure due to hydrogen embrittlement that was related to case hardening.

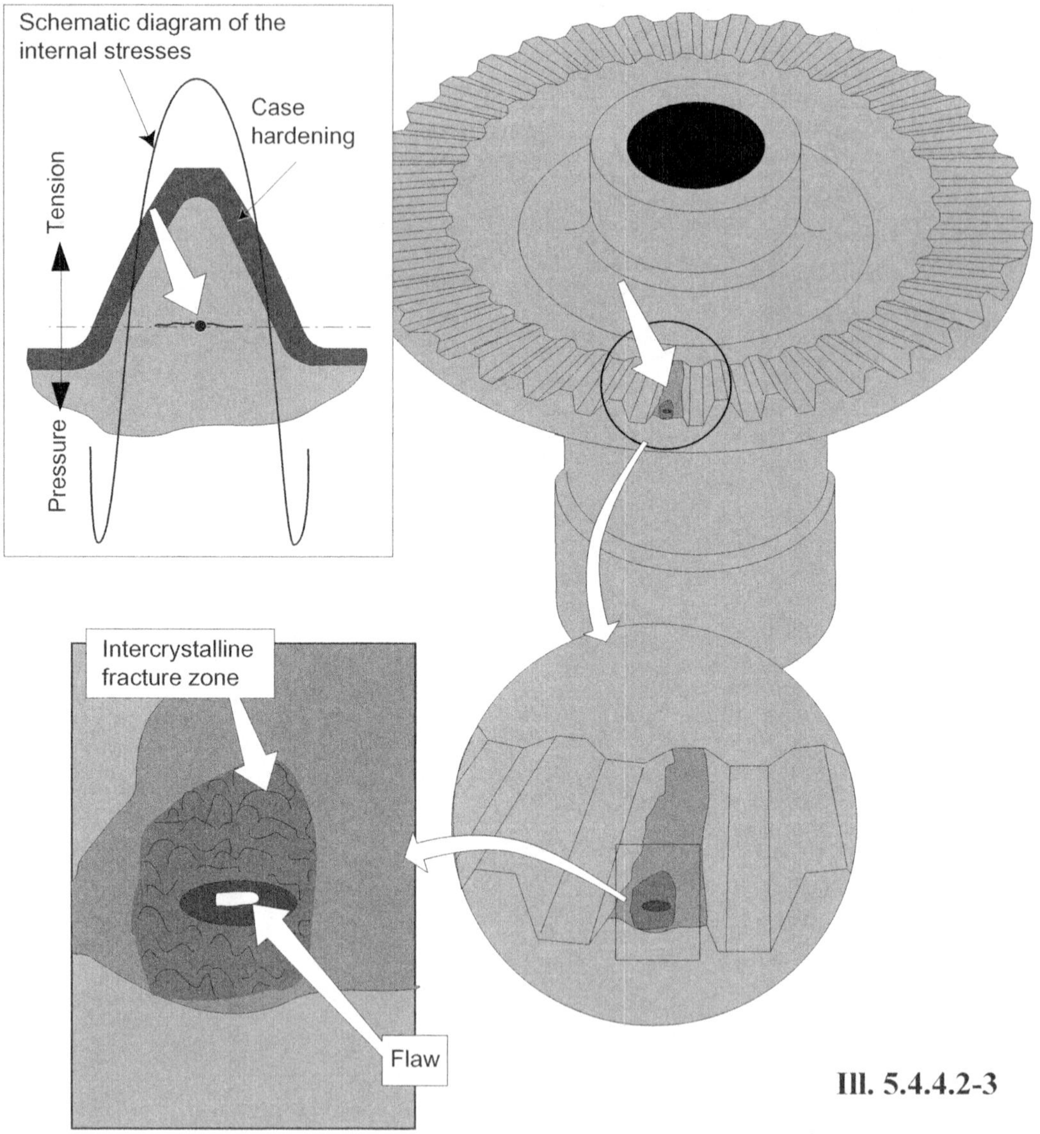

Ill. 5.4.4.2-3

Illustration 5.4.4.2-3: *Teeth fractures on case-hardened gears due to inner crack development following hydrogen embrittlement. The gears evidently absorb hydrogen during case-hardening (see page 5.4.4-8). The damage then seems to occur as follows (Ref. 5.4.4.2-11): the high internal stresses in the case-hardened layer at the tooth surface induce high tension residual stresses in the tooth core (top left diagram). The smaller and more filigreed the tooth, the greater the danger that a circular hydrogen embrittlement crack will develop at internal notches within several hours. An intercrystalline hydrogen-induced brittle fracture then originates in this crack, and can later lead to a dynamic fatigue fracture under the operating loads, destroying the gearing.*

Hydrogen embrittlement in weld seams of compressor and turbine housings made from low-alloy steels.

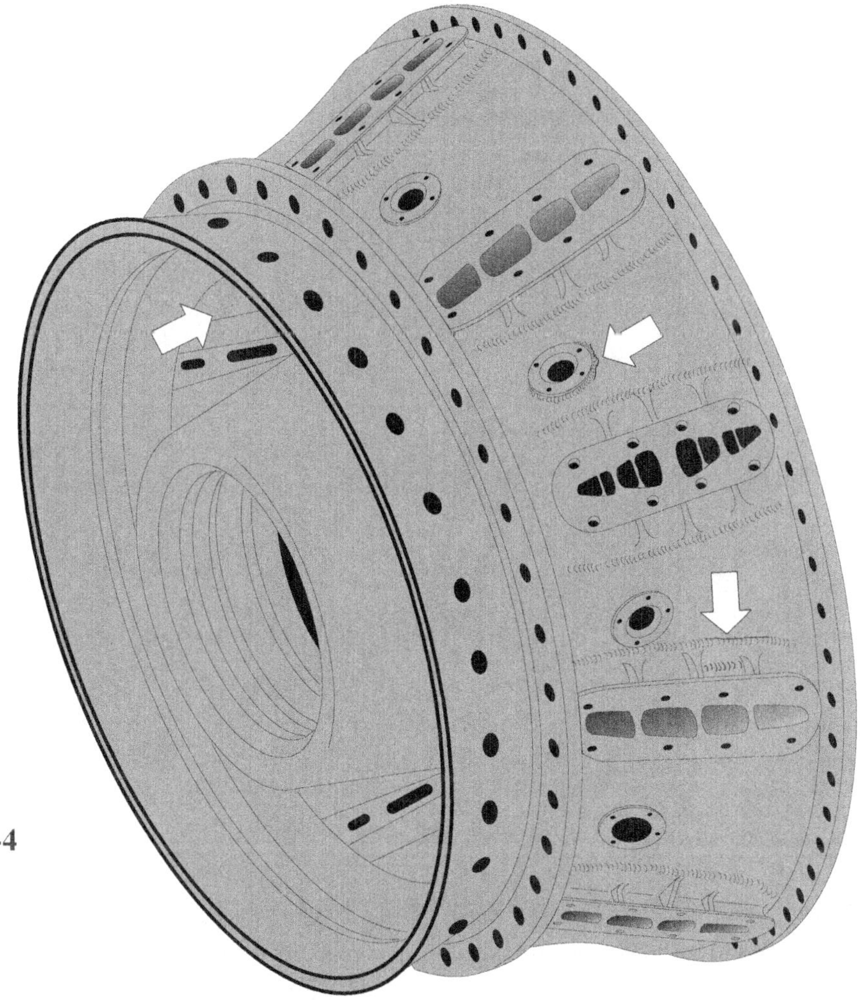

Ill. 5.4.4.2-4

Illustration 5.4.4.2-4: *Older engine types often have compressor housings that are welded constructions made from low-alloy heat-treated steels. Similar housings are found in modern engines near the outlet of the low-pressure turbine. The complex shape of these housings causes them to develop high tension residual stresses that are induced by welding during new part manufacture or repair procedures. If moisture (e.g. in the shielding gas or as conden-sation water) is present during the welding process (see Ill. 5.4.4.1-3), then hydrogen from the moisture can be absorbed by the melt (Ref. 5.4.4-12). This can lead to delayed crack initiation in the weld seam area during storage, manufacturing processes, or operation (arrows show typical crack-prone part zones). For this reason, if this type of crack is suspected, REM inspections of the fracture surface should be conducted first of all, in order to confirm characteristic signs of hydrogen embrittlement (see Ill. 5.4.4.1-2).*

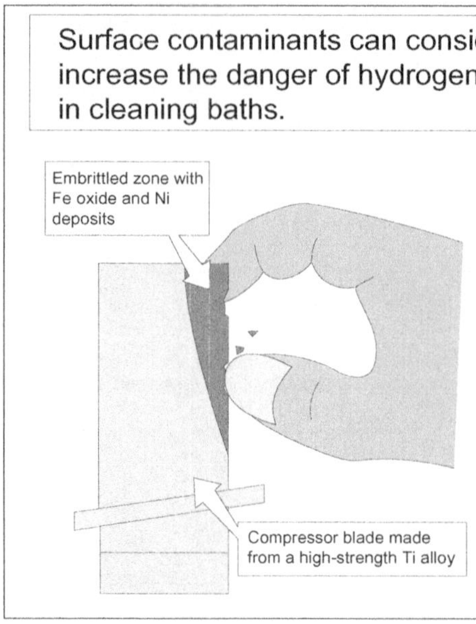

Ill. 5.4.4.2-5

Illustration 5.4.4.2-5: *It has been discovered that, under certain circumstances, hydride formation can cause massive hydrogen embrittlement in high-strength titanium alloys even in oxidizing etching baths (such as HNO_3). Important influences on this include:*

- Deposits of iron oxide. Ni-deposits may also play a part.

- Structural changes due to overheating, such as can occur at the edges of compressor rotor blades during a serious stall (heating-up due to agitation losses).

- Roughening of the surface due to erosion.

In light of this, there is a question as to dangerous hydrogen embrittlement on titanium part surfaces in cleaning baths. Steel shot peening or machining with steel tools (e.g. polishing brush) can leave iron wear products/iron oxide on the part surface.
However, no such cases have been reported yet.

5.4.4.3 Measures Against Damage due to Hydrogen Embrittlement:

There are a number of preventive and corrective measures available to prevent damage due to hydrogen embrittlement:

- **Avoiding hydrogen sources:**

Use of suitable baths (etching baths, cleaning baths, galvanic baths); e.g. "stable" Cd baths or Zn baths (Ref. 5.4.4-12). Testing the baths for any changes (e.g. contaminants, composition) that could cause hydrogen embrittlement. This could be done, for example, by determining the hydrogen content in part-relevant samples or experimentally through suitable toughness tests (Ill. 5.4.4.1-6,).

Preventing hydrogen absorption in metal melts during welding and casting by ensuring a dry environment: drying the welding electrodes (especially with covered electrodes), using dry linings that come into contact with the melt.

Use of dry shielding gas. Avoiding use of water permeable gas feed hoses. Certain synthetic hoses have been proven to be unsuitable.

Drying or sufficiently heating (e.g. above 100 °C) the parts to be welded immediately before welding in order remove any possible condensation water from the surface.

- **Hardening the surfaces before coating** (shot peening, rolling).

- **Optimized coating buildup**, e.g. through preparatory sulfimate nickel-plating followed by application of the actual functional coating.

- **Avoiding coatings** with application processes that have a risk of water absorption (e.g. avoiding Cd coatings).

- Special care **when coating high-strength materials**, especially steels.

- If possible, **no maximization of the hardness/strength potential**, e.g. sufficiently high tempering temperatures with heat-treated steels (Ill. 5.4.4.1-5).

- **Estimation of the embrittlement danger** through testing suitable representative samples (e.g. notch impact sample tests or slow tensile tests).

- **Degasifying ("disembrittling"):** heating the material to about 200-250°C over several hours (one hour to several hours; Ill. 5.4.4.1-7) **immediately after it is charged with hydrogen** (Ill. 5.4.4.1-1, Ref. 5.4.4-10). If, during or after the hydrogen absorption phase, largely sealing **"barrier coatings"** have been separated, then degasification through effusion of the material (escape through the surface) will at least be somewhat impeded.

The degasification effect can then only be expected to balance the hydrogen concentration in the volume of the part.

Chapter 5.4.4 References

5.4.4-1 M.R. Louthan Jr., G. Caskey Jr., "Hydrogen Transport and Embrittlement in Structural Metals", periodical "International Journal of Hydrogen Energy", Vol 1. Pergamon Press, 1976, pages 291-305

5.4.4-2 J.M. Bernstein, A.W. Thompson, "Effect of metallurgical variables on environmental fracture of steels", periodical "International Metals Reviews", December 1976, pages 269 to 287.

5.4.4-3 Tetelman, Mc Evily; "Bruchverhalten technischer Werkstoffe", Stahl und Eisen Publishers mbH, Düsseldorf, 1971, pages 414 to 433.

5.4.4-4 L.Engel, H.Klingele, "Beitrag des Rasterelektronenmikroskops zur Beurteilung wasserstoffinduzierter Brüche", periodical "Archiv Eisenhüttenwesen" 48 (1977) Nr. 10 (October), pages 555 to 560.

5.4.4-5 H.Simon, "Oberflächenreaktionen an Titanwerkstoffen", periodical "Metalloberfläche" 4-1982, pages 211-217.

5.4.4-6 A.W. Thompson, "Microstructural Factors in Hydrogen Embrittlement and Stress Corrosion Cracking" from " Environmental Engineering Material", 1977, pages 3 to 17.

5.4.4-7 V. Jürgens, "Beitrag zur Deutung der Rißentstehung bei verzögerten Brüchen im Schweißgut von Verbindungsschweißungen höherfester Stähle", Dr.-Eng. dissertation, Technical University Braunschweig, 1975.

5.4.4-8 B.F. Brown, NRL Report 6041, Washington D.C. (1063).

5.4.4-9 M.R. Louthan Jr., "Role of Hydrogen in Stress Corrosion Cracking", "Review in Coatings and Corrosion", 2/3 (1977), pages 103 to 119.

5.4.4-10 G.E. Lukas,"Hydrogen Embrittlement in Medium and High Carbon Steels", periodical "Finishing Industries", Jan. 1977, pages 50-52.

5.4.4-11 R.Weiner, H.Fuchs, "Untersuchung zur Wasserstoffversprödung in galvanischen Bädern", periodical "Metalloberfläche", 25. 1971, Volume 1

5.4.4-12 T. Günther, H.Gräfen, "Wasserstoffversprödung von Feinkornbaustählen in Abhängigkeit von der Legierungszusammensetzung, der Gefügeausbildung und der mechanischen Belastung", periodical "Werkstofftechnik", 10, (1979) pages 373 to 390.

5.4.4-13 M.Peters, C. Leyens, J. Kumpfert, "Titan und Titanlegierungen!", DGM, Informationsgesellschaft publishers.

5.4.4-14 S.Beyer, "Wasserstoffversprödungsanfälligkeit hochfester Werkstoffzustände am Beispiel einsatzgehärteter und angelassener Schrauben", Darmstadt, Technical University dissertation 1997, Shaker Publishing GmbH, ISBN 3-8265-2917-0, pages 1-202.

5.4.4-15 "ASM Handbook", Formerly Ninth Edition, Metals Handbook, Volume 13, Corrosion, pages 1037 to 1090.

5.4.5 High-Temperature Corrosion

5.4.5.1 Fundamentals of High-Temperature Corrosion

5.4.5.2 Damage due to High-Temperature Corrosion

5.4.5.3 Measures Against High-Temperature Corrosion Damage

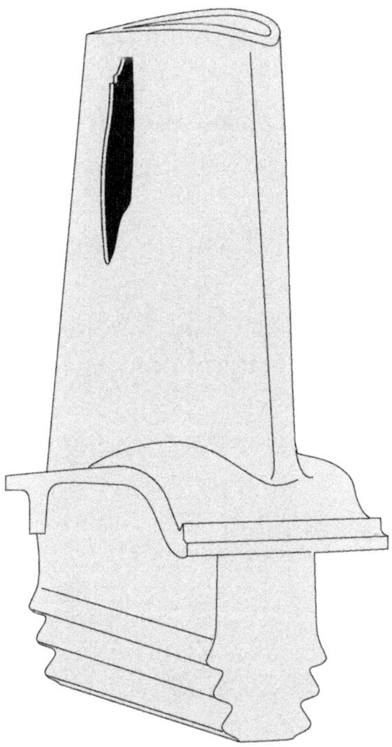

High-temperature corrosion is understood to be the oxidative damaging of metals and metal alloys through hot gases, fluids, and solid materials. It includes all damage caused by reactions with the surrounding atmosphere itself or foreign particles in the atmosphere (Ref. 5.4.5-1, Ill. 5.4.5.1-1). Oxidation and hot gas corrosion (**HGC**, sulfidation) are important life-span reducing damage mechanisms that occur primarily in hot parts in the hot gas flow, such as combustion chambers and turbine bladings. Contaminants in the intake air, in the main air flow in the engine, and in the air system (e.g. cooling air, barrier air, leakage air) not only affect the outer surfaces of cooled parts that are struck by hot gas, but also affect the back and inside surfaces of parts via the cooling air. In modern engines, which use largely corrosion-resistant materials such as titanium and nickel alloys in the compressor area, the corrosion problems that affected the compressor areas of older engine types have shifted into the hot part zone.

5.4.5.1 Fundamentals of High-Temperature Corrosion

Air contaminants in all states of aggregation affect the hot parts through various corrosion-based damage mechanisms. Corrosion is dominant in the lower temperature range up to about 950°C. Above this temperature, oxidation dominates. This results in flat pittings, pitting-like corrosion notches, grain boundary damage, and deep surface changes. This can alter the geometry through material removal or buildup, reducing the material strength through damage, affecting the plastic strain behavior, and shortening the operating life span. One of the specific damage types in hot parts is damage that can be grouped under the term high-temperature corrosion (HTC).

The hot gas typically contains further oxidation-promoting components from the fuel. Damage due to oxidation can be controlled by applying oxidation-resistant coatings (such as Al diffusion coatings, MCrAlY coatings). If aggressive deposits have collected on the affected surface, then rapidly advancing damaging of the base material occurs. Sulfidation has been shown to be especially aggressive (Ill. 5.4.5.1-4). This type of HGC has unique progress dynamics based on what seem to be relatively minor amounts of sulfur.

If ceramic thermal insulation layers are used on hot parts (Ills. 5.4.5.2-4 and 5.4.5.2-5), then special attention must be given to (substrate) oxidation and/or reactions with deposits.

Types of high-temperature corrosion relevant to engines, and their technical terms:

Oxidation: With the operating temperature ranges of modern engines and their typical structural materials, oxidation leads to creation of conversion coatings made from oxides of the base material, such as Al oxide and Cr oxide.

If these brittle oxides are erosively removed or damaged by strain (e.g. heat strain and volume changes, cracks and spalling; Ill. 5.4.5.1-2) then oxidation can progress until it unallowably weakens cross-sections and causes the affected part to fail.

Catastrophic oxidation is a type of oxidation with material loss that occurs due to the creation of liquid or volatile oxides. A typical example of a liquid oxide is vanadium pentoxide with a melting point of about 660 °C. Above 1000°C with high oxygen partial pressure, Cr oxides will largely evaporate. Al oxides, on the other hand, will not noticeably evaporate.

Coking: coking is an internal corrosion type that is based on the formation of carbides that result in a lack of those alloy components that create protective coatings. Damage through coking can occur around injection nozzles and the surrounding combustion chamber areas, especially underneath soot and coke deposits.

Metal Dusting: this special form of coking has been observed in Fe-, Ni- and Co-based alloys under constant coking conditions. It can cause heavy material removal. For this reason, this damage type is also called **catastrophic coking** (Ref. 5.4.5-1). In an intermediate stage, carbides form as internal corrosion products. Oxides, which are the end product of the damage process, typically fall off of the surface as dust.

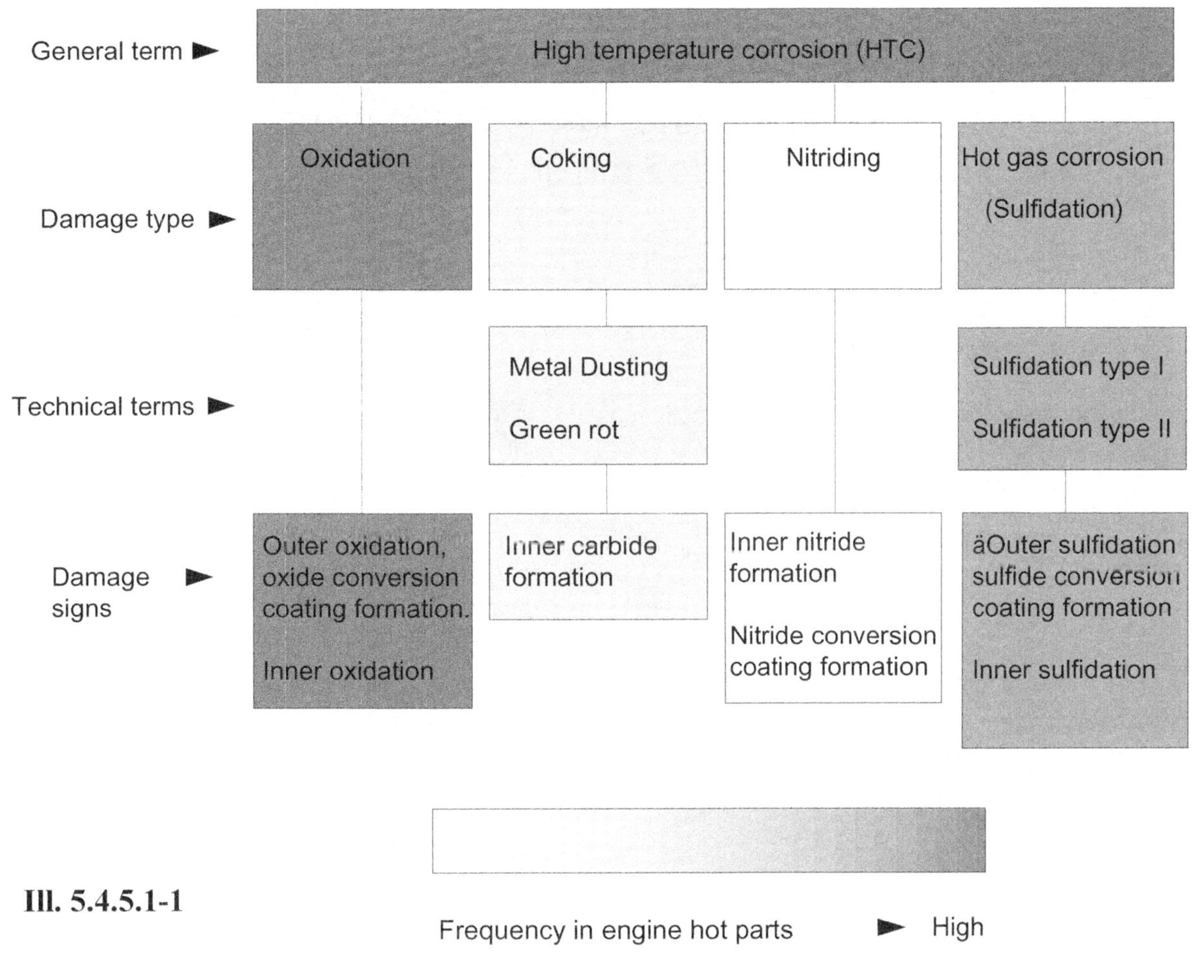

Ill. 5.4.5.1-1

Green Rot is also a type of coking which occurs, however, under alternating coking and oxidizing conditions (Ref. 5.4.5-1). The bonding of Cr prevents the creation of a protecting Cr oxide conversion coating, leading to heavy oxidation of Fe components. Extreme material removal was observed in one case in which, under low oxygen exchange, a Ni based alloy was bolted down with bolts made from Fe based material. This resulted in the Ni based alloy being seriously damaged through material removal with clear green discolorations.

Sulfidation (hot gas corrosion HGC, sulfidation): this oxidation process occurs under the effect of sulfur and, in oxidation-resistant materials, this process is several orders of magnitude faster than normal oxidation (Ill. 5.4.5.1-4). Sulfidation can be initiated and further accelerated by small amounts of silver (Ref. 5.4.5-2, Example 5.4.5.2-1). For this reason, some manufacturers only allow the use of non-silver-coated bolts and nuts in certain hot parts.

Oxidation acts as a protective conversion coating to extend the life span of hot parts, but in its non-protecting state it can act to reduce the life span.

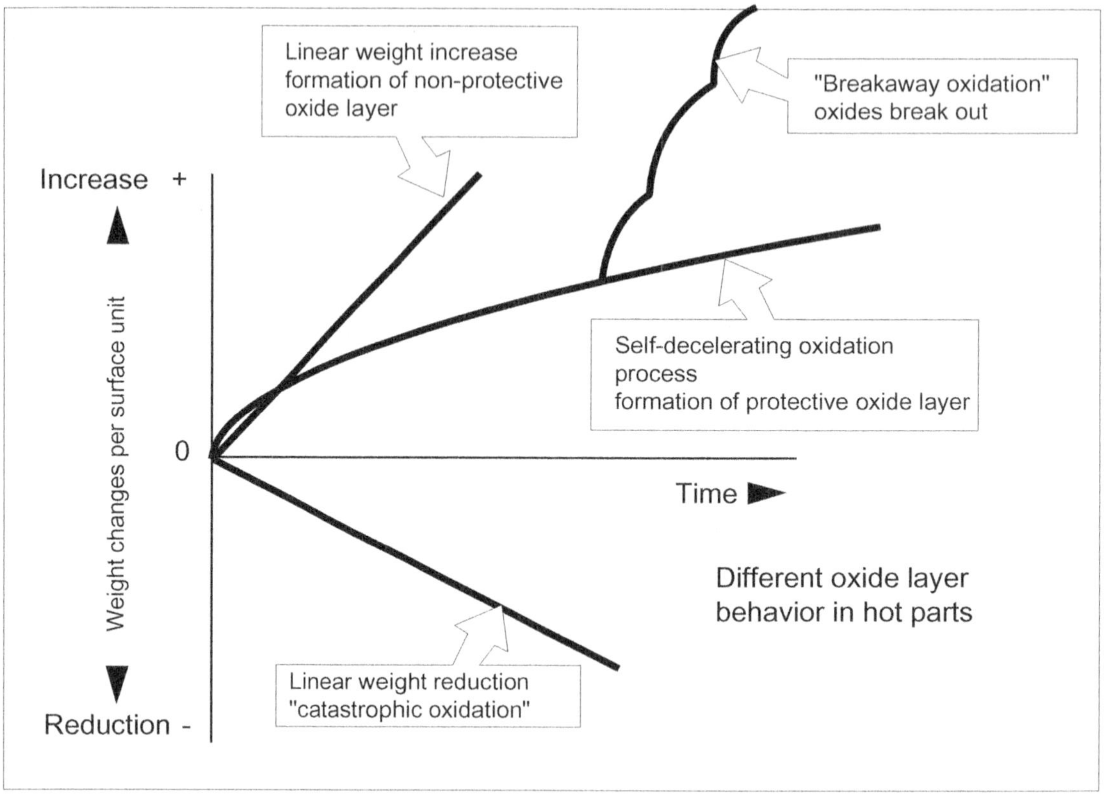

Different oxide layer behavior in hot parts

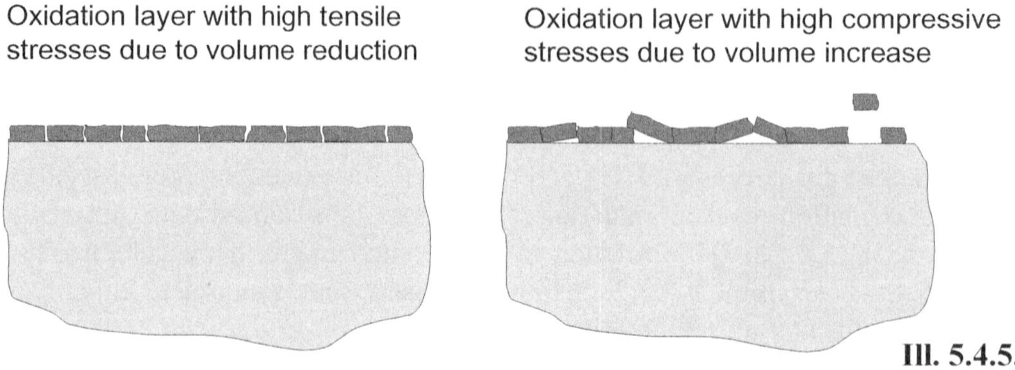

Ill. 5.4.5.1-2

Diagrams of non-protective oxide layers due to crack formation

Illustration 5.4.5.1-2 *(Ref. 5.4.5-1): Based on the mass changes, conclusions can be made regarding protecting or damaging properties of oxide coatings (top diagram). Mass increases that do not level off indicate progressive oxidation due to an oxygen-permeable (diffusion, porosity, and/or cracking) oxide layer.*

The lower diagrams show models of the different damaging mechanisms of brittle oxide layers, where either volume increases induce compressive stress or volume losses induce tensile stress that exceeds the coating strength. As soon as crack initiation occurs in the oxide coating, there is a danger of substrate oxidation, which causes the oxide coating to break off, even if there are tension residual stresses present.

Illustration 5.4.5.1-3 *(Ref. 5.4.5-3): There are three primary damage mechanisms in hot parts, and they can occur in any combination:*

- Solid and liquid deposits build up and cause geometric changes, affecting the aerodynamic performance of profiles or block cooling air bores. There is no shrinkage of supporting cross-sections. The built-up deposits cause a surface-dependent mass increase in the part.

- Solid and liquid particles cause erosive material removal and geometric changes combined with weakened supporting sections and changed aerodynamic properties. The part experiences a mass increase due to the "loading effect (Ill. 5.3.1-7). Progressive mass loss occurs after an incubation period.

- The reaction with gases and solid and liquid deposits (melts) causes strength losses without necessarily reducing the size of sections or changing their geometry. The reaction with surrounding media is accompanied by an almost linear reduction in mass due to a relatively slow loss of surface material.

Combinations of these damage mechanisms can considerably increase the damage relative to a single effect. If, for example, a damaging reaction that creates erosion-sensitive surface layers is combined with erosive loads, then fresh metal surfaces are continually being created, which in turn can promote reactions. This can result in a self-increasing damage process.

The most important types of atmosphere-related damage mechanisms in hot parts.

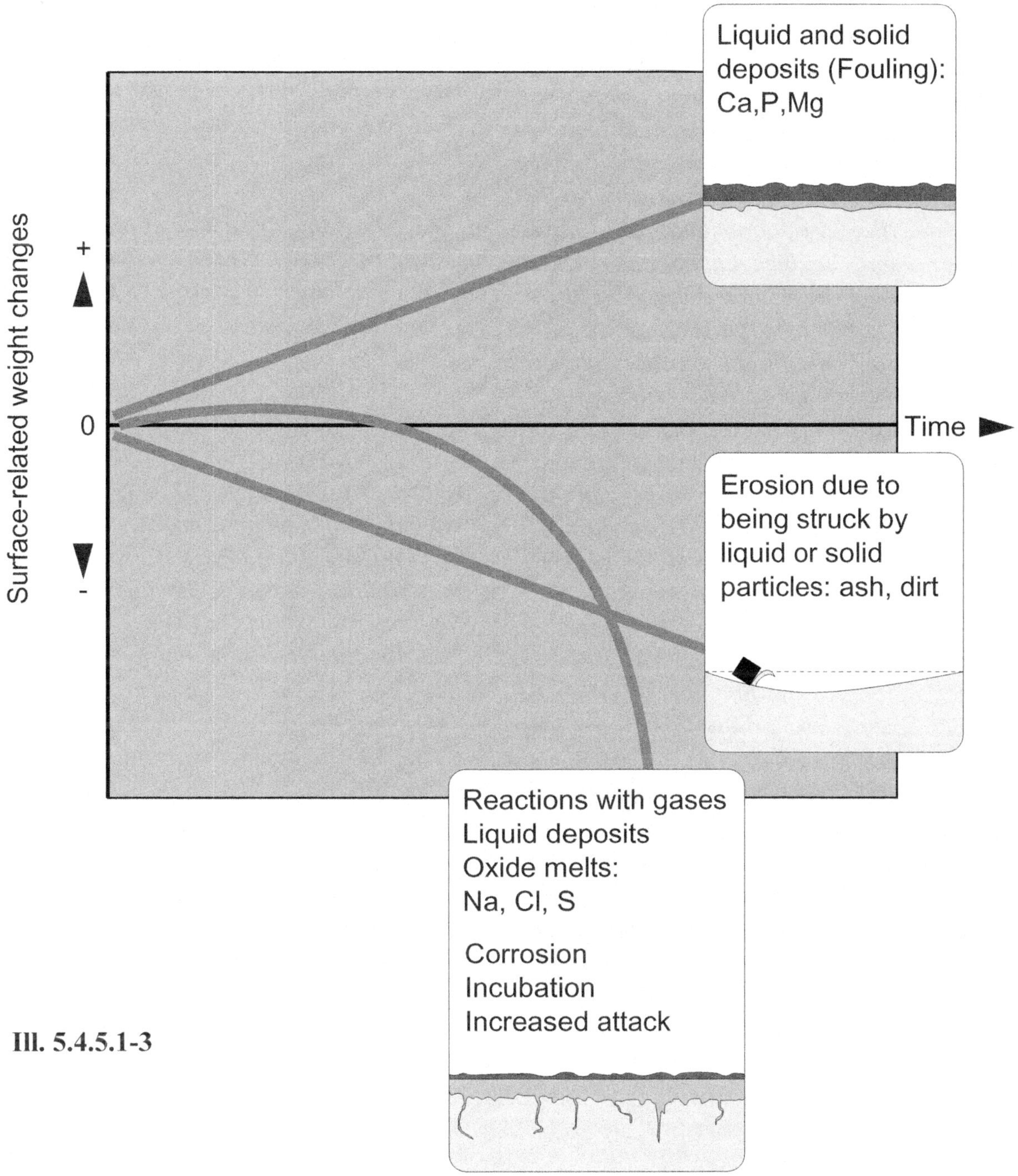

Ill. 5.4.5.1-3

External Factors
High-Temperature Corrosion

Illustration 5.4.5.1-4: *There are two different types of damage processes with sulfidation. Type I occurs above 800°C and Type II occurs between about 600°C to 800°C. These temperatures are not determined by the affected base material, but by the composition of the aggressive deposits. The primary damage mechanism is destruction of the protective oxide conversion coating, which lets in damaging salt melts. The components of damaging salt melts are sodium, vanadium, magnesium, calcium, and sulfur. Because engines burn relatively clean fuel, corrosion-inducing deposits generally come from contaminants in the intake air. Because sodium and sulfur have a special influence, engines that are in a marine atmosphere and/or being acted upon by dusts containing sulfur (such as dusts containing gypsum) are especially threatened. Experience has shown that silver, such as that on silver-plated bolts, can seriously accelerate the sulfidation process (Example 5.4.5.2-1), whereby it probably acts as a catalyst. Silver deposits can evidently be created during standstill in the gas turbine through condensation water causing watery corrosion and acting on surface contaminations. These watery solutions are then transported to other part zones (e.g. flange lugs of the turbine disks). Evidently, corrosion-initiating silver again precipitates from these solutions. Experience has shown this corrosion process to be decisively dependent on the surrounding conditions. Pronounced damage of this type has only been reported in engines that were run in testing rigs in the vicinity of a chemical factory. Here, as well, it should be said that consciously introduced contaminants in the intake air (pollutant burning) must be specifically tested for potential effects such as the ones mentioned. The* ***sulfidation damage mechanism*** *can be explained as follows:*

- *The base material (e.g. Ni-based) is normally protected by a compact Cr_2O_3 coating, which sulfur cannot penetrate.*

- *However, different mechanisms, such as the chemical reaction of the oxide coating with salt melts, can allow sulfur to penetrate the coating.*

- *In this process, the metal ions in the salt melts change places with the Cr and "loosen" the Cr_2O_3 coating, allowing the sulfur to diffuse inward.*

- *Once the sulfur has entered into the base material, it prevents regeneration of the compact protecting Cr oxide coating, so that the salt melt is no longer necessary as a "door opener".*

- *As a result, the base material is destroyed by accelerated oxidation.*

Evidently even very small amounts of sulfur are enough for this damage process, since it only acts on a narrow zone at the transition to the base material and promotes the development of large amounts of Ni oxide. Even small amounts of nickel sulfide are sufficient for transporting the oxygen. Metallographic verification is correspondingly difficult. A non-destructive method for verifying sulfidation in exclusively externally accessible hollow structures (e.g. turbine rotor blades and stator vanes in their assembled state, Ill. 5.4.5.2-1) is ***magnetoscopic testing****. It detects larger amounts of magnetic depleted base material which consists up to 90% from Ni and Co. Nonmagnetic is instead the dense oxide and the unaffected base material.*

External Factors
High-Temperature Corrosion

Even small amounts of sulfur are sufficient to cause catastrophic sulfidation attack.

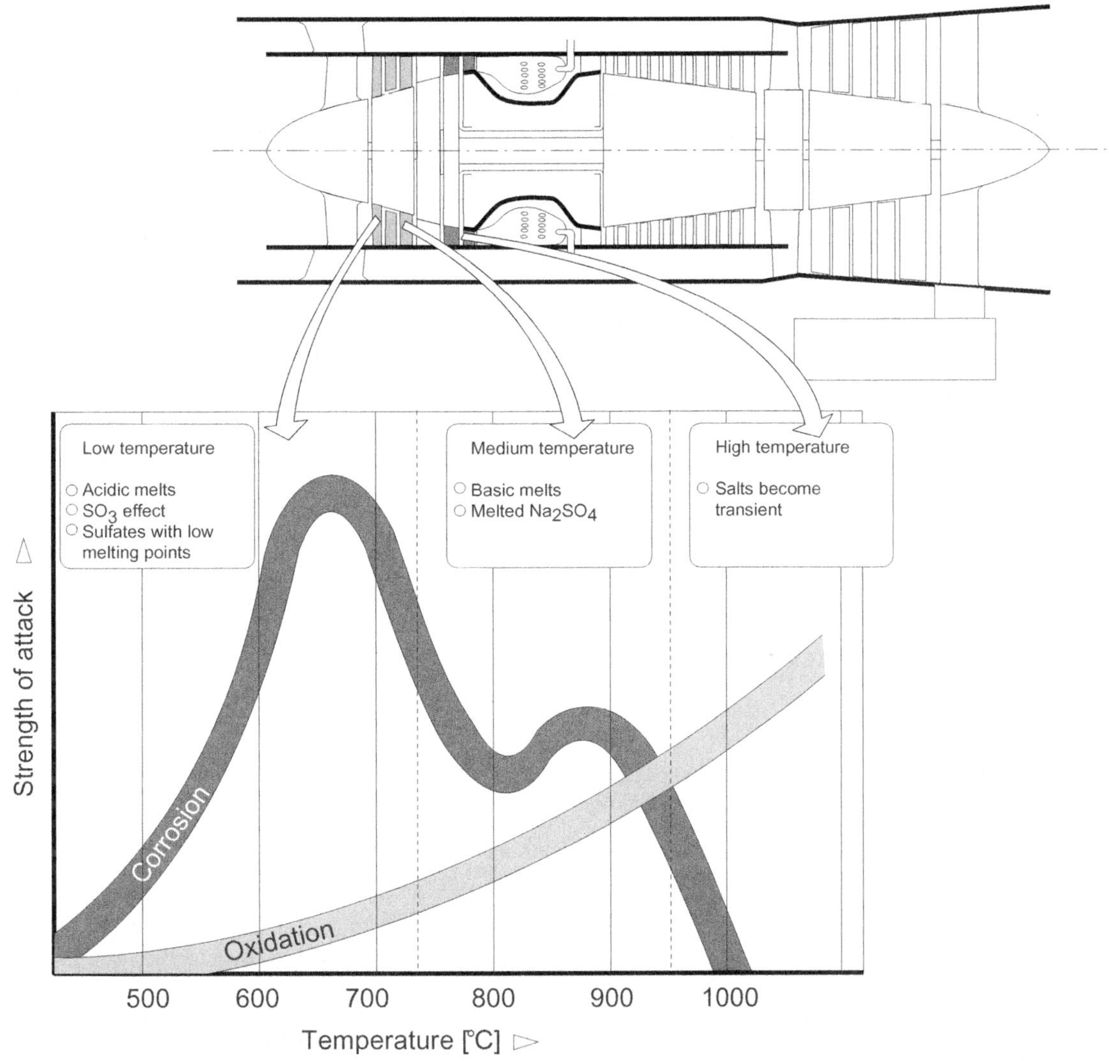

Typical sulfidation symptoms in a metallographic section ▷

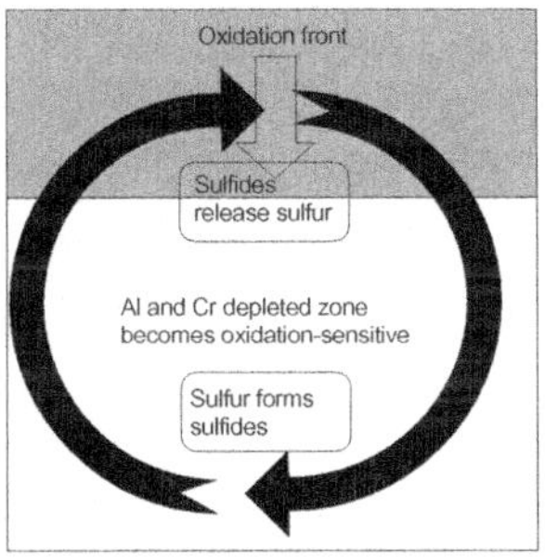

◁ Schematic depection of the sulfidation processes

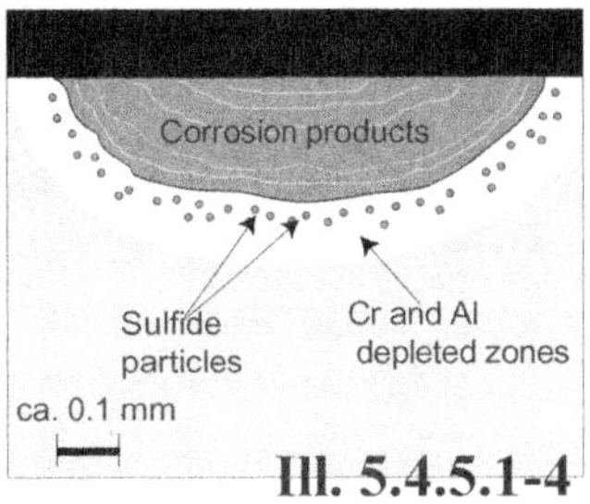

Ill. 5.4.5.1-4

5.4.5.2 Damage due to High-Temperature Corrosion (HTC)

HTC in turbine rotor blades

Experience has shown that overly high temperatures cannot be completely avoided in limited areas of the hot parts. Even the normal temperature fluctuations in the gas flow over the circumference and in radial direction in the high-pressure turbine (HPT) can be considerably greater than 100°C. This puts extreme thermal stress especially on the inlet stator vanes (Ill. 5.4.5.2-1). But the downwind HPT rotor blades with high local temperatures will need oxidation protection, despite the temperature-averaging effect of rotation. The life span of these oxidation-resistant coatings, and therefore also the parts, depends on the emaciation due to oxidation. If the oxidation-resistant coating is worn through, the accelerated damaging of the base material ensues.

Sulfidation (Ill. 5.4.5.1-4) occurs in part zones with temperature ranges suitable for these damage types. It can affect entire blade cross-sections and thoroughly damage them (Ill. 5.4.5.2-2). There is an especially high risk in poorly ventilated hollow structures (Ref. 5.4.5-4) such as, for example, are typical in thick blades which can have "windows" in order to reduce weight. Other susceptible surfaces are those on which dust can easily settle, such as leading edges of stator vane shrouds.

If ceramic thermal barrier coatings are used, they are prone to spalling after longer run times due to substrate oxidation. These coatings, which have necessary functional cracking, can also be burst open and spall by melted deposits that enter into them an then cool (Ill. 5.4.5.2-5).

HTC on honeycomb seals

A special problem is long-time damaging of honeycomb seals. These are used in housings across from rotor blade tips and on the inner stator vane shrouds across from the intermediate stage labyrinths of the rotors. They consist of a thin, honeycomb-like structure made from sheet metal and are designed to abrade upon contact with the labyrinth fins in order to minimize clearances, at least during certain operating conditions. The sheet metal honeycombs are only a few tenths of a millimeter thick and subject to long-term influences of hot gas; especially the seals across from the rotor blade tips can be extremely damaged due to oxidation and lose their strength. This causes large surface areas to break out around the circumference, leading to increased clearances and leakages. If this type of damage seems possible, then suitable materials and/or coatings must be used on the honeycomb structure. A suitable pre-oxidation of the new parts offers a certain amount of improvement. Whether or not this is sufficient, however, needs to be determined in each specific situation.

Typical oxidation symptoms on turbine blades in the high-pressure zone

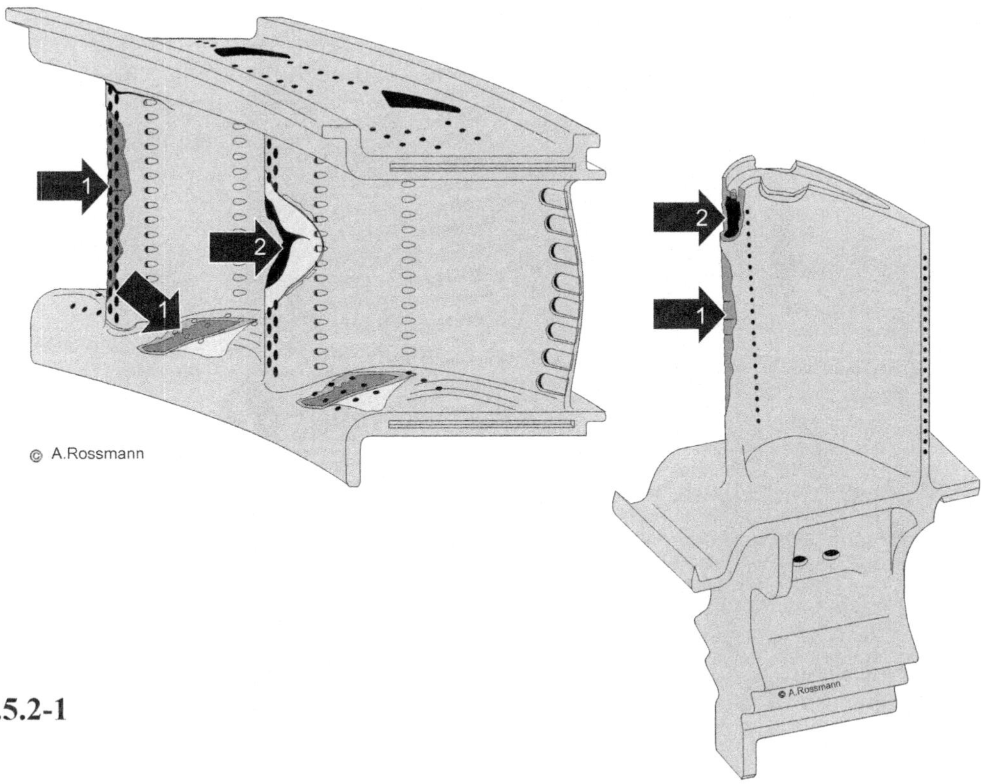

Ill. 5.4.5.2-1

Illustration 5.4.5.2-1: *The inlet stator vanes of the high-pressure turbine and the blades of the first rotor stage are usually subject to extreme temperatures, despite intense cooling measures. If the normal oxidation-resistant coating is emaciated ("3") or has suffered considerable thermal damage (e.g. spalling and rolling up), then an accelerated attack on the base material occurs, and can be seen in a rough and cracked surface ("orange peel effect", arrows "1") that appears dark and/or green, depending on the oxides that form. If the blade wall has been oxidized through ("2"), hot gas can enter into the cooling air ducts, causing cooling air losses and the danger of serious consequential damages not only to the directly affected part, but also to other parts (such as those that are connected to the same cooling air supply).*

Typical internal sulfidation on turbine blades

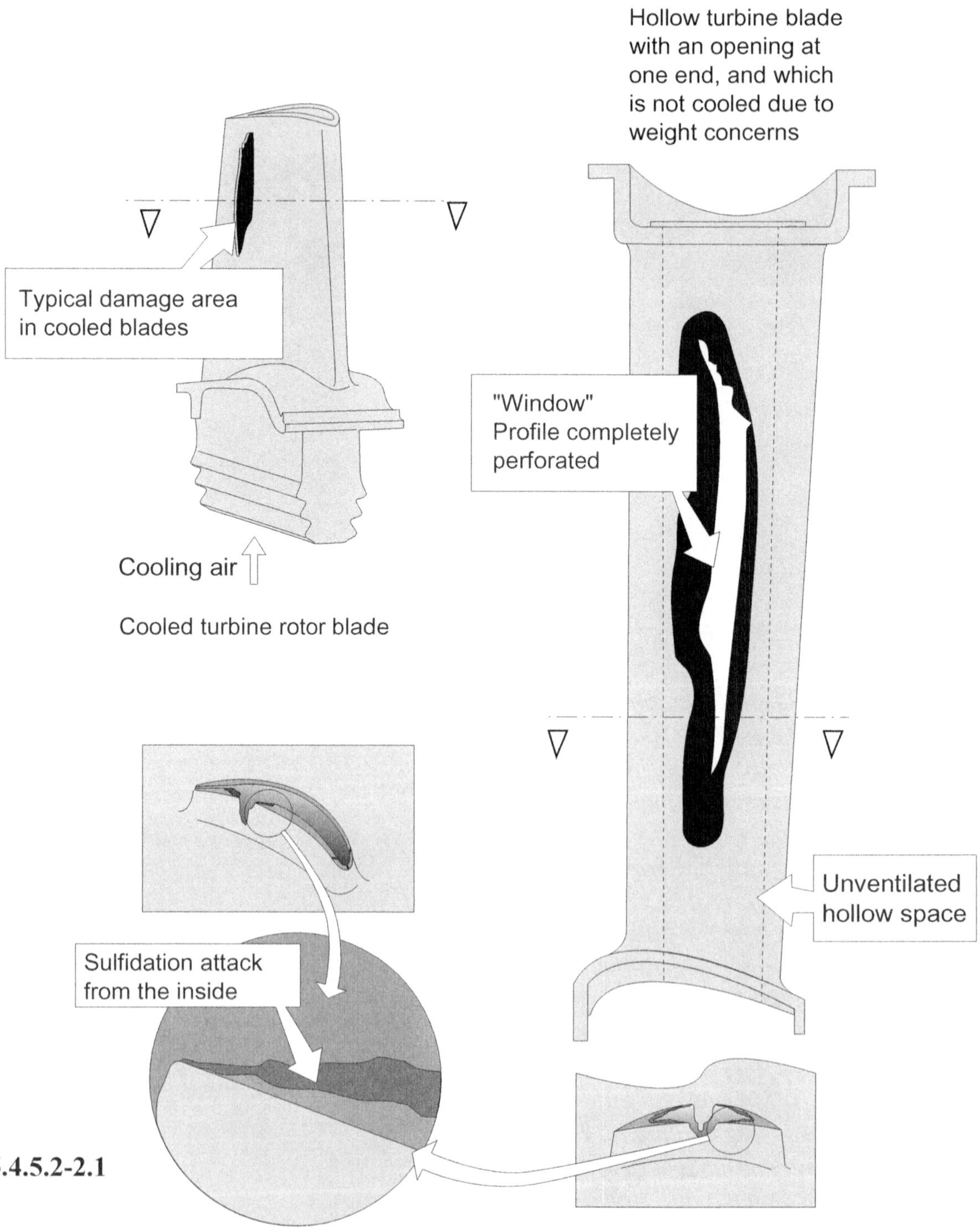

Ill. 5.4.5.2-2.1

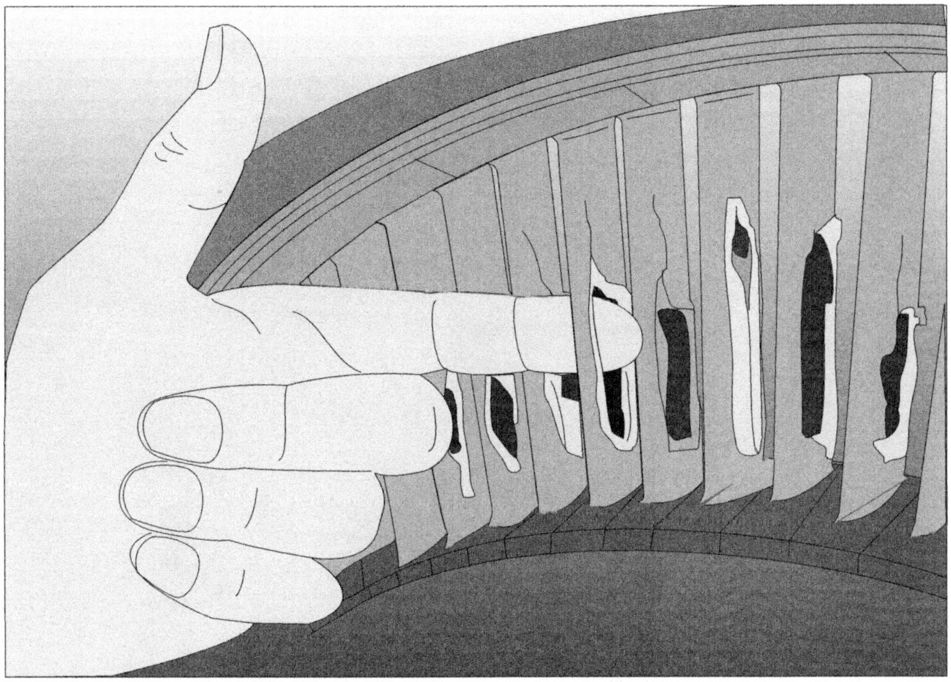

Low pressure turbine staor vanes, perforated by sulfidation.

Ill. 5.4.5.2-2.2

Illustration 5.4.5.2-2.1: *Sulfidation is especially pronounced when the oxygen supply is not sufficient to allow a dense protective oxide layer to develop. This is evidently the case in hollow blades (weight reduction) that are closed on one end. In extreme cases, this results in "windows" opening up in the blades, i.e. both blade walls are broken through. This damage process is made more problematic by the fact that the extent of the damage can only be seen at a very late stage. In blades with cooling air through-flow, there seem to be limited areas inside the blade where aggressive dusts tend to deposit (Ref. 5.4.5-7), and the part temperature is suitable for this phenomenon (Ref. 5.4.5-4). The temperature ranges necessary for sulfidation to occur determine the location of damages on the engine parts.*

Bild 5.4.5.2-2.2 *(Lit 5.4.5-8): This picture according to an photo demonstrates the extreme deterioration of a LP-turbine stator after some 1000 operation hours. The sulfidation attack (Ill. 5.4.2.2-2.1) starts in the hollow vane profiles. If the stator has an „* ***intermesh function*** *" (Ill. 4.5-9) the question for the remaining function arises.*

Silver plated bolts and nuts in hot parts can be accelerated deteriorated by the action of sulfur as also trigger failures in contact surfaces of components.

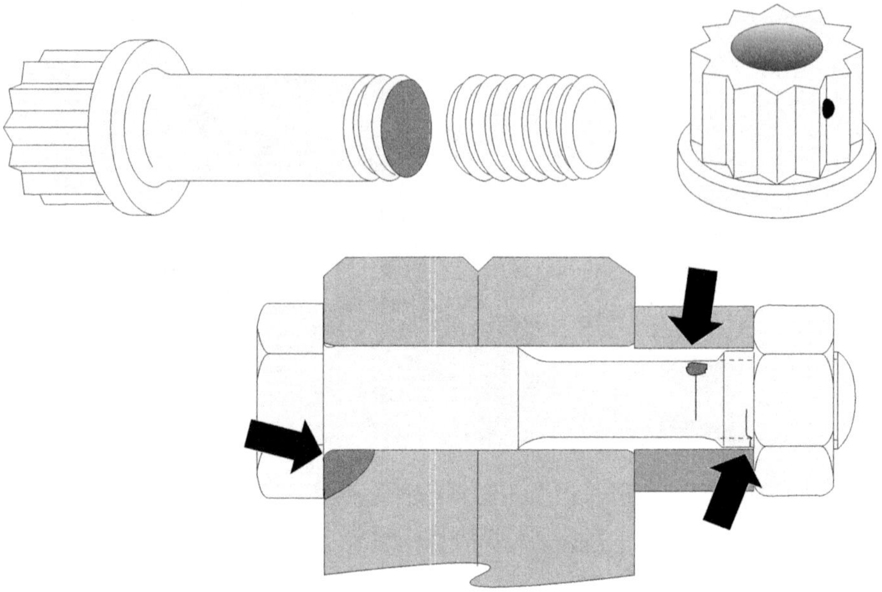

Example 5.4.5.2-1 (Ref. 5.4.5-2):

Excerpt: "...is planning to replace a silver coating on (a fighter) engine high pressure turbine disk front and rear fasteners after an inspection revealed that **small cracks** had developed in the disk during intensive testing....

The engine had been run more than 4,300 cycles, which is equivalent to about 2,000 flight hours, before it was torn down for the inspection....

The **silver coating** used on the nuts and bolts securing the **disk's front and rear retainer** had reacted with **sulfur in the engine's jet fuel**, creating a sulfate compound that exposed the disk surface to acid. The acid led to corrosion and about half a dozen cracks in the disk..."

Comments: In this case it is unclear whether corrosive attack occurred due to the creation of sulfuric acid from condensation water during standstill (volume 3 Ill. 12.4-14, volume 4 Ill. 16.2.2.3-2.2), and/or if it involved sulfidation in dry, hot conditions. Both mechanisms are described in the literature and have been observed in engines.

Illustration 5.4.5.2-3: *Sulfidation is limited to very specific temperature ranges (Ill. 5.4.5.1-4) in which stable aggressive salt melts exist. For this reason, external sulfidation zones develop in very typical and part-specific zones. Aside from suitable temperature, these zones are also determined by a location suitable for deposits. These areas are usually on the pressure side of blades. In turbine blades, these zones are usually behind the bores of the film cooling (top left diagram); in uncooled blades with high edge temperatures, the zones are narrow fields parallel to the edges (top right diagram). Oxides created by the sulfidation are often emaciated*

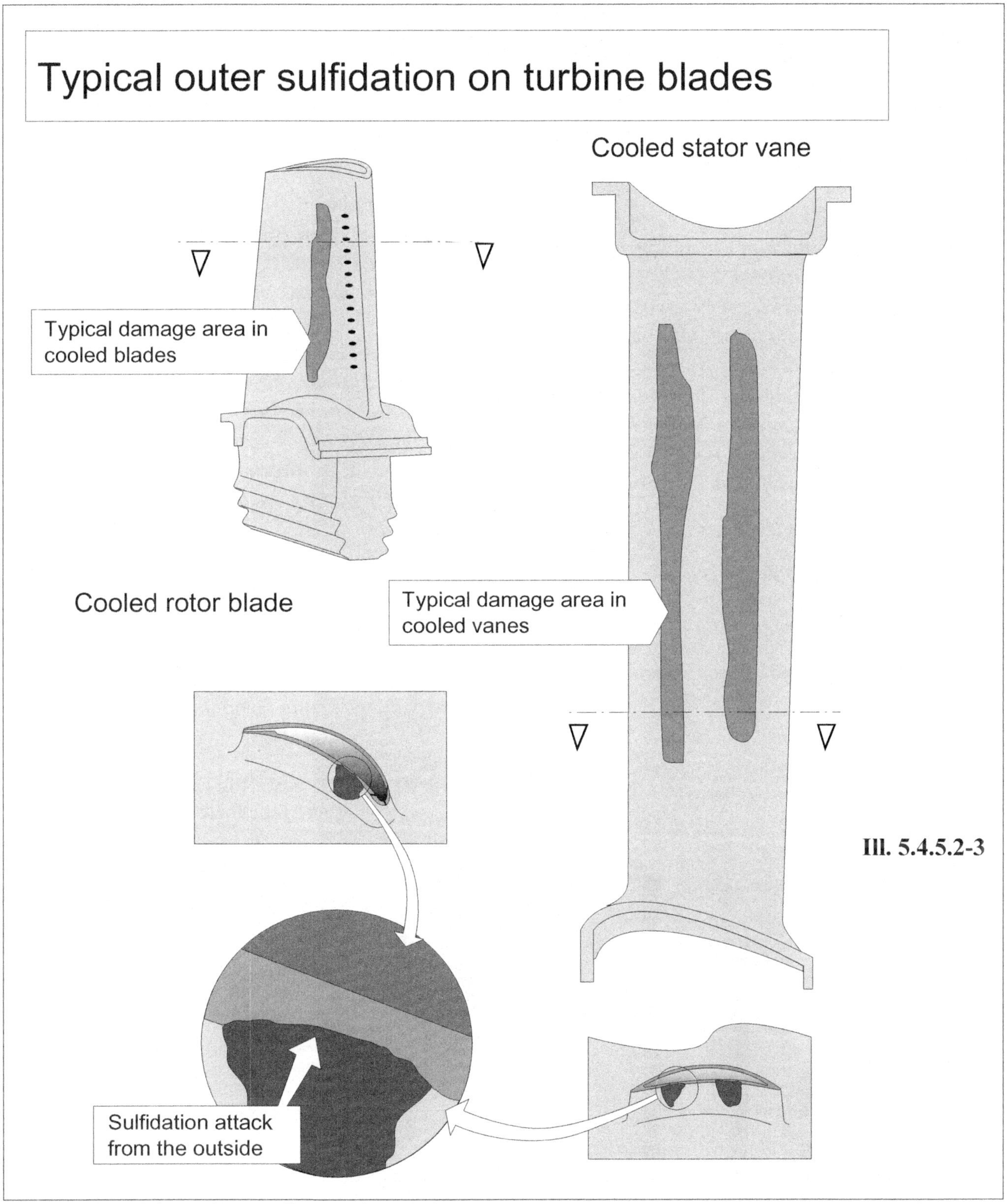

Ill. 5.4.5.2-3

due to erosion from the gas flow and particles it carries (bottom left detail), resulting in visible indentations.

If this is the case, it must be assumed that only a very short usable life span remains, relative to the operating time that has already passed.

External Factors
High-Temperature Corrosion

Illustration 5.4.5.2-4.1 and illustration 4.2 *(Ref. 5.4.5-5): The operating damages from atmospheric plasma-sprayed YSZ (yttrium-stabilized zirconium oxide) thermal barrier coatings (TBC) are primarily dependent on the surface temperature of the coating.*
Damages in the left part of the diagram, especially those that lead to premature failures, were traced back to manufacturing problems. Curves ***"1","2", and "3"*** *show the failure behavior due to dust. These dust deposits include, for example, FeO+NiO from seal wear products and MgO+ CaO,* Al_2O_3, SiO_2 *from dusts ingested from outside the engine. Erosion is more pronounced at low operating temperatures.*
During very long operating hours, effects occur that are based on changes in the coating structure:

- ***Phase changes*** *despite the stabilizing yttrium*

*A special type of phase reaction is the „**mineralization**". It depends on a phase, which is mechaniclly stressed, or which is not in the equilibrion condition and migrates into the equilibriun condition.*

.

- ***Sintereffects*** *influencing the* ***segmentation****. At temperatures above 1093°C fresh sprayed thermal barrier coatings from zirconium oxide can sinter fast (Ref. 5.4.5-9). With this a markedly change of volume and material properties is triggered. Through sintering a shrinkage occurs, which leads to tensile stresses and releases cracks perpenticular to the surface (engl. („mudflat cracks"). The* ***E-modulus*** *(rigidity) of the sintered coating increasesn and it gets more brittle. Also the* ***thermal conductivity*** *rises, so that during constant cooling the component temperatur rises. Examples for this property changes shows Ill. 5.4.5.2.4.2.*

During operation important properties of a ZrO_2 thermal barrier coating can serious downgrade.

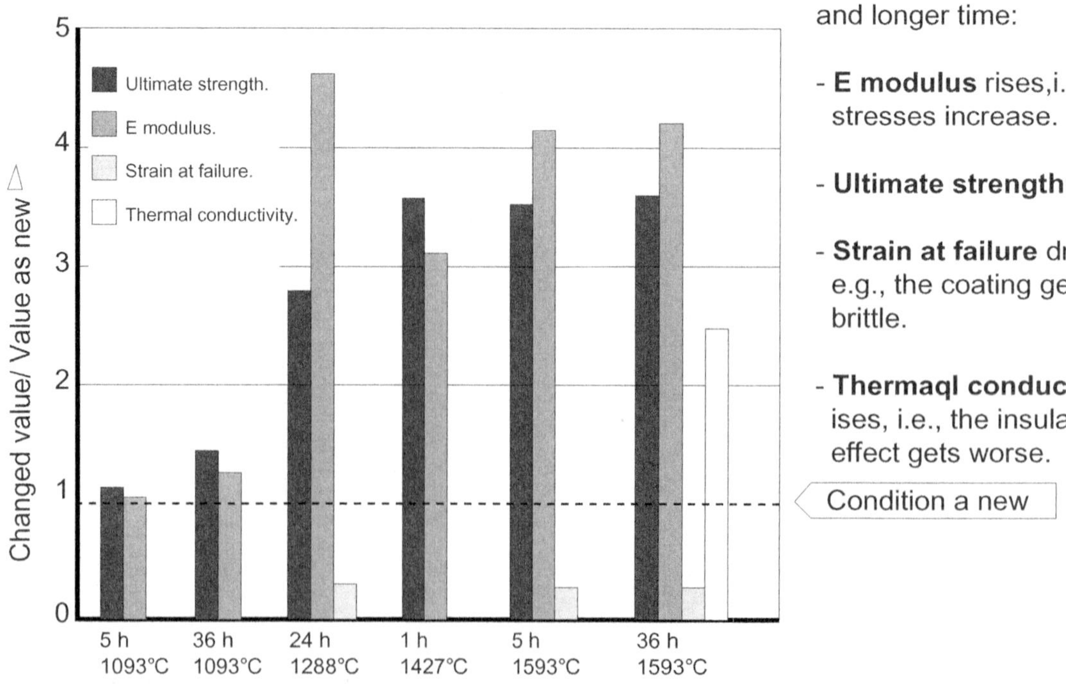

Recognizable tendencies against the new condition during rising temperature and longer time:

- **E modulus** rises,i.e., stresses increase.

- **Ultimate strength** rises.

- **Strain at failure** drops, e.g., the coating ges more brittle.

- **Thermaql conductivity** ises, i.e., the insulating effect gets worse.

Ill. 5.4.5.2-4.1

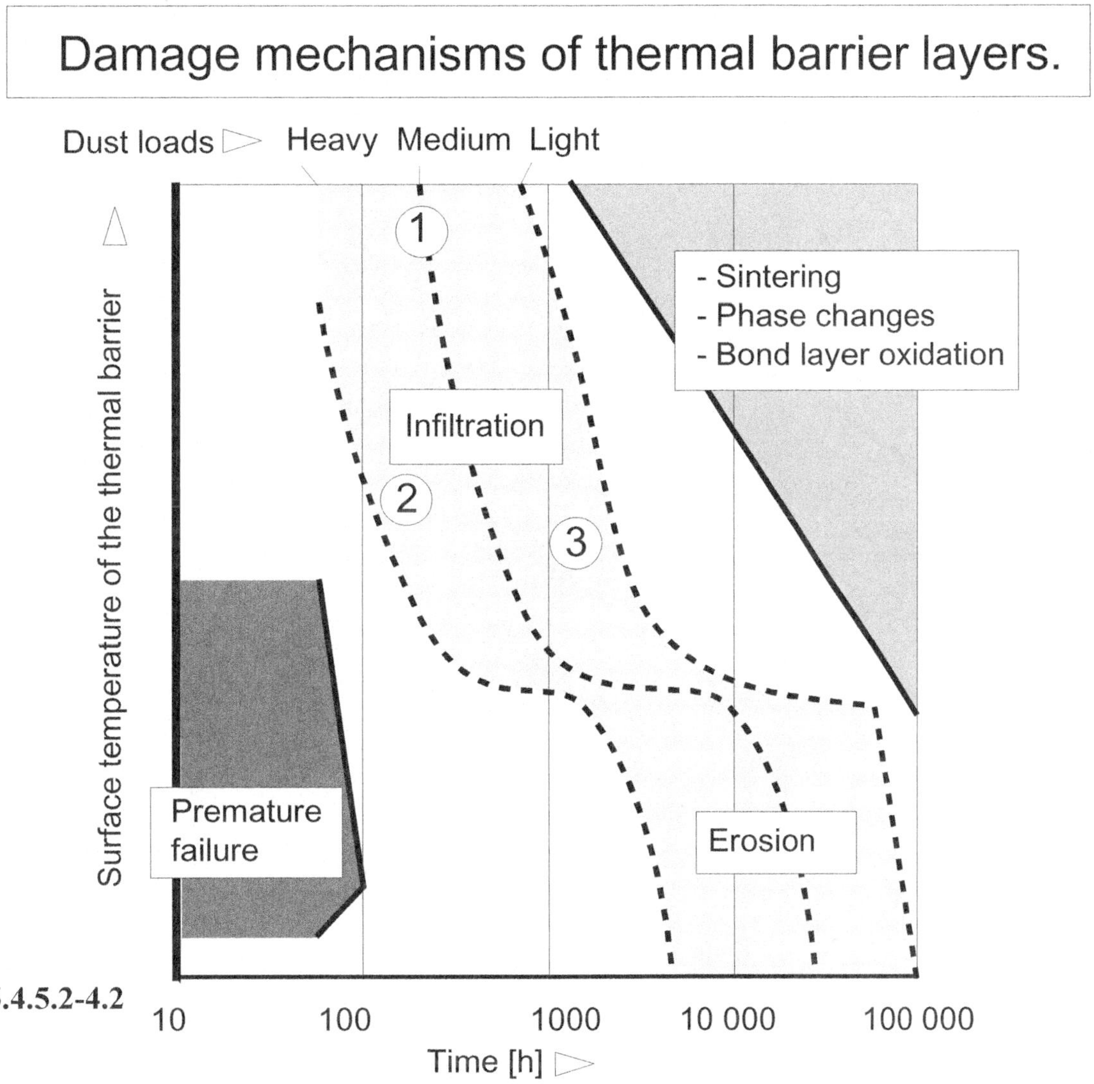

Ill. 5.4.5.2-4.2

*- **Oxidation of the undercoating** (adhesion coating) triggers the so called „**thermal grown oxide scale**" (= TGO) between undercoating to the TBC (Ill. 5.4.5.2-5, Ref. 5.4.5-11). The zirconium-oxide layer is penetrated to the undercoating by oxide. The oxidation, triggered by this process in many cases is the **lifetime limiting effect**. With increasing thickness of the TGO the adherence trength drops and spalling occurs together with a markedly rise of the temperature of the concerned hot part.*

*A special problem which gets more and more attention is the **erosion of the thermal barrier coatings** (TBC). Especially concerned are obviously thermal sprayed coatings. Concerned are break-outs of small particles from the surface. Besides a deterioration of the TBC itself, the life time of the aeroengine can be unacceptable concerned. The **ceramic coating particles have an erosive effect at the blading, following in the gas stream**. So for example the typical thin Al-diffusion coatings (ca. 0,050 mm) can be removed in few hundred operation hours. With this they offer no more an oxidation protection. The result is an accelerated oxidation of the blading which makes a premature exchange necessary.*

Typical operating problems of thermal barrier layers

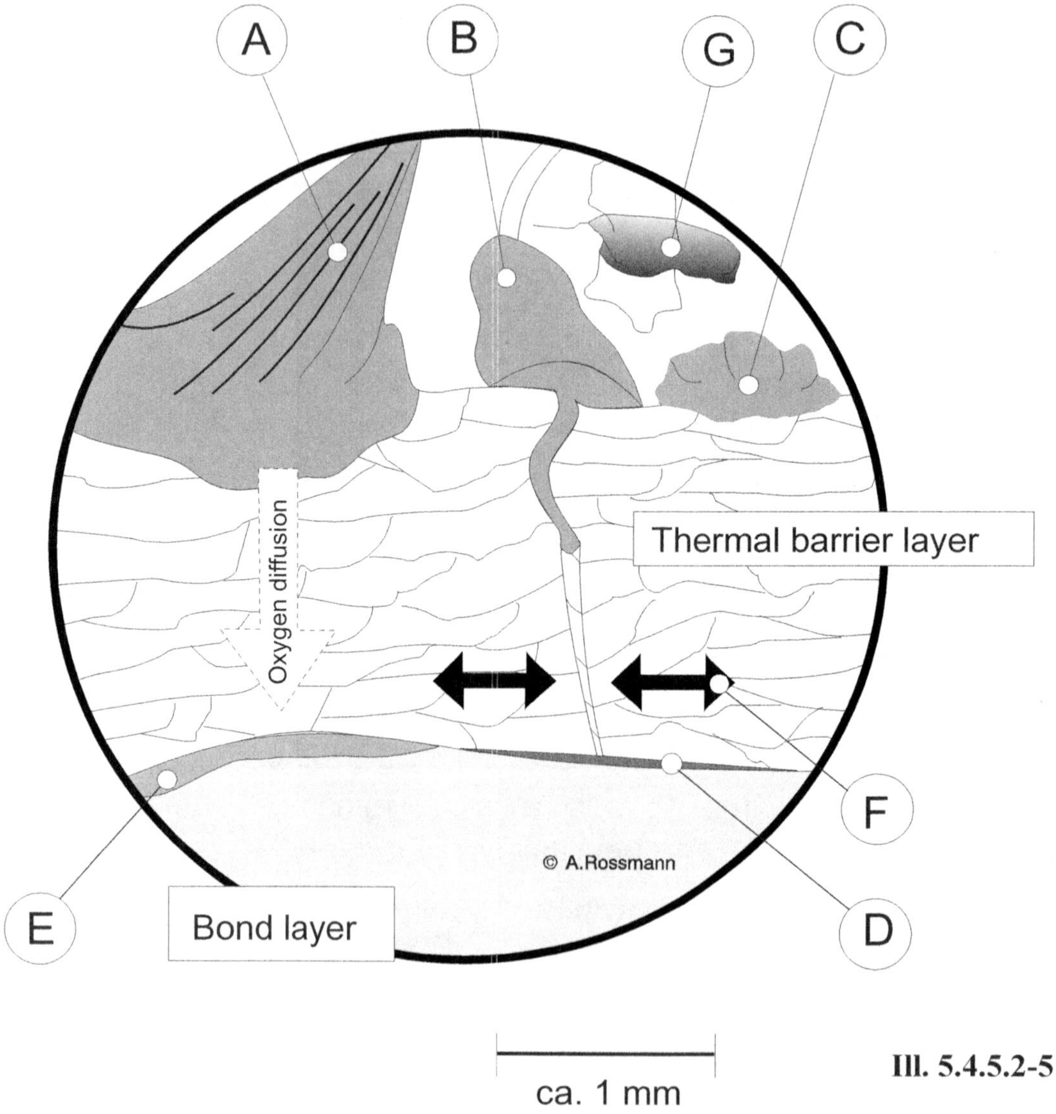

Ill. 5.4.5.2-5

***Illustration 5.4.5.2- 5** (Refs. 5.4.5-6 and 5.4.5-6): The following are typical damage mechanisms in ceramic thermal barrier coatings made from ZrO_2:*

***"A": Erosion** through particles and/or the gas flow. Damages to the high-pressure turbine segments are typical (housing-side seal surfaces across from the HPT rotor blade tips).*

***"B": Melts** of dust deposits can foorm in the combustion chamber and/or at the hevily heated thermal barrier coating (TBC) of the blading.For example such melts can be low melting silicates from ingested dust (Ref. 5.4.5-10). These can enter into the segmentation cracks (Ill. 5.452-4.2) and create a bursting effect in the frozen state after cooling.*
***Salt melts** may show this physical effec less prnounced. They shrink, for example vanadium-sulfate, during the freezing process in a manner, that dangerous high compressive stresses are not to be expected*

***"C": Chemical reactions** with deposits (e.g. fuel remnants). From tests can be identified, that compositions of the elements sodium, sulfur, vanadium and probably also lead and phosphor are primarily responsible for the hot gas corrosion of thremal barrier coatings. The aggressivity develops especiaqlly during **oxidation of the fuel contaminations** in thre combustion in the combustion chamber. Thereby develop intense acid ore alkaline acting salt melts (Ill. 5.4.5.1-4).*

***"D": Delamination of the coating** due to poor bonding (e.g. caused by manufacturing problems)*

***"E": Oxidation of the contact surface**, e.g. the undercoating, caused by oxygen ions being conducted through the hot ceramic coating. This failure mechanism has a relatively long incubation period (up to several 1000 operating hours), making it especially problematic. Also corrosive gases can penetrate the TBC. For example NACL vabpour can let the oxidation layer faster grow at the adhesion layer. The so developimg thicker oxide layer promotes the spalling.*
Against corrosion ba salt melts, usually the zirconium-oxide of the TBC is more resistant as iver the highly oxidation resistant MCrAlY adhesion coatings. Observations bring close, that even thin ZrO_2 coating remains prevent an attack of the adhesion coating by salt melts.
Also solid organic oxides can deteriorate TBCs by acting acid.

***"F": Thermal stresses** from manufacture and operation can cause spalling, especially at convex radii and edges. At convex surfaces like at the suction side of turbine blades the high compression stresses in the ceramiclayer act especially flaking. At concave surfaces the contray is the case (Ref. 5.4.5-9). Simplified the relation is applied: The stress perpenticular to the surface curvature rises direct radius of the surface curvature (i.e., **the more even the curvature, the moree susceptible for spallings!**), the coating thickness and the stress in the coating parallel to the surface.*

***"G": Superficial crumbling** after longer run times in a hot gas flow. These particles can considerably reduce the life spans of diffusion-coated hot parts (such as turbine blades).*

5.4.5.3 Measures Against Damage due to High-Temperature Corrosion

- If possible, sulfidation-prone hot parts should be made from nickel alloys with a high Cr content. If this is not possible due to the required strength properties, the only remaining option is a suitable surface coating.

- Protecting the hot parts with coatings systems that are selected according to the dominant damage mechanism.

- The selection of suitable coating systems for each specific situation is of vital importance. Desirable oxide coatings provide a good seal, are well bonded, grow slowly, and are chemically and mechanically (against erosion and thermal fatigue) stable. The most commonly used coating systems rely on creation of Cr_2O_3 or Al_2O_3. These coatings are produced either through diffusion annealing (homogenizing) in special media (powder, gas), or through thermal spraying (usually) as lay-on coatings. The thicker the coating, the more it affects the strength of the base material. This requires compromises. As protection against hot gas corrosion (sulfidation), Cr_2O_3 conversion coatings are most suitable. The most commonly used coating types are alitized coatings, which are often modified with a Pt intermediate layer on the base material and then referred to as PtAl coatings. Pt is supposed to act as a diffusion barrier, and also improves the protection against Type I hot gas corrosion because it also accumulates in the outer coating zone during operation.

It must be noted that diffusion coatings are relatively brittle at low temperatures, which means that heat strain during startup and shutdown of the engine can cause coating cracks. In this respect, plasma-sprayed coatings such as Type MCrAlY (M stands for a "metal" such as Fe, Co, Ni) perform better than diffusion protective coatings. This is because in MCrAlY coatings, the transition from brittle to ductile behavior occurs at relatively low temperatures.

- Protection against sulfidation can be attained through suitable coatings, preferably with a high Cr content (e.g. Cr diffusion coatings). Oxidation-resistant coatings with a high Al content, such as Al diffusion coatings, only offer limited protection.

- Avoiding silver as a coating material (e.g. on bolts) or as a contaminant in the hot parts.

- Preventing soot and coke creation in areas where hot gas corrosion can be expected.

- Not combining Fe alloys with Ni alloys in hot parts due to the risk of green rot.

- Sufficiently early inspection of the state of the engine parts:

Oxidation damages are evident due to the "orange peel effect" (Ill. 5.4.5.2-1). These damages can be detected with boroscopic inspections and, with sufficient experience with the affected engine type, rough estimates of the remaining life span are possible.

If the external damage has not noticeably progressed, a cross-section of a representative part must be taken for metallographical inspection in order to determine the degree of emaciation.

Large amounts of ferromagnetic oxide (nickel oxide) are created during sulfidation. These can be externally detected with suitable magnetic probes (fluxgate magnetometers), even if they are on the inside of a hollow blade. This allows non-destructive inspection of the blades if they are accessible on the engine, if not, it can be done during an overhaul.

- Non-ventilated hollow spaces, such as hollow blade profiles with one closed end in low-pressure turbines, should be avoided. Experience has shown that even if both ends are closed, problems can arise during operation if cracking allows oxygen and corrosive media to enter the blade.

From this standpoint, solid profiles are the safest. If this is not possible, the inside of the hollow space should be coated. This coating should contain Cr (chromized), and not be a normal Al diffusion coating, as these do not offer optimal protection against sulfidation.

- When washing the compressor to remove salt deposits and dust, it must be ensured that no contaminants are washed into the hot parts and deposit there. For this reason, the manufacturer must set suitable guidelines which must, in turn, be followed by the operator.

Chapter 5.4.5 References

5.4.5-1 Ralf Bürgel, "Handbuch der Hochtemperatur-Werkstofftechnik", Friedr. Vieweg & Sohn Publishing Co., 1998 ISBN 3-528-03107-7, pages 247 to 305.

5.4.5-2 "F110 Disk Crack Traced to Coating", periodical "Aviation Week & Space Technology". August 6, 1984, page 18.

5.4.5-3 R.A. Rudey, J.S. Grobman, "Impact of Future Fuel Properties on Aircraft Engines Fuel Systems", AGARD-LS-96, Agard Lecture Series No. 96 "Aircraft Engine Future Fuels and Energy Conservation", October 1978, Chapter 6, page 19.

5.4.5-4 A.J.A. Mom, J.A.M. Bogers, "Simulated Service Test Behaviour of Various Internal and External Coatings Applied on CF6-50 First Stage Turbine Blades", Proceedings of the conference "Coatings for Gas Turbines and Other Applications", Liège, Belgium, 6-9 October 1986, Chapter 11.

5.4.5-5 P. König, A. Rossmann, "Ratgeber für Gasturbinen-Betreiber", Vulkan Publishing Essen, 1999, ISBN 3-8027-2545-X, pages 95 and 114.

5.4.5-6 R.L.Jones, "Some Aspects of the Hot Corrossion of Thermal Barrier Coatings", periodical, Journal of Thermal Spray Technology", Volume 6 (1) March 1997, pages 77 to 83.

5.4.5-7 M.M Ratwani, A.K. Koul, J-P. Immarigeon, W. Wallace, "Aging airframes and Engines", Proceeding Paper in AGARD-CP-600 Vol. 1, of the AGARD symposium on " Future Aerospace Technology in the Service of the Alliance", Palaiseau, France, 14-17 April 1997, pages A18-1 to Q18-16.

5.4.5-8 K. Steffens, „Technik der Luftfahrtantriebe", Vorlesung 2002/03 an der TH-Aachen.

5.4.5-9 D.M. Nissley, „Thermal Barrier Coating Life Modeling in Aircraft Gas Turbine Engines", Zeitschrift „Journal of Thermal Spray Technology" Volume 6 (1), March 1997, page 91-97.

5.4.5-10 R.L.Jones, „Some Aspects of the Hot Corrosion of Thermal Barrier Coatings", Zeit-schrift „Journal of Thermal Spray Technology" Volume 6 (1), March 1997, page 77-84.

5.4.5-11 E. Berghof-Hesselbächer, H. Echsler, P. Gawenda, M. Schorr, M. Schütze, „Time and Temperature Dependent Development of Physical Defects in Thermal Barrier Coating Sys-tems", Zeitschrift „Praktische Metallografie", 40 (2003) 5, Carl Hanser Verlag, München, page 210-230.

5.5 Special Media (Unusual Influences)

5.5.1 Measures for Avoiding Unusual Influences

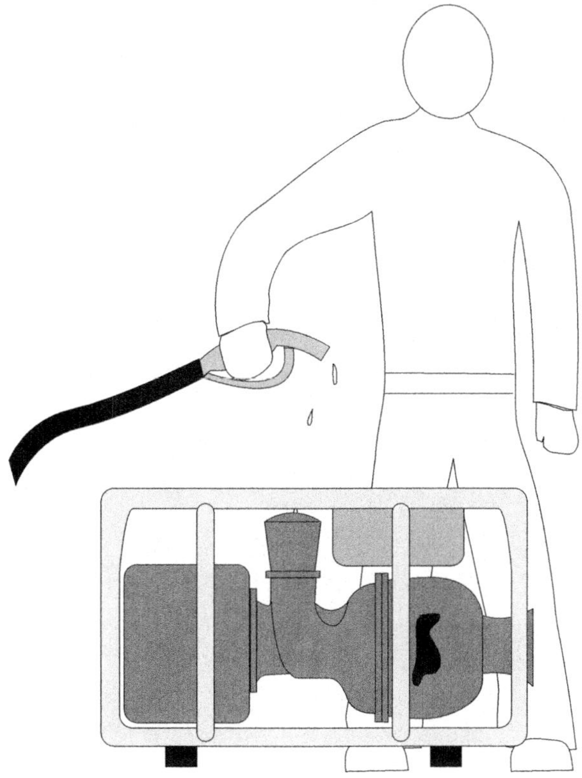

This chapter deals with **foreign materials** and air changes that can dangerously **affect engine operation**, but can not easily be categorized into any of the preceding chapters. These influences include, for example, the **ingestion of fuel** (Ill. 5.5-3) or insufficient oxygen in the **intake air** (Ill. 5.5-6).

One unique damage type is overheating and/or a runaway turbine due to **insufficient drainage of unburned fuel** after an aborted start. If fuel collects in the turbine casing and does not drain in accordance with regulations, it can cause overheating which destroys the hot parts and/or makes the turbine rotor reach **overspeed** and **burst**. Insufficient drainage can be caused by malfunctioning of the **drain system** (the drain valve, etc.) or an unusual engine angle that prevents proper fuel drainage. In one reported case, a helicopter engine under development was restarted in a vertical position after an **aborted start** during a test run. The location of the drainage opening prevented fuel from draining when the engine was in a vertical position. When this fuel burned off, the engine reached uncontrollable overspeed and the turbine disk burst, which was a very dangerous situation for the testing rig personnel.

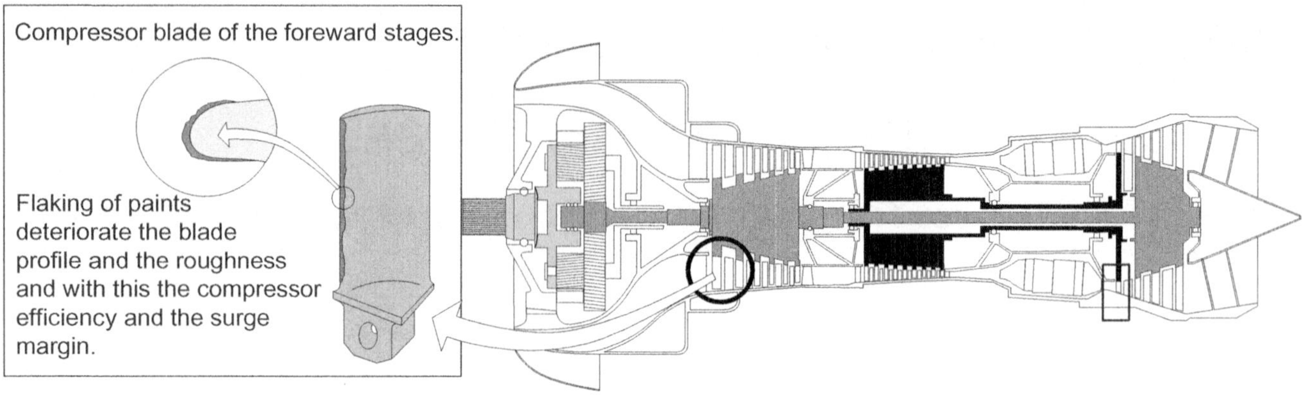

Example 5.5-1 (Ill. 5.5-1, top diagram):

During an acceptance run of a mid-size engine, the operating behavior deteriorated within minutes. Unallowable performance losses and compressor surges occurred. Inspection of the engine revealed a greenish, lacquer-like coating on the compressor blading. Investigation and research followed by a damage analysis determined that painting work was being done in an open area during the acceptance run.

The **paint mist** traveled over about 50 meters into the 8 meter-high compressor intake chute, causing damages of several hundred thousand German Marks (disassembly, overhaul, repeated acceptance run; 2 Marks = 1 Euro). (**A**) source, (**B**) intake flow, (**C**) engine.

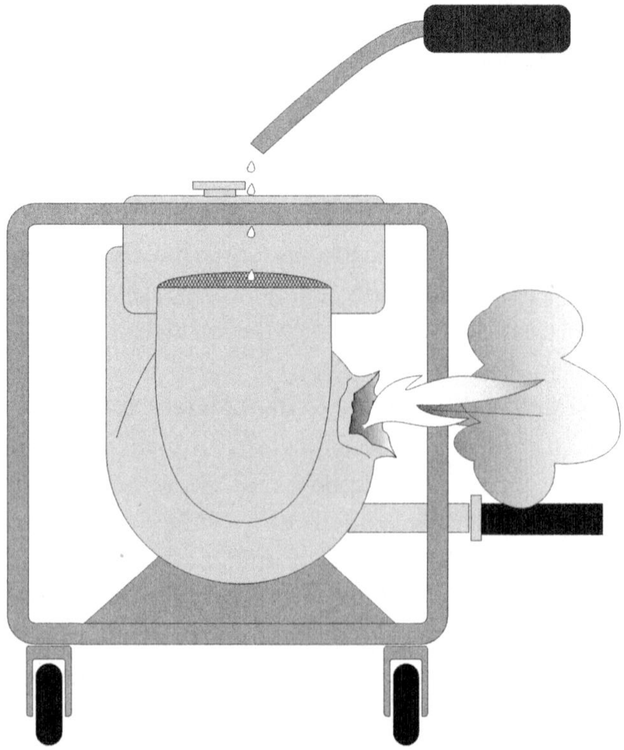

Example 5.5-2 (Ill. 5.5.3):

When fueling up an air pressure-inducing starter unit powered by a small gas turbine, a few droplets of fuel splashed into the air intake chute when the fuel hose was removed from the tank. Reconstruction showed that the ingested fuel amount would have filled a small shot glass. Due to the short time, however, this corresponded to an extremely high fuel-throughflow rate, and the regulator was no longer able to sufficiently limit the RPM. Within seconds, the rotor accelerated to high overspeed like an expanding spring, causing the relatively heavy radial turbine disk to burst. The person who was fueling the tank was seriously injured, and a rotor fragment was found roughly 250 meters away.

! Air impurities must always be taken seriously, even when they are far away.

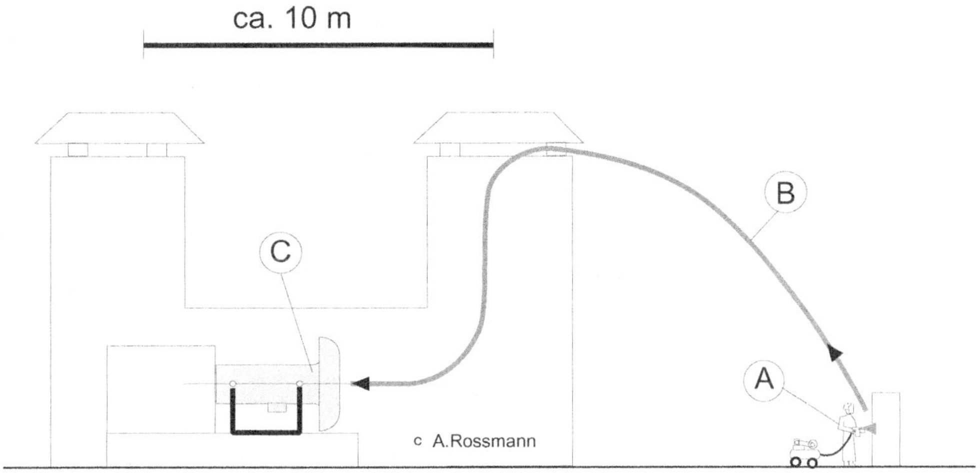

Ill. 5.5-1

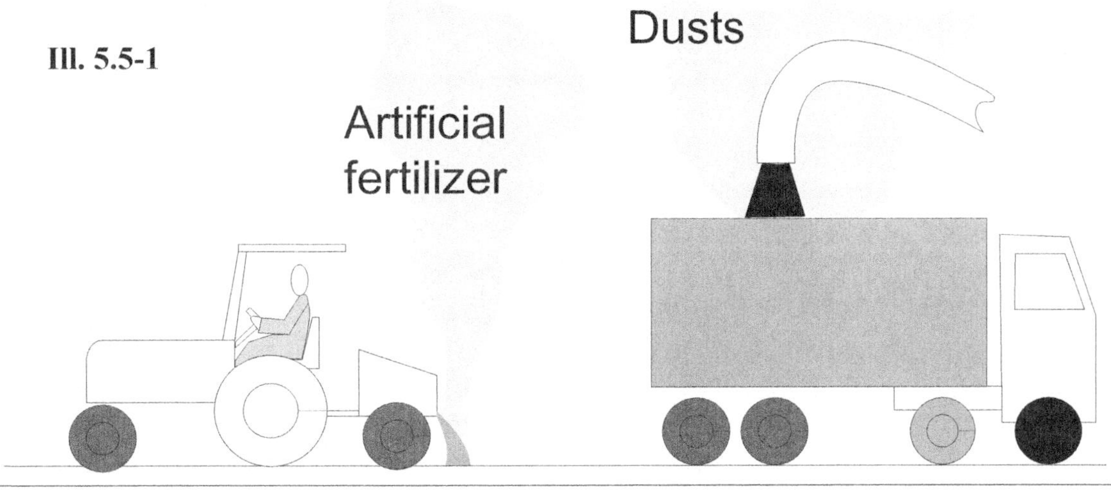

Illustration 5.5-1: *During operation on or near the ground (taxiing, takeoff, landing), or during test and control runs in **testing rigs** or on the aircraft, there is a special danger of damaging foreign materials being **ingested**. It is difficult to imagine the long distances across which engines can suck up foreign materials from a source. The potential damaging effect of foreign media is also frequently underestimated.*

*The top diagram shows the effects of **gaseous media** and **dusts** that are sucked into engines, even over apparently safe distances (Example 5.5-1)*

*Typical examples include **lacquer mists**, dust created during manufacturing processes, dusts created during loading procedures in industrial* *zones, sprayed materials, and artificial* ***fertilizer dusts*** *in agricultural areas.*

Fertilizer dusts: *the spreading of artificial fertilizers results creates dusts with components such as* ***sulfur*** *and* ***phosphorus compounds*** *that can damage hot parts through HGC (Chapter 5.4.5).*

Lacquer mists *can foul (rough bonding deposits) compressor rotor blading and unallowably worsen the operating behavior of the compressor (bringing the surge limit closer and decreasing compressor efficiency). In addition, these pigments often contain heavy metal oxides that can corrosively react with the hot parts.*

Dusts from industrial processes: *Iron oxides, for example, tend to block tight cooling air ducts in the turbine blading.*

Illustration 5.5-2: *The activation of* ***fire extinguishing*** *units takes place in the engine nacelles, the engine area, testing rigs, or on the ground (e.g. during a tailpipe fire).*
In this case, various extinguishing materials are sprayed onto and/or into the engine. These materials may contain components that can ***cause corrosive*** *damage in the compressor or in the hot parts during continued operation. For this reason, it is important that the extinguishing materials used are documented in order to identify engine parts that, based on their composition, could potentially suffer corrosion damage. Often, this information must be used to design an overhaul plan with corresponding cleaning and evaluation procedures.*
External aggregates such as gearings typically have housings made from ***corrosion-sensitive light metal alloys*** *(Al and Mg alloys) or components such as* ***gears*** *and* ***roller bearings*** *made from non-stainless steels. Even* ***stator vane*** *adjusters, actuators, and* ***feedback lines*** *can contain corrosion-sensitive steels. If the fire extinguishing material is not removed, these parts can be damaged corrosively in combination with condensation water over longer operating periods.*
An important factor is whether the extinguishing material entered the engine while it was running ***at operating temperatures****, already considerably cooled, or still hot. At high temperatures, remnants of extinguishing materials can already break down into aggressive components and* ***react with the part surfaces****. In running engines, there is the additional danger of the cooling air system carrying remnants of extinguishing material into the cooling air ducts of the turbine blading, where they can block up both the ducts and the bores of the cooling air film. In one reported case, only 10 operating hours after fire extinguishing material entered the compressor, a boroscopic inspection of the turbine rotor blades revealed considerable signs of overheating due to* ***blocking of the cooling air ducts****.*

Extensive maintenance work may become necessary if fire extinguishing media enter into an engine.

Influencing of compressor components such as blades, spacer rings, and housings:

- Damage to porous coatings such as damage to abradable coatings in the housing and on spacers
- Damage to titanium parts through stress corrosion cracking
- Damage to anti-wear coatings

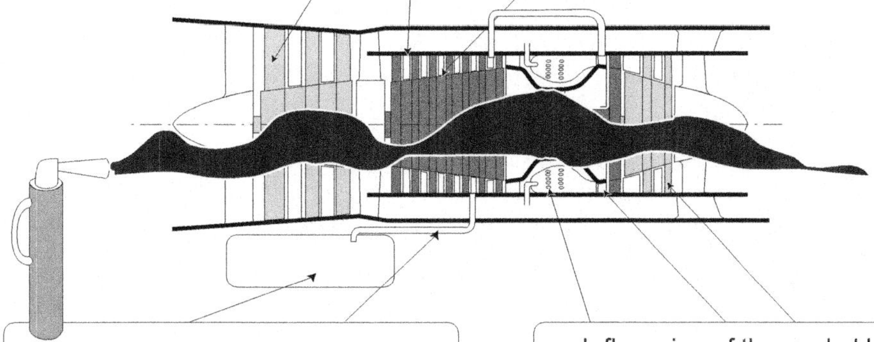

Ill. 5.5-2

Influencing of aggregates and air systems such as air motors, ducts, and vents:

- Corrosion from the inside: damage steel parts and light metals
- Corrosion from the outside: corrosion of light metals

Influencing of the cooled hot parts such as combustion chambers and turbine blades:

- Blocking up of hot parts
- Oxidation of cooling air ducts with worsening of the heat dispersion
- Sulfidation
- Damage to thermal barrier layers

Additionally, chemical reactions can attack the part from the inside and/or worsen the heat transfer into the cooling air by creating heat-insulating oxide coatings. This creates a risk of overheating and premature failure.

Typical extinguishing materials and halons, and their potentially corrosive components *(Ref. 5.5-1):*

Halon 1301 contains CF_3Br, i.e. the halogneides fluorine and bromium which at least temporarily create an aggressive acid during the extinguishing process. This presents at least potential danger of wet corrosion on light metals and steels. Further research is necessary to determine the extent in which titanium alloys are damaged.

Halon 1211 contains $CBrClF_2$, i.e. also contains Cl, the various corrosive effects of which are covered in section 5.4 (pitting corrosion, SCC, corrosion fatigue). Remnants containing Cl can also threaten titanium alloys (Ill. 5.4.2.2-2).
***Modern foam extinguishing systems** contain NH_4HSO_4. After long run times, the sulfur in these deposits can **cause sulfidation** (Ill. 5.4.5.2-3) in combination with **alkali salts** (NaCl) in marine atmospheres. If the foam strikes hot surfaces, such as those of turbine blades, glass-like deposits are created that can be removed from uncoated parts through gentle abrasive blasting (in the disassembled part). Other extinguishing foams contain up to 5% alkali sulfates and **fluorine** compounds (up to 5%). This means that all ingredients for sulfidation attack are present.*

Example 5.5-3.1 (Ill. 5.5-3):

In one aircraft type, the tank opening was located in the inlet area of one of the engines. The fueling personnel evidently forgot to close the tank properly. This allowed the engine to continually **suck fuel out of the tank opening** after startup, and even cutting off the normal fuel feed was not sufficient to stop the engine. The engine showed extreme damage from a fire and unallowable overspeed.auf.

Comment: Unfortunately details of this failure are not known. Also answer to the question, if a 'ground vortex' (Ill. 5.2.1.2-4) to the fuselage region before the aeroengine was enough for the necessary suction.

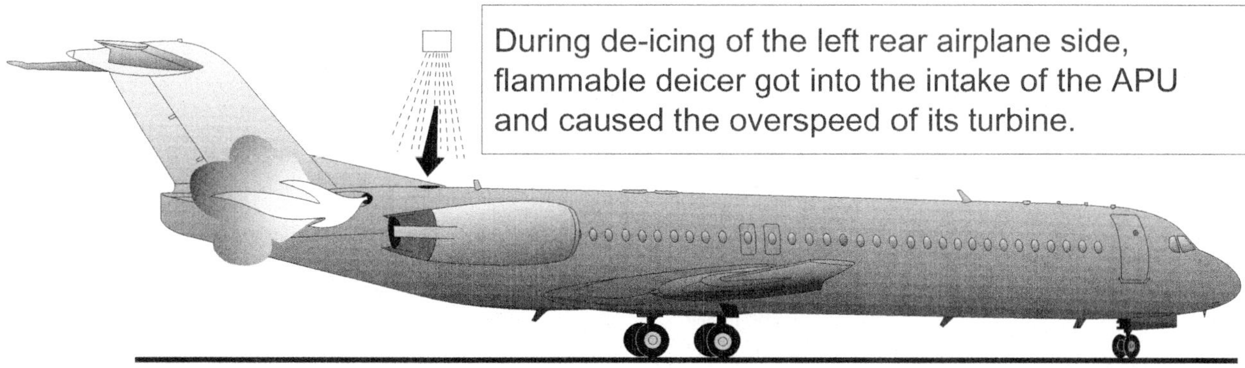

Example 5.5-3.2 (Ref. 5.5-6 and Ref. 55-7):

Excerpt: "...actions follow **uncontained failures** of the ....APU on three ...(airplanes) during ground deicing operations...In (the last)... incident uncontained fragments from the failed **APU** not only cause damage to the ...(aircraft's) tail cone, but one fragment also penetrated the aircraft's aft pressure bulkhead and became embedded in the first aid kit stored behind the aft flight attendant's jump seat at the rear of the cabin.

In its investigation into the March 2001 incident, the NTSB found that if the ...APU **ingests enough Type 1 icing fluid**, an ethylene glycol solution that is combustible when compressed, it will sustain combustion even if its electronic control unit senses an overspeed as a result of the ingestion and cuts off fuel to the APU.

Then, since the ECU no longer has command of the APU's rotor speed, the rotor will continue to accelerate as the **icing fluid keeps burning, until the wheel bursts**.

Comment: Alltogether seems that 3 similar failures occurred. Obviously the unfavourable position **of the APU intake duct without a liquid deflector at the fuselage upper side** of this aircraft type is known. The maintenance manual demands appropriate caution during de-icing. Usually this will be considered. As deicer is used the type 1 is a flammable **mixture of ethylene-glycol**. The operator kept the **APU running during the de-icing**. This promoted the **ingestion of the de-icing liquid**. After this incident it was instructed to shut the APU down before de-icing.

Illustration 5.5-3: Unintentional fuel feed *into the engine can be occur in various different ways:*

Inside the engine:

- Media confusion *between water and the water-methanol for the injection system in the compressor (Ill. 5.5-4). Extreme overheating damage with catastrophic engine failure is the result.*

Insufficient drainage *(top diagram) of unburned fuel after an aborted engine start. The fuel collects in the combustion chamber housing and ignites during an attempted restart. This results in serious consequential damages such as fire damage on the hot parts and/or runaway turbines bursting.*

Fuel entering the engine from outside:

- During fueling *(left diagrams), such as on the ground as depicted here, carelessness by the fueling personnel can result in fuel remnants being ingested by the engine. Surprisingly small amounts are sufficient to cause overheating of the hot parts and/or dangerous overspeed (Example 5.5-2). The bottom diagram shows mid-air refueling during which the tank probe broke off and allowed fuel to escape from the fuel pipe.*

- Ingestion of fuel *by the engine (middle right diagram). This can occur if tank valves in front of the engine intake are not properly closed (Example 5.5-3.1).*

- Ingestion of flammable icing fluid: *Example 5.5-3.2 describes cases in which this resulted in overspeed and bursting of the rotor of an APU turbine.*

- Ingestion of fuel that was released in flight *when flying through ones own "fuel cloud" (bottom right diagram). There has been at least one reported case in which fuel was released before an emergency landing, and the airplane circled around and passed back through its own fuel cloud, resulting in an explosion and crash.*

Fuel that is not supposed to be in the engine can cause extensive damage.

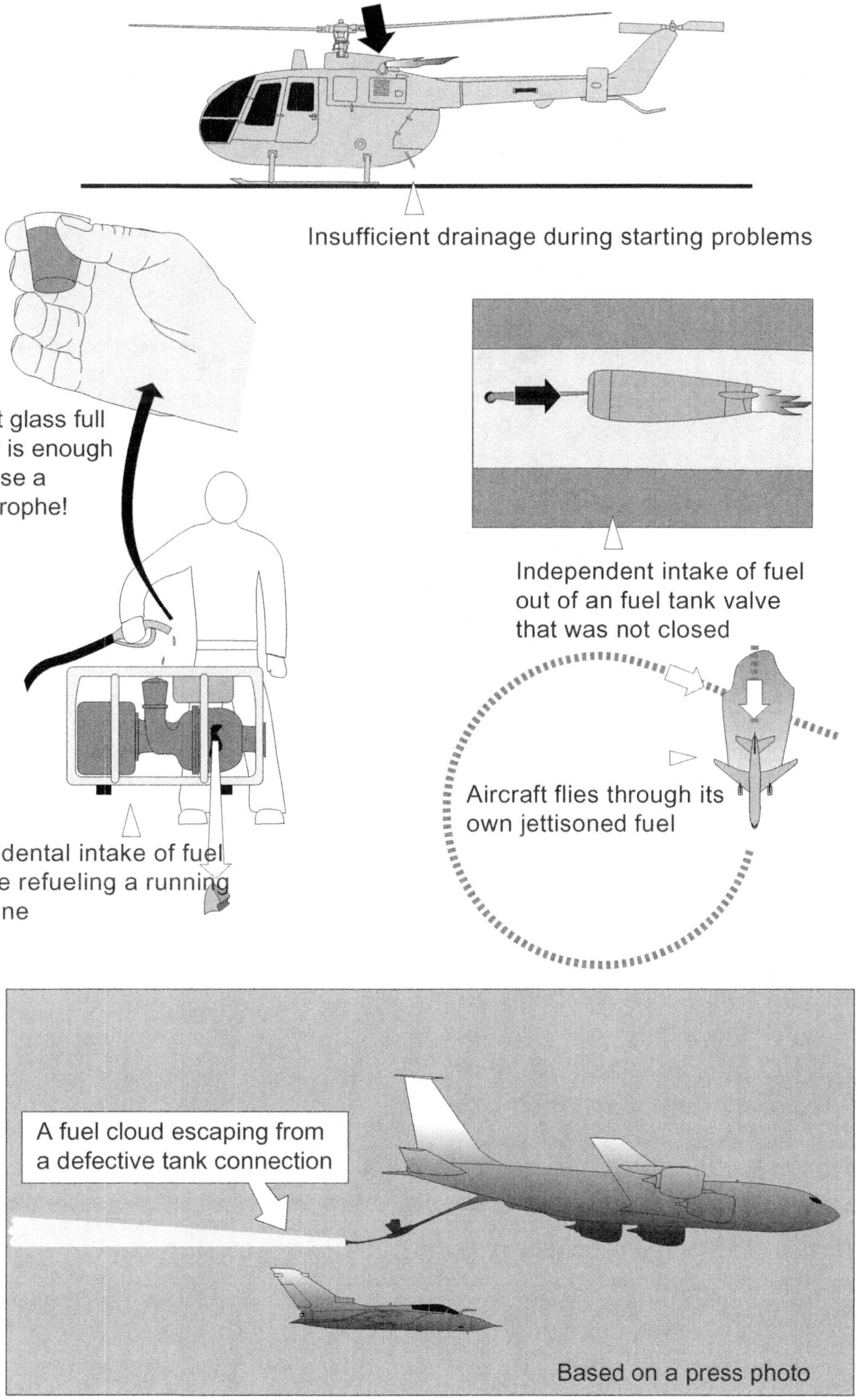

Ill. 5.5-3

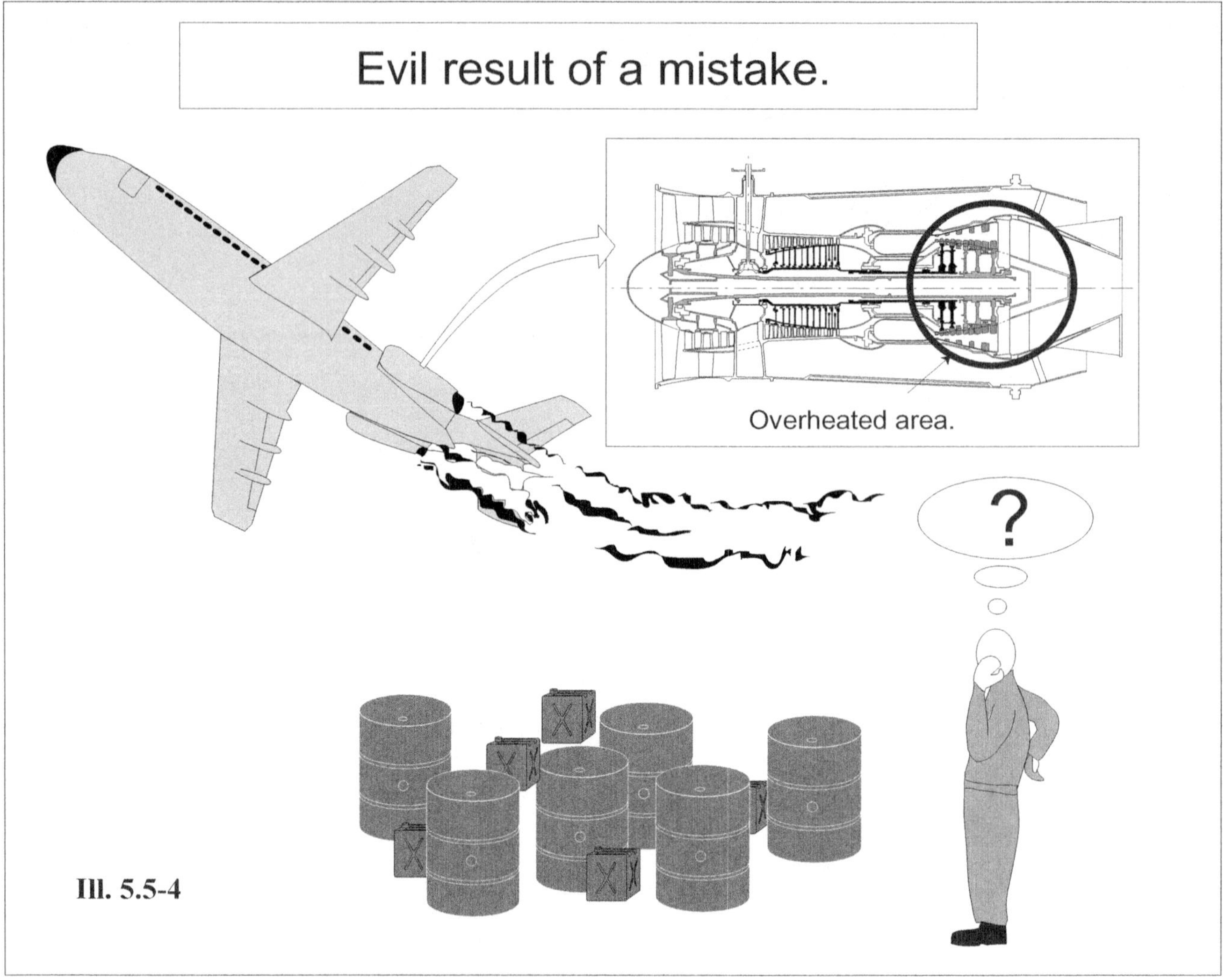

Ill. 5.5-4

Illustration 5.5-4 *(Ref. 5.5-2): An aircraft accident in the early 1970s was caused by a maintenance error (Example 5.5-4). The two engines of the affected aircraft type were outfitted with an injection unit for* ***water into the compressor*** *in order to increase thrust on hot days (Chapter 5.1.5). A mistake resulted in the unit being filled with* ***kerosene****. When the injection system was used during startup, the additional fuel caused* ***extreme overheating*** *of the turbines and both engines failed. Several people were killed during the following attempted emergency landing on a highway.*

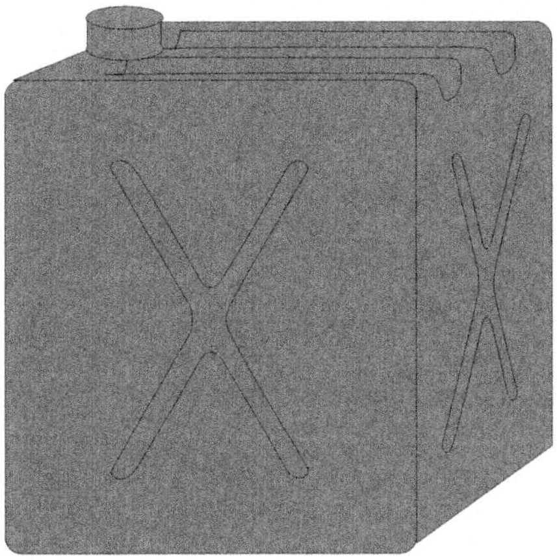

Never use blank containers! Containers must meet specific instructiions. Control the content critically, if there are doubts, no use.

Example 5.5-4 :

Excerpt (Ref. 5.5-2): "Cause of crash... twinjet transport on takeoff ... tentatively has been laid to contamination of water used for the Rolls-Royce Spey engine injection system by kerosene.

German investigation authorities have found indications that two of the five tanks, which are supposed to contain demineralized water, contained pure kerosene. Fluid in the five tanks is pumped to an injection tank in the tail when the pilot selects water for takeoff, as was done on the ....flight when both engines failed at low altitude.

Excerpt (Ref. 5.5-4): "...the five unmarked containers, which were stored in the cargo hold of the (aircraft) and supposedly contained "demineralized water", were standing there...at least one container was filled with pure kerosene, a second contained a mixture of kerosene and water, and a third one was filled with pure water. In Hamburg the deadly mixture was pumped into the injection tank...a worker (apparently said)...to the copilot in the airport..."there's fuel in there," whereupon...(the copilot replied) "no, everything around here smells like fuel."

Because the lighter kerosene content floated at the top of the aircraft's water tank, the accident occurred as follows: during the startup and climbing phases, the pump (taking liquid from the bottom) first sucked up water and then pure kerosene. Experts were able to see the sudden extreme engine temperature increase in the crash fragments: deep burn marks...were found on the injection nozzles of the cooling system; the turbine blading had melted.

Comments: This extreme **maintenance mistake** indicates fundamental **flaws in the whole quality assurance system**. This applies not only to the maintenance personnel, but also the management and logistics.

The reaction of the copilot to the warning indicates insufficient risk-consciousness and mis-estimation of the technical relationships.

The damage process was surely incomprehensible at first because overheating occurred only relatively late during takeoff. This shows how important it is for specialists who are conducting investigations to have, along with their engine-specific technical knowledge, a practice-oriented broad technical education. This of course includes an understanding of physical effects and relationships.

Lack of oxygen, caused by fire, as well as hot gases can trigger a surge in the compressor and/or lower the performance of the aeroengine dangerously.

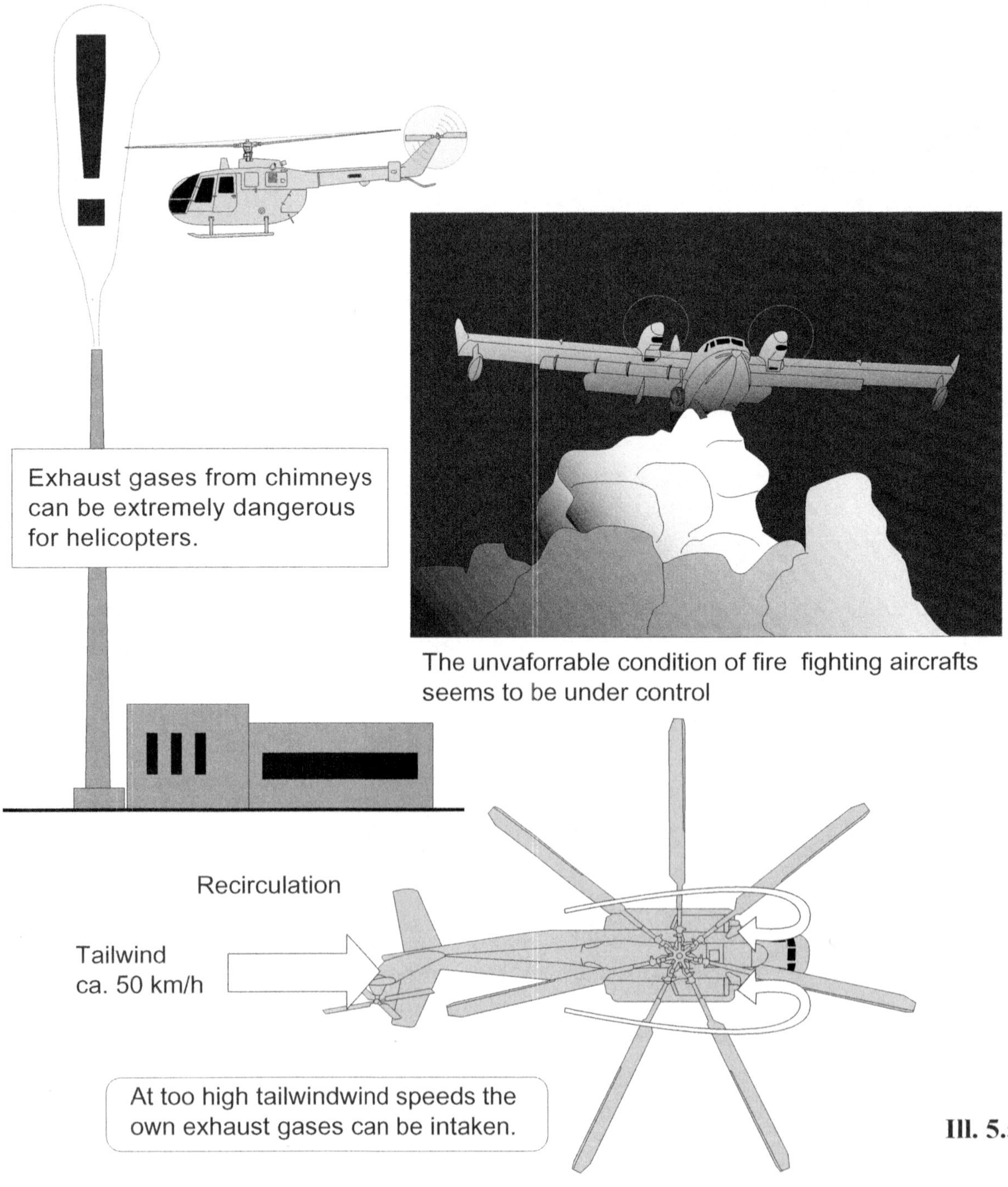

Ill. 5.5-6

***Illustration 5.5-6** (Ref. 5.5-3): The operation of **aircraft over fires** or **chimneys of power- and heating plants** can result in unallowable performance decreases or, in extreme cases, flame-out due to the **increased inlet air temperature and decreased oxygen supply** (Example 5.5-5).*
Interestingly, the lack of oxygen is considered to be the dominant influence (Ref. 5.5-5).
*Through suitable measures (such as, for example, selectable **continuous ignition** and sufficient performance excesses, i.e. distance to surge limit), it is evidently possible to safely operate at relatively low altitudes over expansive and intense fires while conducting firefighting operations.*
*Additionally a **permanent drop of the efficiency and a proneness for surging of the compressor** caused from depositions at the blades **(fouling)** can be expected. This demands the **washing of the compressor** (volume 5 Ill. 19.2.3-3).*
*A special case is the **recirculation** of the own exhaust gases.der eigenen Abgase. This is known from **trust reverser operation** of 4-engined airplanes and **unfavourable wind regimes during stand still** (volume 3 Ill. 11.2.1.2-12). For **vertical take-off aircrafts** this situation is especially problematic (volume 3 Ill. 11.2.1.2-11).*

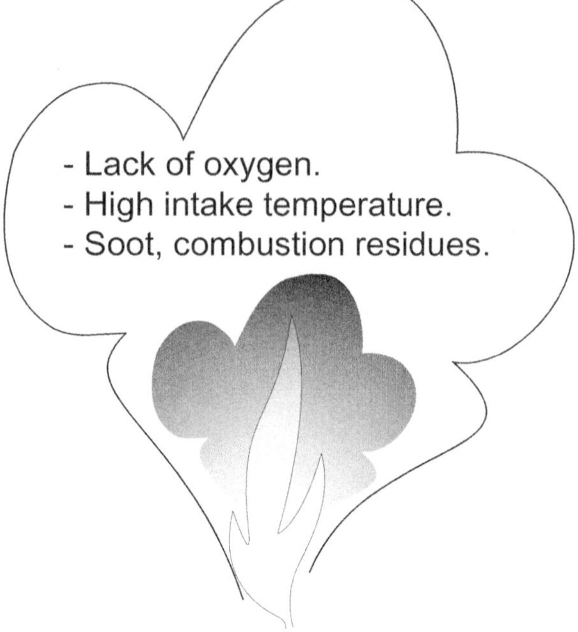

Example 5.5-5 (Ref. 5.5-3, Ill. 5.5-6):

Excerpt 1:

"The first accident in 1974 attracted considerable attention because the exhaust fumes of an incineration plant caused an engine to fail. The following parameters were determined during the accident investigation.

Exhaust gas temperature = + 132°C and exhaust gas composition = 79% nitrogen, 5% oxygen, 10% carbon dioxide, and 6 % steam.

The **high temperature of the gases only caused a drop in performance**. The **low oxygen content** of 5% is the same as the oxygen content of air at 11,000 meters, where no (helicopter) engine can operate, they stop functioning. A further problem is the invisibility of these exhaust fumes; the shape and size of a visible exhaust gas cloud merely show the distribution of the steam, and not the heavier components."

Excerpt 2:

"Due to the above case, in their aviation accident information from November 1996, the aviation accident division of the national aviation authority addresses the dangers to helicopter crews during flights in the immediate vicinity of chimneys. As early as 1974, three occupants of a helicopter were killed when the engine failed during hovering flight around a chimney. Autorotation failed in this densely built-up area."

Comment: The dropout of an aeroengine can be caused by different reactions of compressor and combustion chamber:

- **Surge** because the **surge limit/margin is exceeded** (volume 3 Ill. 11.2.1.1-8). Possible causes are a too high **fuel flow** (high pressure in the combustion chamber) as compensation of the power drop, too **low mass flow** because of the low density of the warm air, **deposits** (profile changes, roughness) on the blading.

- **Quenching/flame of the combustion chamber** because **lack of oxygen**. Causes can be the consumption of oxygen by the fire and an insufficient compression of the compressor.

- **Power drop of the aeroengine** because of too low compressor performance, which can not be compensated with more fuel because of high turbine inlet temperature.

A further possibility is the unsuitable **reaction of the control unit** at the behaviour of the aeroengine.

Potential deteriorating media from the maintenence process and influenced regions of the aeroengine.

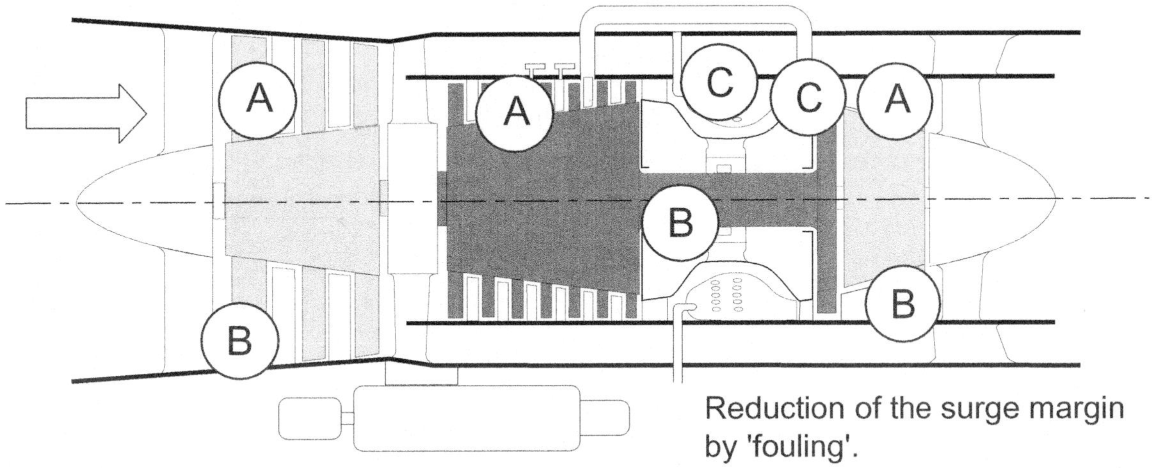

Ill. 5.5-5

Illustration 5.5-5: Media from a maintenence process *can influence, respectively deteriorate an aeroengine. Usually this happenes if the work did* ***not follow the specifications****, respectively* ***not approved media*** *have been used.*
In the following, typical potential deteriorations are compiled.

*- („****A****"),* ***abrasive cleaning media*** *(volume 5 Ill. 19.2.3-1 and Ill. 19.2.3-2).*
- Erosion of soft rub in coatings.
- Roughening of the compressor blading.
- Damage of a paint layer at the blades (aluminium, steel) from elder aeroengine types (Bild 5.3.2-6).
- Clogging of the cooling channels in hot parts like high pressure turbine blades and vanes.

*- („****B****"),* ***washing agents*** *(volume 5 chapter 19.2.3, Ill. 19.2.3-1).*
- Corrosion of sensitive coatings.
- Contamination of the lubrication oil if penetrating into a bearing chamber because of too little sealing air. This situation occurs during the washing process because of too low rotation speeds.
- Contamination of the turbine (danger of high temperature corrosion) through separating dirt from the compressor which accumulates at the turbine surfaces.

*- („****C****"),* ***De-icing fluid*** *(example 5.5-3.2). damages by overheating, run through of the aeroengine with fracture of the rotor.*

Fouled injection water will quickly ruin a compressor.

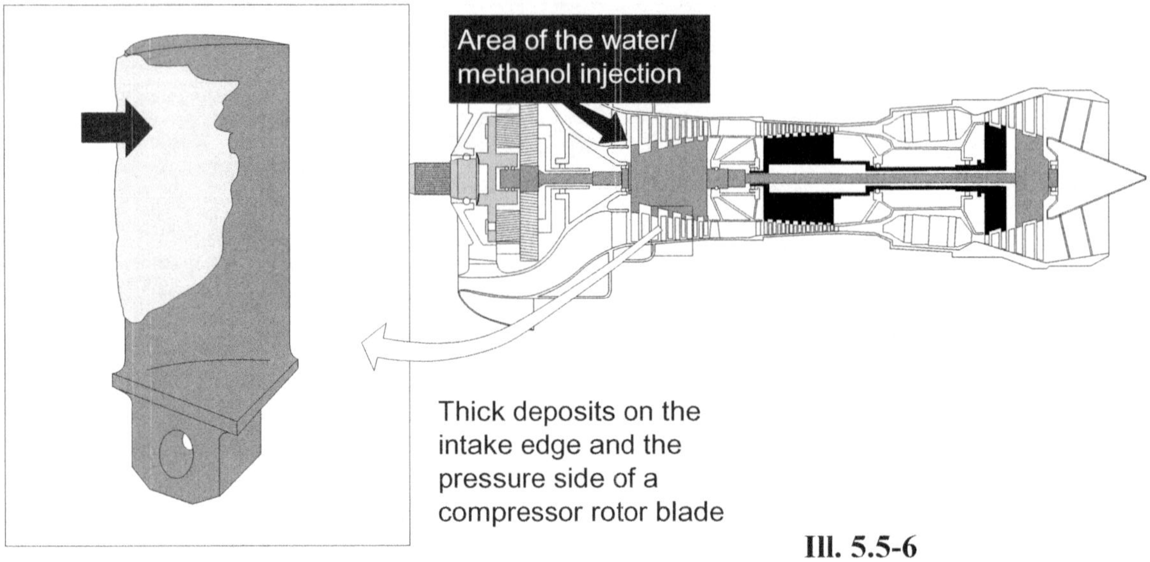

Ill. 5.5-6

***Illustration 5.5-6:** After only several startups following an overhaul, it was noticed that the performance of the engine and the **behavior of the compressor had considerably worsened**. Inspection of the compressor revealed that the compressor rotor blades had a several-millimeter thick, rough, brittle coating on the pressure side and the leading edge. Microanalysis revealed that the primary components of this coating were Ca and Si. Research showed that the water used for the **water-methanol injection** had ten times the allowable conductivity values (a measure of water impurity). The deposits were most likely caused by injection of contaminated water into the compressor on a hot day. It is interesting to note that this amount of coating buildup could be caused by one or few injections.*

***Illustration 5.5-7: Deposits on the conpressor blades** (fouling, volume 5 chapter 19.2.3) lower tht compressor efficiency respectively the aeroengine performance. To compensate this, the fuel supply must be rised. This leads to **increasing hot parts temperatures** with shorter life time, higher repair effort/costs respectively fuel*

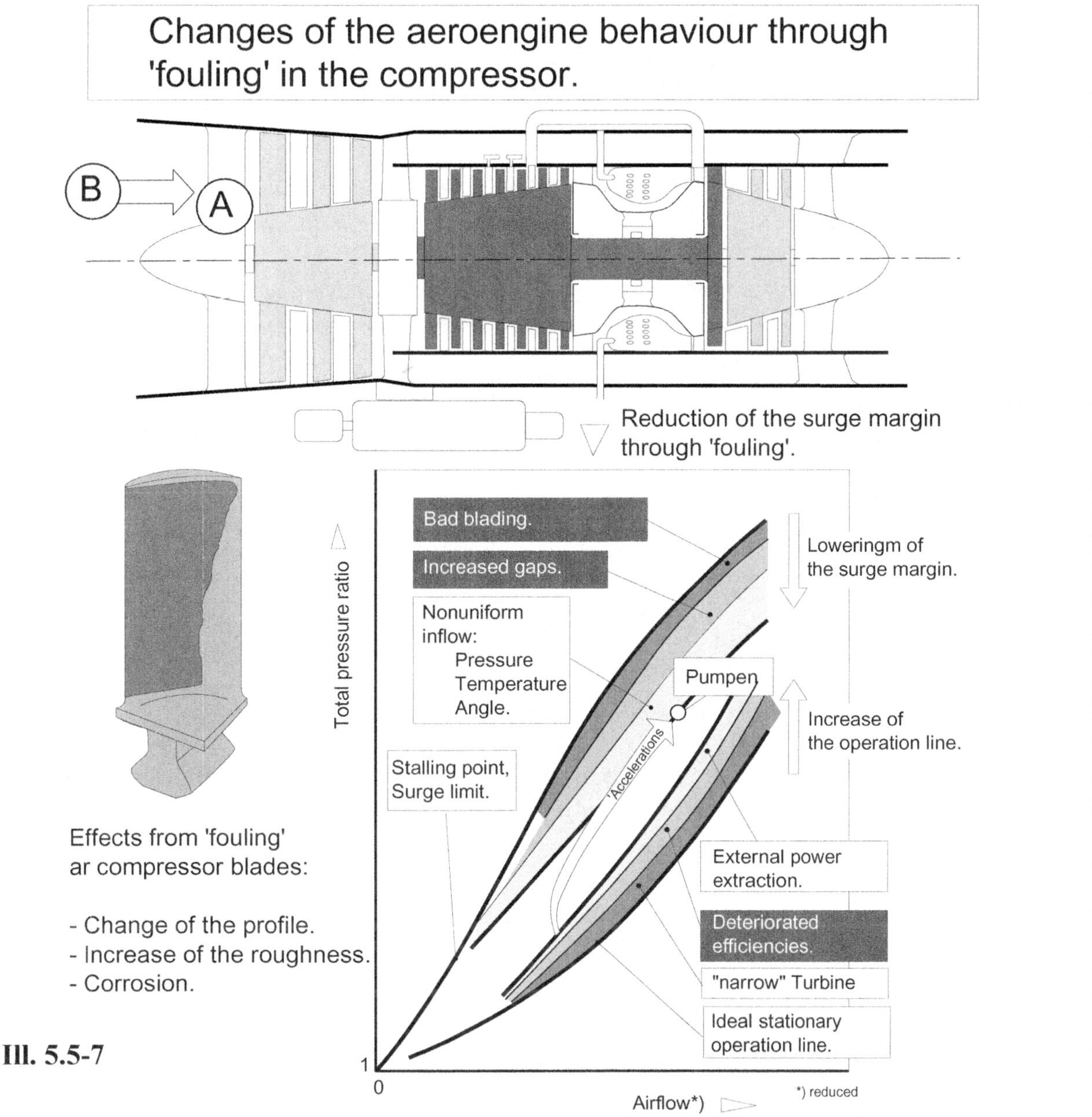

Ill. 5.5-7

consumption and with this also costs (volume 2 Ill. 7.0-2).

*The **surge margin** will be markedly smaller (volume 5 Ill. 19.2.3-13) with the danger of a drop out of the aeroengine and extensive failures (volume 3 Ill. 11.2.1.2-1).*

*The **aeroengine performance drops**. If this is compensated with more fuel, shortens the hot parts lifetime and this increses the **fuel and repair costs**.*

*Fouling develops, when not stocking particles like dust combine with **leakage oil/oil vapour** (e.g., from the front compressor main bearing „**A**“). **Hygroscopic dust** with sea salt can be itself sticky („**B**“).*

*Also the so called **„insecte roughness“ in an extreme case has a similar effect**. Seemingly small aeroengines with radial(!) compressors, which are already operated near the surge margin, are especially affected.*

5.5.1 Measures for Avoiding Unusual Influences

- Preventing accidental fuel infeed:

- Drainage of **unburned fuel remnants** must be done in accordance with regulations, especially after aborted starts.

- The **drainage opening** must be installed so that safe fuel drainage is ensured in every allowable engine position.

- In ground aggregates, a **deflector** should be attached to the **fuel valve neck** along with a warning label with clear instructions for use. If possible, fueling should be avoided while the engines are running.

- The **tank valves** should be in locations that ensure that fuel cannot be ingested in case of valve leaks.

- No engine operation **above larger fires**:

Caution in the vicinity of fires and incinerators (see Example 5.5-5 and Ill. 5.5-6).

If the aircraft is explicitly intended for operation over fires (e.g. for **firefighting**), then appropriate measures are necessary to ensure safe operating behavior of the engines (Ill. 5.5-6). For example, auto re-ignition systems are necessary (Ref. 5.5-5).

- Preventing i**ngestion of foreign materials:**

- If possible, no operation of engines on the ground if **dusts** (such as **fertilizers**) or **lacquer mists** are being released into the air in the vicinity. This is especially true for **field testing rigs** or test runs on the aircraft.

- **Fire extinguishing materials**:

- After fire extinguishing materials have entered into and/or struck the outside of an engine, the **composition of the extinguishing media** must be considered in order to determine which engine parts might be damaged. The compressor and hot parts should be subject to **boroscopic inspections** to help determine future steps. This may result in the realization that cleaning and analysis are unavoidable for estimating continued usability, at least for individual components.

Chapter 5.5 References

5.5-1 W.Rudolph, M. Rieland, "Synthesis and Properties of Various Alternative Fire Extinguishing Agents", AGARD-CP-587, Proceedings of the conference "Aircraft Fire Safety", Dresden, Germany, 14-17 October 1996, Chapter 13, page 13-2.

5.5-2 "BAC 111 Crash", periodical "Aviation Week & Space Technology, September 20, 1971, page 24.

5.5-3 "In der Nähe von Kaminen: Gefahr für Hubschrauber", periodical "Rotorblatt" 4/96, page 41.

5.5-4 "Tödliche Mischung", periodical "Der Spiegel", Nr. 41, 1971, pages 182 and 184.

5.5-5 P.Gilles, "Flameout", periodical "Aviation Maintenance", May 2001, pages 38 to 42.

5.5-6 C. Kjelgaard, Washington DC, "FAA to ban Fokker 100&F28-4000 APU use during deicing", "Air Transport Intelligence News"., 29 March 2002.

5.5-7 „Inadvertent application of deicing fluid in APU", www.fokkerpilot.net, www.ntsb.gov, Fokker 100 N1425A, March 06,2002, page 1-3.

Index

A

Aborted start, undrained fuel 5.5-1
Abradable coatings
- corrosion 5.4.1.2-8
- erosion 5.3.2-9
  - in compressor housings/casings 5.3.2-20

Abrasive
- cleaning material, deteriorations 5.5-15
- wear 5.3.1-5

Acrylglass, erosion 5.3.1-16
Adjustable guide vanes
- bird strike 5.2.2.1-38
- mechansim 5.2.2.1-30

Aeration cell 5.4.1.1-18
Aeroengine
- influence of auxilary media 5.5-15
- measures for firefighting use 5.5-18

Air
- contaminations in the environment, test rigs 5.2.1.3-19
- exchange, corrosion by 5.4.1.1-8
- flow contamination 5.2.1.3-14

Aircraft/airplane
- characteristics by volcanic ash 5.3.2-26
- on the ground, endangered by vulcano ash 5.3.2-27
- operation above chimneys of power- and heating plants 5.5-13

Al/aluminium alloys
- corrosion 5.4.1.2-5, 5.4.2.2-3
  - SCC 5.4.2.2-3

Al-diffusion layer, FOD 5.2.1.1-23
Alkali brittleness 5.4.2.2-11
Altitudes, vulcano ash clouds 5.3.2-29
Angle of attack, influencing wear 5.3.1-9
Ash concentrations, visibility 5.3.2-33
Assembly/installation work, FOD 5.2.1.3-17
Aassessment of vulcano ash influence 5.3.2-33
Auxiliary
- materials
  - corrosion 5.4.1.1-2
  - FOD 5.2.1.3-19
  - triggering SCC 5.4.2.2-1
- units, especially prone to SCC 5.4.2.2-5

Axial face seals, cavitation 5.3.1-23
Axial-radial compressor
- erosion 5.3.2-9
  - zones 5.3.1-19

B

Bathtub curve 5.2.1.3-10
Bearing
- chamber, pressure distribution against FOD 5.2.1.3-19
- front (at the intake) 5.2.1.1-5
- struts, bird strike 5.2.2.3-16

Bending
- moments 5.2.2.3-2
- of a blade 5.2.1.1-10

Bird
- diameter 5.2.2.1-8
- impact
  - at guide vanes 5.2.1.1-5
  - casing struts in the inlet 5.2.2.2-17
  - containment of fragments 5.2.2.3-20
  - high-pressure compressor 5.2.1.1-5
  - impact process on a rotating blade 5.2.2.2-8
  - leading edge stress 5.2.2.2-19
  - one-sided mounted beam 5.2.2.2-6
  - reaction forces on the blade root 5.2.2.2-6
- substitute 5.2.2.2-13
- bird weight, damage at strike 5.2.2.1-7

Bird strike/impact 5.2.1.1-1, 5.2.1.1-5, 5.2.2.1-1, 5.2.2.1-3, 5.2.2.1-30, 5.2.1.3-1, 5.2.2.3-24
- adjustable guide vanes 5.2.2.1-38
- alternative matters/media 5.2.2.2-2
- altitudes 5.2.2.1-6
- at a spinner 5.2.1.1-5
- at fan blading stiffened by clappers 5.2.2.1-26
- at the fuselage 5.2.2.1-41
- bearing struts 5.2.2.3-16
- bird
  - substitute for tests 5.2.2.2-13
  - types 5.2.2.1-16
- blade materials 5.2.2.3-14
- blockage of valves and vents 5.2.2.1-33
- by flocks 5.2.2.1-21
- canada goose 5.2.2.1-25
- casing damage 5.2.2.1-30
- causing oscillations 5.2.2.2-14
- certification 5.2.2.1-2
  - regulations 5.2.2.1-1
  - requirements 5.2.2.3-24
- changes to the blade angle 5.2.2.1-30
- collecting in a s-duct 5.2.2.3-22
- combustion chamber damage 5.2.2.1-33

- compressor surging 5.2.2.1-35
- constructive factors 5.2.2.1-2
- constructive/design influence 5.2.2.3-1
- contact of rotor and staor blades 5.2.2.1-30
- curved inlets, 5.2.2.3-2
- damage 5.2.2.1-1, 5.2.2.1-3
  - depending from weight 5.2.2.1-7
  - distribution at the entire aircraft 5.2.2.1-17
  - in the turbine area 5.2.2.1-33
  - mechanisms 5.2.2.1-1, 5.2.2.2-1
  - of stator vanes 5.2.2.2-14
  - symptoms in the compressor 5.2.2.1-37
  - types 5.2.2.1-32
  - weigh influence 5.2.2.1-7
- damaging rotor shaft 5.2.2.1-10
- deflection 5.2.2.3-12
  - of blades 5.2.2.2-13
- deflector plates during test 5.2.2.1-41
- design without front bearings 5.2.2.1-30
- effect the clapper 5.2.2.1-39
- embrittlement effect 5.2.2.2-3
- energy absorber 5.2.2.3-10
- engine location 5.2.2.3-1
- engine with front bearing 5.2.2.1-30
- fan blades 5.2.1.1-5
  - with clappers 5.2.2.3-20
- fiber reinforced plastics 5.2.2.2-2
- flight
  - phase 5.2.2.1-14
  - situations 5.2.2.1-6 speed 5.2.2.1-2
- flock of birds 5.2.1.1-5
- fracturing of fan blades f 5.2.2.1-25
- fragments of the fuselage 5.2.2.1-41
- friction with the wall 5.2.2.3-2
- front bearing 5.2.2.1-25
- fuselage 5.2.2.1-15
- grills in the engine inlets 5.2.2.3-6
- guide vane 5.2.2.1-40
- haircut 5.2.2.3-15
- helicopters 5.2.2.1-5
- high speed deformation effects 5.2.2.2-3
- high-pressure compressor 5.2.2.3-21
- impact conditions 5.2.2.1-3
- impact forces 5.2.2.1-8
- influence
  - of inlet cross-sections 5.2.2.1-17
  - of S- shaped intake channel 5.2.2.1-20
  - of size 5.2.2.1-9
  - of surrounding an airport 5.2.2.1-16
- inlet
  - cross-section 5.2.2.3-1
  - guide vanes 5.2.2.3-16
  - screen 5.2.2.3-6
- interpretation of allowable damages 5.2.2.3-24
- liner from the air intake duct 5.2.2.1-35
- location of the inlet 5.2.2.3-8
- low speed impact 5.2.1.1-5, 5.2.2.1-10,. 5.2.2.1-20, 5.2.2.1-36, 5.2.2.3-10
  - rotor blades 5.2.2.1-39
- main bearings 5.2.1.1-5
- measures against 5.2.2.1-1, 5.2.2.3-24
  - protection grill 5.2.2.3-2
- metallic melts 5.2.2.1-33
- migratory bird routes 5.2.2.1-16
- military aircraft 5.2.2.1-2, 5.2.2.1-15
  - transport 5.2.2.1-20
- multiple engines 5.2.2.3-4
- multiple engines failing 5.2.2.1-21
- nose cone 5.2.2.1-2
- plastic deformation
  - progress 5.2.2.2-5
  - of the blades 5.2.2.1-28
- power after 5.2.2.1-1
- pressure
  - fluctuations 5.2.2.2-4
  - of the bird mass 5.2.2.1-28
  - schock pressure 5.2.2.2-2
- protection
  - grills 5.2.2.3-7, 5.2.2.3-10
  - studies 5.2.2.3-11
- reaction of a blade 5.2.2.3-12
- resultant vibrations 5.2.2.2-10
- retention chamber' 5.2.2.3-10
- risk
  - around airports 5.2.2.1-5
  - influenced by altitude 5.2.2.1-14
- safer intake design 5.2.2.3-8
- sequence of events 5.2.2.2-1
- schock pressure 5.2.2.2-2
- several aeroengines of a single aircraft 5.2.2.1-18
- shutdown of the engine 5.2.2.1-1
- simulation 5.2.2.2-1
- speeds 5.2.2.1-14
- spinner 5.2.2.1-30, 5.2.2.1-35
  - deformations 5.2.2.1-36
- statistical analysis 5.2.2.1-5
- stresses at leading edge 5.2.2.3-2
- substitute mass 5.2.2.1-2
- swan neck channel/duct 5.2.2.1-10, 5.2.2.3-2, 5.2.2.3-20
- technological influences 5.2.2.3-1
- tests 5.2.2.1-2
  - if constructive changes. 5.2.2.1-2
- the same engine by several birds 5.2.2.1-21
- threaded/bolted connections 5.2.2.1-30
- titanium fires 5.2.2.1-1
- torsion
  - angle change 5.2.2.1-32

fracture of the gas producer shaft 5.2.2.3-8
turboprop engines 5.2.2.1-20, 5.2.2.3-8
typical
damage at large fan engines 5.2.2.1-42
endangered parts 5.2.2.1-30
VTOL 5.2.2.1-32
wide chord blade 5.2.2.1-36, 5.2.2.2-10, 5.2.2.3-19
weight, damage at strike 5.2.2.1-7
definition of 5.2.2.1-3
intake guide vanes 5.2.1.1-5
Blade/blading
Al alloy, corrosion 5.4.1.1-20
bending by FOD 5.2.1.1-10
blending of notches 5.2.1.1-15
edges, sensitivity for FOD 5.2.1.1-16
erosion 5.3.1-1
influenced by bleed valves 5.3.2-13
FOD 5.2.1.1-9
robust against 5.2.1.3-18
fouling 5.2.1.1-2
influence of the location of a notch 5.2.1.1-16
materials, bird strike behaviour 5.2.2.3-14
notches by particles 5.2.1.1-6
plastic collapse 5.2.2.2-5
profile changes 5.3.2-5
shapes, influence on FOD 5.2.1.2-1
wear products 5.3.1-6
Blasting device, removing dust layers 5.3.2-24
Bleed valves, influencing erosion on blades 5.3.2-13
Blending of notches 5.2.1.1-15
Blind holes, danger of SCC at bolts 5.4.2.2-19
Blisk, suitable intake design 5.2.2.3-8
Blistering, of coating due to corrosion 5.4.1.2-23
Blockages in the cooling air
dust 5.2.1.1-5
Blocking of air bores/channels by vulcano ash 5.3.2-26
Bolts /boltings
corrosion 5.4.1.1-7
preventing metallic contacts 5.4.1.1-25
Boltings,
Bond layer, corrosion 5.4.1.2-8
Borescope inspections
compressor 5.2.1.1-15
FOD 5.2.1.3-18
Braze, corrosion at the edge 5.4.1.2-2
Brittle coating, erosion behaviour 5.3.1-9
Burn-in, bath tub curve 5.2.1.3-10
Burnishing, SCC 5.4.2.2-11

C

Cable connections, corrosion 5.4.1.2-18
Canada goose, bird strike 5.2.2.1-25
Carbon fiber
cause short circuits 5.2.1.1-2
triggering corrosion 5.4.1.3-4
Carbon particles from combustion chamber
erosion by 5.3.1-7
impact *5.2.1.1-22*
Casing/housing
abrasive dust 5.3.3-9
wear products 5.3.1-6
Catalog parts, sensitivity for SCC 5.4.2.2-5
Cathodic corrosion protection 5.4.1.2-2
Cavitation 5.3.1-1, 5.3.1-5. 5.3.1-20
axial face seals 5.3.1-23
gear pump in the fuel system 5.3.1-5,
mechanism 5.3.1-20, 5.3.1-22
oil dampened anti friction bearings 5.3.1-23
transmission gears 5.3.1-5
Cavitation corrosion 5.3.1-5, 5.3.1-22
Cavitation erosion 5.3.1-5
Cell action 5.4.1.1-15
corrosion 5.4.1.1-20
Centrifuging of sand in the high pressure compressor 5.2.1.1-5
Certification and rating regulations
bird strike 5.2.2.1-1, 5.2.2.1-2
requirements 5.2.2.3-24
CFRP (carbon fiber reinforced plastic), triggering corrosion 5.4.1.1-15, 5.4.1.1-18
Chimneys of power- and heating plants, aircraft operation above 5.5-13
Chromium steels, corrosion 5.4.2.2-2
Circumferentially-traveling trapped particles, erosion 5.3.2-4
Cirrus cloud, containing vulcano ash 5.3.2-33
Clamping bolt, fracture by corrosion 5.4.1.2-16
Cleaning
baths
corrosion 5.4.1.1-3
containing Cl, inducing SCC 5.4.2.1-17
medium, deterioration by erosion 5.5-15
rag as foreign object 5.2.1.3-6
Clearance increase at blade tips by erosion 5.3.2-9
Clogging, turbine nozzle 5.3.2-21
Coating
against erosion 5.3.1-5
components, corrosion 5.4.1.1-3
containing Cl, danger of SCC in tatanium alloys 5.4.2.1-17
corrosion 5.4.1.1-15
symptoms at sprayed 5.4.1.2-8

crack initiation when brittle 5.3.3-1
delaminateby particle impact 5.3.2-12
particles from armored areas 5.3.2-21
rolling-up 5.3.2-12
Cockpit glass, rain erosion 5.3.1-20
Coke
impact/carbon impact 5.2.1.1-22
particles
in fuel 5.3.1-7
wear by 5.3.1-6
Collecting/accumulating of bird parts at struts 5.2.1.1-5
Combustion chamber
damage by bird strike 5.2.2.1-33
erosion by coke 5.3.2-21
particles from 5.3.1-6, 5.3.2-20
Composition of the extinguishing media 5.5-18
Compressor
blades/blading
compressor
deposits 5.5-16
erosions 5.2.1.1-5
corrosion
at vanes 5.4.1.2-2
adjustment systems 5.4.1.2-18
endangered components 5.4.1.2-2
fouling 5.2.1.1-2
hard facing 5.3.3-6
SCC 5.4.2.2-5
borescope inspections 5.2.1.1-15
cleaning, triggering corrosion 5.4.1.1-3
corrosion
blades 5.4.1.1-5
disks 5.4.1.1-5, 5.4.1.2-11
efficiency, influenced by erosion 5.3.2-6
FOD 5.2.1.1-14
fouling 5.5-13, 5.5-16
by leakage oil 5.5-17
influencing
hot parts temperatures 5.5-16
surge margin 5.5-17
brazed guide vanes, corrosion 5.4.1.2-2
insect roughness 5.5-17
particle trajectories in 5.3.1-13
SCC systems 5.4.2.1-2
surge/-ing
bird strike 5.2.2.1-35
from vulcano ash 5.3.2-27
washing 5.5-13
Condensate formation (water) 5.4.1.2-12
corrosion 5.4.1.1-2, 5.4.1.2-11, 5.4.1.3-2
Conservation oils, corrosion 5.4.1.1-4
Construction/design
changes, bird strike test 5.2.2.1-2
design of the inlet area, erosion 5.3.3-1
factors, bird strike 5.2.2.1-2

measures against erosion 5.3.3-1
Contact corrosion 5.4.1.1-17, 5.4.1.1-18
by fiber-reinforced plastic 5.4.1.1-18
Containment of fragments by bird impact 5.2.2.3-20
Contaminants
ingested 5.2.1.1-7
of air and oil flow 5.2.1.3-14
water-methanol injection 5.5-16
Continuous ignition 5.5-13
Cooling air ducts, blocking by vulcano ash 5.3.2-29
Corroded cross-sections, repairing 5.4.1.3-5
Corrosion 5.2.1.1-2, 5.4.1.1-1, 5.4.1.1-1, 5.4.1.1-3, 5.4.1.1-4, 5.4.1.1-5, 5.4.1.1-7, 5.4.1.1-12, 5.4.1.1-15, 5.4.1.2-2, 5.4.1.2-5, 5.4.1.3-1
abradable coating 5.4.1.2-8
attack at Mg alloy 5.4.1.2-22
blades made from an Al alloy 5.4.1.1-20
blistering 5.4.1.2-23
bond layer 5.4.1.2-8
brazed compressor guide vanes 5.4.1.2-2
by air exchange 5.4.1.1-8
corrosion
by auxiliary materials 5.4.1.1-2
by deposits 5.4.1.1-3
by exhaust gases from munitions 5.4.1.1-3
by fire extinguisher 5.5-4
by fused salt deposits 5.4.1.2-2
by hard facing 5.4.1.1-20
by hot salt, in titanium alloys 5.4.2.1-17
by ingestion 5.4.1.1-8
by peening 5.4.1.1-4
by sea salt 5.4.1.1-3
by splash water in nacelles 5.4.1.2-24
by cell action 5.4.1.1-20
cable connections 5.4.1.2-18
clamping bolt fracture 5.4.1.2-16
cleaning baths 5.4.1.1-3
coatings 5.4.1.1-15
components 5.4.1.1-3
compressor
cleaning 5.4.1.1-3
disk 5.4.1.2-11
stator blades/vanes 5.4.1.2-2
adjustment systems 5.4.1.2-18
condensate formation (condensation water) 5.4.1.1-2, 5.4.1.2-11, 5.4.1.3-2
conservation oils 5.4.1.1-4
conserving compressors as protection 5.4.1.3-1
construction/design, influence at 5.4.1.1-4
crack-inducing mechanisms 5.4.2.1-3
cracks in threaded connections 5.4.2.1-13
damage 5.4.1.1-15
corners/edges 5.4.1.2-5
mechanisms 5.4.1.1-1
deposits 5.4.1.1-15

- during
  - manufacture 5.4.1.3-4
  - overhaul 5.4.1.1-22
- edge of a braze 5.4.1.2-2
- endangered compressor components 5.4.1.2-2
- erroneous fire alarms 5.4.1.2-24
- etching procedures 5.4.1.1-22
- fading/polishing out 5.4.1.3-5
- fiber-reinforced plastics 5.4.1.1-15
- fundamentals 5.4.1.1-1, 5.4.1.1-15
- gearbox housing 5.4.1.1-7
- grain boundaries 5.4.1.1-22
- hand sweat 5.4.1.1-18
- helicopter engines 5.4.1.1-8
- high strength steel 5.4.1.2-24
- hinge
  - bearing 5.4.1.1-7
  - eyes 5.4.1.2-18
- hollow areas 5.4.1.3-4
- hot parts 5.4.1.1-7
- in titanium alloys by hand sweat 5.4.2.1-17
- influenced by
  - design 5.4.1.3-4
  - maintenance problems 5.4.1.2-12
  - surface area ratio 5.4.1.1-17
  - tools 5.4.1.3-5
- intergranular 5.4.1.1-10
- jamming variable guide vane actuation 5.4.1.2-24
- local cell 5.4.1.1-17
- lubricants, unsuitable 5.4.1.1-4
- media
  - from inside the engine itself 5.4.1.1-4
  - ingestion during flight5.4.1.1-8
  - manufacture and overhaul 5.4.1.1-3
  - SCC 5.4.2.1-3
- maintenance caused 5.4.1.3-5
- markers/writings 5.4.1.1-4
- measures/remedies 5.4.1.1-15, 5.4.1.3-1
- metallic contact 5.4.1.3-4
- military aircraft 5.4.1.1-2
- mounting bolts 5.4.1.2-24
- not caused by mechanical loads 5.4.1.2-1
- of hollow structures 5.4.1.1-17
- parked/stored airplanes 5.4.1.1-8
- pittings 5.4.1.1-15
- porous spray coatings 5.4.1.2-8
- prevention, development, testing, and verification 5.4.1.3-4
- processes baths 5.4.1.3-4
- production 5.4.1.1-2
- promoted by silver 5.4.1.1-7
- protection
  - by air-conditioning/dehumidifying unit 5.4.1.3-1
  - by dehumidification 5.4.1.3-1, 5.4.1.3-2
  - cathodic 5.4.1.2-2
  - coatings 5.4.1.3-4
- repair 5.4.1.1-2, 5.4.1.3-4, 5.4.1.3-5
- resistant materials 5.4.1.3-4
- rings 5.4.1.1-18
- roller bearing 5.4.1.1-7
- rub coatings 5.4.1.1-15, 5.4.1.2-8
- sea atmospheres 5.4.1.1-16
- sensitivity 5.4.1.1-17
  - in elder engine types 5.4.1.2-5
    - steels 5.4.1.1-1
- sensitized material 5.4.1.1-22
- splash water in nacelles 5.4.1.2-24
- spline couplings 5.4.1.1-7
- standstill 5.4.1.3-1
  - times/periods 5.4.1.1-2, 5.4.1.2-11
- storage 5.4.1.1-2
  - during manufacture 5.4.1.3-4
- symptoms at sprayed coatings 5.4.1.2-8
- tarnishing 5.4.1.1-12
- tension bolts through rotor disks 5.4.1.2-1
- testing/tests 5.4.1.2-8, 5.4.2.1-14
  - methods 5.4.1.3-5
  - samples 5.4.2.1-14
    - CT 5.4.2.1-17
- triggered
  - by carbon fibers 5.4.1.3-4
  - by TiN 5.4.1.1-15
  - by WC/Co 5.4.1.1-15
- types, crack-initiating 5.4.1.1-10
- washing compressors 5.4.1.3-5
- wear products 5.4.1.1-4
- welded constructions made from steels 5.4.1.1-25
- without mechanical loading 5.4.1.1-15, 5.4.1.2-1

Corrosion fatigue 5.4.1.1-1, 5.4.1.1-10, 5.4.1.1-14

Corrosion system 5.4.1.1-10

Corrosive
- deposits 5.4.1.1-3
- ingestion
  - from fire extinguisher 5.4.1.1-8
  - ground running 5.4.1.1-8
  - low altitudes 5.4.1.1-8
  - media 5.4.1.1-3

Cr coating/layer, SCC under 5.4.2.2-11

Cr Ni steels, SCC 5.4.2.2-2

Cr- steels
- disks made from corrosion-sensitive 5.4.1.1-1
- SCC 5.4.2.2-2

Crack corrosion 5.4.1.1-17

Crack
- growth
  - below the threshold 5.2.1.1-20
  - rate at SCC 5.4.2.2-1
- initiation

brittle coatings 5.3.3-1
corrosion types 5.4.1.1-10
by FOD 5.2.1.1-10
Critical fracture toughness 5.2.1.1-10
Critical impact speed 5.2.1.1-10
FOD 5.2.1.1-21
CT samples for SCC testing 5.4.2.1-17

D

Damage
indirectly by erosive particles 5.3.2-3
influence of mechanism 5.2.1.1-20
mechanisms, foreign objects 5.2.1.1-8
Damping foils 5.3.2-12
Deflection by bird strike 5.2.2.3-12
Deflector at fuel valve neck 5.5-18
Deformation speed, causing embrittlement 5.2.2.3-15
Dehumidification of air, corrosion protection 5.4.1.3-1, 5.4.1.3-2
Deicing medium/fluid 5.5-15
Delamination, coating 5.3.2-12
Delivery flight, FOD danger 5.2.1.1-14
Deposits
corrosion 5.4.1.1-15
on the conpressor blades 5.5-16
vulcano ash 5.3.2-27
Desert environment
hot parts 5.3.2-24
problems on engine 5.3.2-24
Design/construction, influencing corrosion 5.4.1.3-4
Deterioration
by abrasive cleaning media 5.5-15
by vulcano ash 5.3.2-33
by washing fluids 5.5-15
Development phase, erosion tests 5.3.3-1
DOD (domestic object damage = own object damage) 5.2.1.1-1
Drainage/drain system
insufficient 5.5-8
malfunction 5.5-1
Droplet impact/strike 5.3.1-5
at glass 5.3.1-5
at polycarbonate 5.3.1-5
erosion 5.3.1-20
mechanism 5.3.1-21
Dry corrosion from fused salt 5.4.1.2-2
Dust
erosion by 5.3.1-1, 5.3.1-5
blockages in the cooling air 5.2.1.1-5
deposits, filter of a helicopter engine 5.3.2-9
glassy depositions of molten 5.3.2-29
ingestion 5.5-3
from industrial processes 5.5-4
in guide vane bearings 5.3.2-4
layers, removing by blasting device 5.3.2-24
particles, effect of melting 5.3.2-21
Dynamic fatigue, starting from FOD 5.2.1.1-9

E

Economical consequences of vulcano ash 5.3.2-33
Efficiency of particle separator/filter 5.3.3-4
Elder aeroengine types, corrosion-sensitive materials 5.4.1.2-5
Electrical systems, damage by vulcano ash 5.3.2-27
Embrittlement
by strain 5.2.1.1-17
effect during bird strike 5.2.2.2-3
influenced by deformation speed 5.2.2.3-15
under high-speed 5.2.2.2-6
Enamel coating, erosion 5.3.1-16
Energy absorber, bird strike 5.2.2.3-10
Engine/aeroengine
problems through desert environment 5.3.2-24
inlet/intake
design against FOD 5.2.1.3-19
filter devices at helicopter 5.3.3-2
underneath the fuselage, FOD problem 5.2.1.2-6
filters 5.3.3-2
location, influence at bird strike risk 5.2.2.3-1
mounted on the rear, FOD 5.2.1.1-12
ER-coating, erosion resistant coating 5.3.3-8
Erosion 5.2.1.1-2, 5.2.1.1-5, 5.3.1-1, 5.3.3-1
abradable coatings 5.3.2-9
in compressor housings/casings 5.3.2-20
acrylglass 5.3.1-16
axial-radial compressors 5.3.2-9
behaviour of brittle coating 5.3.1-9
blading/blades 5.3.1-1
hooking 5.3.2-13
by abrasive cleaning media 5.5-15
by carbon particles 5.3.1-7
by cavitation 5.3.1-5
by circumferential-traveling trapped particles 5.3.2-4
by coke from the combustion chamber 5.3.2-21
by particles from water ice 5.3.2-4
by rain 5.3.1-5
drops 5.2.1.1-6, 5.3.1-1
by sand 5.3.2-15
by vulcano ash 5.3.2-26
ceramic thermal barriers 5.3.1-2
characteristic zones 5.3.1-19
clearance increases 5.3.2-9
compressor
blades 5.2.1.1-5
rotor spacer 5.3.2-18
constructive

design of the inlet area 5.3.3-1
measures 5.3.3-1
damage 5.3.1-1, 5.3.2-1, 5.3.2-11, 5.3.2-12
of the spinner 5.3.3-5
droplet strikes 5.3.1-20
dust oncentrates in the hub area 5.3.1-19
effective parameters 5.3.1-3
enamel coating 5.3.1-16
fan design 5.3.3-5
fiber-reinforced materials 5.3.1-2, 5.3.1-13, 5.3.1-20
galvanic coatings 5.3.3-1
glass 5.3.1-9, 5.3.1-16
hard facing 5.3.3-1
helicopters 5.3.1-2
influencing singularities 5.3.2-17
inner casing 5.3.3-9
in labyrinths 5.2.1.1-5
intensifying oxidation 5.3.2-4
incubation period 5.3.1-24
influenced
by inlets configuration 5.3.2-14
inlet duct 5.3.1-20
by S-shaped ducts 5.3.2-17
by side-mounted inlets 5.3.2-9
erosion
influencing compressor efficiency 5.3.2-6
inspections 5.3.3-1
labyrinths 5.3.1-16
lacquers 5.3.3-1
loading effect 5.3.1-16
material parameters 5.3.1-16
measures/remedies 5.3.3-1
mechanisms, 5.3.1-1, 5.3.1-3
energy transfer 5.3.1-9
multi layer coating 5.3.3-8
on blades near bleed valves 5.3.2-13
optimizing operation 5.3.3-1
oxidation combination effect of 5.3.2-21
particles
erosive 5.3.1-6
from the thermal barrier coating 5.3.2-20
in the inlet air 5.3.1-6
selecting for tests 5.3.1-11
strength and hardness 5.3.1-11
size 5.3.1-15
speed 5.3.1-15
parameters of operation 5.3.1-20
pattern on rotor spacer 5.3.2-18
pipe bends 5.3.2-4
plexiglass 5.3.1-16
properties of the eroded part 5.3.1-3
protection/protective
coating 5.3.1-5
fiber-reinforced plastics 5.3.1-16
filter cartridges 5.3.3-3
hard facing of blades 5.3.3-6
titanium nitride (TiN) 5.3.3-2, 5.3.3-8
wear-resistant, hard coatings 5.3.3-2
wire mesh/sieve 5.3.3-4
rate 5.3.1-24
influenced by hardness 5.3.1-11
particle size 5.3.1-15
radial compressors 5.3.2-9, 5.3.2-14
removing of protective oxide coatings 5.3.2-21
rubber or PU coatings 5.3.1-11
sand storm 5.3.2-14
spacer ring 5.3.2-18
spalled ceramic thermal barriers 5.3.1-2
stator vanes 5.3.1-13
steel plates 5.3.1-11
tests 5.3.3-1
selecting particles 5.3.1-11
TiN coating 5.3.1-13
titanium alloys 5.3.1-9
under influence of liquids 5.3.1-20
VTOL aircraft 5.3.1-2
wear, by ingested dust 5.2.1.1-2
resistant, hard coatings 5.3.3-2
= ER-coating 5.3.3-8
Wire mesh/sieve, erosion protection 5.3.3-4
zones
axial-radial compressor 5.3.1-19
radial compressor 5.3.1-19
Erroneous fire alarms due to corrosion 5.4.1.2-24
Etching procedures, corrosion 5.4.1.1-22
Excavation of material by FOD 5.2.1.1-10
Exhaust gases from munitions, corrosion by 5.4.1.1-3
Extinguishing materials/media 5.5-5
composition of the 5.5-18
systems, foam type 5.5-6
Extrusions on material surface 5.4.2.1-7

F

Fan area
blading/blades FOD 5.2.1.1-5
bird strike 5.2.1.1-5
fracturing 5.2.2.1-25
plastic deformations 5.2.2.1-28
with clappers, 5.2.2.3-20
stiffened 5.2.2.1-26
designed against erosion 5.3.3-5
Fastening devices as foreign objects 5.2.1.3-1
Fatigue
crack, caused by FOD 5.2.1.1-11
fractures from FOD 5.2.1.1-12
strength, drop due to rain erosion 5.3.1-20
Fertilizer dusts, ingestion 5.5-4
Fiber reinforced plastics (FRP)

bird strike 5.2.2.2-2
corrosion 5.4.1.1-15
contact corrosion by 5.4.1.1-18
erosion 5.3.1-2, 5.3.1-13, 5.3.1-20
by rain 5.2.1.1-6, 5.3.1-5
protection 5.3.1-16
hail 5.2.1.1-6
Field testing rigs
FOD risk 5.2.1.3-19
vicinity 5.5-18
Fighter aircraft, FOD 5.2.1.2-6
Filter
cartridges, erosion protection 5.3.3-3
engine intake 5.3.3-2
helicopter engine inlet 5.3.3-2
particles, cleaning effect 5.3.3-1
systems, helicopter 5.3.3-4
Fine erosion 5.3.1-3
Fire
alarms due to corrosion 5.4.1.2-24
extinguisher
corrosive ingestion 5.4.1.1-8
foam 5.5-6
trigering corrosion 5.5-4
fighting, measures at the engine 5.5-18
ingestion of media from 5.5-4, 5.5-18
Flame out due to vulcano ash 5.3.2-27
Flight speed at bird strike 5.2.2.1-2
Flue gases from amunition 5.4.1.1-3
Foam extinguishing systems 5.5-6
FOD 5.2.1.1-1, 5.2.1.1-5, 5.2.1.1-6, 5.2.1.3-17
assembly/installation work 5.2.1.3-17
at Al-diffusion layer 5.2.1.1-23
at small engines 5.2.1.3-6
auxiliary materials 5.2.1.3-19
bearing chamber pressure distribution 5.2.1.3-19
blading 5.2.1.1-9
design 5.2.1.3-18
boroscopic inspections 5.2.1.3-18
by personal equipment 5.2.1.3-2
cause/caused by
from landing gear 5.2.1.1-12
groundvortex 5.2.1.2-1
hammer shock' 5.2.1.2-21
maintenance 5.2.1.3-9
thrust reverser 5.2.1.2-1, 5.2.1.2-9, 5.2.1.3-18
causing
fatigue crack 5.2.1.1-11
mechanical overstress 5.2.1.1-9
cleaning rag 5.2.1.3-6
compressor 5.2.1.1-14
consequences at rotor blades 5.2.1.1-11
crack initiation 5.2.1.1-10
critical impact speed 5.2.1.1-21
cylindrical 5.2.1.3-8
damage mechanisms 5.2.1.1-8
danger
at delivery flight 5.2.1.1-14
by pop rivet 5.2.1.2-21
dynamic fatigue 5.2.1.1-9
engine inlet design 5.2.1.3-19
engine
inlets underneath the fuselage 5.2.1.2-6
mounted on the rear 5.2.1.1-12
excavation of material 5.2.1.1-10
fastening devices 5.2.1.3-1
fatigue fractures from 5.2.1.1-12
field testing rigs 5.2.1.3-19
fighter aircraft 5.2.1.2-6
fire at titanium compressor blades 5.2.1.1-15
from adjustable shutter and ramps systems 5.2.1.2-20
geometry of the impact marks 5.2.1.3-13
height of an inlet above the ground 5.2.1.2-14
identifivation 5.2.1.3-1, 5.2.1.3-9
intake 5.2.1.1-8
impact
direction 5.2.1.1-22
marks 5.2.1.3-7
imprints 5.2.1.3-8
influence/influenced
by vortex diameter 5.2.1.2-12
of impact speed 5.2.1.1-21
of the location of the engines 5.2.1.2-5
of blade shapes 5.2.1.2-1
inlet/intake design 5.2.1.3-17
ingestion/intake 5.2.1.2-1
high on the wing, 5.2.1.2-12
inlet ducts 5.2.1.3-12
insensitive materials 5.2.1.3-18
labyrinth seals 5.2.1.3-19
location at turbine blades 5.2.1.1-22
manufacturing residue 5.2.1.3-19
measures against 5.2.1.3-17
metric or non-metric thread 5.2.1.3-8
minimizing by inlet shutter mechanism 5.2.1.2-19
notch (-es) 5.2.1.1-9
effect 5.2.1.1-15
oscillating 5.2.1.3-12
periphery of running engines 5.2.1.3-17
pop rivet 5.2.1.2-21
preventing/preventive
further 5.2.1.3-9
ingestion 5.2.1.2-1
measures 5.2.1.3-1, 5.2.1.3-17
projectiles 5.2.1.3-1
protecting/protection
grills 5.2.2.3-7
shutter mechanism 5.2.1.2-14
rear mounted engines 5.2.1.2-5

FOD
FOD

ricochet 5.2.1.3-8
remnants point at source 5.2.1.3-9
residual stress 5.2.1.1-9
rust 5.2.1.3-19
sealants 5.2.1.3-19
sensitivity of blade edges 5.2.1.1-16
FOD
sucked up from the ground 5.2.1.2-1
surface, EDX analyzed 5.2.1.3-9
TBC spalling by 5.2.1.1-24
thermal barrier coating 5.2.1.1-23, 5.2.1.1-24
thread imprints 5.2.1.3-8
titanium alloy 5.2.1.1-21
    blades 5.2.1.1-10
        sensitivity 5.2.2.3-19
tools 5.2.1.3-1
turbine rotor blades 5.2.1.1-23
types 5.2.1.1-8, 5.2.1.3-1
typical characteristics 5.2.1.3-13
    cylindrical 5.2.1.3-8
    geometry of the impact marks 5.2.1.3-13
    human-related 5.2.1.3-2
    identifivation 5.2.1.3-1, 5.2.1.3-9
        intake 5.2.1.1-8
    imprints 5.2.1.3-8
Fouling 5.2.1.1-2
  by hygroscopic dust 5.5-17
  compressor 5.5-13, 5.5-16
    by leakage oil 5.5-17
Fracture toughness, critical 5.2.1.1-10
Fretting (wear), increased by erosion 5.3.2-4
Front bearing 5.2.1.1-5
  bird strike 5.2.2.1-25, 5.2.2.1-30
Fuel
  fueling, ingestion into the engine 5.5-8
  nozzle, clogging by dust 5.3.2-21
  released in flight 5.5-8
  salt deposits, corrosion by 5.4.1.2-2
  system, cavitation 5.3.1-5
    valve neck, deflector 5.5-18
Fuselage, bird strike 5.2.2.1-15

G

Galvanic coatings, erosion 5.3.3-1
Gaseous media, ingestion 5.5-3
  from vulcano clouds 5.3.2-27
Gear pump
  cavitation 5.3.1-5, 5.3.1-22
Gearbox housing, corrosion 5.4.1.1-7
Gears, SCC 5.4.2.2-11
Glass
  droplet impact 5.3.1-5
  erosion 5.3.1-9, 5.3.1-16
Glass beeds, shot-peening with 5.4.1.2-2
Glassy depositions of molten dust 5.3.2-29
Grain boundaries, corrosion 5.4.1.1-22
Ground
  running, corrosive ingestion 5.4.1.1-8
  vortex/inlet vortex formation, specific conditions 5.2.1.2-10
    causing FOD 5.2.1.2-1
Guide vanes, bird strike 5.2.1.1-5, 5.2.2.1-40

H

Hail 5.2.1.1-6
  damaging fiber-reinforced plastic (FRP) 5.2.1.1-6
Haircut by bird strike 5.2.2.3-15
Hammer shock 5.2.1.2-21
  causing OOD (own foreign objects) 5.2.1.2-21
Hand sweat
  corrosion 5.4.1.1-18
    SCC in titanium alloys 5.4.2.1-17, 5.4.2.2-15
Hard coatings/hard facing,
  against erosion/as erosion protection 5.3.3-1, 5.3.3-2
    at compressor blades 5.3.3-6
  corrosion triggered by 5.4.1.1-20
  hard particles impact 5.2.1.1-10
  WC/Co 5.3.3-1
Hardness, influencing SCC 5.4.2.2-2
HCF fractures 5.2.1.1-20
Helicopter engines
  air inlet ducts 5.2.1.3-12
  birdstrike 5.2.2.1-5
  corrosion 5.4.1.1-8
  dust filter 5.3.2-9
  erosion 5.3.1-2
    by rain 5.3.1-20
  filter systems 5.3.3-4
  sand erosion 5.3.3-7
  separation systems 5.3.3-2, 5.3.3-4
High pressure compressor
  bird impact /bird strike 5.2.1.1-5, 5.2.2.3-21
  centrifuging of the sand 5.2.1.1-5
High-pressure water jet for cleaning 5.3.1-21
High speed
  deformation, effect on blade material 5.2.2.2-3
  embrittlement due to 5.2.2.2-6
  impact 5.2.2.2-17
High strength material
  sensitive for SCC 5.4.2.1-1
  steel, corrosion 5.4.1.2-24
High temperature corrosion 5.4.1.1-1, 5.4.1.1-16
Hinge
  bearing, corrosion 5.4.1.1-7
  eyes, corrosion 5.4.1.2-18
Hollow areas/spaces/volumes, corrosion 5.4.1.3-4
Hooking, erosion shape on blades 5.3.2-13

Hot gas recirculation, vertical take-off aircrafts 5.5-13
Hot parts
corrosion and sulfidation 5.4.1.1-7
in desert environment 5.3.2-24
MoS2 triggering SCC 5.4.2.2-16
SCC 5.4.2.2-16
temperatures by compressor fouling 5.5-16
vulcano ash burning in 5.3.2-27
Hot salt
corrosion in titanium alloys 5.4.2.1-2, 5.4.2.1-17, 5.4.2.2-3, 5.4.2.2-14
Hub area of a compressor rotor, erosion by dust to concentration 5.3.1-19
Hugoniot pressure (bird impact) 5.2.2.2-4
Hydrogen-induced cracks and fractures 5.4.1.1-1
Hygroscopic dust, 5.4.2.1-2
compressor fouling by 5.5-17

I

Anti cing/deicing fluid, ingestion into the engine 5.5-8
Identification of foreign objects 5.2.1.3-1, 5.2.1.3-9, 5.2.1.3-17
Ignition, activated if vulcano ash danger 5.3.2-28
continuous 5.5-13
Impact
conditions during bird strike 5.2.2.1-3
of hard partic 5.2.1.1-10
direction, FOD 5.2.1.1-22
forces at bird strike 5.2.2.1-8
marks of a FOD 5.2.1.3-7
geometry of the foreign object 5.2.1.3-13
speed
critical 5.2.1.1-10
FOD 5.2.1.1-21
Impregnation systems against porosity 5.4.1.3-5
Increased fretting wear, erosion 5.3.2-4
Injection system, media confusion 5.5-8
Inlet air temperature, influencing performance 5.5-13
Incubation period, erosion 5.3.1-24
Industrial atmosphere 5.4.1.1-3
Infant mortality (bath tub curve) 5.2.1.3-10
Influence of a notch location on a blade 5.2.1.1-16
Ingestion/intake
at testing rigs 5.5-3
corrosion 5.4.1.1-8
dust 5.5-3
erosive wear 5.2.1.1-2
fertilizer 5.5-4
from industrial processes 5.5-4
fire extinguishing materials 5.5-18
foreign objects 5.2.1.2-1
fuel released in flight 5.5-8
gaseous media 5.5-3
lacquer mists 5.5-3, 5.5-18
media of fire extinguishing 5.5-4
Injection (into the compressor)
maintenance error 5.5-10
water 5.5-10
water-methanol 5.5-8
Inlet/intake
air, foreign materials 5.5-1
containing erosion particles 5.3.1-6
configuration, influencing erosion 5.3.2-14
cross-section, risk of bird strike 5.2.2.3-1, 5.2.2.3-8, 5.2.2.1-17
design, influencing FOD 5.2.1.3-17
duct
helicopter 5.2.1.3-12
influence on erosion 5.3.1-20
guide vanes, bird strike 5.2.2.3-16
location against bird strike
oxygen supply 5.5-13
screen against bird strike 5.2.2.3-6
vortex / ground vortex 5.2.1.2-10
Insect roughness 5.2.1.1-2
in the compressor 5.5-17
Inspection
rivets 5.2.1.3-17
threaded fasteners 5.2.1.3-17
erosion 5.3.3-1
Insufficient drainage of unburned fuel 5.5-1
Intake
design, suitable for minimum FOD on blisk 5.2.2.3-8
duct, S- shaped bended 5.2.2.1-20
grill-like structures in the engine inlet 5.2.2.3-2
guide vanes, bird strikes 5.2.1.1-5
minimum distance of persons 5.2.1.3-18
stator, FOD 5.2.1.1-5
Intergranular corrosion 5.4.1.1-10, 5.4.1.1-14, 5.4.1.1-22

K

L

Labyrinth seals
erosion 5.2.1.1-5, 5.3.1-16
FOD 5.2.1.3-19
wear products 5.3.1-6
Lacquer mists 5.5-4
erosion 5.3.3-1
ingested 5.5-3, 5.5-18
Landing gear
FOD caused from 5.2.1.1-12
mud guards 5.2.1.3-17
Laser peening 5.2.1.1-15
Leading edge (blade), stresses at bird strike 5.2.2.3-2

Leakage oil, triggering fouling 5.5-17
Loading effect, erosion 5.3.1-16
Local cell (corrosion) 5.4.1.1-17
Location
position of the engines, influence on FOD 5.2.1.2-5
of the inlet from propeller engines 5.2.2.3-8
Low speed impact 5.2.2.2-17, 5.2.2.1-20
bird strike 5.2.1.1-5,5.2.2.1-10, 5.2.2.1-36,5.2.2.3-10
at rotor blades 5.2.2.1-39
Lubricants, causing corrosion 5.4.1.1-4

M

Machined carbon fiber surfaces 5.4.1.1-18
Magnesium alloys, corrosion 5.4.1.2-5
Main bearings, influence of bird strike 5.2.1.1-5
Maintenance
corrosion 5.4.1.3-5
causing FOD 5.2.1.3-9
error, injection fluid 5.5-10
media, influence at the aeroengine 5.5-15
problems, influencing corrosion 5.4.1.2-12
Manufacture(-ing) process
corrosion 5.4.1.3-4
media 5.4.1.1-3
SCC corrosion in titanium alloys 5.4.2.1-17
at Ti-alloys 5.4.2.2-11
steps or parameters 5.4.2.2-1
residue, causing FOD 5.2.1.3-19
Markers/writings, corrosion 5.4.1.1-4
Materials
insensitive against FOD 5.2.1.3-18
notch impact strength 5.2.2.3-14
Measures
against erosion damage 5.3.1-1
against vulcano ash 5.3.2-33
Mechanical overstress by foreign objects 5.2.1.1-9
Media (liquid), confusion at injection system 5.5-8
Metallic contact, corrosion 5.4.1.3-4
Mg (magnesium) alloy, corrosion attack 5.4.1.2-22
Military aircraft
bird strike 5.2.2.1-2, 5.2.2.1-15
corrosion 5.4.1.1-2
Military engine, overhauls 5.2.1.1-2
Mineral wear 5.3.1-5
MoS_2, triggering SCC in hot parts 5.4.2.2-16
Mounting bolts (shear bolts), corrosion 5.4.1.2-24
Mud guards at landing gear 5.2.1.3-17
Multiple
engines bird strike 5.2.2.1-21, 5.2.2.3-4
splined shafts, SCC 5.4.2.2-11

N

Ni-alloys, SCC by sulfur 5.4.2.1-2
Nitrogen-hardened stainless 18 8 steel, SCC 5.4.2.2-8
Nose cone or spinner, bird strike 5.2.2.1-2
Notch
at blading 5.2.1.1-6
by FOD 5.2.1.1-9
impact strength of materials 5.2.2.3-14
kinds 5.2.1.1-18
on the blade, influence of the location 5.2.1.1-16
smooth out /blending 5.2.1.1-15
Nozzle (fuel), clogging by dust 5.3.2-21

O

Objects, caused by thrust reverser 5.2.1.2-9
Oil dampened anti friction bearings, cavitation 5.3.1-23
Oil flow, contamination 5.2.1.3-14
OOD (own object damage = DOD domestic object damage) 5.2.1.1-1
Operating
aircraft over fires (e.g., fire fighting) 5.5-13
temperatures triggering SCC 5.4.2.2-1
Oscillating/oscillations
by bird strike 5.2.2.2-14
of foreign object 5.2.1.3-12
Overhaul
corrosion 5.4.1.1-22
media 5.4.1.1-3
military engine 5.2.1.1-2
philosophy (e.g., peace oriented) 5.2.1.3-10
Overspeed (of an aeroengine)
by undrained fuel 5.5-1
by deicer fluid 5.5-15
Own object damage = OOD 5.2.1.1-1
Oxidation
effect of combination with erosion 5.3.2-21
intensified by erosion 5.3.2-4
Oxygen supply/content at infake 5.5-13

P

Parked/stored airplanes, corrosion 5.4.1.1-8
Particle(-s)
causing notches and burrs 5.3.2-5
erosive 5.3.1-6
filters 5.3.3-1
from combustion chamber 5.3.1-6
from labyrinths 5.3.1-6
from rub coatings 5.3.1-6
from thermal barrier coatings 5.3.1-6
from the combustion chamber 5.3.2-20

from water ice, erosion 5.3.2-4
impact, coating delamination 5.3.2-12
separator/filter, efficiency 5.3.3-4
types at helicopters 5.3.3-2
size
erosion 5.3.1-15
following the air flow 5.3.1-18
impact speed 5.3.1-15
trajectories in compressor 5.3.1-13
separators 5.3.3-1
strength and hardness influencing erosion 5.3.1-11
vulcano ash 5.3.2-26
Peace oriented military overhaul philisophy 5.2.1.3-10
Peening
corrosion caused from 5.4.1.1-4
wear 5.3.1-5, 5.3.1-9
Perchloroethylene degreasing baths, inducing SCC 5.4.2.2-19
Performance
drop due to recirculation 5.5-13
due to increased inlet air temperature 5.5-13
Personal equipment as foreign object 5.2.1.3-2
Pipe bends, erosion 5.3.2-4
Pitting corrosion 5.4.1.1-11
Plastic/ductile
collapse of a blade 5.2.2.2-5
deformed zones, SCC 5.4.2.2-13
Plexiglass (polyxmethylmethacrylate), erosion 5.3.1-16
Polycarbonate, droplet impact 5.3.1-5
Pop rivet, FOD-danger of 5.2.1.2-21
Porosity, impregnation systems 5.4.1.3-5
Porous spray coatings, corrosion 5.4.1.2-8
Power (of aeroengine) after bird strike 5.2.2.1-1
Preventing
damage from volcanic ash 5.3.2-27
further foreign object damage 5.2.1.3-9
measures against FOD 5.2.1.3-1
metallic contacts at bolts 5.4.1.1-25
the ingestion of foreign 5.2.1.2-1
Process baths, triggering corrosion 5.4.1.3-4
Production, causing corrosion 5.4.1.1-2
Projectiles as foreign object 5.2.1.3-1
Propeller engines, bird strike 5.2.2.3-8
Protection/protective
coating against erosion 5.3.1-5
problems due to removing oxide by erosion 5.3.2-21
grills against bird strike 5.2.2.3-7, 5.2.2.3-10

R

Radial compressor
erosion 5.3.2-9, 5.3.2-14
zones 5.3.1-19
Rain 5.2.1.1-6
Rain erosion 5.2.1.1-6, 5.3.1-5, 5.3.1-20, 5.3.1-25, 5.3.1-26
at fiber-reinforced plasics 5.2.1.1-6
at helicopter rotor blades 5.3.1-20
fiber-reinforced plastics 5.3.1-5
incubation period 5.3.1-24
influence 5.3.1-25
at fatigue strength 5.3.1-20
droplet size 5.3.1-26
diameters 5.3.1-26
impact angle 5.3.1-28
impact speed 5.3.1-27
on cockpit glass 5.3.1-20
surface roughness 5.3.1-28
Random failures 5.2.1.3-10
Reaction layers, Cl triggering SCC 5.4.2.2-11
Realistic testing 5.4.1.3-1
Rear mounted engines, FOD 5.2.1.2-5
Remained parts/remnants of a foreign object 5.2.1.3-7
source 5.2.1.3-9
Repair
corrosion 5.4.1.1-2, 5.4.1.3-4
of corroded cross-sections 5.4.1.3-5
Residual stress by FOD 5.2.1.1-9
Retention chamber against bird strike 5.2.2.3-10
Recirculation, drop of performance 5.5-13
Rinsing, SCC 5.4.2.2-19
Roller bearing, corrosion 5.4.1.1-7
Rolling-up of a coating/paint 5.3.2-12
Rollover (plastic deformation) 5.3.2-11
Rotor
blades, consequences from FOD 5.2.1.1-11
disks, corrosion at holes for tension bolts 5.4.1.2-1
fracture due to deicing fluid 5.5-15
shaft, damage by bird strike 5.2.2.1-10
Rub coatings
corrosion 5.4.1.1-15, 5.4.1.2-8, 5.4.1.1-5
particles from 5.3.1-6
Rubber coatings
from PU, erosion 5.3.1-11
wear behaviour 5.3.1-9
Rust as FOD 5.2.1.3-19

S

S-duct (shaped housing/casing ducts)
collecting bird fragments 5.2.2.3-22

influencing erosion 5.3.2-17
Sample for corrosion tests 5.4.2.1-14
Sand and dust erosion 5.2.1.1-5, 5.3.1-5, 5.3.2-15
at helicopter engines 5.3.3-7
Sand storm, erosion 5.3.2-14
SCC see also 'stress corrosion cracking'
Al alloys 5.4.2.2-3
auxiliary units especially prone 5.4.2.2-5
belowm Cr coating 5.4.2.2-11
by boiling condesate during operation 5.4.2.2-9
by burnishing 5.4.2.2-11
by reaction layers containing Cl 5.4.2.2-11
by rinsing 5.4.2.2-19
by unallowable deviations 5.4.2.2-1
cathodic protective effects 5.4.2.2-19
compressor blading 5.4.2.2-5
corrosive media 5.4.2.1-3
Cr Ni steels 5.4.2.2-2
Cr steels 5.4.2.2-2
crack 5.4.2.2-13
growth rate 5.4.2.2-1
crack-inducing mechanisms 5.4.2.1-3
danger of SCC at bolts in blind holes 5.4.2.2-19
during welding 5.4.2.2-11
electrochemical testing method 5.4.2.2-19
fractured surface, typical characteristics 5.4.2.1-10
gears 5.4.2.2-11
hot salt on titanium alloys 5.4.2.1-2
in hot parts 5.4.2.2-16
in plastic deformed zones 5.4.2.2-13
in process baths
cleaning baths containing Cl 5.4.2.1-17
degreasing baths by perchloroethylene 5.4.2.2-19
influenced
by anufacturing steps or parameters 5.4.2.2-1
by hardness 5.4.2.2-2
mechanism at titanium alloys 5.4.2.1-9
MoS_2 triggering SCC in hot parts 5.4.2.2-16
multiple-splined shafts 5.4.2.2-11
nitrogen-hardened stainless 18 8 steel 5.4.2.2-8
of titanium alloys
by hot salt corrosion 5.4.2.2-14
by manufacturing processes 5.4.2.2-11
preventing 5.4.2.2-18
by shot peening 5.4.2.1-5
reaction layers containing Cl 5.4.2.2-11
reproducing 5.4.2.1-17
screws/bolts for pipe and hose 5.4.2.2-5
sensitivity of catalog parts 5.4.2.2-5
sensitization of material during operation 5.4.2.2-7
step-by-step crack growth 5.4.2.1-9
stress
above a threshold 5.4.2.1-3
induced 5.4.2.1-9
systems in compressors 5.4.2.1-2
tension bolt 5.4.2.2-7
tests 5.4.2.1-14
of Ti alloys 5.4.2.2-19
titanium (TI) alloys 5.4.2.1-17, 5.4.2.2-3, 5.4.2.2-11
triggered
by auxiliary materials 5.4.2.2-1
by operating temperatures 5.4.2.2-1
by trichlorethylene 5.4.2.2-11
by watery media 5.4.2.2-2
variable vanes 5.4.2.2-8
Schock pressure (see also hygoniot pressure) 5.2.2.2-4
bird strike 5.2.2.2-2
Screws/bolts for pipe and hose, SCC 5.4.2.2-5
Sea
atmospheres, corrosion 5.4.1.1-16
salt corrosion by **5.4.1.1-3**
Sealants,
as FOD 5.2.1.3-19
for porous materials 5.4.1.3-5
Segregation corrosion 5.4.1.1-12
Seizing of contacting surfaces by vulcano ash 5.3.2-26
Sensitization
corrosion 5.4.1.1-22
of material, SCC duing operation 5.4.2.2-7
Separation
Separators for particles 5.3.3-1, 5.3.3-2
systems at helicopters 5.3.3-4
Short circuits by carbon fiber 5.2.1.1-2
Shot peening
preventing SCC 5.4.2.1-5
with glass beads 5.4.1.2-2
Shutdown of the engine after bird strike 5.2.2.1-1
Side-mounted inlets, influencing erosion 5.3.2-9
Silver promoting corrosion 5.4.1.1-7
Soot
deposits in the compressor 5.4.1.1-3
wear by 5.3.1-6
Sound at tapping a part, dull sound 5.4.1.1-22
Spacer ring (at rotor)
erosion 5.3.2-18
pattern 5.3.2-18
wear products 5.3.1-6
Spalling, TBC by FOD 5.2.1.1-24
Stress corrosion cracking (see SCC) 5.4.2.2-1, 5.4.2.2-2, 5.4.2.2-3, 5.4.2.2-18
Special media to be avoided 5.5-1
Spinner
bird strike 5.2.1.1-5, 5.2.2.1-30, 5.2.2.1-35
deformations by bird strike 5.2.2.1-36
designed against erosion 5.3.3-5
Spline couplings, corrosion 5.4.1.1-7
Sprayed coatings, corrosion symptoms 5.4.1.2-8

Stabilized steels 5.4..1.1-22
Standstill, corrosion 5.4.1.1-2, 5.4.1.2-11, 5.4.1.3-1
Stator vane
adjuster 5.4.1.2-20
malfunction 5.4.1.2-20
erosion 5.3.1-13
Sticking parts of the foreign object 5.2.1.3-7
Storage
aeroengine with corrosion protection 5.4.1.3-1
corrosion 5.4.1.1-2
during manufacture, corrosion 5.4.1.3-4
stored airplanes, corrosion 5.4.1.1-8
Stossziffer ('impact/shock number') 5.2.2.2-13
Strain
causing embrittlement 5.2.1.1-17
induced corrosion cracking 5.4.2.1-4
SCC 5.4.2.1-9
Stress Corrosion Cracking (see SCC) 5.4.1.1-1, 5.4.1.1-10, 5.4.1.1-14, 5.4.2.1-1
damages 5.4.2.2-1
fundamentals 5.4.2.1-3
preventing 5.4.2.1-2
stress-induced 5.4.2.1-9
system 5.4.2.1-1
Sucking foreign objects by vortexes 5.2.1.2-1
Sulfidation 5.4.1.1-7
by vulcano gases and ash 5.3.2-27
producing corrosion pits 5.4.2.2-16
Sulfur
and sulfuric acid in vulcano clouds 5.3.2-27
SCC at Ni-alloys 5.4.2.1-2
Surface corrosion 5.4.1.1-13
area ratio influencing corrosion 5.4.1.1-17
Surge margin, influenced by compressor fouling 5.5-17
Swan neck (channel, see als S-shaped duct), influence af bird strike 5.2.2.1-10, 5.2.2.3-2, 5.2.2.3-20, 5.3.2-17
Sweat, SCC in production 5.4.2.2-15

T

Tank filler, location endangering fuel suck in 5.5-18
Tapping a part, dull sound 5.4.1.1-22
Tarnishing, tiggering stress corrosion 5.4.1.1-12
TBC (thermal barrier coating), spalling by FOD 5.2.1.1-24
Tension bolt, SCC 5.4.2.2-7
Test/testing rigs
danger of ingestion 5.5-3
air contaminations 5.2.1.3-19
erosion 5.3.3-1
development phase 5.3.3-1
methods for corrosion 5.4.1.3-5
SCC 5.4.2.1-14
Thermal barrier coating (TBC) 5.2.1.1-23
erosion particles from 5.3.2-20
FOD 5.2.1.1-23, 5.2.1.1-24
particles from 5.3.1-6
Thermal expansion, balancing 5.4.1.1-1
Thread imprints at FOD 5.2.1.3-8
Threaded connections
corrosion cracks 5.4.2.1-13
fasteners inspection 5.2.1.3-17
Thrust reverser
causing FOD 5.2.1.2-1, 5.2.1.3-18
by objects from the runway 5.2.1.2-9
influencing vortex formation 5.2.1.2-17
Ti (titanium) **alloys**
compressor blades, fire by FOD 5.2.1.1-15
corrosion 5.4.2.2-3
by hot salt 5.4.2.1-17
erosion 5.3.1-9
FOD 5.2.1.1-21
at blades 5.2.1.1-10
hot salt corrosion 5.4.2.2-3, 5.4.2.2-14
SCC 5.4.2.1-17, 5.4.2.2-3, 5.4.2.2-11
by coatings containing Cl 5.4.2.1-17
by hand sweat 5.4.2.1-17
by hot salt 5.4.2.1-2
by manufacturing process 5.4.2.1-17
mechanism 5.4.2.1-9
testing 5.4.2.2-19
wear behaviour 5.3.1-13
Titanium fire 5.2.1.1-15
after bird strike 5.2.2.1-1
Titanium nitride (TiN) coating
erosion 5.3.1-13
protection 5.3.3-2
triggering corrosion 5.4.1.1-15
erosion protection 5.3.3-8
Tools
as foreigen object 5.2.1.3-1
influencing corrosion 5.4.1.3-5
Torsion fracture of the gas producer shaft by bird strike 5.2.2.3-8
Transmission gear, cavitation 5.3.1-5
Trichlorethylene, triggering SCC 5.4.2.2-11
TC/Co (tungsten carbide in cobalt matrix)
hardfacing 5.3.3-1
triggering corrosion 5.4.1.1-15
Turbine rotor blades, FOD 5.2.1.1-23
location of FOD 5.2.1.1-22
Turboprop engines, bird strike 5.2.2.1-20, 5.2.2.3-8

U

Unburned fuel, insufficient drainage 5.5-1

Unusual Influences 5.5-1
avoiding 5.5-18

V

Vapor lock 5.3.1-1
Variable guide vane actuation, jamming by corrosion 5.4.1.2-24
SCC 5.4.2.2-8
Vertical take-off aircrafts (VTOL), hot gas recirculation 5.5-13
Vicinity of field testing, influence 5.5-18
Visibility if ash concentrations in the air 5.3.2-33
Volcanic ash
characteristics at an airplan 5.3.2-26
dangers 5.3.2-26
influenced by thrust reversers 5.2.1.2-17
Vortex (ground vortex) 5.2.1.2-11
at high located engines 5.2.1.2-12
diameter influencing FOD 5.2.1.2-12
formation 5.2.1.2-11
measures against 5.2.1.3-18
maximum ingestible FOD 5.2.1.2-13
parameters that influence the formation 5.2.1.2-12
phenomena 5.2.1.2-11
sucking foreign objects 5.2.1.2-1
suppression (Vortex killer) 5.2.1.2-19
devices against FOD 5.3.3-1
triggering influences 5.2.1.2-11
wind conditions influence formation 5.2.1.2-17
Vortex killer (vortex prevention) 5.2.1.3-18, 5.3.3-1
VTOL aircraft
bird strike 5.2.2.1-32
erosion 5.3.1-2
Volcano/volcanic ash
blocking
air bores 5.3.2-26
air ducts 5.3.2-29
the flow cross-sections in the turbine 5.3.2-27
burning into the hot parts 5.3.2-27
clouds,
altitudes 5.3.2-29
cirrus 5.3.2-33
gases 5.3.2-27
sulfur and sulfuric acid 5.3.2-27
compressor surge 5.3.2-27
damage of lectrical systems 5.3.2-27
deposits 5.3.2-27
deteriorations by 5.3.2-33
dust, glassy depositions of molten 5.3.2-29
economical consequences 5.3.2-33
endangering aircraft on the ground 5.3.2-27
erosion 5.3.2-26
flame out 5.3.2-27
glass-like coating 5.3.2-27
identification with onboard radar 5.3.2-27
influence assessment 5.3.2-33
measures 5.3.2-33
melting 5.3.2-27
particles, hard and sharp-edged 5.3.2-26
preventing damage by 5.3.2-27
seizing of contacting surfaces 5.3.2-26

W

Washing
compressors 5.4.1.3-5, 5.5-13
media,
deterioration by 5.5-15
remains 5.4.1.1-3
Water
injection 5.5-10
water-methanol .5-8
contamination 5.5-16
Water jet, high-pressure cleaning 5.3.1-21
Watery media
containing chlorine 5.4.2.1-2
triggering SCC 5.4.2.2-2
Wear (caused by particles 5.3.1-1)
behaviour
of rubber coatings 5.3.1-9
of titanium materials 5.3.1-13
by chipping 5.3.1-9
by coke particles 5.3.1-6
in fuel 5.3.1-7
by soot 5.3.1-6
influenced by angle of attack 5.3.1-9
products
corrosion 5.4.1.1-4
from blading 5.3.1-6
from casings 5.3.1-6
from rotor spacer rings 5.3.1-6
Wear-in (bath tub curve) 5.2.1.3-10
Wear-out (bath tub curve) 5.2.1.3-10, 5.2.1.3-10
Welded constructions made from steels corrosion 5.4.1.1-25
Welding, SCC by reaction layers 5.4.2.2-11
Wide chord blade, bird strike 5.2.2.1-36,. 5.2.2.2-10, 5.2.2.3-19
Wind influencing vortex formation 5.2.1.2-17
Wire mesh/sieve, erosion protection 5.3.3-4

Printed in Dunstable, United Kingdom

91338429R00246